Soft Matter under Exogenic Impacts

NATO Science Series

A Series presenting the results of scientific meetings supported under the NATO Science Programme.

The Series is published by IOS Press, Amsterdam, and Springer in conjunction with the NATO Public Diplomacy Division

Sub-Series

I. Life and Behavioural Sciences	IOS Press
II. Mathematics, Physics and Chemistry	Springer
III. Computer and Systems Science	IOS Press
IV. Earth and Environmental Sciences	Springer

The NATO Science Series continues the series of books published formerly as the NATO ASI Series.

The NATO Science Programme offers support for collaboration in civil science between scientists of countries of the Euro-Atlantic Partnership Council. The types of scientific meeting generally supported are "Advanced Study Institutes" and "Advanced Research Workshops", and the NATO Science Series collects together the results of these meetings. The meetings are co-organized bij scientists from NATO countries and scientists from NATO's Partner countries – countries of the CIS and Central and Eastern Europe.

Advanced Study Institutes are high-level tutorial courses offering in-depth study of latest advances in a field.
Advanced Research Workshops are expert meetings aimed at critical assessment of a field, and identification of directions for future action.

As a consequence of the restructuring of the NATO Science Programme in 1999, the NATO Science Series was re-organised to the four sub-series noted above. Please consult the following web sites for information on previous volumes published in the Series.

http://www.nato.int/science
http://www.springer.com
http://www.iospress.nl

Series II: Mathematics, Physics and Chemistry – Vol. 242

Soft Matter under Exogenic Impacts

edited by

Sylwester J. Rzoska

Department of Biophysics and Molecular Physics,
Institute of Physics, Silesian University, Katowice, Poland

and

Victor A. Mazur

Department of Thermodynamics, Institute of Energy and Ecology,
State Academy of Refrigeration,
Odessa, Ukraine

Springer

Published in cooperation with NATO Public Diplomacy Division

Proceedings of the NATO Advanced Research Workshop on
Soft Matter under Exogenic Impacts: Fundamentals and Emerging Technologies
Odessa, Ukraine, 8–12 October 2005

A C.I.P. Catalogue record for this book is available from the Library of Congress.

ISBN 978-1-4020-5870-7 (HB)
ISBN 978-1-4020-5872-1 (eBook)

Published by Springer,
P.O. Box 17, 3300 AA Dordrecht, The Netherlands.

www.springer.com

Printed on acid-free paper

TABLE OF CONTENTS

Part I: General Issues

Part II: Glass forming liquids

PARTICIPANTS

Directors:

Sylwester J. Rzoska
Department of Biophysics and Molecular Physics,
Institute of Physics, Silesian University, Katowice, Poland.

Victor M. Mazur
Department of Thermodynamics, Institute of Energy and Ecology,
State Academy of Refrigeration, Odessa 65026, Ukraine.

Key speakers:

Gyan P. Johari
Department of Materials Science and Engineering, McMaster University, Canada.

Christiane M. Alba-Simionesco
Lab. Chimie Physique, C.N.R.S, Université Paris-Sud Orsay, France.

Marie-Claire-Bellisent
Laboratoire Leon Brillouin, Gif-Dur-Yvette, France.

Thomas Kraska
Institute for Physical Chemistry, University at Cologne, Köln, Germany.

George Floudas
Department of Physics, Univ. of Ioannina, Ioannina, and FORTH-BRI
Greece.

Attila R. Imre
KFKI Atomic Energy Institute, Budapest, Hungary.

Simone Capaccioli
Istituto Nazionale per la Fisica della Materia and Dip. di. Fisica Univ. di Pisa, Italy.

Roelof D. Schuiling
Utrecht University, Utrecht, The Netherlands.

Aleksandra Drozd-Rzoska
Institute of Physics, Silesian University, Katowice, Poland.

Marian Paluch
Institute of Physics, Silesian University, Katowice, Poland.

Stefan Jurga
A. Mickiewicz University, Department of Macromolecular Physics, Poznan, Poland.

Jerzy Zioło
Institute of Physics, Silesian University, Katowice, Poland.

Samo Kralj
Univerza v Mariboru, Oddelek za fiziko, Maribor, Slovenia.

Josep Lluís Tamarit
ETSEIB, Universitat Politècnica de Catalunya, , Barcelona, Catalunya, Spain.

Jichen. C. Li
Dept. Physics, Biomolecular Sci, Manchester University, Manchester, UK.

H. Eugene Stanley
Center for Polymer Studies, Dept. Phys., Boston University, Boston, USA.

Mikhail A. Anisimov
Dept. Chem. Engn. & Inst. for Phys. Sci. & Tech. Univ. of Maryland College Park, USA.

C. Mike Roland
Naval Research Laboratory, Washington, USA.

Kia L. Ngai
Naval Research Laboratory, Washington, USA.

Eugene Yakub
Computer Science Dept., State Economic University, Odessa, Ukraine.

Leonid Bulavin
Taras Shevchenko Kiev National University, Ukraine.

Nikolay Malomuzh
Odessa National University, Odessa, Ukraine.

Aleksandr V. Chalyi
Physics Dept., National Medical University, Kiev, Ukraine.

Kirill Schmulovich
Institute of Experimental Mineralogy, Chernogolovka, Russia.

Akira Onuki
Department of Physics, Graduate School of Science, Kyoto University, Kyoto, Japan.

Other Participants:

George Cordoyiannis
National Centre for Scientific Research "Demokritos", Aghia Paraskevi, Greece.

Michał Mierzwa
Institute of Physics, Silesian University Katowice, Poland.

Wiesław Sułkowski
Institute of Chemistry, Silesian University, Katowice, Poland.

Sebastian Pawlus
Institute of Physics, Silesian University, Katowice, Poland.

Małgorzata Janik
Institute of Physics, Silesian University, Katowice, Poland.

Sławomir Maślanka
Institute of Chemistry, Silesian University, Katowice, Poland.

Milan Svetec
Regional Dev. Agency and Fac. of Educ., University of Maribor, Maribor, Slovenia.

Josef Bartos
Polymer Inst., Slovak Academy of Sciences, Bratislava, Slovakia.

Ricardo Casalini
Naval Res. Lab., Washington DC and George Mason University, Fairfax, USA.

Vlad Popa-Nita
Faculty of Physics, University of Bucharest, Romania.

Dmitry Yu. Ivanov
St. Petersburg State Uni. of Refrig. and Food Engn, Saint Petersburg, Russia.

H. Schvets
Odessa Natl. University, Odessa, Ukraine.

Vitaly Rogankov
Odessa Natl. University, Odessa, Ukraine.

Vladimir Kulinski
Odessa Natl. University, Odessa, Ukraine.

Sergey Artemenko
Odessa Natl. University, Odessa, Ukraine.

Participants of the ARW NATO "*Soft Matter under Exogenic Impacts: Fundamentals and Emerging Technologies*" Odessa, Ukraine, 2005 at the terrace of Hotel "Morskoy". Arrows indicate directors of the workshop. Note the Black Sea in the background.

PREFACE

Strong sensitivity even to moderate external impacts is a canonical feature of soft matter systems, the novel category covering critical liquids, polymers and blends, liquid crystals, glass forming liquids, bioliquids and biopolymers, . . . The dominance of mesoscale structures, rich phase structure, and complex dynamics are their subsequent common features, which can be strongly influenced by external impacts. This book presents the hardly available so far comprehensive presentation of high pressures, negative pressures, random constraints, and strong electric field exogenic impacts on various soft matter systems. Exogenic means derived or originating externally. It is used in social sciences, geology, biology, or medicine but hardly, if at all used in physics so far. Using this term we would like to point out strong and comprehensive influence of various external factors on soft matter system. The most classical example of exogenic actions is an asteroid impact, the problem from the discussion presented in this volume begins.

This book was prepared due to the Advanced Research Workshop NATO *"Soft Matter under Exogenic Impacts: Fundamental and Emerging Technologies"*, a brainstorming discussion which took place in Hotel Morskoy, Odessa, Ukraine, 8–12 October 2005. r.

The book was prepared to the cooperation of the prominent specialist focusing on soft matter and exogenic impacts, the direction which joins unusual experimental and theoretical difficulties with the promise of milestone results. It can serve as a guide in the novel and still puzzling research area. However, it also contains new results, appearing just in this volume. Identification of new types of physical behavior, new technological materials, ultimate verification of condensed and soft matter physics models, new applications in geophysics, biophysics, biotechnology – all these are topics of this book.

Directors of the ARW NATO and all participants are very grateful to the Science NATO Programme which support made it possible the brainstorming discussion in superior conditions.

The research activity of some of polish participants was possible due to the Polish Committee for Sci. Res. (KBN, grant no 2PO3B 025 34, for years 2003–2003, grant. resp. S. J. Rzoska).

Sylwester J. Rzoska (Poland)

Victor A. Mazur (Ukraine)

PART I: GENERAL ISSUES

ASTEROID IMPACT IN THE BLACK SEA: A BLACK SCENARIO

ROELOF DIRK SCHUILING*
Faculty of Earth Sciences, Utrecht University, P.O. Box 80.021, 3508 TA Utrecht, The Netherlands, e-mail: schuiling@geo.uu.nl

RICHARD B. CATHCART
Geographos, 1300 West Olive Avenue, Suite M, Burbank, California 91506-2225, USA, e-mail: rbcathcart@charter.net

VIOREL BADESCU
Candida Oancea Institute of Solar Energy, Faculty of Mechanical Engineering, Polytechnic University of Bucharest, Spl. Independentei 313, Bucharest 79590, Romania,
e-mail: badescu@theta.termo.pub.ro

Abstract. Discussions of the consequences of an impact of an extraterrestrial body in a particular region often focus only on very large asteroids, neglecting many phenomena attributable to small asteroid impacts. For impacts at sea it is normally the beach run up of impact-generated waves, which attracts most interest. Here, however, the vulnerability of the Black Sea region (comprised of Romania, Ukraine, Russia, Georgia, Turkey, and Bulgaria) will be examined, if it becomes subject to toxic and flammable gases that may someday be emitted by a catastrophically overturned Black Sea as a consequence of the impact of a medium-sized asteroid.

Keywords: Black Sea, asteroid impact, overturn, hydrogen sulfide, immobilization of metals.

1. Introduction

Studies of meteorite interactions with human societies have chiefly resulted from the public's increasing interest in the current and future threats posed by asteroid impacts, as well as the possible effects of past asteroid impacts on land and into the ocean. Similar in concept to the Richter scale used to rank earthquakes, the Torino Impact Hazard Scale ranks the energetic input of

---

* To whom correspondence should be addressed.

S. J. Rzoska and V. A. Mazur (eds.), Soft Matter under Exogenic Impacts, 1–8.

asteroids colliding with our planet.[1] Swiss Reinsurance (CF) and Munich Reinsurance Company (FRG) researchers are trying to calculate the pay-off costs ensuing from small and large asteroid impacts upon human infrastructures – the risk assessors are monitoring science periodicals to provide data that will enable them to devise mathematical models putting a price tag on unforeseen tragedies.[2] Scientists, too, seek to assess the damage to infrastructures from asteroid impacts.[3] Events of great magnitude and low frequency are not easily or accurately defined, yet science's ongoing investigations of extant and impending environmental hazards lead to a focused attention on major, sudden, disastrous event-processes at various places.

Neal Ascherson's *The Black Sea* (1995)[4] elucidates a nightmarish prediction of what might happen to the Black Sea: "It is a Black Sea apocalypse. This nightmare is known by the harmless word 'turnover', a sudden rolling-over of water layers, when the balance of densities that had kept the heavier mass below the lighter is reversed. This phenomenon... [that] has been observed and studied in lakes whose depths are anoxic and charged with hydrogen sulfide, or carbon dioxide. The catastrophic release of carbon dioxide gas by the overturn of Lake Nyos in Cameroon on 21 August 1986, which asphyxiated 1,746 persons, vividly illustrates the danger of a turnover to humans."[5]

The Black Sea is the largest anoxic water body. If "turnover" were to take place in the Black Sea, it would be the worst natural cataclysm to strike the Earth since the last Ice Age, more devastating in its human consequences than the eruption on Thera in about 1500 BC or than the Krakatoa eruption in Indonesia in 1883. Since the break-up of the USSR, there is no longer a single planning authority in the region surrounding the Black Sea, which could develop a contingency plan in case a Black Sea overturn would take place. There is some technical cooperation in the region, fostered by the Convention on the Protection of the Black Sea Against Pollution (1992), the Odessa Ministerial Declaration (1993) and the Black Sea Strategic Action Plan (1996).

2. Contemporary Black Sea conditions

The Black Sea's formation remains a matter of controversy,[6,7] but it seems to to have taken on its present-day oceanographic characteristics some 7,600 years ago. The Black Sea is an oval-shaped body of seawater, situated between 40° 55′ and 46° 32′ North latitude by 27° 27′ and 41° 42′ East longitude; it has a coastline of ~4,090 km and a maximum water depth of ~2,200 m. The distinguishing peculiarity of the Black Sea is the presence of a "permanent" halocline situated at a depth of 100–200 m, a stratification generated mostly by freshwater inputs from rivers flowing into the Black Sea. Of its total volume of water – approximately 547,000 km^3 – some 475,000 km^3, or ~87%, is anoxic, without dissolved oxygen and therefore lifeless except for anaerobic bacteria. The Black Sea is Earth's biggest single reservoir of hydrogen sulfide (H_2S), with a total quantity of dissolved H_2S of 5.3×10^9 tons. H_2S is generated by bacterial reduction of sulfate both in the water column and in the Holocene seafloor sediments deposited since the connection of the Black Sea with the Mediterranean Sea around 7,600 years ago. There are indications that the halocline, owing to human interventions (dams on rivers, pollution of the seawater) may rise even further, resulting in H_2S bubbles entering the atmosphere.[8] The summer season outbursts of H_2S already consistently cause mass mortality of benthic organisms and huge fish kills floating on shelf zone waters, which eventually reach and rot on beaches everywhere. For submarine archaeologists, this toxic water zone represents a potential mine of historical data on our ancestors since, as Bascom[9] noted: "A wooden ship ... [settling] in an anaerobic area, such as the Black Sea, should endure indefinitely"; metallic objects, such as modern ships, however, won't fare so well because of corrosion caused by H_2S-laden seawater.

3. Hydrogen sulfide gas (natural and anthropogenic)

Hydrogen sulfide is present in Earth's air.[10] Globally, anthropogenic emissions – mostly from industrial chemical factory processing and urban sewage treatment plants – yearly amount to >3 Mt, while normal natural emissions are ~100 Mt/yr. H_2S has a residence time in the atmosphere from ~18 hours to less than two days. It is a colorless gas slightly soluble in seawater and occurs in volcanic terrains, in gases released from underground natural gas fields and oilfields in addition to being produced by bacteria through the direct reduction of sulfate. When released in the atmosphere, it will form sulfur dioxide and sulfuric acid (H_2SO_4).

For humans, hydrogen sulfide is a broad-spectrum poison, meaning that it can poison several different sub-systems in the body. Accidental occupational H_2S poisoning is observed rarely in workers at industrial sites. Hydrogen sulfide poisoning occurs by inhalation. High-level exposure, 1500–3000 mg/m^3 (1000–2000 ppm, where 0.706 ppm = 1 mg/m^3), induces hyperpnoea (rapid breathing) followed by respiratory inactivity (apnoea) and death. Chronic low-level exposure to H_2S gas from the ~25 Mwe Puna Geothermal Venture (established 1991) near Hilo, Hawaii, has reportedly been linked to neurotoxicity symptoms exhibited by neighborhood residents.

Hydrogen sulfide has a density of 1.393 g/L at 25°C, making it 18% heavier than air. Its explosive limits range from 4–44% in air, producing a combustion product of sulfur dioxide that is also toxic to exposed humans; ignition temperature is ~260°C, usually producing a blue flame. H_2S saturated seawater at atmospheric pressure at ~20°C contains the equivalent of 2.9 volumes of gas per volume of water. The Black Sea is strongly undersaturated with H_2S as its average concentration is around 10 mg/l. Kump et al. (2003)[11] proposed "Death by Hydrogen Sulfide: A Kill Mechanism for the End-Permian Mass Extinction". They asserted that a global extinction of biota could have happened following a massive flux of H_2S gas to the Earth-atmosphere. They suggested a new global anoxia mechanism, "… the possibility that the chemocline separating anoxic, sulfidic deep water from the atmosphere, originally at the base of the wind-mixed [topmost photic] layer of the ocean, collapsed catastrophically to the surface as sulfide levels built up in the deep ocean …. Widespread outcropping of sulfidic waters creates a flux to the atmosphere that quickly depletes the troposphere of hydroxyl radical, thereby allowing a hydrogen-sulfide rich plume to waft across the continents. During the late Permian, such a cloud would poison terrestrial organisms in the way that aqueous H_2S had already caused the demise of marine organisms."

Contrary to this scenario Ryskin proposed that the End-Permian Extinction was caused by methane-driven oceanic eruptions.[12]

4. H_2S acidic cloud generation

The total energy released by the Earth, through heat conduction, volcanism and earthquakes is ~1.3×10^{21} Joules/yr, or ~310,000 Mt annually of TNT.[13] If the top 100 m of the Black Sea were suddenly vaporized by an instantaneous application of 140×10^{18} Joules provided by a falling asteroid, it would expose a deadly reservoir of seawater on the verge of exsolving a huge quantity of hydrogen sulfide gas and methane. What makes the Black Sea an especially noteworthy—Damoclean in scope—threat to people living on its shores is its unusual concentration of H_2S. Currently, a Black Sea with a significant concentration of H_2S poses an acid trauma threat to humans nearby that is generally similar to the actual acid trauma suffered by animals and plants that lived during the Cretaceous when an asteroid striking the Earth created Chicxulub Crater near Yucatan, Mexico.[14] The large asteroid struck a terrain especially rich in sulfur-bearing rocks (calcium sulfate), and its impact vaporized the sulfate rock and deposited many millions of tons of sulfur dioxide gas in the Earth-atmosphere. Next, sulfuric acid and nitrogen oxides formed clouds that enveloped much of the planet; clouds were composed of pyrolized organic material (soot) and sulfuric acid aerosol haze.[15] The large Cretaceous asteroid impact was extraordinary in its global effect; a future small asteroid impact during the Holocene would have extraordinary regional and international geopolitical repercussions. Melosh has discussed the relation between impacts and tsunamis (Impact-generated Tsunamis: An Over-rated Hazard) in which he stated, with categorical certitude, that only "... asteroids >100 m in diameter may pose a serious hazard to humanity and could require substantial expansion of the current efforts to identify Earth-crossing asteroids >1 km in diameter." Citing the 1968 conclusions of Van Dorn, Melosh[16] thinks small asteroids impacting the ocean pose only a modest threat to coastal shipping and that beach run up waves will not be especially noteworthy compared to earthquake-generated tsunamis.[17] Earthquake-generated tsunamis have a long history of occurrence on the strands of the Black Sea.[18]

5. Impacts on Black Sea region's residents

The Black Sea is lodged in a basin that is almost surrounded by mountains with the Carpathians on the west, the Caucasus on the north and east, and the Pontides to the south. Its shoreline hosts the populations of Turkey, Bulgaria, Romania, Ukraine, Russia, and Georgia. Many millions of persons live in these six countries, but it is difficult to estimate how many of these persons live within one kilometer from the coast. A *minimal* estimate for the human population at risk nowadays to a future event of the type we postulate here is ~300,000 persons—figuring ~70 persons/km^2 multiplied by the 4,090 km length of the Black Sea's coastline. This *minimal* estimate is far too low; because coastal regions are always more densely populated than inland areas. We have only to think of coastal cities such as Odessa, Suchumi, Batum, Trabzon, Samsun, Zonguldak, Varna and Constantza.

Shortly after the 11 September 1927 "Great Crimean Earthquake" gigantic fiery flares ranging from 20–500 m high and covering large areas were observed erupting from the Black Sea.[19] Probably, these flames were caused only in part by the spontaneous combustion of hydrogen sulfide that had exsolved from below the halocline. Mostly, the flames were caused by methane eruptions triggered by the earthquake. Even in normal times, dense bubble curtains or plumes (also flares) of methane can attain altitudes of 200 m in the water column or even reach the surface where methane can inflame. A "burning sea" is not unusual during thunderstorms over the Black Sea. Above many methane seeps the bottom waters must be (super-)saturated with methane. If there were a turnover, and these waters are no longer pressurized, methane will quickly exsolve. The methane will act as a carrier gas for the H_2S that is dissolved in these same waters. Even out of water that is undersaturated, H_2S will diffuse into the atmosphere when the water is splashed into the atmosphere.

6. Use of H_2S as a fuel?

Macro-engineering experts have proposed a means of mitigating this environmental hazard. Rustam Akhmedov, sometime during the early 1980s, first realized (in a slightly quixotic fashion) that H_2S might become an economically valuable energy resource that is energetically only slightly inferior to widely harnessed natural gas extracted from subterranean petroleum deposits. Use of this aqueous H_2S resource for energy production, as Akhmedov desires, will also remove an environmental hazard – any future Black Sea overturning – that just may be initiated sometime in the indefinite future by a small asteroid impact. An extraction scheme for the elimination of hydrogen sulfide from the Black Sea has been proposed by Cazacu[20] and Cazacu and Iancu.[21,22] If the Black Sea's hydrogen sulfide was efficiently mined and burnt in electricity-generating stations, then an impending ecological macro-problem could be mitigated and the six nations surrounding the Black Sea would have a vast energy resource. The extant H_2S reservoir/resource of the Black Sea is estimated to be $\sim 5.3 \times 10^9$ tons. When burnt, $\sim 1.56 \times 10^{10}$ Joules is liberated per ton–but at the cost of creating two tons of unwanted/undesirable SO_2. This calls for some viable industrial-scale use for this toxic byproduct if H_2S is used for commercial energy production. Burning all the H_2S resident in the Black Sea today (i.e., 5.3×10^9 tons) in a 100% efficient commercial electricity power plant would produce $\sim 8.3 \times 10^{19}$ Joules. Using H_2S as a fuel will require the development of a new type of power plant. Deep-lying pipelines like the large-diameter seawater intake pipeline in Hawaii, built by the Natural Energy Laboratory of Hawaii Authority[23] can draw the gas-loaded seawater upwards (contained artificial upwelling) where degasifiers, compressors and H_2S gas storage units will control the production and flow of a potentially useful flammable gas.[24] Most likely, degassing the deep waters will also release some methane, which could add to the energy production.

7. Immobilization of metal-rich industrial wastes

It has also been suggested to use the Black Sea as a georeactor.[25–27] The Black Sea is the world's largest body of reducing, sulphide-rich waters. Suspended in these waters are very fine metal sulphide particles. If metal-rich sludges or metal-rich industrial effluents would be pumped into the deep waters of the Black Sea (below 200 m), these would be converted to metal sulphides and slowly settle to the bottom, where they would join a large amount of natural metal sulphides already present in the bottom muds. The reducing water body as well as the low-permeability mud itself will insure that reducing conditions will be maintained over geological periods, forming the modern equivalent of metal-rich "black shales". The conditions under which black shales are deposited are aptly called "euxinic facies" after the Black Sea's name in Antiquity. This identifies certain black shales, high in organic carbon, sometimes having concentrations of metal sulfides that make them even mineable. If such a scheme for the immobilization of toxic wastes rich in heavy metals is adopted, it may become a source of income for the countries surrounding the Black Sea, as they can levy a tax on the use of their natural resource. Although, compared to the amount of H_2S dissolved in the Black Sea, the fraction that becomes fixed as metal sulfides will be minor, it will alleviate the dangers of a Black Sea turnover somewhat.

8. Potential post-impact recovery problems

In the foregoing we have sketched the potential dangers of the H_2S and partially also methane rich waters of the Black Sea for the surrounding population.[28] If this dangerous natural resource is used, for example for energy production, or as a chemical for the immobilization of hazardous metal waste streams, its potential danger may be slowly diminished. It is unlikely, however, that such measures, even if applied on a grand scale, will in the short term be sufficient to avert the dangers posed by a sudden release of the gases. The impact of a 100 meter asteroid in the Black Sea may cause a catastrophic overturn. This may result in a geographically large depopulated region—like that expected after detonation of a "neutron bomb". That is, the infrastructure will remain relatively intact, but the inhabitants will be missing. It is difficult to imagine what this will mean for the eventual repopulation of the affected coastal zones.

References

1 Binzel, R. P. (2000) The Torino Impact Hazard Scale, Planetary and Space, *Science* **48**, 297–303.
2 Poveda, A. et al. (1999) The expected frequency of collisions of small meteorites with cars and aircraft, *Planetary and Space Science* **47**, 715–719.
3 Remo, J. L. (1998) An approach to assessing the technological cost/benefits of critical and sub-critical cosmic impact prevention, *Journal of the British Interplanetary Society* **51**, 461–470.
4 Ascherson, N. (1995) The Black Sea (Hill and Wang, NY), 267–268.
5 Cotel, A. J. (1999) A trigger mechanism for the Lake Nyos disaster, *Journal of Volcanology and Geothermal Research* **88**, 343–347.
6 Sperling, M. et al. (2003) Black Sea impact on the formation of eastern Mediterranean sapropel S1. Evidence from the Marmara Sea, *Palaeogeography, Palaeoclimatology, Palaeoecology* **190**, 9–21.

7. Spada, G. et al. (1999) Chandler wobble excitation by catastrophic flooding of the Black Sea, *Annali di Geofisica* **42**, 749–754.
8. Humborg, C. et al. (1997) Effect of Danube River dam on Black Sea biogeochemistry and ecosystem structure, *Nature* **386**, 385–388.
9. Bascom, W. (1971) Deep-water archeology, *Science* **174**, 61–269.
10. Akimoto, H. (2003) Global air quality and pollution, *Science* **302**, 1716–1718.
11. Kump, L. R. et al. (2003) Death by hydrogen sulfide: a kill mechanism for the endpermian mass extinction, lecture at: *Annual Meting of the Geological Society of America,* Seattle, USA.
12. Ryskin, G. (2003) Methane-driven oceanic eruptions and mass extinctions, *Geology* **31**, 741–744.
13. Fowler, C. M. R. (1993) The Solid Earth: An Introduction to Global Geophysics, *Cambridge UP*, NY, 472–480.
14. Brett, R. (1992) The Cretaceous-Tertiary extinction: A lethal mechanism involving anhydrite target rocks, *Geochimica et Cosmochimica* **56**, 3603–3606.
15. Pope, K. (2002) Impact dust not the cause of the Cretaceous-Tertiary mass extinction, *Geology* **30**, 99–102.
16. Melosh, H. J.: http://www.lpl.arizona.edu/~jmelosh/ImpactTsunami.pdf
17. Dalton, R. (2003) Long-lost wave report sinks asteroid impact theory, *Nature* **421**, 679.
18. Nikonov, A. A. (1997) Tsunami occurrence on the coasts of the Black Sea and the Sea of Azov, Izvestiya, *Physics of the Solid Earth* **33**, 77–87.
19. Kondorskaya, N. V. and Shebalin, N. V. (1982) New Catalog of Strong Earthquakes in the Territory of the USSR from Ancient Time to 1975, *SE-31, United States Department of Commerce*, Boulder, Colorado.
20. Cazacu, M. D. (1999) A survey of the techniques which might preserve the biosphere reservation Danube Delta, lecture at: *The 8th Symp. Technologies, Installations and Equipments for Improvement of Env. Quality,* Bucharest, Romania.
21. Cazacu, M. D. and Iancu, R. (2001) Elimination of hydrogen sulfide from deep water of Black Sea by a gas-lift plant, lecture at: *11th Conf. Sperin,* Bucharest, Romania.
22. Cazacu, M. D. and Iancu, V. R. (2003) Advantageous technologies for Black Sea water problems and solutions, lecture at: *Conf. Int. Environment and Energie,* Mangalia–Romania.
23. Shuster, L. A. (2003) Thinking deep, *Civil Engineering* **73**, 47–53.
24. Spiridonov, A. (1990) Saving the Black Sea, *Science in the USSR* **6**, 51–58.
25. Schuiling, R. D. (2005) Immobilisation of metal wastes by reaction with H_2S in anoxic basins. Concept and elaboration, *Proc. MedCoast Conference,* October 2005.
26. Schuiling, R. D. (1996) Geochemical engineering: principles and case studies *in Geochemical Approaches to Environmental Engineering of Metals.* Ed. R. Reuther, Springer-Verlag, 3–12.
27. Mazur, V. A., Tsykalo A. L., and Schuiling, R. D. (1997) Biogeochemical processes in the depths of the Black Sea as a model of wastes transforming technology (in Russian), *Emergency Situations and Civil Defense* **2**, 40–509.
28. Neretin, L. N. et al. (2004) Advancing knowledge of anoxic systems of the world ocean, *EOS* **85**, 47 & 5.

THE CONDUCTIVITY OF HYDROGEN IN EXTREME CONDITIONS

VALERIY T. SHVETS*
Odessa State Academy of Refrigeration, Dvoryanska 1/3, 65082 Odessa, Ukraine

SVYATOSLAV V. SAVENKO
Debye Institute, Soft Condensed Matter, Utrecht University, Princetonplein 5, 3584 CC Utrecht, The Netherlands

JEVGENIY K. MALYNOVSKI
Special Designer Office National Academy of Science of Ukraine, Pushkinska 37, 65011 Odessa, Ukraine

Abstract. We consider the electroresistivity of metallic hydrogen within the framework of perturbation theory in electron–proton interaction. To this end we employ the Kubo linear response theory while using the two-time retarded Green functions method to calculate the relaxation time. The expressions for the second and third order contributions are calculated, which serve as a basis for the estimation of the sum of the series. To describe the electron subsystem the random phase approximation is used at that the exchange interactions and correlations are taken into account in a local field approximation. The structure of the proton subsystem is assumed to be determined by the Percus-Yevick equation with the effective hard sphere diameter being the only parameter of the theory in this case. The effective pair proton interaction is used to identify the effective hard sphere diameter as a function of the temperature and the density of the system. The behavior of the second and the third order contributions is investigated in a wide range of the temperature and the temperature. In the whole range the values of the parameters of the theory are closed to those indicating the limits of nearly free model applicability.

Keywords: Metallic hydrogen, conductivity.

* To whom correspondence should be addressed. Valeriy T. Shvets, e-mail: valtar@paco.net

S. J. Rzoska and V. A. Mazur (eds.), Soft Matter under Exogenic Impacts, 9–21.

1. Introduction

The interest to hydrogen can be partly explained by the fact that it is the most widely spread element in the universe. Hydrogen in metallic state contributes a major portion to the mass of the planets in the solar system. It is also the unique object from the theoretical point of view, as it is the only substance for which the exact electron–ion interaction is known. This allows to calculate the properties of the metallic hydrogen using the minimal number of approximations, and thus with the maximum reliability.

However the fact that the Coulomb interaction is not weak and fast decreasing one complicates the theoretical investigations. The workaround developed in the electron gas theory is the random phase approximation, which corresponds to approximate summation of the perturbation theory series in coulombic electron– electron interaction.

The possibility for the metallic phase of hydrogen to exist was predicted in 1935.[1] It was suggested that at high densities hydrogen should exhibit the transition from diatomic molecular state, with insulator properties, to an atomic state with metallic conductivity. The actual discovery of metallic hydrogen and the detailed investigation of conductivity as a function of the pressure and the temperature followed at 1996.[2] In this study the molecular hydrogen in a liquid state was subjected to shock compression up to 1.80 Mbar at the temperatures in the 2200–4400 K range. The dielectric-metal transition was observed at the pressure 1.4 Mbar and the temperature 3000 K, at that the resistivity of metallic phase measured 500 μΩ cm. In the pressure and the temperature range 1.4–1.8 Mbar and 3000–4400 K the electroresistivity remained virtually unchanged. As a matter of fact the examined transition was of a metal–semiconductor type, as the band-gap did not vanish completely, but decreased from 15 to 0.3 eV, which is almost equal to the temperature of a sample. It should be noted that the earlier experimental and theoretical studies of the conjectured metallic state exist. In the work[3] the electric conductivity of molecular hydrogen was measured at substantially lower pressure 0.1–0.2 Mbar, the results followed the exponential law with the rate specific to semiconductors with the 12 eV band-gap. The discovery of metallic hydrogen was first reported at 1978,[4] the authors reported the discovery of metallic hydrogen at the pressure 2 Mbar with the resistivity 1000 μΩ cm.

When the new substances are discovered, the examination of the equilibrium properties usually precedes the studies of transport phenomena. At present the research on the static properties of metallic hydrogen is carried out intensively.[5–10] Yet, the theoretical investigations of the electron transport phenomena in metallic hydrogen are far less advanced. For one, in the paper[11] the resistivity of metallic hydrogen was calculated in a wide range of the temperature and the pressure, but using the fixed value of the packing fraction, typical for the liquid metals close to the melting point, and, what is more important, only the first order contribution of the perturbation series in electron-proton interaction. As we show in the current paper it is this approximation which is responsible for almost an order of magnitude difference between the experimental and theoretical values. Of some interest is the computer modeling of the metallic hydrogen properties including the electroresistivity.[12] Unfortunately, the magnitude of the resistivity obtained from simulations differs from the experimental values by the order of magnitude as well. The distinctive feature of the treatises on static properties of metallic hydrogen is adoption of the nearly free electron model. Here we are going to employ this model for calculation of electroresistivity of metallic hydrogen. At that we assume that the hydrogen is in the true metallic state neglecting the band gap of 0.3 eV observed in experiments. This state can be realized either at a higher pressure, or at a higher temperature. We like to note that the Jupiter core with the radius equal to half the radius of the planet consists of hydrogen at the pressure 3–40 Mbar and the temperature 10,000–20,000 K.

2. Hamiltonian

We assume the hydrogen to be in disordered atomic state with all the electrons collectivized. In this case the Hamiltonian of the electron subsystem of metallic hydrogen can be taken in the form similar to that of the liquid simple metals

$$H = H_0 + H_{ie}. \tag{1}$$

This model is applicable even in the presence of the molecular hydrogen along with the atomic one. Owing to its electroneitrality the scattering of electrons on the hydrogen molecules falls far short of the scattering on protons. Again, just like in the case of the liquid simple metals, the electron gas is considered degenerated, and the ionic subsystem to be disordered and static one. Using the nearly free electrons approximation, the Hamiltonian of noninteracting electron gas can be written as

$$H_0 = \sum_{\mathbf{k}} \varepsilon_k a_{\mathbf{k}}^{+} a_{\mathbf{k}}, \tag{2}$$

where $a_{\mathbf{k}}^{+}, a_{\mathbf{k}}$ are the operators of nucleation and annihilation of electrons with the wave vector $\mathbf{k}$,

$$\varepsilon_k = \hbar^2 k^2 / 2m \tag{3}$$

is the energy of a free electron and m is its mass.

The Hamiltonian of the electron–proton interaction can be taken in line with the diffraction model of a metal, which incorporates the electron–electron interaction via screening of the electron–ion interaction

$$H_{ie} = V^{-1} \sum_{\mathbf{q}} W(q) \rho^i(\mathbf{q}) \rho^e(-\mathbf{q}). \tag{4}$$

Here V is the volume of the system,

$$W(q) = -V(q)/\varepsilon(q) \tag{5}$$

is the screened potential energy of electron–proton interaction,

$$V(q) = 4\pi e^2/q^2 \tag{6}$$

is the Fourier transform of the Coulomb interaction,

$$\varepsilon(q) = 1 + [V(q) + U(q)]\pi_0(q) \tag{7}$$

is the dielectric permittivity of the electron gas in the random phase approximation with taken into account the exchange interaction and correlations of electron in local field approximation,

$$U(q) = -2\pi e^2/(q^2 + \lambda k_F^2) \tag{8}$$

is the potential energy of the exchange interaction and the correlations of the electron gas with $\lambda \approx 2^{13}$, k_F is the Fermi wave vector,

$$\pi_0(q) = \frac{mk_F}{\pi^2\hbar^2}\left(\frac{1}{2} + \frac{4k_F^2 - q^2}{8k_F q}\ln\left|\frac{2k_F + q}{2k_F - q}\right|\right) \tag{9}$$

is the polarization function of a free electron gas,

$$\rho^e(\mathbf{q}) = \sum_{\mathbf{k}} a_{\mathbf{k}}^{+} a_{\mathbf{k}+\mathbf{q}} \tag{10}$$

is the Fourier transform of the density operator of the electron gas, and

$$\rho^i(\mathbf{q}) = \sum_{n} \exp(-i\mathbf{q}\mathbf{R}_n) \tag{11}$$

is the Fourier transform of the density of ion subsystem with $\mathbf{R}_n$ denoting the radius-vector of the n-th ion.

We like to stress that the metallic hydrogen is the only system with exactly known unscreened potential of the electron–ion interaction, which is the Coulomb law. This fact significantly simplifies the calculation of various properties of metallic hydrogen, as long as the problem of modeling the electron–ion interaction, leading to introduction of additional fitting parameters, is avoided.

Just like in the theory of many of the disordered simple metals, the ratio of the potential energy of the electron–ion interaction to the Fermi energy is the small parameter of the theory not for all values of the wave vector. This situation is illustrated on the Fig. 1.

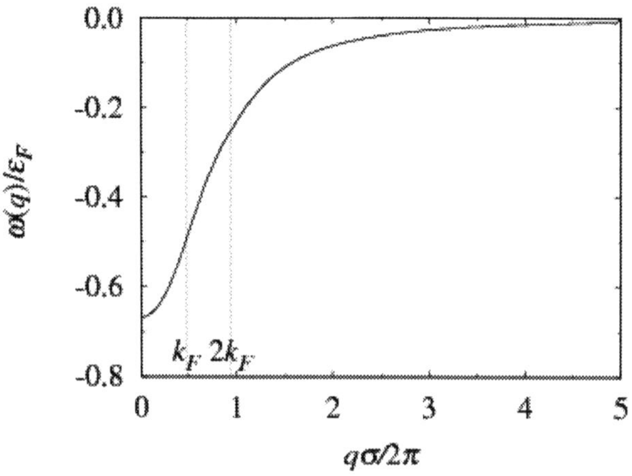

Figure 1. The small parameter of the theory. Two vertical lines denote the values of the wave vector equal to k_F and $2k_F$ correspondingly. These values correspond to $\sigma = 4$ a.u. and $\eta = 0.45$.

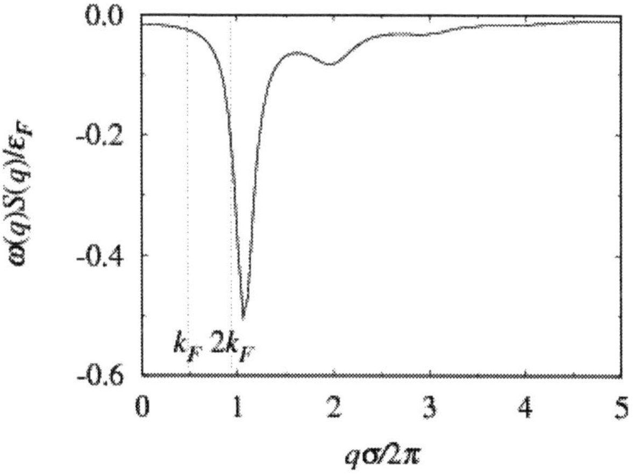

Figure 2. The small parameter of the theory. Two vertical lines denote the values of the wave vector equal to k_F and $2k_F$ correspondingly. These values correspond to $\sigma = 4$ a.u. and $\eta = 0.45$.

The actual small parameter of the theory is the product of the above mentioned ratio on the structure factor of the ion subsystem. From the Fig. 2

it can be seen that this parameter again cannot be considered small for every value of the wave vector.

Inasmuch as these values of the wave vector provide substantial contribution to the corresponding integrals, one can expect the perturbation series in this parameter of the theory to converge slowly. This circumstance does not allow us to restrict ourselves by only the first term of the perturbation series when calculating the resistivity.

3. Resistivity

The electroresistivity of well conducting disordered simple metals can be found using the well known Drude formula

$$R = \frac{m}{ne^2}\frac{1}{\tau}. \tag{12}$$

Here n is the electron gas density and τ is the relaxation time for the electroconductivity process. We are going to exploit this formula to calculate the resistivity of metallic hydrogen as well. At that we will consider only the electron contribution to the resistivity.

Within the framework of the linear response theory of Kubo and the method of two-time retarded Green functions[13–15] the inverse relaxation time can be written in the form of the series in electron–proton interaction

$$\tau^{-1} = \sum_{n=2}^{\infty} \tau_n^{-1}. \tag{13}$$

The general term of this expansion is of the form

$$\tau_n^{-1} = \frac{N}{V^n} \sum_{\mathbf{q}_1,\ldots,\mathbf{q}_n} W(\mathbf{q}_1)\cdots W(\mathbf{q}_n) S(\mathbf{q}_1,\ldots,\mathbf{q}_n) \Gamma(\mathbf{q}_1,\ldots,\mathbf{q}_n). \tag{14}$$

Here $S(\mathbf{q}_1,\ldots,\mathbf{q}_n)$ is the n-particle structure factor of the ion subsystem, N is the number of ions in the system, $\Gamma(\mathbf{q}_1,\ldots,\mathbf{q}_n)$ is the electron multipole for the electric conductivity process.

The second order contribution to the inverse relaxation time for the disordered simple metals, first obtained by Ziman,[16] is well explored and reads

$$\tau_2^{-1} = \frac{m}{4\pi\hbar^3 v k_F^3} \int_0^{2k_F} W^2(x) S(x) x^3 dx, \tag{15}$$

where v is the inverse proton number density. We now examine the third order contribution in more detail. It has the form

$$\tau_3^{-1} = \frac{N}{V^3} \sum_{\mathbf{q}_1,\mathbf{q}_2,\mathbf{q}_3} W(\mathbf{q}_1)W(\mathbf{q}_2)W(\mathbf{q}_3)S(\mathbf{q}_1,\mathbf{q}_2,\mathbf{q}_3)\Gamma(\mathbf{q}_1,\mathbf{q}_2,\mathbf{q}_3). \quad (16)$$

For the noninteracting electron gas the electron multipole, obtained by using the kinetic equation and characterizing the electric conductivity, has the form

$$\Gamma(\mathbf{k}_1-\mathbf{k}_2,\mathbf{k}_2-\mathbf{k}_3,\mathbf{k}_3-\mathbf{k}_1) = \frac{\pi\hbar}{3mNk_BT}(\mathbf{k}_1-\mathbf{k}_2)^2 n(\mathbf{k}_1)[1-n(\mathbf{k}_1)]\frac{\delta(\varepsilon_{\mathbf{k}2}-\varepsilon_{\mathbf{k}1})}{\varepsilon_{\mathbf{k}2}-\varepsilon_{\mathbf{k}3}}, \quad (17)$$

where $n(k)$ is the Fermi–Dirac distribution function. With some manipulation[14,15] the third order contribution can be expressed as a principal value integral

$$\tau_3^{-1} = \frac{m^2}{24\pi^5\hbar^5 k_F^2}\int_0^\infty \frac{f(k)}{k_F-k}dk. \quad (18)$$

As long as the electron–electron interaction is exactly known, the major approximation we use is the geometric approximation for the three-particle structure factor[14–18]

$$S(\mathbf{q}_1,\mathbf{q}_2,\mathbf{q}_3) = S(\mathbf{q}_1)S(\mathbf{q}_2)S(\mathbf{q}_3). \quad (19)$$

As a consequence of this approximation we obtain the following expression for the function $f(k)$:

$$f(k) = \frac{1}{k_F+k}\sum_{n=0}^{\infty}(2n+1)A_n B_n^2(k), \quad (20)$$

with

$$A_n = \int_0^{2k_F} W(q)S(q)P_n\left(\frac{2k_F^2-q^2}{2k_F^2}\right)q^3 dq, \quad (21)$$

$$B_n(k) = \int_{|k-k_F|}^{k+k_F} W(q)S(q)P_n\left(\frac{k^2+k_F^2-q^2}{2kk_F}\right)q\,dq, \quad (22)$$

where $P_n(x)$ is the Legendre polynomial of order n.

4. The effective pairwise proton interaction

The knowledge of the effective pair proton interaction is essential for estimation of the conductivity of metallic hydrogen. In the context of the current work it is a particularly valuable feature as the only parameter it depends on is the number density of the system. Since the effective pair interaction allows to calculate the effective hard sphere diameter for a given temperature, it enables us to determine the hard sphere diameter as a function of density and temperature.

The expression for the effective pair potential is known from the second order perturbation theory in electron–ion interaction and reads

$$V_{ef}(R) = \frac{e^2}{R} - \frac{v}{\pi^2} \int_0^\infty Q(q) \frac{\sin(qR)}{qR} q^2 dq, \tag{23}$$

where

$$Q(q) = \frac{v}{2} V^2(q) \frac{\pi(q)}{1 + V(q)\pi(q)}. \tag{24}$$

The polarization function of electron gas with exchange interaction and correlations taken into account in a local field approximation can be written as

$$\pi(q) = \frac{\pi_0(q)}{1 + V(q)\pi_0(q)}. \tag{25}$$

This expression does not contain any fitting parameters characterizing the ion subsystem and depends only on the Fermi wave vector, i.e., on the number density of the system. The only approximation it involves is the random phase approximation for the electron subsystem with accounting for the exchange interaction and electron correlations in a local field approximation. Figure 3 provides the plot of the effective pair ion interaction for a number density n=0.001 gm/cm^3.

Now we can introduce the effective hard sphere diameter as the minimum ion approach distance for a given temperature. This distance can be found from equating the total energy of protons to zero at the closest approach

$$V_{ef}(\sigma) = 3k_B T / 2. \tag{26}$$

Here the reference point for potential energy is the bottom of the potential well.

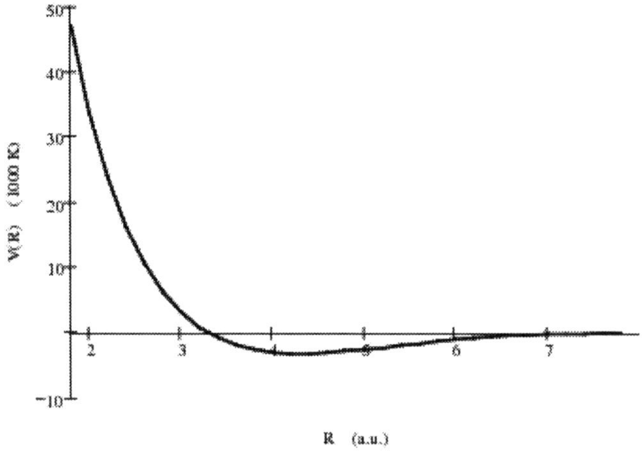

Figure 3. The effective pair interaction, $n = 0.001\text{gm/cm}^3$.

5. Results and discussion

If for the structure factor of the ion subsystem we use the Percus-Yevick expression, then the resistivity depends on three parameters. Namely these are the hard sphere diameter, the packing fraction and the Fermi wave vector (see Fig. 4).

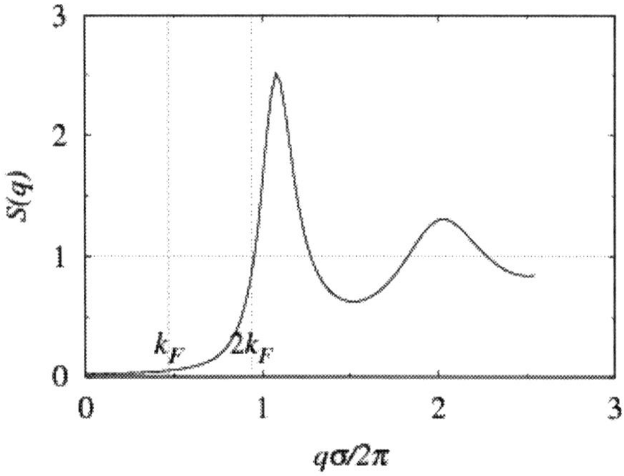

Figure 4. The pair structure factor of the ion subsystem. Two vertical lines denote the values of the wave vector equal to k_F and $2k_F$ correspondingly. These values correspond to $\sigma = 4$a.u. and $\eta = 0.45$.

If all the electrons are considered collectivized then the Fermi wave vector can be related to the number density through $n = k_F^3 / 3\pi^2$, while on the other hand $v\eta = \pi\sigma^3/6$. It is convenient to choose the hard sphere diameter as the independent parameter. Provided the temperature is known this parameter can be uniquely defined as well as we show in the previous section.

As long as liquid hydrogen exists in nature in a wide pressure and temperature range it is instructive to investigate the behavior of resistivity as a function of these parameters.

First we consider the resistivity using only the first nonvanishing term of the perturbation series in electron–proton interaction. As illustrated in Fig. 5 the second order contribution decays as the density grows and increases as the temperature goes up. At the density 0.32 mol/cm^3 and the temperature 3000 K, i.e., under conditions corresponding to the transition to metallic state, the resistivity comprises 20 μΩ cm if all electrons are metallized.

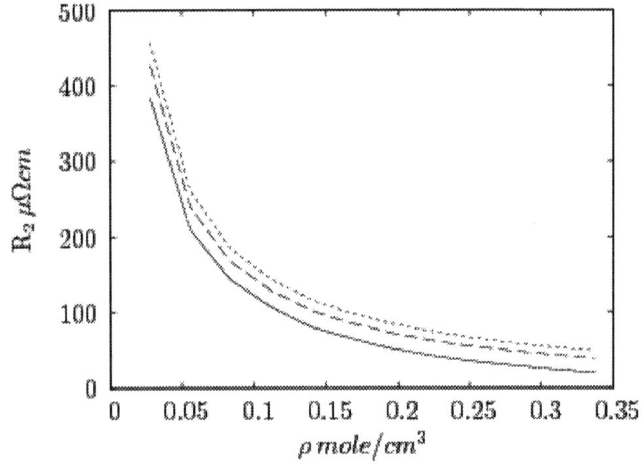

Figure 5. The resistivity of metallic hydrogen calculated using only the contribution of the second order in electron–proton interaction. --- T=3000 K, - - T=6000 K, ...T=12000 K.

This value is in the good agreement with the results obtained in work[11] using the same approximation, as well as with the simulation data.[12] If we assume that only a part of the hydrogen changed to an atomic state, which corresponds to smaller electron gas densities, then the resistivity reaches the experimental value of 500 μΩ cm at the density 0.2 mol/cm^3. This corresponds approximately to 5% of the full amount of hydrogen transformed into atomic state. This result agrees with[2] conjectured that only 5% of the hydrogen at the dielectric-metal transition is in the atomic state.

The Fig. 6 represents the resistivity of metallic hydrogen with the higher order terms of perturbation theory taken into account. The third order contribution is not small which requires some approximation to estimate the sum of the perturbation theory series. Here we assume that those are the series with the geometric ratio equal to ratio of the third to the second order contributions.

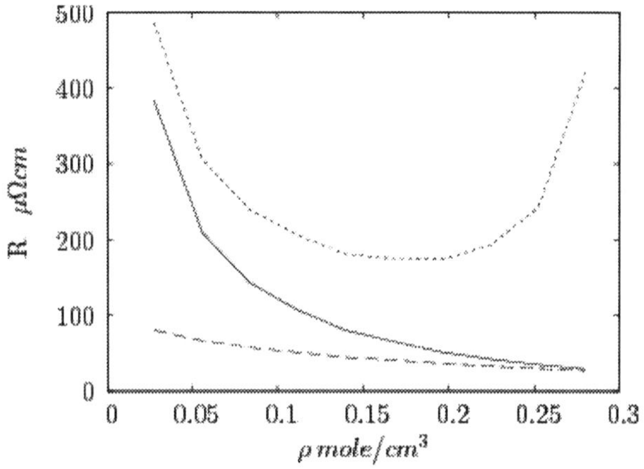

Figure 6. The resisitivity of metallic hydrogen using the contributions of a higher order in electron–proton interaction at the temperature T=3000 K. --- R_2, - - R_3, … $R_2/(1–R_2/R_3)$.

When the hydrogen in atomic state makes up to 5% of the whole amount, the third order contribution comes up to only 20% of the second order contribution. As the density of the electron gas increases, which corresponds to increase of the atomic hydrogen fraction, both second and third order contributions decay, but the second order term decays faster. As a result, at the density 0.3 mol/cm^3 the second and the third order contributions become equal, and the series diverge. Close to this point the resistivity reaches experimentally observed value. At that it is in excess of 170 μΩ cm in the whole range, while despite the value of the second order contribution is rather small in the most part of the range. This raises an issue of applicability of the nearly free electron model used here. The validity criterion is the smallness of dimensionless parameter $\gamma = \hbar / \varepsilon_F \tau$. The Fig. 7 illustrates the dependence of this parameter on the density of the electron gas. It suggests that for the most of the densities it is but little less than one, and for the large densities it can exceed one. Thus, we have used the nearly free electron model close to its applicability limit.

Another parameter which has an impact on the solution is the degree of degeneracy of the electron gas $\alpha = k_B T / \varepsilon_F$. This parameter is small in the whole range and the electron gas is strongly degenerated. At the higher temperatures the validity of the nearly free electron model is impaired, which is illustrated on the Fig. 8.

An alternative description of the electron subsystem is the tight binding approximation, however the experimental values of the conductivity are out of the limits of its applicability as well. The experimentally observed resistivity 500 μΩ cm was not the goal of this work, however the values obtained are reasonably close to it.

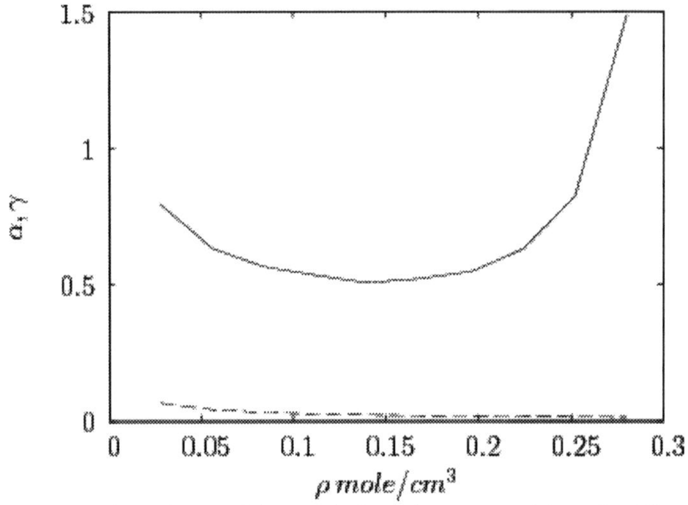

Figure 7. The density dependence of the nearly free electrons model applicability criterion and degeneracy parameter at the temperature $T = 3000$ K.

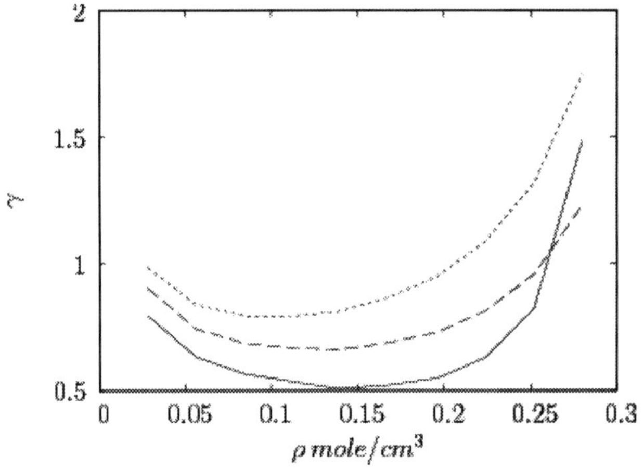

Figure 8. The temperature and density dependence of the nearly free electrons model applicability criterion. ---$T = 3000$ K, - - $T = 6000$ K, and … $T = 12000$ K.

Our theory do not incorporate two factors: the existence of the band-gap in the energy spectrum of the electrons and the presence of the hydrogen in molecular state at the pressure 1.4 Mbar and the temperature 3000 K. The scattering of the conduction electrons on the hydrogen molecules can provide a noticeable contribution to the resistivity. The width of the band-gap decays quickly as the pressure (density) grows and it's unlikely that it produces a considerable contribution to the resistivity at the pressures exceeding 3 Mbar.

References

1. Wigner, E. and Huntington, H. B. (1935) *J. Chem. Phys.* **3**, 765.
2. Weir, S. T., Mitchell, A. C. and Nellis, W. J. (1996) Metallization of Fluid Molecular Hydrogen, *Phys. Rev. Lett.* **76**, 1860–1864.
3. Nellis, W. J., Mitchell, A. C., McCandless, P. C., Erskine, D. J. and Weir, S. T. (1992) Electronic Energy Gap of Molecular Hydrogen from Electrical Conductivity Measurements at High Shock Pressures, *Phys. Rev. Lett.* **68**, 2927–2931.
4. Hawke, P. S., Burgess, T. J., Duerre, D. E., Huebel, J. G., Keeler, R. N., Klapper, H. and Wallace, W. C. (1978) Observation of Electrical Conductivity of Isotropically Compressed Hydrogen at Mbar Pressures, *Phys. Rev. Lett.* **41**(14), 994–998.
5. Bonev, S. A. and Ashcroft, N. W. (2001) Hydrogen in Jellium: First-principle Pair Interactions, *Phys. Rev. B* **64**, 224112 (1)–(9).
6. Nagao, K., Bonev, S. A. and Ashcroft, N. W. (2001). Cusp-condition Constraints and the Thermodynamic Properties of Dense Hot Hydrogen, *Phys. Rev. B* **64**, 224111 (12).
7. Richardson, C. F. and Ashcroft, N. W. (1997) High Temperature Superconductivity in Metallic Hydrogen: Electron–Electron Enhancements, *Phys. Rev. Lett.* **78**(1), 118–122.
8. Maksimov, E. G. and Shilov, Yu. I. (1999) Hydrogen at High Pressure, *Phys. Usp.* **42**(11), 1121–1138.
9. Pierleoni, C., Ceperley., D. M. and Holzmann, M. (2004) Coupled Electron–Ion Monte Carlo Calculations of Dense Metallic Hydrogen, *Phys. Rev. Lett.* **93**, 146402(1)–(4).
10. Filinov, V. S., Bonitz, M., Fortov, V. E., Ebeling, W., Levashov, P. and Schlanges M. (2004) Thermodynamic Properties and Plasma Phase Transition in Dense Hydrogen, Contrib. *Plasma Phys.* **44**(5–6), 388–394.
11. Stevenson, D. J. and Ashcroft, N. W. (1974) Conduction in Fully Ionized Liquid Metals, *Phys. Rev. A* **9**(2), 782–788.
12. Kwon, I., Collins, L., Kress, J. and Troullier, N. (1996) Electrical Conductivities for Hot, Dense Hydrogen, *Phys. Rev. E* **54**(3), 2844–2851.
13. Geldart, D. J. M. Vosko, S. H. (1966) The Screening Function of an Interacting Electron Gas, *Can. J. Phys.* **44**(96), 2137–2171.
14. Shvets, V. T. (2000) The Choice of the Pseudopotential and the Electrical Resistivity of Simple Disordered Metals, *Phys. Met. Metallogr. (Russia)*, **89**(3), 211–216.
15. Shvets, V. T. and Belov, E. V. (1999) Choice of Peudopotential and Electro-resistance of Simple Metals, *Acta Phys. Pol. A* **96**(6), 741–750.
16. Ziman, J. M. (1961) A Theory of Electrical Properties of Liquid Metals. I. The Monovalent Metals, *Phil. Mag.* **6**(68), 1013–1034.
17. Shih, W. H. and Stroud, D. (1985) Thermolynamic Properties of Liquid Si and Ge, *Phys. Rev. B* **31**(6), 3715–3724.
18. Shvets, V. T. (2002) *Green's Function Method in the Theory of Metals (Latstar, Odessa)*, pp. 293–301.

DYNAMIC CROSSOVER AND LIQUID–LIQUID CRITICAL POINT IN THE TIP5P MODEL OF WATER

PRADEEP KUMAR, S. V. BULDYREV, H. E. STANLEY
Center for Polymer Studies and Department of Physics, Boston University, Boston, MA 02215, USA; Department of Physics, Yeshiva University, 500 West 185th Street, New York, NY 10033, USA

Abstract. Water is hypothesized to have a low temperature phase transition line which separates a high density water at high temperatures from a low density water at low temperatures. This negatively sloped first order liquid-liquid phase coexistence line terminates at a critical point known as the liquid-liquid critical point. This critical point is hypothesized to exist in a deeply supercooled region of the phase diagram of water. Recent experiments have given an indirect indication of this phase transition. Neutron scattering and NMR experiments on water confined in hydrophilic nanopores revealed that the dynamics of water changes from the dynamics of a high density liquid (non-Arrhenius) at high temperatures to that of a low density liquid (Arrhenius) at low temperatures. Motivated Motivated by these experiments we investigated the TIP5P model of water. We find that the dynamic transition appears to be related to crossing the specific heat maxima line (Widom line) which emanates from the liquid-liquid critical point. Moreover, we find that water not only undergoes dynamical change but also structural changes on crossing the Widom line. The structure of the liquid changes from a high density liquid to a low density liquid.

Keywords: liquid-liquid critical point, fragile-to-strong transition, supercooled water, TIP5P, Widom line.

S. J. Rzoska and V. A. Mazur (eds.), Soft Matter under Exogenic Impacts, 23–33.

1. Introduction

Water is the most abundant liquid on earth. This occupies almost 70% of the earth's surface and 90% of our bodies by volume. It is not only the abundance of water and its importance in biology, atmospheric, and engineering sciences that makes it interesting but rather a large set of its unusual properties that makes it unique among liquids. Response functions of water such as isobaric heat capacity, isothermal compressibility, etc. seem to diverge as temperature is decreased.[1,2] These and many other unusual properties of water, also referred to as water anomalies, were hypothesized to be the consequence of a first order phase coexistence line between liquids of two different densities which may exist at low temperatures.[3,4] This coexistence line terminates at a critical point which is located in the deep supercooled region of water known as the "no man's land" because it is difficult to do measurements on liquid water in this region as it freezes spontaneously (see Fig. 1). Recent experiments on confined water, where it was possible to avoid freezing of water by confining it to hydrophilic nanopores, indicated that there might be a possibility of a liquid–liquid critical point at low temperatures. Specifically it was found that the dynamics of water changes from non-Arrhenius to Arrehenius behavior upon cooling isobarically. Since non-Arrehenius behavior is a characteristic of high density liquid where the activation energy increases as the temperature decreases and Arrehenius behavior is a characteristic of a low density liquid where the activation energy is a constant, it was speculated that this dynamic transition was somehow associated with the liquid–liquid phase transition. In order to correctly interpret the interesting finding, we need to understand the behavior of the system in the vicinity of a critical point.

By definition, in a first order phase transition, thermodynamic state functions such as density ρ and enthalpy H discontinuously change as we cool the system along a path crossing the equilibrium coexistence line (Fig. 2(a), path β). In a real experiment, this discontinuous change may not occur at the coexistence line since a substance can remain in a supercooled metastable phase until a limit of stability (a spinodal) is reached[1] (Fig. 2(b), path β).

If the system is cooled isobarically along a path above the liquid–gas critical pressure P_c (Fig. 2(a), path α), the state functions continuously change from the values characteristic of a high-temperature phase (gas) to those characteristic or a low-temperature phase (liquid). The thermodynamic response functions which are the derivatives of the state functions with respect to temperature (e.g., isobaric heat

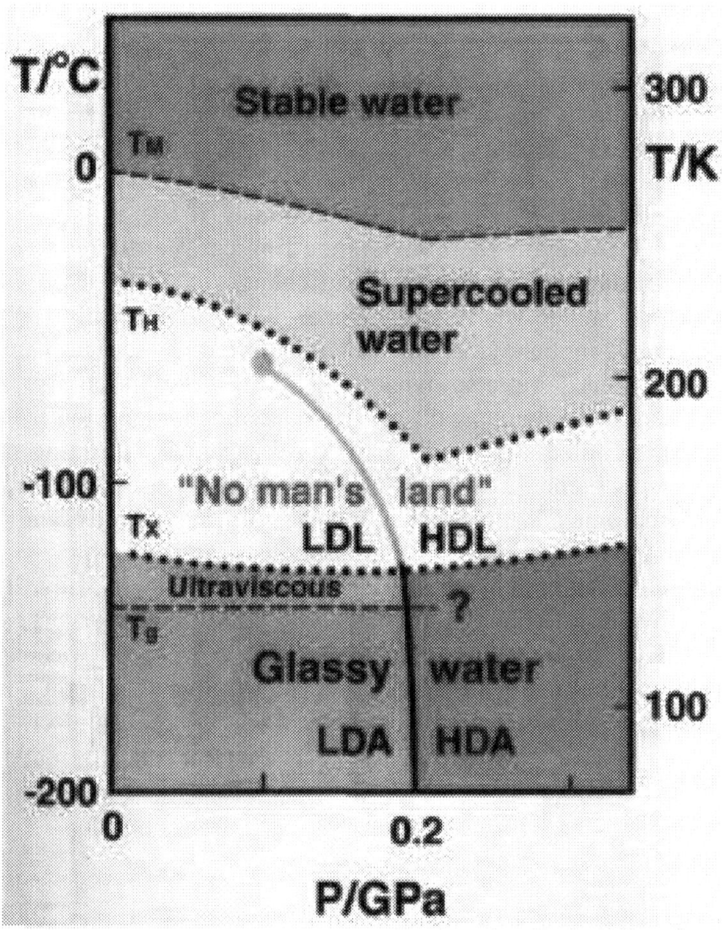

Figure 1. Phase diagram of liquid water. The liquid–liquid critical point lies in the experimentally inaccessible "no man's land." Courtesy of O. Mishima.

capacity $C_P = (\partial H/\partial T)$) have maxima at temperatures denoted by $T_{max}(P)$. Remarkably these maxima are still prominent far above the critical pressure,[5–8] and the values of the response functions at $T_{max}(P)$ (e.g., C_P^{max}) diverge as the critical point is approached. The lines of the maxima for different response functions asymptotically approach one another as the critical point is approached, since all response functions become expressible in terms of the correlation length. This asymptotic line is sometimes called the Widom line, and is often regarded as and extension of the coexistence line in to the "one-phase regime" (Fig. 2). If the system is cooled at constant pressure P_0, and P_0 is not too far from the critical pressure P_c, then there are two classes of behavior possible. (i) If $P_0 > P_c$ (Fig. 2(b), path α) then experimentally–measured quantities will change dramatically but continuously in the vicinity of the Widom line (with huge fluctuations as measured by, e.g., C_P). (ii) If $P_0 < P_c$ (Fig. 2(b), path β), experimentally-measured quantities will change discontinuously if the coexistence line is actually seen. However, the coexistence line can be difficult to detect in a pure system due to metastability, and changes will occur only when the spinodal is approached where the gas phase is no longer stable.

The changes in behavior may include not only static quantities like response functions[5–8] but also dynamic quantities like diffusivity. In water, a significant change in dynamical properties has been suggested to take place in deeply supercooled states.[2,9–15] Unlike other network forming materials[16] water behaves as a fragile liquid in the experimentally accessible window.[2,17,18] Based on analogies with other network forming liquids and with the thermodynamic properties of the amorphous forms of water, it has been suggested that, at ambient pressure, liquid water should show a dynamic crossover from fragile behavior at high T to strong behavior at low T.[14,19–23] Using Adam–Gibbs theory,[24] the dynamic crossover in water was related to the C_P^{max} line.[11,12,25] In this work, we interpret recent experiments on water[9–11] as arising from the presence of the hypothesized liquid–liquid critical point, which gives rise to a Widom line and an associated fragility transition (Fig. 2(c), path a).

2. Methods

Using molecular dynamics (MD) simulations,[26] we study the TIP5P model of water which accurately represents water behavior at ambient conditions and has a liquid–liquid critical point. This model[27] treats water as a multiple site rigid body, interacting via electrostatic site–site interactions complemented by a Lennard-Jones potential.[28] We evaluate the loci of maxima of the relevant response functions, compressibility, and specific heat, which coincide close to the critical point and give rise to the Widom line. We find evidence that, for both potentials, the dynamic crossover occurs just when the Widom line is crossed.

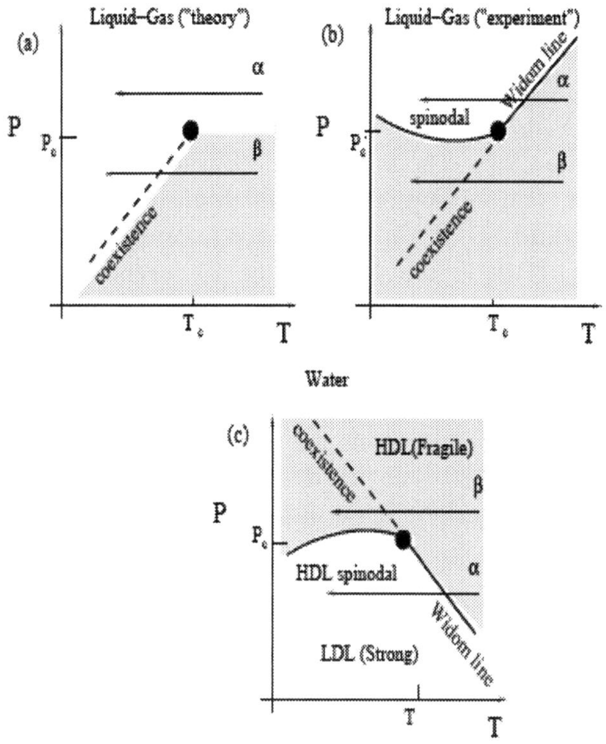

Figure 2. (a) Schematic phase diagram for the critical region associated with a liquid–gas critical point. Shown are the two features displaying mathematical singularities, the critical point (closed circles) and the liquid–gas coexistence (bold dashed curve). (b) Same as (a) with the addition of the gas–liquid spinodal and the Widom line. Along the Widom line, thermodynamic response functions have extrema in their T dependence. The path α denotes a path along which the Widom line is crossed. Path β denotes a path meeting the coexistence line. (c) A hypothetical phase diagram for water of possible relevance to the recent neutron scattering experiments by Chen *et al.*[9,10] on confined water. The liquid–liquid coexistence, which has a negative sloped coexistence line, generates a Widom line which extends below the critical point, suggesting that water may exhibit a fragile-to-strong xs transition for $P < P_c$ (path α), while no dynamic changes will occur above the critical point (path β) (adapted from Ref. 12).

Our results for the TIP5P potential are based on MD simulations of a system of $N = 512$ molecules, carried out both in the NVT and NPT ensembles using the techniques described in Refs. 26 and 29.

3. Results

Figure 3 shows the relevant portion of the P–T phase diagram for the TIP5P model. A liquid–liquid critical point is observed,[29,30] from which the Widom line develops. The coexistence curve is negatively sloped, so the Clapeyron equation implies that the high-temperature phase is a high density liquid (HDL) and the low-temperature phase is a low density liquid (LDL). Figure 4 shows the T dependence of the diffusion coefficient D, evaluated from the long time limit of the mean square displacement along isobars. The isobars crossing the Widom line (path α) show a crossover (i) from a non-Arrhenius behavior at high T (which can be fit by a power law function $D \sim (T - T_{\mathrm{MCT}})^{\gamma}$) where T_{MCT} is the crossover temperature predicted by the mode coupling theory[31], (ii) to an Arrhenius behavior at low T (which can be described by $D \sim \exp(-E_a/T)$). The crossover between these two functional forms takes place at the Widom line.

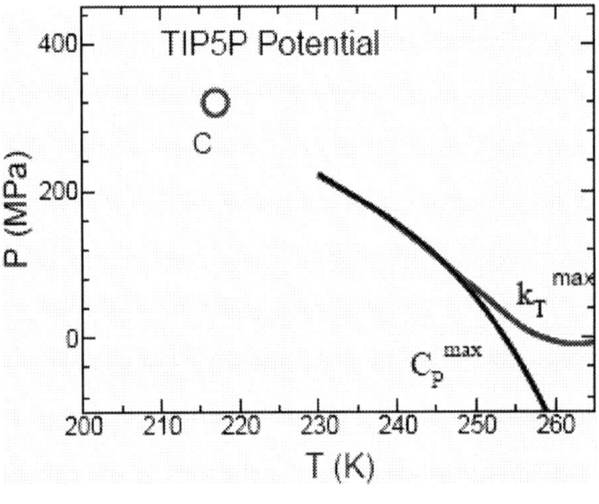

Figure 3. Relevant part of the phase diagram for the TIP5P potential, showing the liquid–liquid critical point C at P_c = 320 MPa and T_c = 217 K, the line of maximum of isobaric specific heat Cp^{max} and the line of maximum of isothermal compressibility KT^{max}.

Thus, we see that the simulations for TIP5P water model support the connection between the Widom line and the dynamic fragility crossover. Crossing the Widom line, not only gives a dynamic crossover but also the structural changes where water goes from a less ordered, "high-density-like" water at high temperatures to more ordered, "low-density-like" water at low temperatures. In Fig. 5, we show the oxygen–oxygen pair correlation functions $g(r)$ corresponding to two temperatures, one above the Widom line and another below the Widom line. Above the Widom line at $T = 250$ K, water is less ordered as seen from the smaller first and second peaks of $g(r)$, while below the Widom line at $T = 230$ K, water is more ordered, reflected in the increase in the amplitude of the first and the second peaks of $g(r)$. The area between the first and the second peak of $g(r)$ decreases as the Wisdom line is crossed, suggesting that the number of bifurcated bonds decreases and hence

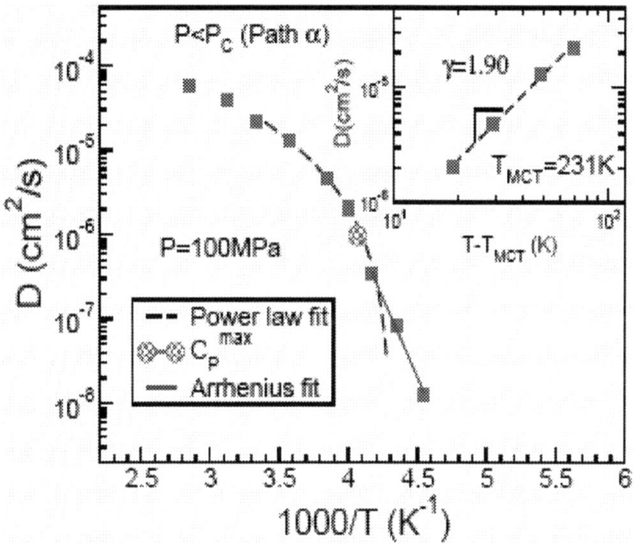

Figure 4. At high temperatures, D behaves like that of a non-Arrhenius liquid and can be fit by $D \sim (T - T_{MCT})^{\Gamma}$ (also shown in the inset) where $T_{MCT} = 231$ K and $\gamma = 1.90$, while at low temperatures the dynamic behavior appears to change to that of a liquid where D is Arrhenius.

fifth neighbors (which is a feature of high density water) also decreases upon crossing the Widom line from the high-temperature side to the low-temperature side. Figure 6 shows the oxygen–oxygen structure factor of water on two sides of the Widom line at $P = 100$ MPa. The structure of the liquid on high-temperature side of the Widom line is less tetrahedral compared to the structure of the liquid on the low-temperature side of the Widom line. In Fig. 7, we show the corresponding hydrogen bonds formed by the water molecules with a hydrogen bond-length cut-off distance of 0.3 nm and angle width of 10 degrees.

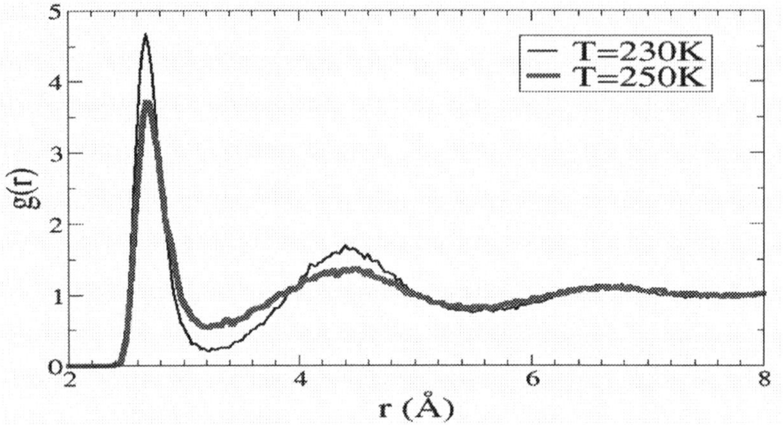

Figure 5. Oxygen–oxygen pair correlation function above and below the Widom line at $P = 100$ MPa. Above the Widom line, water is less ordered while upon crossing the Widom line, water becomes more ordered.

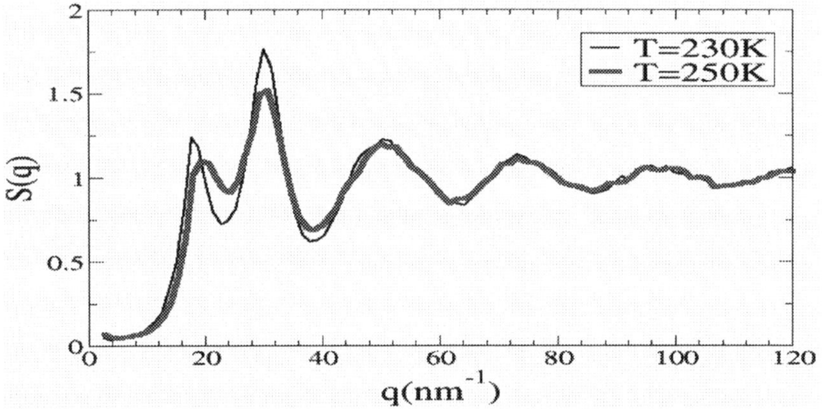

Figure 6. Structural changes on crossing the Widom line at $P = 100$ MPa. (a) Oxygen–oxygen structure factor $S(q)$ above the Widom line (see Fig. 5). (b) Structure factor $S(q)$ below the Widom line. Tetrahedrality of the water molecules increases dramatically upon crossing the Widom indicated by a sharp increase in the first and second peaks of the structure factor.

For the temperature below the Widom line water is more structured locally and hence more hydrogen bonds with a strict constraint of the hydrogen-bond angle are formed compared to water above the Widom line temperature.

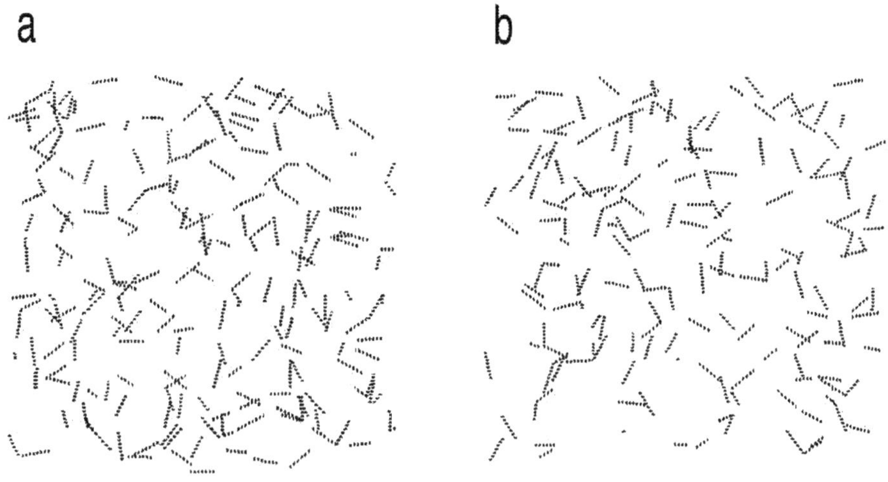

Figure 7. Structural changes on crossing the Widom line at P = 100 MPa. (a) Hydrogen bonds below the Widom line (see Fig. 5). (b) Hydrogen bonds above the Widom line. Number of the hydrogen bonds with strict tetrahedral structures increases on crossing the Widom line upon cooling.

4. Discussion and summary

Our findings are consistent with the possibility that the observed dynamic crossover along path *a* is related to the behavior of C_P, suggesting that enthalpy or entropy fluctuations may have a strong influence on the dynamic properties.[15,32] Indeed, as the thermodynamic properties change from the high-temperature side of the Widom line to the low-temperature side, $(\partial S/\partial T)_P = C_P/T > 0$ implies that the entropy must decrease. The entropy decrease is most pronounced at the Widom line when $C_P = C_P^{max}$. Since the configurational part of the entropy, S_{conf}, makes the major contribution to S, we expect that S_{conf} also decreases sharply on crossing the Widom line.

D changes from non-Arrhenius ("fragile") to Arrhenius ("strong") behavior, while the structural and thermodynamic properties change from those resembling HDL to those resembling LDL, due to the *negative* slope of the Widom line.

In summary, our results are consistent with thc cxpcrimcntal observation in confined water of (i) a fragility transition for $P < P_c$,[9,10,11] and (ii) a peak in C_P upon cooling water at atmospheric pressure.[33] Thus, our work offers a plausible interpretation of the results of Refs. 9–11 as supporting the existence of a hypothesized liquid–liquid critical point.

Acknowledgments

We thank C. A. Angell, S.-H. Chen, G. Franzese, J. M. H. Levelt Sengers, L. Liu, M. G. Mazza, P. H. Poole, S. Sastry, F. Sciortino, F. W. Starr, B. Widom, L. Xu, and Z. Yan for helpful discussions and NSF grant CHE 0096892 for support. We also thank the Boston University Computation Center, Yeshiva University Office of Academic Affairs, and StFX hpcLAB for allocation of CPU time.

References

1. Debenedetti, P. G., and Stanley, H. E. (2003) *Physics Today* **56** (6), 40–46.
2. Debenedetti, P. G. (2003) *J. Phys.: Condens. Matter* **15**, R1669–R1726.
3. Mishima O., and Stanley, H. E. (1998) *Nature* **392**, 164–168.
4. Poole, P. H., Sciortino, F., Essmann, U., and Stanley, H. E. (1992) *Nature* **360**, 324–328.
5. Anisimov, M. A., Sengers, J. V., and Levelt Sengers, J. M. H. (2004) in *Aqueous System at Elevated Temperatures and Pressures: Physical Chemistry in Water, Stream and Hydrothermal Solutions* eds. Palmer, D. A., Fernandez-Prini, R., and Harvey, A. H. (Elsevier, Amsterdam).
6. Levelt, J. M. H. (1958) in *Measurements of the compressibility of argon in the gaseous and liquid phase,* Ph.D. Thesis (University of Amsterdam, Van Gorkum & Co., Assen).
7. Michels, A., Levelt, J. M. H., and Wolkers, G. (1958) *Physica* **24**, 769–794.
8. Michels, A., Levelt, J. M. H., and de Graaff, W. (1958) *Physica* **24**, 659–671.
9. Liu, L., Chen, S.-H., Faraone, A. Yen, C.-W., and Mou, C.-Y. (2005) *Phys. Rev. Lett.* **95**, 117802.
10. Faraone, A., Liu, L., Mou, C.-Y., Yen, C.-W., and Chen, S.-H. (2004) *J. Chem. Phys.* **121**, 10843–10846.
11. Mallamace, F., and Chen S.-H. (2006) *J. Chem. Phys. E* **73**, 020402–8.
12. Xu, L., Kumar, P., Buldyrev, S. V., Chen, S.-H., Poole, P., Sciortino, F., and Stanley, H. E. (2005) *Proc. Nat. Acad. Sci.* **102**, 16558.
13. Chen, S.-H., Liu, L., Fratini, E., Baglioni, E. P., Faraone, E., and Mamontov, E. (2006) *Proc. Nat. Acad. Sci.* (submitted).
14. Kumar, P., Xu, L., Yan, Z., Mazza, Buldyrev, S. V., Chen, S.-H., Sastry, S., and Stanley, H. E. (2006) (submitted).
15. Scala, A., Star, F. W., La Nave, E., Sciortino, F., & Stanley H. E. (2000) *Nature* **406**, 166–169.
16. Starr, F. W., Angell, C. A., and Stanley, H. E. (2003) *Physica A* **323**, 51–66.

17. Angell, C. A. (2004) *Ann. Rev. Phys. Chem.* **55**, 559–583.
18. Angell, C. A. (1993) *J. Phys. Chem.* **97**, 6339–6341.
19. Kumar, P., Buldyrev, S. V., Starr, F. W., Giovambattista, N., and Stanley, H. E. E. (2005) *Phys. Rev. E* **72**, 051503.
20. Horbach, J., and Knob, W. (1999) *Phys. Rev. B* **60**, 3169–3181.
21. Lang E. W., and Ludemann, H. D. (2004) *Angew Chem. Intl. Ed. Engl.* **21**, 315–329.
22. Prielmeier, F. X., Lang, E. W., Speedy, E. J., and Ludemann, H. D. (1987) Bunsenges, Ber. (1987) and *Phys. Rev. Lett.* **59**, 1128–1131.
23. Ito, K., Moynihan, C. T., and Angell, C. A. (1999) *Nature* **398**, 492–495.
24. Tanaka, H. (2003) *J. Phys.: Condens. Matter* **15**, L703-L711.
25. Poole, P. H., Saika-Voivod, I., and Sciortino, F. (2005) *J. Phys.: Condens. Matter* **17**, L431–L437.
26. Rapaport, D. C. (1995) in *The Art of Molecular Dynamics Simulation* (Cambridge University Press, Cambridge).
27. Mahoney, M. W., and Jorgensen, W. L. (2000) *J. Chem. Phys.* **112**, 8910–8922.
28. Stillinger, F. H., and Rahman, A. (1972) *J. Chem. Phys.* 57, 1281–1292.
29. Yamada, M., Mossa, S., Stanley, H. E., and Sciortino, F. (2002) *Phys. Rev. Lett.* **88**, 195701.
30. Paschek, D. (2005) *Phys. Rev. Lett.* **94**, 217802.
31. Goetze, W. (1991) in *Liquids, Freezing, and Glass Transition*, eds. J. P. Hansen, D. Levesque, and J. Zinn-Justin (North-Holland, Amsterdam), p. 287.
32. Saika-Voivod, I., Poole, P. H., and Sciortino F. (2001) *Nature* **412**, 514–517.
33. Maruyama, S. Wakabayashi, K., and Oguni, M. (2004) *AIP Conf. Proc.***708**, 675–676.

AMORPHIZATION OF ICE BY COLLAPSE UNDER PRESSURE, VIBRATIONAL PROPERTIES AND ULTRAVISCOUS WATER AT 1 GPA

G. P. JOHARI*
Department of Materials Science and Engineering, McMaster University, Hamilton, Ontario, L8S 4L7, Canada

OVE ANDERSSON
Department of Physics, Umeå University, 901 87 Umeå, Sweden

Abstract. When subjected to a uniaxial pressure of 0.7 to 1.5 GPa, structures of hexagonal and cubic ices at a temperature below 140 K collapse and the ordered arrangement of molecules is lost. Another well-known (tetrahedrally bonded and open structure) crystal, SiO_2, also collapses and become amorphous but at 25–30 GPa and 300 K. This is known as pressure-induced amorphization of crystals. Here we report, (i) how the vibrational properties, e.g., molar volume V, limiting high frequency permittivity ε_∞, ultrasonic sound velocity, and thermal conductivity κ change during the pressure-amorphization, and (ii) how the amorphized ice relaxes to a lower energy state on heating to 140 K, and becomes ultraviscous water of dielectric relaxation time of ~1 s at 1 GPa pressure. As the extent of amorphization increases on increasing the pressure to 1.5 GPa, V and κ irreversibly decrease and ε_∞ and the ultrasound velocity increase. Amorphization begins at a lower pressure for micron-size ice crystals than for larger crystals. It also begins at a lower pressure at high temperatures of ice. At a fixed pressure and temperature, ice continues to amorphize up to a period of several days according to a stretched exponential kinetics and a pressure- and temperature-dependent rate constant. It is proposed that lattice faults, which are also produced during pressure-deformation of ice cause a distribution of the Born instability pressures, and the amorphization process becomes pressure- and time-dependent. Pressure-induced amorphization of ice at 77 K produces kinetically unstable high energy amorphs in the same manner as mechanical deformation of other crystals produces kinetically unstable, high energy amorphs which, on heating, become an ultraviscous liquid. But, in contrast,

*To whom correspondence should be addressed. G. P. Johari, Department of Materials Science and Engineering, McMaster University, Hamilton, Ontario, L8S 4L7, Canada; e-mail: joharig@mcmaster.ca

S. J. Rzoska and V. A. Mazur (eds.), Soft Matter under Exogenic Impacts, 35–74.

the ice amorphs are denser than the parent ices, and bulkier than ice VI the stable phase, and ice XII the metastable phase.

Keywords: Pressure-induced amorphization, amorphous and crystalline ices, thermal conductivity, dielectric spectroscopy, high pressure measurements.

1. Introduction

One of the most important exogenic effects on materials is that of pressure. The pressure may result from natural or man-made impact, from a shock wave generated by it, or from a sustained stress as in geological occurrences. Here we describe an unusual phenomenon of irreversible collapse of crystals under pressure, that also destroys their structural order. Historically, as early as 1963, Skinner and Fahey[1] had reported that stishovite, a form of crystalline silica, becomes amorphous by the simple procedure of mechanical grinding, and others had reported similar observations for other crystals soon thereafter.[2] In his compilation of the properties of silica, Primak[3] had noted that under a shock pressure of more than 20 GPa, SiO_2 becomes amorphous. In 1972, Brixner[4] reported that $Gd_2(M_2O_4)$ crystals become amorphous when subjected to pressure. Since 1981, there have been a number of systematic studies of shear-induced mechanical amorphization of metal crystals in a high speed ball mill, in which crystals were subjected to high impact.[5,6] Hemley et al.[7] reported that α-quartz and coesite crystals transform to amorphous solids at 25–35 GPa and 300 K, indicating the thermoelastic instability of tetrahedral network structures at high compression. The amorphous state is recognized by the loss of sharp scattering features in their x-ray and neutron diffraction features, or as broadening of the peaks in the vibrational spectra. Briefly, collapse of a crystal's structure has now been discovered in materials with covalent, ionic, and hydrogen bonding and in metals. Some of further examples are: calcium silicate perovskite,[8] α-berlinite,[9] α-quartz[7,10] porous silicon,[11] zeolites,[12] various aluminophosphates and aluminosilicates, β-eucryptite ($LiAlSiO_4$),[13] ZrV_2O_7,[14] ZrW_2O_8,[15] TiP_2O_7,[16] and most recently,[17] natrolite (Na_2 $[Al_2SiO_3O_{10}]\cdot 2H_2O$) and edingtonite (Ba $[Al_2SiO_3O_{10}]\cdot 4H_2O$). The process of collapse of crystals under pressure and the consequent loss of characteristic x-ray diffraction features has been called pressure-induced amorphization, and for several crystals it has been related to their negative coefficient of thermal expansion over a certain temperature range. Sharma and

Sikka[18] and Greaves[19] have reviewed some of the earlier studies on the subject. This exogenic effect is relevant to geological formations as a result of meteoritic impact. It is also relevant to plate tectonics, when friction between the continental plates produces an enormous stress, and which on sudden release produces shock waves in the earth's crust. Most recently, the subject of pressure-amorphization has received attention in a different context: the manufacture of ceramic armor plates whose crystalline material becomes partly amorphous on impact and hence absorbs energy.

One of the most discussed solids amongst the pressure-collapsed states is one that was accidentally produced by uniaxially pressurizing a common solid, hexagonal ice, at 77 K in a piston cylinder apparatus to ~1.5 GPa. The tetrahedral (hydrogen) bonding in the ice's open structure is much weaker than the covalent bonding in SiO_2, and as a result, hexagonal ice at 77 K began to collapse under a relatively low pressure of ~1 GPa. The density increased from 0.93 to 1.31 g cm^{-3} at 77 K and 1 GPa and it transformed to an amorphous solid.[20] The collapsed state has been called the high-density amorph. It has been thought that under pressure at 77 K, hexagonal ice slowly melts irreversibly at ~1 GPa pressure, making it a first example of a slow crystal-liquid (thermodynamic) transformation upon isothermal pressurization. However, studies by Johari and Jones,[21] who attempted to find similar amorphization by presumed melting of tetrahydrofuran ice clathrate at a much lower pressure of ~0.4 GPa, found no indication of such a collapse. They concluded that, "... the collapse of crystalline order and of the hydrogen bonded structure under a uniaxial stress is a more likely cause of the pressure-induced amorphization than the proposed melting along the extrapolated phase boundary."[21] The high-density amorph has been recovered at ambient pressure at 77 K, and various properties of its recovered state have been studied at ambient pressure. Its structure has been determined from x-ray and neutron diffraction studies in the recovered state. The diffraction features, or the details of the structure factor obtained from studies by the same group and different groups have been found to differ significantly. In a discussion of the subject of multiplicity of high-density amorphs formed by pressure-collapse,[22] it has been concluded that lack of control of pressure, temperature, and time conditions has led to such differences, and that there is no single high-density amorph of ice. Cubic ice also transforms to a high-density amorph but at somewhat different pressures.[23, 24]

When the high-density amorph obtained by pressure-amorphizing hexagonal and cubic ices at 77 K is heated at a fixed pressure in the 0.1 MPa–0.07 GPa range, it gradually, and irreversibly, expands on heating to a form known as low-density amorph,[25] which apparently has the same density as hexagonal and cubic ices. The temperature of this transformation also varies with the procedure used to amorphize hexagonal and cubic ices. Moreover, qualitative features of this transformation kinetics are debated. Thus, not only the mechanism of pressure-amorphization, but also the structure and the behavior of the pressure-amorphized ice have been debated. Johari[26] has critically reviewed developments in this subject until the year 2000. Gromnitskaya et al.[27] have provided a review of the ultrasonic properties of the high-density and low-density amorphs, and of the changes in the properties of hexagonal ice during the amorphization. Debenedetti[28] has published a general review, "Supercooled and glassy water" in which pressure-amorphized ice has also been discussed. Andersson and Inaba[29] have provided a detailed review of the changes in the thermal conductivity of the amorphized state and of crystalline ices. These reviews may be consulted for the advances made in the pressure-induced amorphization of ice.

Loerting et al.[30] discovered that pressure-amorphized ice becomes further ~5% denser when heated from 77 to 168 K at a fixed pressure in the range ~0.8 to 1 GPa, and thereafter the still denser state can be recovered at 0.1 MPa and 77 K. This discovery has caused an ambiguity in the earlier interpretation of the properties of the pressure-amorphized ice when it was thought that there was only one high-density amorph and it was believed by various groups that this amorph had been produced by pressurizing to 1.5 GPa at 77 K in a short time. Since the pressure, temperature, and time conditions in the earlier studies were not controlled, variation in the properties of the high-density amorphs formed could not be attributed to any one, or all these conditions. Here we describe studies of the formation of the high-density amorph by pressure-collapse of cubic and hexagonal ices under precise conditions at the highest temperature of 130 K, by using extremely slow pressurization rates over a longer period of time. In particular, we describe, (i) how a vibrational property, namely the thermal conductivity, κ, changes as hexagonal and cubic ice collapse under different pressure, temperature and time conditions and become amorphous, and (ii) how heating of the amorph formed by collapse leads to an ultraviscous liquid. We also describe a mechanism for the pressure-collapse of the ices that produces a high-density amorph. It is based on the Born criteria for elastic instability, but, in our description, the lattice defects in the sample modify the instability conditions and produce a distribution of the collapse, or amorphization pressures. Since Loerting et al.'s[30] plot of piston displacement against temperature (Fig. 1, Ref. 30) obtained by heating at 4–6 K/min shows that the transformation is halfway through the transformation to the densest amorph at 130 K and 1.1 GPa, the slow pressurizing rate at 130 K used here produces the amorph that they had obtained by heating to higher temperatures.

Properties of water and of its solid phases at high pressures are of general interest. These properties are also of much importance for cryobiology,[31] astrophysics, and geophysics of planets and of their satellites.[32–36] There are 13 known crystalline phases of ice.[37] Many of these crystalline and amorphous phases are apparently present in large quantities in satellites and giant planets such as Europa, Callisto, and Ganeymede, as well as in the nuclei of comets and interstellar dust.[32–36] As the focus of this NATO Advanced Workshop is on experimental aspects of exogenic effects, we have limited our discussion here to three aspects: (i) the experimentally observed changes in the properties during the pressure-induced amorphization of ice, (ii) the mechanism of the pressure-induced amorphization of ice, and (iii) the recovery of normally unachievable ultraviscous water under pressure by an indirect route of pressure collapsing hexagonal ice. We propose that there is only one (ultimate) high-density amorph. It forms by thermally-activated relaxation at pressures near 1 GPa or by slow pressurization at high temperatures, and, like the mechanically amorphized crystals, the various structurally and kinetically different amorphs produced in earlier studies correspond to (nonergodic) states of different fictive temperatures at high pressures. Therefore, this article is based mainly on experimental studies performed since the last review in the year 2000.[26] Some of the aspects described here have been reported earlier.[38]

2. Experimental methods

We have chosen to describe here measurements of three vibrational properties, (i) the volume, (ii) the limiting high frequency dielectric permittivity ε_∞, i.e., the sum of the contribution to the dielectric permittivity from electronic and infrared polarization, and (iii) the thermal conductivity, κ. These properties were measured in real time during the course of pressure-amorphization at a fixed temperature, T, and at a fixed pressure, p, isothermally as a function of time. We have also chosen to describe measurements of a configurational property, namely, the dielectric relaxation time, by using a different technique and a separate set of experiments. It is conceivable that in the future, all three measurements can be performed simultaneously in a single experimental procedure.

For thermal conductivity measurements, the principle underlying the hot-wire method was used. This principle and the method have been described in detail earlier.[39] Briefly, the hot-wire sample cell for high pressure measurements of κ is shown in Fig. 1(A). The cell is made of Teflon, which is an ideal material for high-pressure studies since it deforms easily under pressure and it is easy to machine. The hot-wire probes used in the investigation of the solids were Ni-wires of about 40 mm long and 0.1 mm diameter, but Pt-wires have also been commonly used as probes. The reason for using Ni and Pt is that their electrical resistance varies relatively strongly with temperature. This makes wires of these two metals relatively well suited in experiments such as ours in which the wire itself is used both as heater and temperature sensor. Since the Teflon cell had to be small enough to occupy the limited space in the high pressure vessel, the wire was placed horizontally in a ring of constant radius within the Teflon cell. This method yields more accurate values than one that uses thin planar heaters, because planar heaters require a substrate for their physical support when the starting sample is a liquid, as is the case in our study. Since a part of the heat flux goes through the supporting substrate, the substrate decreases the signal associated with the sample. For accurate measurements with planar heaters, this correction needs to be made by using thermal properties of the planar heaters' substrate as a function of p and T, which is often not the case.

The Teflon cell was filled with ~25 ml of tissue-culture grade pure water supplied by Sigma Chemicals, or by water purified by using Milli-Q® Ultrapure WaterSystems. The cell was sealed with a tightly fitting, 5 mm thick Teflon cover and the piston inserted. The assembled Teflon cell filled with water was mounted in a piston-cylinder type of pressure vessel of 45 mm internal diameter, as shown in Fig. 1(B), and the load was applied using a hydraulic press. In such measurements on a solid sample, a small pressure must be maintained to ensure good thermal contact between the wire and the sample. The pressure was determined from the load/area ratio with an empirical correction for friction which had been established by a calibration curve in which the resistance of a manganin wire was measured as a function of pressure during the pressure cycling. The inaccuracy in pressure is estimated as ±40 MPa at 1 GPa and 298 K and, due to increased friction, ±60 MPa at 40 K and 1 GPa. The temperature was varied by cooling or warming the entire pressure vessel by means of a closed cycle helium cryostat. To do so, the pressure vessel was contained inside a box under vacuum, in thermal contact with the cryostat's head. The temperature of the sample was measured using a chromel–alumel thermocouple which was placed inside the vessel. The inaccuracy in temperature is estimated as ±0.5 K. The apparatus used for the purpose has been described in detail elsewhere.[40] Its drawing is shown in Fig. 1(B).

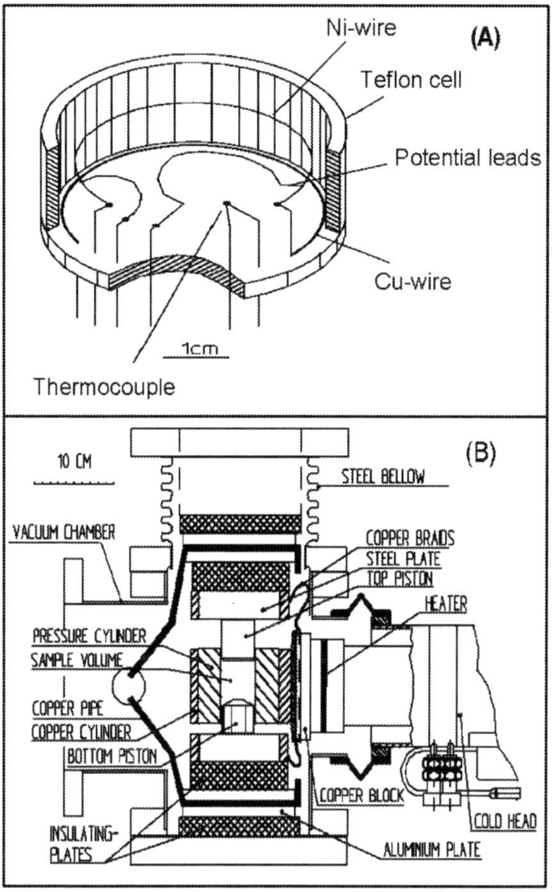

Figure 1. (A) The high pressure cell assembly for measuring the thermal conductivity and heat capacity. (B) The arrangement of the high pressure vessel showing the application of the uniaxial stress and arrangement that allows the use of helium cryostat in a massive assembly.

To determine κ, the hot-wire probe embedded in the sample was heated by a 1.4 s duration electric pulse of almost constant power, and the wire's resistance was measured as a function of time. The temperature rise of the wire was calculated by using its electrical resistance–temperature relation, i.e., the wire working as both heater and sensor for the temperature rise. This method as well as other dynamic methods for measuring κ are based on solutions of the time-dependent equation for heat conduction. The exact solution of the temperature rise ΔT of an infinitely long, infinitely conducting wire immersed in an infinitely large specimen is [41]

$$\Delta T = \frac{2q\alpha^2}{\pi^3\kappa}\int_0^\infty \frac{1-\exp(-\beta u^2)}{u^3\{(uJ_0(u)-\alpha J_1(u))^2+(uY_0(u)-\alpha Y_1(u))^2\}}du, \quad (1)$$

where q is the constant heating power supplied per unit length of the wire, $\alpha = 2\rho C_p/(\rho_w C_w)$, $\beta = \kappa t/(\rho C_p r^2)$, t is the time, r is the radius of the hot-wire, ρ and C_p are the density and specific heat capacity of the specimen, ρ_w and C_w are the density and specific heat capacity of the hot-wire, J_0 an J_1 are Bessel functions of the first kind of zero and first order, and Y_0 and Y_1 are Bessel functions of the second kind of zero and first order. Equation (1) was fitted to the data points, and κ was thus determined. For temperatures above 100 K, the inaccuracy in κ is estimated as ±2%. As the sensitivity of the hot-wire (i.e., the temperature derivative of resistance) decreased with decrease in the temperature, the inaccuracy in the measured κ increased, and it became ±4% at 40 K. The standard deviation in the measurements was found to be an order of magnitude smaller than the inaccuracy.

The limiting high frequency permittivity was measured[21] by compressing hexagonal ice contained in a piston cylinder, concentric electrode dielectric cell made from Be–Cu alloy. This cell was used in the dielectric measurements of ice VI, where details may be found.[42, 43]

For dielectric spectroscopy study, two dielectric cells were used, one a parallel plate capacitor of nominally 139 pF air capacitance, and another a concentric electrode capacitor of 19 pF. The change in geometry of the latter up to a pressure of 1 GPa is insignificant when the pressure is hydrostatic.[44] The capacitor was placed inside a Teflon container identical to the one used for the thermal conductivity and specific heat measurements, and the same pressure assembly was used, with the same pressure and temperature controls.[40] Electrical leads of the dielectric cell were connected to a Solartron 1260 impedance analyser. The dielectric spectra were measured in real time as the pressure and temperature of the sample was cycled at a very slow rate.

The Teflon cell was filled with ca. 25 ml of pure water. The cell was sealed with a tightly fitting, 5 mm thick Teflon cover and the piston inserted. The whole assembly was placed in a vacuum chamber and load was applied. The pressure in the cell was determined from the ratio of load to area to which a correction for friction was applied. This correction had been previously established in a separate, *in situ*, experiment by using the pressure dependence of the resistance of a manganin wire. As for the thermal conductivity measurements, the temperature was varied by using a built-in helium cryostat equipped with heater, and the temperature, and pressure on the sample, were computer-monitored continuously. The accuracy here was also ±0.5 K for the temperature and ±0.06 GPa for the pressure at 1 GPa. The high pressure vessel, containing water at ~0.04 GPa, was cooled from 298 K to about 100 K. Thereafter, its temperature was raised and stabilized at 115–130 K, which took a total of 10 h. The pressure was then raised at an average rate of 0.1 GPa/h. The heating rate was typically 15 K/h and cooling rate, 10 K/h, near 150 K. Thus, one set of experiments took a continuous period of several days. Measurements were also performed by using water doped with 1.7 $\times 10^{-4}$ mole fraction of KOH. The additional results of KOH doped pressure amorphized ice together with results of pure samples are useful for confirming the amorph's relaxation behavior. This is particularly so because the KOH-doping has been found to decrease the dielectric relaxation time in hexagonal ice due "to the release of water molecules from the ice rules",[45] therefore, allowing also measurements of the relaxation time of hexagonal ice at a lower temperatures than that of pure hexagonal ice. As the results on hexagonal ice are less relevant here they are excluded.

The temperature difference between the sample and the Teflon cell wall was also measured continuously in order to determine any rapid transformation to another crystalline or amorphous state. This measurement is equivalent to differential thermal analysis performed at a very slow heating rate. Because of the long duration of the experiments and slow rate of pressure and temperature changes the ice frequently crystallized rapidly during its pressurizing from 0.8 to 1.1 GPa, i.e., in the pressure-amorphization range. This has not been the case in the studies by other groups who have used ~10 times higher pressurization and cooling rates–in some studies raising the pressure from 0.1 MPa to 1.5 GPa in 5 min. For caution in studies that may use slow pressurization and slow heating and cooling rates in the future, we note that the failure rate in our amorphization experiments is high. On average, one in five experiments was successful in providing the required data.

3. Results and discussion

3.1. CHANGE IN THE VIBRATIONAL PROPERTIES ON COLLAPSE OF ICE UNDER PRESSURE

During the course of the collapse of a crystal, topological rearrangement of atoms or molecules is forcibly altered by the pressure and the energy increases. If the change in the arrangement leads to total disorder, the collapsed solid appears amorphous in the x-ray and neutron diffraction spectra. However, if the change in the topological arrangement leads to nanometer size crystals of high pressure phases with a very large net surface area, the product formed after the collapse still appears amorphous to x-rays and neutrons. In either case, the collapse, occurring at low temperatures when the product is a rigid solid, produces a structure of fixed configuration like that of a nonergodic state, and therefore change in only those properties that are vibrational (non-relaxing) in origin can be observed during and after the collapse of a crystal's structure. These are determined by measurements of the volume, the ultrasonic sound velocity, the magnitude of the infrared-red polarization and the thermal conductivity during the course of collapse of a crystal. (Mechanical deformation of crystals in a high-speed ball mill also amorphizes a crystal, to a solid amorph if vitrification temperature of the amorph is much lower than the ball-milling temperature, and to an ultraviscous liquid, as was found for aspirin,[46] if it is higher.) Among those thermodynamic properties that arise from vibrational motions, the volume and the heat capacity are usually measured at different temperatures and pressures. We also recall that the pressure-collapse of hexagonal ice was accidentally discovered by Mishima et al.[20] by detecting the volume change, when they measured the piston displacement during the uniaxial compression of hexagonal ice at 77 K,[20] and found that a large decrease in its volume occurred. It should be noted that similar attempts for determining phase transformation of tetrahydrofuran ice clathrate at low temperatures had been made by Ross and Andersson a few years before but they did not study the volume change.[47] The original plots by Mishima et al.[20] had shown an inverted and broad sigmoid shape decrease in the piston displacement and an increase in the density and they stated that "... a new amorphous state of density of about 1.31 g cm^{-3} is produced by the transformation of ice Ih at 77 K and 10 kbar."[20] Mishima et al.[20] explained the pressure-collapse as melting of ice under pressure and refreezing to a solid amorph at 77 K, but Johari and Jones[21] argued that if the collapse of hexagonal ice was caused by pressure-melting then tetrahydrofuran ice clathrate, whose melting point is more sensitive to pressure than that of hexagonal ice, should collapse at a much lower pressure of 0.3 to 0.4 GPa. But experiments on the ice clathrate did not yield a sigmoid-shape displacement against pressure plot, nor could its pressure-collapsed solid be recovered at ambient pressure.

Studies of pressure-collapse of hexagonal ice were later performed by using samples prepared in three different ways,[26] (i) by freezing water, which produced up to 0.5 mm size crystals, (ii) by transforming cubic ice to microcrystalline hexagonal ice, and (iii) by allowing the micron-size crystals of hexagonal ice formed from the cubic ice to grow to a relatively large size. The original plots of the piston displacement against pressure were of course obtained by Mishima et al.,[20] but here in Fig. 2(A) we show the corresponding plots obtained by Johari.[26] The plots show that the mm-size crystals begin to collapse at ~1 GPa at 77 K and micron-size, crystals at ~0.7 GPa. This shows that the pressure needed to collapse the structure of ice decreases as the crystal size is decreased or as the ratio of surface to bulk energy is increased. He also reported[26] that his crude attempt to collapse a 2 mm diameter 3 mm long single crystal of hexagonal ice at 77 K by raising the pressure to ~2 GPa did not succeed.

Another vibrational property, namely, the limiting high-frequency permittivity, ε_∞, which is measured under conditions when dipolar relaxational contribution is absent, also changes during the pressure-collapse. The permittivity ε' of hexagonal ice at 77 K measured at 1 kHz frequency has no relaxational contribution and the dielectric loss is less than 1×10^{-6}.[21] Its ε' measured at 1 kHz frequency is therefore equal to its ε_∞ at 77 K. Its plot against the pressure at 77 K is shown in Fig. 2(B). On compression of hexagonal ice at 77 K, ε' (= ε_∞) increases initially slowly from its value of 3.1. Near the onset of the pressure-collapse, it increases rapidly and then approaches a constant value of 3.4 in a sigmoid-shape manner. Decompression of the sample does not restore the original value and ε' (=ε_∞) of the pressure-amorphized solid remains 3.36.[21] The increase in ε_∞ from 3.14 to 3.40 on pressure-collapse is partly due to the increase in the square of the optical refractive index from 1.73 to 1.79, as the density increases from 0.93 to 1.17 g cm^{-3}. The remainder of the increase is attributed to the increase in the infrared contribution to ε_∞. Accordingly, it was concluded that the frequency of translational vibrations in the pressure-collapsed solid is lower than in hexagonal ice and/or the absorptivity associated with these vibrations is higher. From that information, it was inferred[21] that the near neighbor distance between the water molecules in the amorphous phase formed by the structural collapse should be greater than in hexagonal ice. This inference was confirmed from direct determination of the x-ray diffraction of the pressure amorphized solid by Bosio et al.[48] In the absence of far-infrared spectra studies during the pressure-collapse of ice, the information obtained from the ε_∞ studies remains significant.

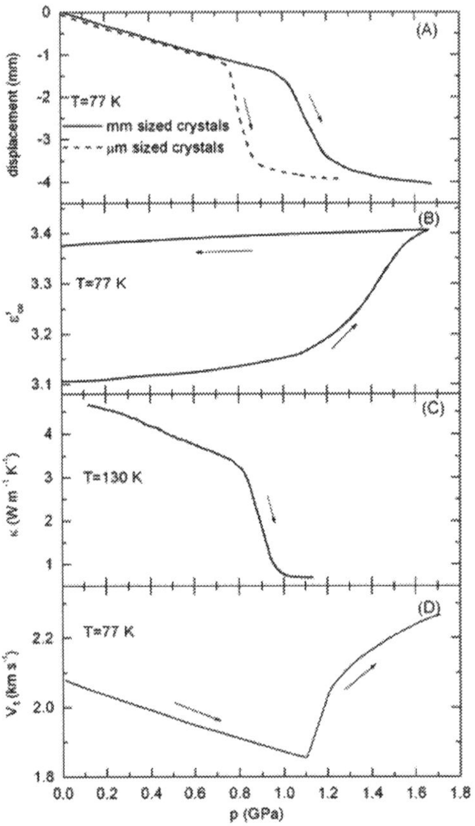

Figure 2. Changes in the vibrational properties of hexagonal ice on increase in the pressure showing the collapse of amorphization of ice. (A) Volume decrease indicated as displacement of the piston in the pressure vessel containing hexagonal ice crystals of mm size and μm size at 77 K. Data taken from Ref. 26. (B) Increase in the limiting high frequency permittivity at 77 K. Data taken from Ref. 21. After pressure-amorphization of hexagonal ice, decreasing the pressure to ambient does not restore the original values of volume, thermal conductivity, or the limiting high-frequency permittivity. (C) Decrease in the thermal conductivity of hexagonal ice at 130 K. Data taken from Ref. 38. (D) The The increase in the velocity of transverse sound wave at 77 K. Data taken from Ref. 27.

Much of the progress in the pressure-collapse and consequent amorphization of hexagonal ice have come from measurements of two properties. The first is the thermal conductivity, which has been studied by Andersson and coworkers.[22,24,29,38,49,50] In their studies on hexagonal and cubic ices, thermal conductivity, κ, decreased on increase in the pressure in an inverted broad sigmoid shape manner, similar to the piston displacement or the volume of the ices. Moreover, studies performed at different temperatures also showed the same feature. A typical plot of κ against pressure at 130 K is shown in Fig. 2(C). Its shape is remarkably similar to the plot of the volume (piston displacement) of hexagonal ice sample containing micrometer size crystals, and the two plots show about the same onset pressure of ~0.8 GPa for the collapse of the structure. This indicates that the effect of increase in the temperature on the onset pressure for collapse is similar to the effect of decrease in the crystal size, or increase in the net surface (interfacial or grain-boundary) energy of the ice sample. This is a remarkable finding in that it suggests that thermal energy has a role qualitatively similar to the surface energy. We conclude that in terms of the high pressure needed, it is easier to collapse hexagonal ice by either decreasing the crystal size at 77 K or by increasing the temperature at which the pressure is raised.

The second property that has led to an understanding of the pressure-collapse of ice is ultrasound velocity. Brazhkin and coworker's[51–53] have measured these properties during the course of pressure-collapse of hexagonal ice at different temperatures. These studies have been reviewed in details by Gromnitskaya et al.,[27] and it has been shown that the velocity of sound waves and shear modulus increase sharply as the hexagonal ice structure collapses, and the bulk modulus increases.[27] A typical plot of the (transverse) ultrasonic velocity against pressure read from Gromnitskaya et al.'s data[27] is shown in Fig. 2(D). Their studies have also provided information on how change in the temperature has an effect on the collapse-pressure of hexagonal ice and the properties of the collapsed state.

However, the ultrasound measurements did not show a sigmoid shape change over a certain pressure range. Instead, the shape of the plot showed a sharp change in the slope, thereby indicating that the pressure-collapse of hexagonal ice is initially very sharp, quite the opposite of the spread-out or broad shape plots observed in the volume, thermal conductivity and the high-frequency permittivity measurements.

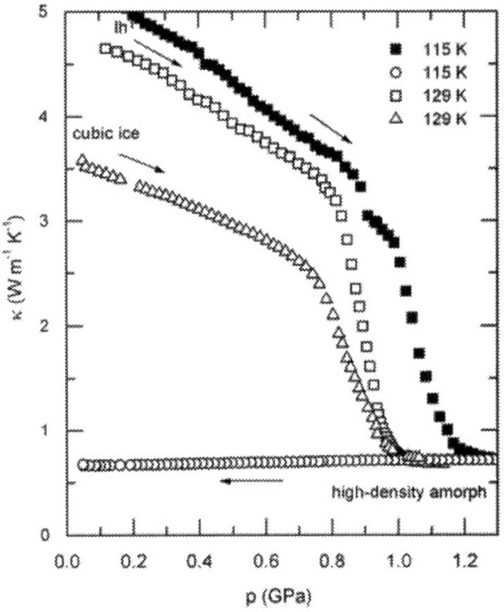

Figure 3. Thermal conductivity of hexagonal ice at 115 and 129 K and of cubic ice at 129 K is plotted against the pressure showing their respective amorphization ranges, and the thermal conductivity of the pressure-amorphized solid on lowering the pressure to ambient. The jog from the smooth sigmoid shape in the plot for 115 K near 0.9 GPa is caused by the inadvertent waiting at this pressure for an unknown period. The plots show that the onset pressure for amorphization decreases with increase in the temperature and is about 0.1 GPa lower for cubic ice at 129 K than for hexagonal ice. Data taken from Ref. 38.

Hexagonal ice crystals are mechanically anisotropic, more easily deformable along the basal plane than along a plane normal to it. It was considered that pressure-collapse of its polycrystalline sample could be due to this anisotropy because mechanical anisotropy can cause a stress concentration at some grain junctions of the randomly oriented crystals. To investigate merits of this possibility, experiments were performed on mechanically isotropic cubic ice in order to determine whether or not it collapses at the same pressure as the anisotropic hexagonal ice and whether or not there is a similar effect of thermal energy on the pressure needed to collapse its structures. Johari et al.[23] and later Mishima[54] showed that cubic ice at 77 K and higher temperatures also collapses under pressure. This is clear evidence for the mechanical anisotropy not being the underlying reason for the collapse of hexagonal ice under pressure. Figure 3 shows some of the studies on hexagonal and cubic ice performed by using thermal conductivity measurements.

Here, κ of hexagonal ice measured at 115 and at 129 K and of cubic ice measured at 129 K are plotted against the pressure. The plots show their respective ranges of pressure-collapse. Remarkably, the onset pressure for collapse of cubic ice at 129 K is ~0.1 GPa lower than that for hexagonal ice, the grain size in both cases remaining large. Moreover, this pressure for hexagonal ice decreases from 0.9 GPa at 115 K to 0.8 GPa at 129 K, i.e., as thermal energy is increased, the onset pressure decreases. This is the opposite of the effect observed for vitrification of liquids because vitrification pressure increases as the temperature is increased. The bottom plot in Fig. 3 shows that κ of the pressure-collapsed solid decreases when the pressure is decreased, and that there is a further effect that appears as a jog for the 115 K plot at p near 0.9 GPa. This jog from the smooth, sigmoid-shape curve was caused by the inadvertent waiting at this pressure for an unknown period. Its occurrence is significant as it indicates that pressure-collapse continues isothermally at a fixed p, and it has led us to study the time-dependence of the pressure-collapse of ice.

Since pressure-collapse of ice has produced an amorphous solid state of water, we henceforth use the term pressure-induced amorphization instead of pressure-collapse.

3.2. TIME-, PRESSURE- AND TEMPERATURE-DEPENDENCE OF THE EXTENT OF AMORPHIZATION

To study the time-dependence of pressure-induced amorphization of hexagonal ice, κ of a sample was measured as its pressure was slowly raised to 0.8 GPa, which is a higher pressure than the onset pressure for amorphization, and kept at 0.8 GPa and 130 K. Its κ was then measured continuously with time for 1 h. The results are shown in Fig. 4, and are indicated by the arrow. After 1 h, the sample's pressure was raised to 0.85 GPa and the measurements made for another 1 h. These data are also shown in Fig. 4 along the side of the arrow. Finally, the pressure was slowly raised to completely amorphize the ice. Another sample was studied by continuously increasing the pressure at the same rate and its results are shown in Fig. 4 for comparison. It is evident that amorphization continues with time at a fixed T and p and that the amount amorphized in 1 h, as indicated by the vertical decrease in κ, is more at 0.85 GPa than at 0.8 GPa. Thus, once hexagonal ice has been brought to a pressure in the amorphization range and kept at that pressure isothermally, the amorphization process becomes time-dependent. This time dependence is an important aspect of the amorphization mechanism and, therefore, it has been studied in detail, as described in the following: In experiments on hexagonal ice under pressure that lasted for several days, measurements of κ were made continuously with time by keeping the ice at several, fixed *p–T* conditions. For example, in one experiment, hexagonal ice was kept at 0.8 GPa and 128 K and its κ was measured over a period of 110 ks (30.6 h). The measured value is plotted against time in Fig. 5(A). It shows a much larger decrease in κ over time and the decrease itself seems to occur in an asymptotic manner, but with little indication of an approach to a limiting value in an experimentally convenient time. Further experiments were performed on hexagonal ice for other *p–T* conditions and for different time periods. The data obtained for six such conditions are shown in Figs. 5(B) and 5(C).

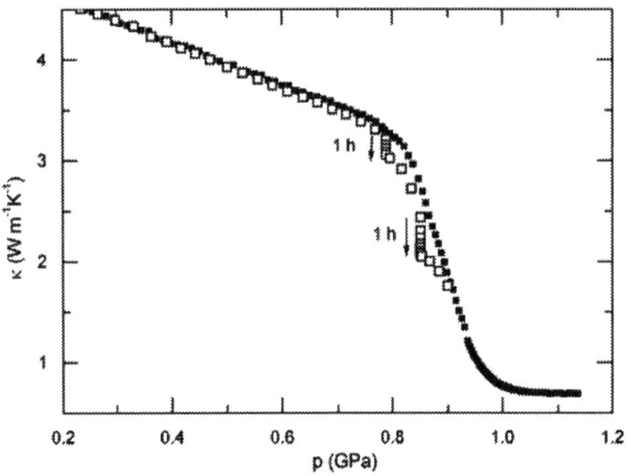

Figure 4. Thermal conductivity of hexagonal ice at ~130 K is plotted against the pressure. The vertical arrows during the course of crystallization are for fixed pressure conditions of 0.8 and 0.85 GPa at which the sample was kept for 1 h after the pressure-amorphization had begun. Data taken from Ref. 22.

The time-dependence of κ at these *p–T* conditions, as shown in Fig. 5, indicates one or several of the four following occurrences: (i) hexagonal ice crystallites in the polycrystalline sample transform at different rates by a mechanism that continuously changes the conditions required for their amorphization with time, (ii) the sample becomes a mixture of hexagonal ice and an amorph (of high density) already at its initial stages, (iii) all the ice sample has amorphized but the amorph is kinetically unstable and is tending toward a more stable state by relaxation, and

(iv) highly deformed crystallites or else a mixture of nanometer-size crystals of high pressure polymorphs have formed whose deformation or composition continuously changes with time. In occurrences (i) and (ii), the rate of approach to a stable state would be very small in the early period immediately after the onset of amorphization because the amount of amorph formed would be negligibly small. It would become significant only when most of the sample is amorphous. (Note that this latter effect appears to have produced amorphs with different x-ray and neutron diffraction and other features[55] and it has, as discussed earlier, also led to further increase in density when an amorph that had been formed by pressurizing hexagonal ice at 77 K to 1.45 GPa was heated to 160 K at ~1 GPa.[30]) In their physical manifestation, occurrence (iii) and (iv) are analogous to structural relaxation of the state obtained by mechanical amorphization of crystals in the high speed ball - milling process, to structural relaxation of a glass formed by hyper-quenching the melt, or to structural relaxation of an amorphous solid made by vapor-deposition. All the three states have high energy and high fictive temperatures. They become denser spontaneously with time, as their states approach asymptotically an amorphous structure of lower energy and lower fictive temperature. In the plots in Fig. 5(B), κ also decreases with time asymptotically toward a limiting value. From these plots, it is evident that κ of the sample as well as the as the rate of the asymptotic decrease in κ with time vary with the p–T conditions.

The manner in which κ decreases with time at a given p–T condition is more clearly expressed by normalizing the $\kappa(t)$ values for each set of measurements given in Fig. 5(B) by the initial value $\kappa(0)$, and then plotting it against the time. Figure 5(C) shows the plots of these normalized, $[\kappa(t)/\kappa(0)]_{p,T}$, values. It is evident that $[\kappa(t)/\kappa(0)]_{p,T}$ decreases with time, and tends toward $[\kappa(\infty)/\kappa(0)]_{p,T}$, that is different for different p–T conditions. The plots in Fig. 5(C) also indicate that $[\kappa(t)/\kappa(0)]_{p,T}$ values at a given time differ even when the p–T conditions are almost the same, as for the plots at 0.86 GPa and 130 K and 0.85 GPa and 129 K. This finding means that the amorphization rate differs even when p–T conditions are closely similar. Moreover, the plot for 0.85 GPa and 129 K and that for 0.88 GPa and 127 K cross over, thus showing that there is a time reached at which $[\kappa(t)/\kappa(0)]_{p,T}$ is the same for different p–T conditions, although the rate of decrease in $[\kappa(t)/\kappa(0)]_{p,T}$ with time differs. Clearly, a given value of κ of the sample can be reached by different p–T–t paths. Altogether these findings indicate that the initial conditions of the ice sample determine the rate of its amorphization and that the degree of amorphization or the κ value reached at a given p, T, and t is determined by more variables than have so far been considered in these studies.

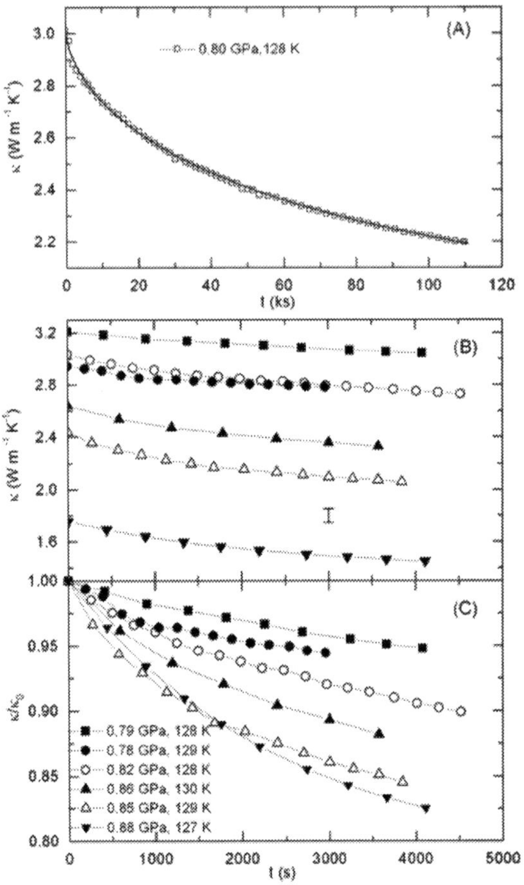

Figure 5. (A) Thermal conductivity of partially amorphized state of hexagonal ice at 0.8 GPa and 128 K. The line is calculated from stretched exponential relation given here. Data are taken from Ref. 49. (B) Thermal conductivity of several samples plotted against time. Data are taken from Ref. 49. (C) Normalized values of the measured thermal conductivity with respect to the initial (zero time) value of the samples kept at different pressures and temperatures are plotted against time. The pressure and temperature conditions are as labeled. Data are taken from Ref. 38.

The asymptotic nature of the decrease in thermal conductivity with time at 0.8 GPa and 128 K seen in Fig. 5(A) can be described by a stretched exponential relation,

$$\kappa(t) = \kappa(\infty) + [\kappa(0) - \kappa(\infty)]\exp[-(t/\tau_0)^{\beta}, \quad (2)$$

where $\kappa(0)$ is the value of κ at the instant when the fixed pressure experiment at 0.8 GPa was begun, $\kappa(\infty)$ the limiting long time value of κ, and τ_0 the characteristic amorphization time. The fit of Eq. (2) to the data is shown by the smooth lines in Fig. 5(A). The normalized value of the same data is plotted against logarithmic time in Fig. 6(A), where the fit is also shown. The values obtained from these fits are: $\kappa(0) = 3$ W m^{-1} K^{-1}, $\kappa(\infty) = 1.8$ W m^{-1}K^{-1}, $\tau_0 = 90.9$ ks (1500 min), and, $\beta = 0.6$. From these parameters we calculate that, at 0.8 GPa and 128 K, it would take 120 h to reach an amorph whose κ value is within 5% of the $\kappa(\infty)$ value of 1.8 W m^{-1} K^{-1}. Moreover, the plot in Fig. 5(A) shows that at 0.8 GPa and 128 K, κ of the amorph formed would not decrease to the same value as that of the amorph formed at 1.15 GPa and 129 K in the plot of Fig. 3. To elaborate, κ of the amorph formed at 1.15 GPa and 129 K is ~0.7 W m^{-1} K^{-1} (Fig. 3), but $\kappa(\infty)$ of an amorph formed at 0.8 GPa and 128 K is 1.8 W m^{-1} K^{-1}, i.e., 1.1 W m^{-1} K^{-1} higher. Therefore, it would seem that either the amorph formed after keeping for a formally infinite time at 0.8 GPa and 128 K would be different from that formed at 1.15 GPa and 129 K, or Eq. (2) does not apply for the entire transformation. In view of the finding that, (i) κ further increases on heating the amorph formed at 1 GPa,[22] which may be only negligibly due to any densification because increase in κ on heating is an amorph's normal behavior, and the density increase that occurs on heating the amorph formed by pressurizing hexagonal ice at 77 K from 77 to 165 K at a fixed p in the 0.8–1.1 GPa GPa range[30] has already occurred here during slow pressurization at 129 K, and (ii) this denser amorph in its recovered state at ambient pressure has different x-ray diffraction features (see Ref. 38 for discussion), we conclude that the final state of an amorph formed on pressurizing hexagonal ice depends upon the p–T conditions.

Generally speaking, Eq. (2) has been fitted to the relaxation spectra in molecular relaxation studies, to the enthalpy and volume changes in structural relaxation studies, and to the extent of transformation in chemical reaction kinetics. In all these studies, the quantity β has been found to be less than 1. In dielectric and mechanical relaxations and in calorimetric structural relaxation studies, it has been interpreted in terms of a broad distribution of relaxation times[56–59] and attributed to possible microscopic heterogeneity in the liquids and solids,[60,61] and in chemical reaction kinetics, it has been interpreted in terms of a diffusion-controlled kinetics, dispersive kinetics, or a distribution of the reaction rate constants. This latter idea was provided and developed by Plonka.[62] According to Plonka,[62] the dispersive kinetics of such (transformation) reactions, means that homogenization (in a viscous liquid and solid) does not occur over the timescale of the transformation. Consequently, a molecular level heterogeneity of the reactants and products develops within the bulk of a sample on the transformation timescale. An extension of Plonka's theory[62] to the amorphization process of hexagonal ice would mean that molecular level regions of the amorph and hexagonal ice exist in the bulk of the sample over the transformation time scale.

The decrease in κ with time for a given p–T condition may be expressed in terms of the extent of the pressure-induced amorphization, α_{amorph}. This extent can be determined by using the $\kappa(0)$, $\kappa(t)$, and $\kappa(\infty)$ values. Here, the extent of amorphization may be calculated for the four cases given above, i.e., (i) by assuming that the sample is a mixture of hexagonal ice and an amorph of $\kappa(\infty)$ of ~1.8 W m^{-1} K^{-1} at 129 K and 0.8 GPa, (ii) by assuming that the sample is a mixture of hexagonal ice and an amorph of $\kappa(\infty)$ of ~0.7 W m^{-1} K^{-1} at 129 K and 0.8 GPa, (iii) by assuming that the sample is an amorph tending toward a more stable, still amorphous state of $\kappa(\infty)$ of ~1.8 W m^{-1} K^{-1} at 129 K and 0.8 GPa, and (iv) by assuming that the amorph is either highly deformed crystallites of hexagonal ice and/or a mixture of nanometer size crystals of high-pressure phases. This may be done by using a single equation,

$$\alpha_{\text{amorph}}(t) = \left(\frac{\kappa(0) - \kappa(t)}{\kappa(0) - \kappa(\infty)} \right)_{p,T}. \quad (3)$$

The calculated value of α_{amorph} for both, $\kappa(\infty)$ of ~1.8 W m^{-1} K^{-1} at 129 K and 0.8 GPa $\kappa(\infty)$ of ~0.7 W m^{-1} K^{-1} at 129 K and 0.8 GPa is plotted against the logarithmic time in Fig. 6(B). (Note that the plot calculated by assuming that the sample is a mixture of hexagonal ice and an amorph of $\kappa(\infty)$ of ~0.7 W m^{-1} K^{-1} does not reach a long time limiting value of 1 because the sample remains a mixture of an amorph and ice at that time.) This plot has an extended sigmoid shape. Although we can obtain only part of this shape from our data, it resembles the shape of the plots usually obtained for slow crystallization, phase transformation, chemical reactions and structural relaxation processes.

Techniques such as infrared and Raman spectroscopy in real time, which may help quantitatively resolve how hexagonal ice amorphizes under pressure, have not so far been used. In the absence of such data, we continue our discussion of the pressure-induced amorphization on the premise that there are, as given here earlier, at least four conditions in which the solid during its continuous amorphization say at 0.8 GPa and 129 K may exist:

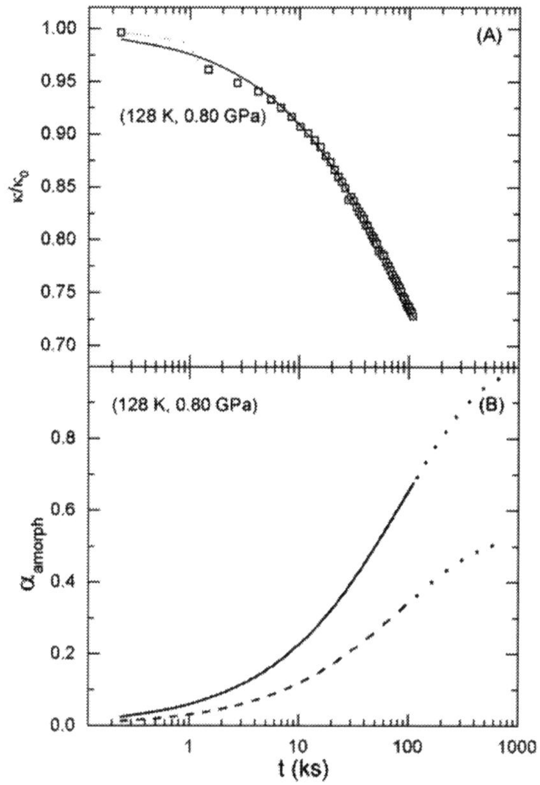

Figure 6. (A) The normalized value of thermal conductivity measured during the amorphization of hexagonal ice at 0.8 GPa and 128 K is plotted against logarithmic time. κ_0 is the value at the beginning of the measurements. (B) The degree of amorphization of the ice at 0.8 GPa and 128 K, calculated from Eq. (3) is plotted against the time on a logarithmic scale: For the solid line κ_∞ is assumed to be 1.8 W m^{-1} K^{-1}, and for dashed line κ_∞ is assumed to be 0.7 W m^{-1} K^{-1}. The dotted lines correspond to extrapolations by using Eq. (2). The plot was calculated by assuming that the sample is a mixture of crystalline and amorphous ice and the amorphs $\kappa(\infty)$ of ~0.7 W m^{-1} K^{-1} does not reach a long time limiting value because the sample remains a mixture of ice and amorph at that time. Data taken from Ref. 38.

(i) The solid is a mixture of hexagonal ice and a certain unknown amount of the amorph whose $\kappa(\infty)$ is ~1.8 W m^{-1} K^{-1}.

(ii) The solid is a mixture of hexagonal ice and a certain unknown amount of an amorph whose $\kappa(\infty)$ is ~0.7 W m^{-1} K^{-1}.

(iii) The solid is entirely an amorph that gradually transforms to its own $\kappa(\infty)$ value of ~1.8 W m^{-1} K^{-1} at 0.8 GPa and 129 K.

(iv) The solid consists of highly deformed hexagonal ice crystals, or else is a mixture of nanometer-size crystals of high pressure polymorphs whose deformation or composition continuously changes with time. In its physical manifestation, it slowly relaxes to a lower energy amorphous solid. This is analogous to structural relaxation of a mechanically amorphized crystal in the high speed ball milling process or else to structural relaxation of a hyperquenched glass. Both procedures produce a solid of high fictive temperature and high energy, which spontaneously densify with time toward a state of lower fictive temperature and lower energy.

Hemley et al.[63] have reported that marked changes in the texture of their ice sample occured as the pressure was increased in their diamond anvil high pressure cell that contained the hexagonal ice. Also extensive fracturing of the sample occurred at ~0.5 GPa and the sample became turbid by the time the pressure reached ~1 GPa, with a new phase appearing along the fracture faults.[63] This indicates possible existence of ice and a texturally different amorph together at least up to a pressure of ~1 GPa. Salzmann et al.[64] had found that x-ray diffraction spectra of the recovered sample at ambient pressure and 77 K contained distinctive features of both the amorph and hexagonal ice, thus implying that the state of the sample may be a mixture of the ice and the amorph, as in (i) and (ii) above, and ignoring the possibility that some of the ice could have also formed during the handling of the samples.

The state described in (iii) above requires that all hexagonal ice samples amorphize abruptly. But no study of ice has shown that complete amorphization occurs abruptly. It also seems inconceivable that near the onset of amorphization, the hexagonal ice and the amorph would have the same thermal conductivity, thereby preventing us from detecting any possible abrupt amorphization. (Note that in one dilatometric study of a sample of 1–1.5 mm thick hexagonal ice at 77 K, amorphization apparently occurred[65] first slowly and then abruptly, reaching completion at ~1.06 Ga.[65]) But we now know that when the sample is kept at a fixed pressure, that is far above the onset pressure, for a short period, the plot of the piston displacement against pressure in the amorphization range of both hexagonal and cubic ices would be vertical. This vertical plot can be mistaken as an indication of a relatively abrupt decrease in the volume. For example, the jog in the plot of κ at 115 K in Fig. 3, or the vertical decrease in κ in Fig. 4 here could have been mistaken as an indication of abrupt amorphization. It is worth noting that the rate of pressurization in the earlier study[65] had not been controlled, and an unknown period of time was allowed between pressure-increase steps, and this could lead to a feature similar to the one seen in Fig. 4. Therefore, we do not consider (iii) above as reliable occurrence. The possibility then remains that the ice crystals begin to deform and/or to recrystallize to nanometer size crystals of high pressure phases at a certain pressure as mentioned in (iv) above, and that this occurrence is followed by the relaxation of the structure with time to an amorphous phase at 0.8 GPa and 129 K. The structural relaxation may occur via a series of sequentially lower, denser states of thermal conductivity until the final state whose characteristic κ is ~1.8 W m^{-1} K^{-1} has been reached.

3.3. MECHANISM OF PRESSURE-INDUCED AMORPHIZATION

Pressure-amorphization was first seen as pressure melting at 77 K at p near 1 GPa by using a nonlinearly extrapolated hexagonal ice-water equilibrium phase boundary line in the p–T plane. But this view has since been abandoned. Therefore, we consider whether an alternative mechanism based on Lindemann theory can be consistent with the above-given findings on pressure-induced amorphization of ice. It is recognized that calculations based on the Lindemann melting theory apply to an ideal single crystal without point defects or lattice faults, and this is not true for pressure-amorphization of a polycrystalline sample which contains point defects, dislocations, and a large grain-boundary area. As mentioned earlier here, compression of a ~2 mm diameter and ~3 mm long single crystal of hexagonal ice embedded in ~5 mm thick-wall indium cup at 77 K had showed no volume decrease up to ~2 GPa,[26] thereby suggesting that a single crystal did not collapse in the same way as polycrystalline ice did. Moreover, Lindemann melting is instantaneous, not time-dependent with pressure- and temperature-dependent kinetics observed for pressure-induced amorphization. Therefore, we consider other manners of melting that are peculiar to a polycrystalline sample. For example, it is also known that arrangement of molecules and atoms at the grain-boundaries in a polycrystalline sample is liquid-like, and significant premelting occurs at the three-grain and four-grain junctions when a polycrystalline sample with submicron size crystallites is heated in the temperature range far below its bulk melting point.[66] In this incipient process, the melt and solid remain at thermodynamic equilibrium as the change in the solid–liquid interfacial energy compensates for the change in the bulk energy. It is also known that the surface layer of ice crystals is disordered or water-like.[67] The amount of water present contained in the grain boundaries and grain junctions of micron size grains in polycrystalline hexagonal ice at ambient pressure has been determined from both experiments[68] and calculations.[66] This amount is expected to change at high pressures, because the relative ratio of the surface energy to bulk energy changes. The pressure-induced amorphization of polycrystalline ice itself has been carried out by uniaxial loading, at a rate of usually ~0.1–0.2 GPa/min in most experiments. In this process, ice crystal splastically deform anisotropically, converting the uniaxial load to a hydrostatic pressure presumably within less than 30 s at 77 K. In this process, ice crystallites reorient, the sample recrystallizes, new grain-boundaries form and the population of the three- and four-grain junctions changes, as would occur for any polycrystalline sample. This in turn would continuously change the incipient melting conditions at the grain junctions and grain boundaries. Although it is still the process of melting, it differs from the Lindemann melting.

It has been occasionally suggested that pressure-induced amorphization of ice is due to compaction via the filling of interstitial voids. Details of the arguments are given earlier.[38] We consider the consequences of our above-described findings of the amorphization kinetics in developing a mechanism for pressure-induced amorphization. Born[69] had mathematically described the conditions for the loss of stability of an ideal crystal lattice, i.e., of a *perfect single crystal*. According to his theory, a crystal lattice becomes mechanically unstable when increase in the *hydrostatic* pressure softens the transverse acoustic phonon modes and the elastic modulus decreases. The collapse occurs homogeneously throughout the crystal lattice, and its pressure is determined by the manner in which the elastic constants change on compression. Born had considered the lattice stability violation conditions in relevance to transformation of one crystal form to another, and not for transformation of a crystal to a disordered structure, which was not known at the time. But his theory has been used to calculate,[70] in a quasi-harmonic approximation, the mechanical collapse pressure of an ideal ice single crystal at different temperatures. This, however, may not be relevant to polycrystalline ice, which contains grain boundaries and three- and four-grain junctions in which impurities segregate, and its individual crystals contain point defects, impurities, and dislocations. Moreover, the state of the crystals in a polycrystalline mass changes as uniaxial loading plastically deforms ice, causes its crystal grains to reorient, dislocations population to increase and the dislocations to move. All of these features, that are characteristic of a sample, are expected to alter the Born instability (or stability violation) pressure and would cause the extent of amorphization to depend upon the time, temperature and pressure, and to a small degree on the sample itself. It has also been found that dislocations in a single crystal of ice have an extended noncrystalline core,[71] which would alter the conditions of pressure-amorphization. Thus it may seem that pressure-induced amorphization of ice also involves two mechanistically distinct processes: (i) incipient melting at the inter-granular regions in a polycrystalline sample, and (ii) increase in the population of dislocation cores containing disordered arrangements of water molecules in the crystallites.

In a mechanical collapse resulting from the Born stability violation in a crystal, lattice faults would lower the collapse pressure, because these faults store energy and in most cases lower the crystal's density from that of an ideal lattice. More importantly, a variation in the population of the lattice faults would cause different crystallites in the sample to collapse at different pressures. Thus, one would expect a distribution of the Born instability pressures (of different crystallites) in a polycrystalline sample, which would broaden the pressure range for the mechanical collapse of the sample. Thus, even at a very slow compression rate, full amorphization would be reached only after the pressure is of a magnitude high enough to collapse the near ideal ice crystal of the highest Born instability pressure. (The situation may be seen as analogous to a multi-component crystalline composite, in which each component would collapses at its own characteristic pressure.) If this occurred then ice would coexist with the amorph at formally infinite annealing time, because its $\kappa(\infty)$ of 1.8 W m^{-1} K^{-1} at 0.8 GPa and 128 K is higher than κ of 0.7 W m^{-1} K^{-1} for the amorph formed at 1.15 GPa at 129 K. This means that the final state achieved would remain a mixture of (stronger) hexagonal ice crystallites that did not collapse at 0.8 GPa and 128 K and the amorph that formed by the collapse of (weaker) ice crystallites. This would also explain the observation that the onset pressure of amorphization decreases with decrease in the crystal grain-size in the sample.[26]

Electrostatic interactions in the structure of crystalline ices have been found to be co-operative in nature,[72,73] with the consequence that breaking of some H-bonds weakens the strength of the others. Thus, breaking of some H-bonds as a result of structural collapse would weaken the neighboring H-bonds in the ice crystal. If that were to occur, pressure-induced amorphization of ice would become easier, i.e., after part the structure has collapsed, further collapse would require a smaller increase in pressure. But the stretched sigmoid shape plots of volume, thermal conductivity and high frequency permittivity against pressure show that this weakening effect is inconsequential. Therefore, the regions of collapse would seem to be small enough not to cause a sudden and rapidly increasing collapse of the whole crystal. Thus, although the Born criterion for crystal stability may remain valid for the mechanical collapse of ice, the features of amorphization of hexagonal ice are altered by a distribution of the mechanical collapse pressure in a polycrystalline mass. Nevertheless, the time-dependent pressure-amorphization of hexagonal ice at a given p–T condition may be reconciled with Born's stability criterion if occurrence of another molecular process that changes the crystallite's microstructure is included. This mechanism would continuously alter, with time, the distribution of the collapse pressure at a given p–T condition. On the basis of our knowledge of plastic deformation of crystals and their recrystallization, we suggest that this process would be diffusion of defects, redistribution of impurities and dislocations and partial melting.

We conclude that the amorphization pressure, the pressure-range for complete amorphization, and the characteristic amorphization time, τ_0, at a given p–T condition are determined by at least four effects: (i) the rate of plastic deformation of crystallites and their recrystallization under a uniaxial stress, (ii) the concentration of lattice faults in the crystallites, (iii) the Born instability pressure of the crystallites and the distribution of this pressure, (iv) the redistribution of impurities and dislocations and partial melting during the period of amorphization, and (v) the pressurizing rate. It is conceivable that in Johari's earlier experiment mentioned in Ref. 26, a single crystal of hexagonal ice did not amorphize because the Born instability pressure for a single crystal is much higher than for polycrystalline samples, and this pressure was not reached in his experiment.[26] On the opposite end, it is conceivable that crystals of ice approaching several nanometers in size may amorphize at a pressure as low as 0.1 GPa. This may be tested by an experiment.

We should mention another mechanism that has been proposed for pressure-induced amorphization of crystals. Sikka[74] has proposed that the development of steric constraints leads to amorphization of crystals under pressure. But this mechanism does not account for the observation that pressure-induced amorphization of hexagonal ice is time-dependent, with a well-defined kinetics with a distribution of times.

3.4. PATH-DEPENDENT STRUCTURE AND THERMODYNAMIC STATE OF THE AMORPH

An open structure, such as that of hexagonal ice and cubic ice lattice, may also gradually elastically distort to a fixed limit determined by the p–T conditions of the sample. Therefore, we consider whether different p–T conditions for amorphization produce the same amorph or produce different amorphs each with a distinct x-ray diffraction and physical property. A difference between the x-ray diffraction features has been found in a number of studies and this has been discussed previously.[22] Reasons for the difference between the various amorphs have been envisaged in terms of a multiplicity of energy minima, each energy minimum corresponding to a certain structure of the high-density amorphs formed at a certain p–T condition,[55] but all minima (like a set of closely-spaced corrugations) are inside a shallow energy minima that represents the energy and the structure of the high-density amorph. But studies of the amorphs produced at a single temperature of 77 K and the same pressure of ~1.5 GPa have shown different x-ray diffraction features after subtracting the diffraction lines from contaminant hexagonal ice and other ices. Also, recent studies have suggested that there may be a continuity of amorphs ranging from the low-density amorph (density 0.92–0.94 g cm^{-3} at 0.1 MPa and 77 K) to the so-called very high-density amorph[30] (density 1.25 g cm^{-3} at 0.1 MPa and 77 K), and the amorph formed at 77 K is just one of the state that is intermediate in this continuity of amorphs. In this sense, an amorph of $\kappa(\infty)$ of ~1.8 W m^{-1}K^{-1} formed at 0.8 GPa and 129 K would also be one of the state in this continuity of amorphs. The differences in the various amorphs may be partly due to the intrinsic nature of the sample in which transformation has occurred to only a limited extent and partly due to the insufficient time allowed for the sluggish transformation to reach completion.

Analysis of the density, ρ, and ultrasonic velocity, v_{trans}, data of the amorph obtained by pressure-amorphization have been useful in determining whether or not thermodynamically identical states are formed by using different pressure and temperature paths.[75] Briefly, ρ and v_{trans} of HDA at 1.6 GPa at different temperatures are available within an accuracy of 1% for ρ and 2% for v_{trans}, in Fig. 2 of Ref. 27. At 1.6 GPa and 77 K, the ρ values of the amorph are spread over 0.05 g cm^{-3} range, from 1.33 to 1.38 g cm^{-3} and the v_{trans} values over 0.07 km s^{-1} range, from 2.21 to 2.28 km s^{-1}. This spread is beyond the quoted error of the measured data. This spread means that the pressure–temperature path for making the amorph determines its ρ and v_{trans} at 1.6 GPa and 77 K. We also note that in Fig. 1 of Ref. 27, hexagonal ice had sharply transformed to the amorph at ~1.06 GPa at 77 K and at ~0.85 GPa at 125 K. Thereafter, the amorph had been pressurized up to 1.65 GPa at 77 K, and to 1.57 GPa at 125 K.[27] Since the ρ (and v_{trans}) values for the amorph are not at the same *T*, it would seem that the data cannot be used for determining whether or not ρ (and v_{trans}) is only a function of state, i.e., ρ (and v_{trans}) is independent of the *p*–*T* path by which a given state has been reached. But we can still use these data by recasting a thermodynamic test as follows: At a given pressure, a property *u* of a *solid* cannot be the same at two temperatures, or alternatively, at a given temperature, *u* of a *solid* cannot be the same at two pressures. (Otherwise, there would be two values of (du/dT) at the same *p* and *u*, or two values of (du/dp) at the same *T* and *u*.) Graphically speaking, this means that the plots of *u* against *p* at different *T* should not cross. (Note that ρ for water is the same at 275 and 279 K, and such plots still do not cross.)

But the plot of ρ for the amorph at 77 K against the pressure in Fig. 1, Ref. 27 approaches the plot of ρ for the amorph at 125 K from below and crosses it at ~1.3 GPa where ρ = 1.33 g cm^{-3}. There is a similar crossing of the plots at v_{trans} = 2.15 km s^{-1}, and at a pressure higher than 1.3 GPa. Thereafter, ρ and v_{trans} at 77 K remain higher than ρ and v_{trans} at 125 K. From these plots in Fig. 1, Ref. 27, we determined two quantities, $\Delta V = (V_{\text{amorph,77 K}} - V_{\text{amorph,125 K}})$ and $\Delta v_{trans} = (v_{\text{trans, amorph,77 K}} - v_{\text{trans amorph,125 K}})$ at several pressures above 1.2 GPa and provided their respective values in Table I, Ref. 75. These calculations showed that, instead of remaining positive at all pressures, ΔV and Δv_{trans} for the amorph became negative as *P* was increased. This also means that compressibility of the amorph at 77 K is *higher* than that at 110 K, which is unusual because compressibility of a solid decreases as the temperature is decreased. For details Ref. 75 may be consulted.

We conclude that the density of the solid produced at low temperatures by rapid increase in pressure is a function of the pressure–temperature path. This is so, even if our recently found time-dependence of thermal conductivity during the amorphization of hexagonal ice is ignored. Temperature, time, and pressurizing rate have an effect on the measured properties. Loerting et al.[30] recently discovered that the amorph formed at 77 K densifies by ~5 % when heated to 160 K at a pressure higher than 0.8 GPa, and we have found that the amorphization process is time-dependent. The temperature–pressure conditions for amorphization of hexagonal ice in earlier studies had varied in the experiments performed by different groups, as well as in the experiments performed by the same group, and therefore the amorphs they had studied are now seen to be different solids out of a multiplicity of amorphous solids that lie in the density range of 0.93 and 1.25 g cm^{-3} at 0.1 MPa and 77 K.[38] As pointed out earlier,[22] this difference has been evident from slight differences in the structure factor of the high-density amorphs, and in the kinetics of their transformation to a state of low density on heating at ambient pressure.[75–78] This structural difference is also evident from our recent (unpublished) C_p measurements which show that C_p at 130 K and 0.1 MPa of the amorph made by collapsing hexagonal ice at 130 K is 4% higher than C_p of the amorph made by collapsing hexagonal ice at 77 K.

3.5. DIELECTRIC RELAXATION OF THE AMORPH AND CONVERSION TO ULTRAVISCOUS WATER

In most of the earlier studies, hexagonal ice was pressure-amorphized at ~77 K, usually by raising the pressure from 0.1 MPa to 1.5 GPa in a duration of 5–20 min. When this solid was heated at a fixed pressure of ~0.8 GPa or higher from 77 to 160 K, it further densified by ~5% (~13 mm^3/250 mm^3 read from Fig. 3, Ref. 30) in a time-dependent manner. Almost 90% of this densification occurred on reaching 150 K temperature at a heating rate of ~5 K/min. (Fig. 3 in Ref. 30.) As mentioned earlier here, pressure-induced amorphization of hexagonal ice in our studies was performed at a fixed temperature of 130 K, and the pressurizing rate was 80 times slower than the rate used in the earlier studies. Also, the sample was heated much more slowly, over a period 20 times longer than in the earlier studies. Therefore, the pressure-amorphized state at 1.2 GPa and 130 K obtained here was almost fully densified and an amorph which was the same or similar to that obtained by Loerting et al.[30] was produced.

For dielectric spectroscopy studies, the pressure on the amorph at 130 K was slowly decreased from 1.2 to 1 GPa and its dielectric relaxation spectra at 130 K was measured in the 0.5 Hz–1 MHz showed. The spectra showed no relaxation peak and only the high frequency part of the peak was observed.[79] The sample at a fixed pressure of 1 GPa was then heated and its permittivity and loss, ε' and ε'', spectra recorded continuously with time and temperature. As the temperature was increased, the data showed indication of the ε''-peak's shift to higher frequencies and ultimately in our spectral range and a corresponding shift occurred in the dielectric permittivity, ε', spectra. Finally, a relaxation peak appeared in our spectral range at a temperature above 140 K. On further heating, particularly beyond 150 K, the relaxation peak height of both the pure and the KOH-doped samples began to decrease with time. This indicated a decrease in the orientational polarization, as the sample had begun to crystallize on heating to a temperature above ~150 K. A simultaneously recorded temperature difference between the sample and the Teflon vessel showed a small, slowly growing exothermic effect which reached a local maximum at ~150 K. This confirmed that slow crystallization had begun and which accelerated at $T > 150$ K. It has been found already that on slow heating at a pressure of ~1 GPa, the fully dense pressure-amorphized state crystallizes at T near 165 K.[80] Typical spectra of ε' and ε'' for the pure and KOH-doped amorphized-ice at 1 GPa are shown in Fig. 7.

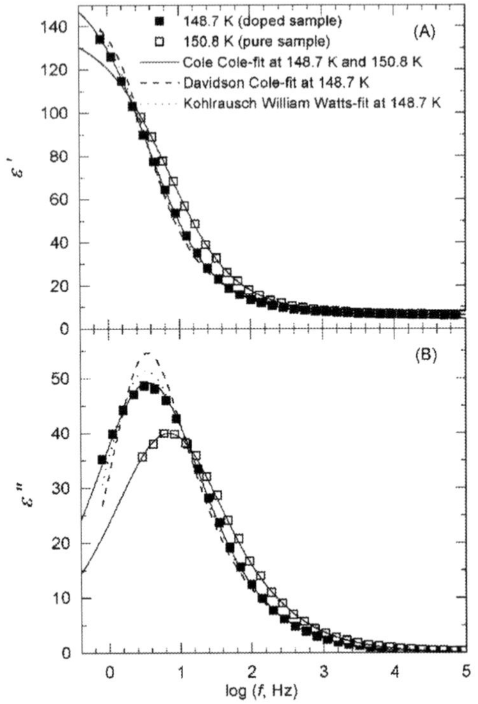

Figure 7. The dielectric permittivity and loss spectra for pure amorph and KOH-doped amorph at 1 GPa at 150.8 and 148.7 K, and its analysis in terms of the Cole–Cole symmetrical distribution of relaxation times. For the KOH-doped sample, the measured values were converted to ε' and ε'' by using the cell constant, $C_g = C_\infty$ (KOH-doped water)/ε_∞ (pure water) at 130 K. The best fit parameters of the Cole–Cole function are: $\alpha = 0.29$ and $\varepsilon_0 = 164$, $\varepsilon_\infty = 6.4$, and $\tau_{CC} = 0.044$ s. The corresponding data for pure water at 150.8 K are: $\alpha = 0.30$ and $\varepsilon_0 = 132$, $\varepsilon_\infty = 6.4$, and $\tau_{CC} = 0.022$ s. Data taken from Ref. 79.

As shown in Fig. 7, the spectra are best described by a symmetrical distribution of relaxation times function given by the Cole and Cole,[81] $\varepsilon^*(\omega) = \varepsilon_\infty + (\varepsilon_0 - \varepsilon_\infty)/$

$[1+(i\omega\tau)^{1-\alpha}]$, where ω $(=2\pi f)$ is the angular frequency and τ is the characteristic relaxation time. (Note that this type of distribution has been found for water also in the range 273–323 K.[82,83] For a detailed discussion see Ref. 79.) The τ value was calculated from the reciprocal of the peak frequency in the ε'' spectrum, i.e., $\tau = (2\pi f_{peak})^{-1}$. This is plotted against the temperature in Fig. 8. At 150 K, τ is 30 ± 5 ms for both pure and KOH-doped VHDA. This shows that this amount of doping has no discernible effect on τ of the amorph at 1 GPa, although it has a very large effect on the relaxation of time of hexagonal ice.[45] This finding is consistent with the general observation that the relaxation time of water is not greatly affected by the addition of small amounts of electrolytes.

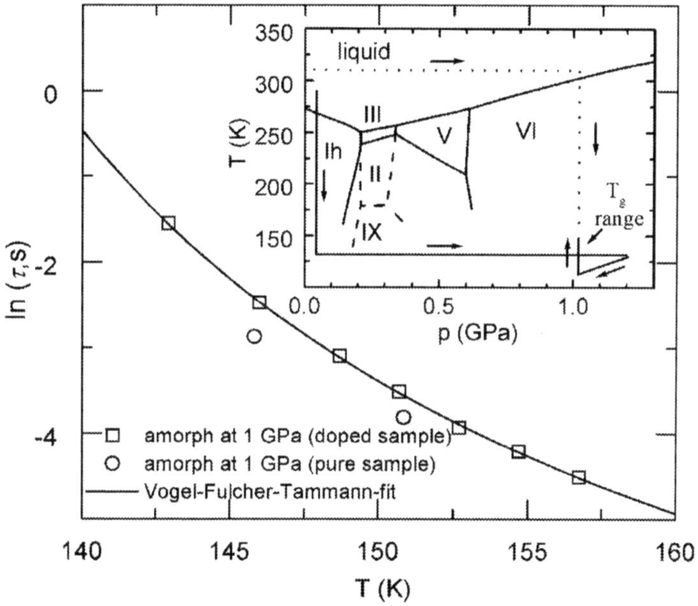

Figure 8. The relaxation time of pure and KOH-doped amorph at 1 GPa plotted against the temperature. The line is the best fit of the Vogel–Fulcher–Tammann equation, $\ln\tau = -9.81 + [204.0/(T - 118.2\ \text{K})]$. Data taken from Ref. 79. Inset is the phase diagram of the ices. The temperature–pressure region in which amorphization studies of hexagonal ice have been performed are indicated. The path that leads to an ultraviscous liquid state is shown by arrows. The path from the liquid water to ultraviscous state shown by the dashed line requires supercooling at 1 GPa. This path has not been experimentally achieved.

Reorientational and translational diffusion motions of H_2O molecules in liquid water with tetrahedral bonding require that hydrogen bond break and then reform with a new neighbor or the same neighbor. The reorientational and translational motions are concomitant, and it has been found that the dielectric relaxation time, the self-diffusion time, and the structural relaxation time, the last determined from the T_g endotherm, are of the same order of magnitude for hyperquenched glassy water and for the annealed vapor-deposited amorphous solid water at 136 K.[84,85] These relaxation times have been estimated as ~35 s at 136 K.[84,85] A short extrapolation of the plot in Fig. 8 yields the dielectric relaxation time for water as ~1 s at 140 K and 1 GPa. This small value of τ shows that the state of the amorph is likely to be ultraviscous water at 1 GPa and 140 K, because τ for an amorphous solid is much greater than 1 ks.

Since the temperature-dependence of the reorientational and translational motions determines the vitrification temperature, T_g, of a supercooled liquid (or the glass-softening temperature on heating a vitreous solid), this means that T_g for water at 1 GPa is below 140 K. Also, as for most ultraviscous liquids, water at 1 GPa crystallizes in its ultraviscous state and not in the rigid glassy state. Based on the dielectric relaxation studies, we conclude that water at 1 GPa is in a metastable equilibrium state at temperatures above 140 K. This state could be achieved if water at 1 GPa could be supercooled through its ice VI phase boundary to ~140 K. Here of course this state has been achieved via an irreversible path of pressure-amorphization of hexagonal ice. This is illustrated in the inset of Fig. 8.

3.6. THERMODYNAMICS AND KINETICS OF PRESSURE-COLLAPSED AMORPH AND OF ULTRAVISCOUS WATER

It is known that in contrast to glasses, which are formed by supercooling a melt, amorphous solids can be formed by using a variety of different techniques; (i) vapor-deposition on a cold substrate,[86] (ii) rapid evaporation of a solution, (iii) chemical reaction that leaves a solid product, (iv) electrodeposition, and (v) mechanical deformation of crystals in a high speed ball mill. Some of these techniques are used in technology. These amorphous solids are in a much higher energy state and, on raising the temperature, they rapidly relax structurally to a lower energy, but still amorphous state.[86] Further heating slowly mechanically softens the solid and it ultimately becomes an ultraviscous liquid. The ultraviscous liquid may crystallize rapidly and in some cases may not show the characteristic glass-softening endotherm.[86] This may also occur as a result of crystal nucleation in local regions of molecular mobility when heat is released during the exothermic structural relaxation.[85] The thermodynamic path along which these changes occur is irreversible. Nevertheless, the path is unidirectionally cyclic in as much as the finally obtained state can be converted to the original states of vapor, solution, chemical reactants, crystal, etc., and the amorph can be reproduced.

In contrast, a glass or a vitrified solid is formed only by cooling a liquid at a fixed pressure or by compressing a liquid at a fixed temperature. It is often a process that can be cycled by heating the glass and cooling the melt, and it is seen to be a reversible process, although the cooling path in a temperature plane differs from the heating path, as structural relaxation occurs during the time taken to cool the liquid or heat the glass. In those cases when a liquid cannot be supercooled by usual methods, the glass has to be formed by hyperquenching the liquid, i.e., by cooling at a rate higher than 10^5 K s^{-1}, as in the case of water.[87] The glass thus obtained has a high fictive temperature, T_f, than a glass formed by normal cooling of a liquid, say at 0.1 K s^{-1}. With time, the hyperquenched glass structurally relaxes to a lower energy state of lower T_f, at a rate that increases with increase in the temperature. (For a discussion of the subject, see for example Ref. 58.)

When such a glass is heated, its energy and T_f decrease. After the state has crossed the equilibrium liquid line in the temperature plane for a certain heating rate, the energy and T_f increase, the glass softens and becomes an ultraviscous liquid. If the liquid crystallizes rapidly, the glass-softening may be immediately followed by crystallization and may not be observed if crystal nuclei form as a result of the heat released during structural relaxation.[85] For a material that does not crystallize in the ultraviscous state, this process is illustrated in Fig. 9(A).

We propose that in terms of its high energy, the pressure-collapsed state of hexagonal and cubic ices is qualitatively similar to the amorphous state formed by mechanical deformation of crystals, to the glassy state formed by hyperquenching of a liquid, or to the amorphous state formed by vapor-deposition. As illustrated in Fig. 9(A), the deformed amorphous state obtained by mechanical deformation of crystals has high internal energy and, as in mechanical amorphization of metal crystals, the self diffusion rate in the deformed state is much faster than in the normal state. The remarkable difference between a mechanically amorphized crystal and the amorph formed by pressure-collapse of hexagonal and cubic ices is that the mechanically amorphized state of crystals is bulkier than the parent crystals, and the amorph formed by pressure-collapse of ice is denser than the parent crystal. But it should be borne in mind that this amorph is denser with respect to hexagonal and cubic ices, which do not survive at a pressure higher than

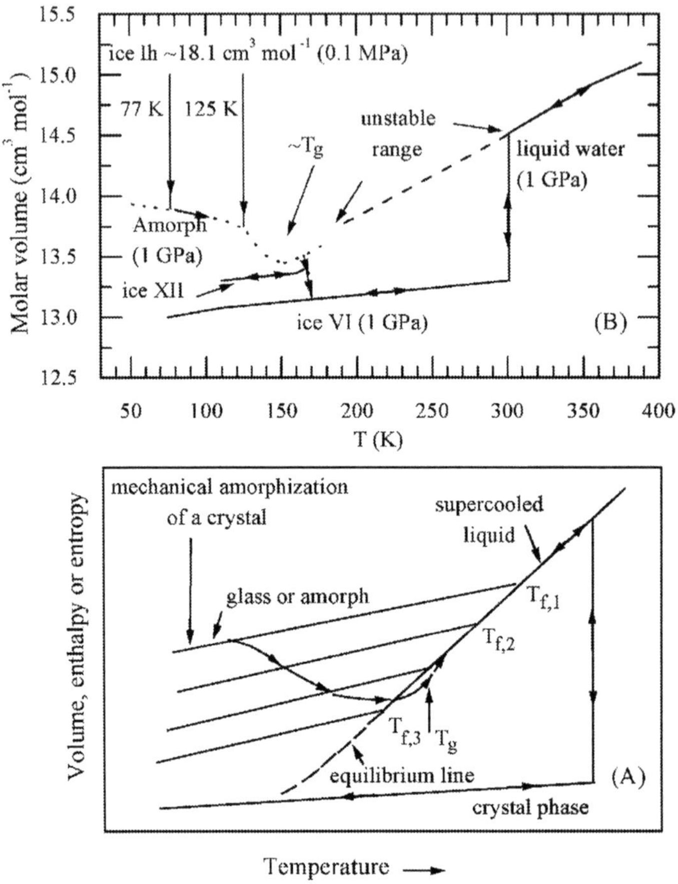

Figure 9. (A) An illustration for the formation of high energy amorphs of high fictive temperatures by mechanical deformation, and/or by rapid cooling. The spontaneous structural relaxation in a heating rate dependent manner to a low fictive temperature, low energy state and then glass transition. In this case the ultraviscous liquid does not crystallize. The segments of the plots are marked. (B) Pressure-collapse of hexagonal ice, its structural relaxation to a denser, low thermal conductivity solid with time and on heating, its gradual transformation to an ultraviscous liquid which crystallizes to ice XII on fast heating and finally to ice VI.

0.8 GPa, and not denser with respect to ice VI which is the stable crystal phase at these pressures. Therefore, a comparison of the density of the amorph needs to be made with respect to the density of ice VI, which is the stable phase at the collapse pressures of the hexagonal ice. In earlier discussion of the properties of the amorph at ambient pressure and 77 K, this aspect has been unfortunately overlooked. It remains to be seen whether an amorph can be produced by mechanical deformation of ice VI at 1 GPa, or else produced by mechanical deformation of recovered ice VI at ambient pressure and 77 K by using a high speed ball mill in the same manner as ordinary crystals.[88]

To illustrate the pressure-induced amorphization of hexagonal and cubic ices, the structural relaxation of the solid on heating, and the conversion of the relaxed state to ultraviscous water, we have plotted the molar volume against the temperature of these various states in Fig. 9(B). Hexagonal and cubic ices have molar volume of ~18.1 ml mol^{-1} and they collapse to a solid of molar volume 13.8 ml mol^{-1} at ~1 GPa at 77 K, and to 13.75 ml mol^{-1} volume at 125 K. The collapsed state becomes denser on heating according to a temperature-dependent rate, and after the equilibrium line has been crossed on heating the volume increases. Further heating transforms it to ultraviscous water whose dielectric relaxation time is ~1 s at 140 K. On further heating to 150-160 K range, the liquid may crystallize to ice XII, which exists in the 0.7–1.5 GPa and 158–212 K range, but also the resulting phase may depend upon the heating rate.[89,90] Ice XII can be thermally cycled at 1 GPa pressure along the path shown by the oppositely pointing arrows. (Note that crystallization has been found to occur also to mixtures of ice VI and XII[80] and mixtures of ice IV and XII,[89] and therefore there is some ambiguity of crystallization in this metastable state.) On heating, ice XII transforms to the stable phase ice VI[90] at 1 GPa as is indicated in Fig. 9(B). Ice VI can be heated at 1 GPa and it melts to water at a temperature slightly above 300 K. Studies of the liquid–ice VI phase boundary have shown that liquid water does not supercool through the ice VI stability region and water at 1 GPa in the 300–140 K range is unstable. Its ultraviscous state can be obtained via the pressure-collapse route of hexagonal ice shown in Fig. 8 inset.

4. Concluding remarks

As an exogenic effect, high pressure has been long known to reversibly transform one crystal form to another form at an equilibrium phase boundary. But recent studies have shown that high pressure also has a prominent role in transforming a crystal permanently to an amorphous solid. This is most easily observed for crystals with open structures that float on their melt. Hexagonal and cubic ices collapse under less than 1 GPa pressure, with a rate that varies with the temperature, crystal size, and pressure. The pressurizing rate, the temperature, and the period of sustained pressure all have an effect on the amorphous state formed by the collapse of ice. The pressure required to collapse hexagonal ice becomes less as the temperature is increased. This is the opposite of the effect observed for vitrification of liquids – vitrification pressure increases as the temperature is increased. This pressure is lower for cubic ice than for hexagonal ice and it is also lower when the ice crystal size is in the micron range. It seems that the higher the thermal energy, or else the higher the surface energy, of a sample, the lower is the pressure that ultimately amorphizes it. Thus, for a given amorphization pressure there is a thermal energy equivalence for the surface energy.

The kinetics of pressure-collapse and amorphization shows a distribution of rates instead of a single rate and this distribution is expressed in terms of the stretched exponential parameter. The width of the pressure range over which the structural collapse occurs, the time-dependence, and the temperature-, and pressure-dependences of structure collapse, can be reconciled with Born's theory of mechanical collapse by transverse phonons softening, only if the effects of lattice faults (point defects, dislocations) and of internal surfaces are included. A single crystal hexagonal ice may require a much higher pressure for structural collapse than has been used for polycrystalline samples. It is also conceivable that crystals of ice approaching several tens of nanometers may amorphize at a much lower pressure. As the nanometer size crystals would be difficult to distinguish from the amorph by x-ray diffraction, volume and thermal conductivity changes on increase in pressure would need to be used. Such effects may be observable for other crystals.

Pressure-collapsed crystals are in a high-energy state and in this state they are analogous to mechanically-amorphized crystals and the amorphs that are produced by vapor-deposition. After their spontaneous relaxation at a relatively high temperature, they become analogous to the hyperquenched glassy state and show a glass-softening range and temperature. The pressure-collapsed state of hexagonal ice at 1 GPa structurally relaxes rapidly on heating. During this occurrence, its density increases and thermal conductivity decreases and the solid become ultraviscous water at 140 K. On further heating the ultraviscous water crystallizes to the denser state of ice XII[89,90] and then to ice VI,[90] or to a mixture of ice XII and VI[80] or of ice XII and ice IV.[89] Thus, at 1 GPa, the amorph is bulkier than ice VI, its stable crystal phase. It is conceivable that ice VI may become mechanically amorphized in a high-speed ball mill, in the same manner as ordinary crystals do, either at 1 GPa at temperatures below 130 K, or in the recovered state at ambient pressure and 77 K. The ultraviscous state of water at 1 GPa, which is difficult to obtain by supercooling water at 1 GPa, can be obtained via the pressure-collapse of hexagonal ice.

Finally, it should be noted that in a recent high resolution x-ray scattering study of the amorph formed by pressurizing hexagonal ice at 77 K to 1.8 GPa, it has been found that the amorph shows crystal-like inelastic response, i.e., "Despite the apparent structural disorder of HDA, its dynamic response appears crystal-like with a close resemblance to ice IX pointing an intriguing high degree of short range order."[91] But from the phonon dispersion data obtained from an inelastic neutron diffraction study of single crystal of ice at 140 K hydrostatically compressed to 0.55 GPa by using fluid nitrogen as pressure transmitting medium,[92] it has been concluded that already at 0.50 GPa pressure, "They reveal a pronounced softening of various low-energy modes, in particular, those of the transverse acoustic phonon branch in the [100] direction and polarization in the hexagonal plane."[92] This has been used to suggest that the lattice instability already at 0.5 GPa leads to amorphization of ice under pressure. In another development, negative thermal expansion of inorganic crystals has been considered as a requirement for their pressure-collapse and to infer that such crystals would amorphize under pressure.[13,15] Strassle et al.[92] have used the negative expansion coefficient of hexagonal ice at low temperatures to determine its Born instability pressure as ~2.5 GPa, meaning that a single crystal of ice may amorphize at a pressure higher than 2.5 GPa. This pressure is higher than that observed here, and consistent with the finding that a single crystal of ice did not amorphize at a hydrostatic pressure of ~2 GPa.[26] Nevertheless, it is important to examine whether the dissolved nitrogen in ice in those experiments in which fluid nitrogen has been used as pressure transmitting medium,[92] has an effect on the phonon dispersion data. Dissolved nitrogen at ambient pressure has been known to decrease the dielectrically measured molecular reorientation time of hexagonal ice by several orders of magnitude,[93] and it is not certain whether it also has an effect on the temperature at which the expansion coefficient of pure ice is negative.

References

1. Skinner, B. J. and Fahey, J. J. (1963) *J. Geophys. Res.* **68**, 5595.
2. Schrader, R. and Dusdorf, W. (1966) *Krist. Tech.* **1**, 59.
3. Primak, W. (1975) *The Compacted States of Vitreous Silica, Studies of Radiation Effects in Solids* (Gordon and Breach, London 1975).
4. Brixner, L. H. (1972) *Mater. Res. Bull.* **7**, 879.

5. Yarmakov, A. Y., Yurchikov, Y., and Barinov, V. A. (1981) *Phys. Met. Metall.* (English Transl.) **52**, 50.
6. Koch, C. C., Cavin, O. B., McKamey, C. J., and Scarborough, J. O. (1983) *Appl. Phys. Lett.* **43**,1017.
7. Hemley, R. J., Jephcoat, A. P., Mao, H. K., Ming, L. C., and Manghnani, M. H. (1988) *Nature* **334**, 52.
8. Liu, L.-G. and Ringwood, A. E. (1975) *Earth Planet Sci. Lett.* **28**, 209.
9. Kruger, M. B. and Jeanloz, R. (1990) *Science* **249**, 647.
10. McNeil, L. E. and Grimsditch, M. (1992) *Phys. Rev. Lett.* **68**, 83.
11. Deb, S. K., Wilding, M., Somayazulu, M., and McMillan, P. F. (2001) *Nature* **414**, 528.
12. Greaves, G. N., Meneau, F., Sapelkin, A., Colyer, L. M., Gwynn, I. A. P., Wade, S., and Sankar, G. (2003) *Nature Mater.* **2**, 622.
13. Zhang, J., Zhao, Y., Xu, H., Zrelinskas, M. V., Wang, L., Wang, Y., and Uchida, T. (2005) *Chem. Mater.* **17**, 2817.
14. Carlson, S. and Krogh Andersen, A. M. (2001) *J. Appl. Cryst.* **34**, 7.
15. Perottoni, C. A. and da Jornada, J. A. H. (1998) *Science* **280**, 886.
16. Lipinska-Kalita, K. E., Kruger, M. B., Carlson, S., and Krogh Andersen, A. M. (2003) *Physica B* **337**, 221.
17. Goryainov, S. V. (2005), *Eur. J. Minerology* **17**, 201.
18. Sharma, S. M. and Sikka, S. K. (1996) *Prog. Mater. Sci.* **40**, 1.
19. Greaves, G. N. (2001) NATO Advanced Research Workshop Frontiers of High Pressure Research III, Application of high pressure to low dimensional electronic materials, Colorado State University (Kluwer, Academic, 2001).
20. Mishima, O., Calvert, L. D., and Whalley, E. (1984) *Nature* **310**, 393.
21. Johari, G. P. and Jones, S. J. (1986) *Phil. Mag. B.* **54**, 311.
22. Johari, G. P. and Andersson, O. (2004) *J. Chem. Phys.* **120**, 6207.
23. Johari, G. P., Hallbrucker, A., and Mayer, E. (1990) *J. Phys. Chem.* **94**, 1212.
24. Andersson, O. and Suga, H. (2002) *Phys. Rev. B* **65**, 140201.
25. Mishima, O., Calvert, L. D., and Whalley, E. (1985) *Nature* **314**, 76.
26. Johari, G. P. (2000) *Phys. Chem. Chem. Phys.* **2**, 1567.
27. Gromnitskaya, E. L., Stal'gorova, O. V., and Brazhkin, V. V. (2001) *Phys. Rev. B* **64**, 094205. Note that for some analysis in this paper and in a paper (Kohl, I., Mayer, E., and Hallbrucker, A. (2001) *Phys. Chem. Chem. Phys.* **3**, 602) density of D_2O ice XII has been mistaken for density of H_2O ice XII. Use of the correct value would change some of their conclusions.
28. Debenedetti, P. G. (2003) *J. Phys. Condens. Matter* **15**, R1669.
29. Andersson, O. and Inaba A. (2005) *Phys. Chem. Chem. Phys.* **7**, 1441.
30. Loerting, T., Salzmann, C., Kohl, I., Mayer, E., and Hallbrucker, A. (2001) *Phys. Chem. Chem. Phys.* **3**, 5355.
31. Mehl, P. and Boutron, P. (1987) *J. Phys. (Paris) Colloq.* **48**, C1-449.
32. Poirier, J. P. (1982) *Nature* 299, 683.
33. Tielens, A. G. G. M. (1987) and Allamandola, L. J. in *Physical Processes in Interstellar Clouds*, Eds. G. E. Morfill and M. Scholer (Reidel, Dordrecht, 1987), p. 223.

34. Mumma, M. J., Weissman, P. R., and Stern, S. A. (1993) in *Protostars and Planets III*, Eds. E. H. Levy, J. I. Lunine, and M. S. Matthews (Univ. of Arizona Press, Tucson, 1993), p. 1177.
35. Jenniskens, P. and Blake, D. F. (1994) *Science* **265**, 753.
36. Jenniskens, P., Blake, D. F., Wilson, M. A., and Pohorille, A. (1995) *Astrophys. J.* **455**, 389.
37. Petrenko, V. F. and Whitworth, R. W. (1999) *Physics of Ice* (Oxford University Press, Oxford, 1999).
38. Johari, G. P. and Andersson, O. (2004) *Phys. Rev. B* **70**, 184108.
39. Håkansson, B., Andersson, P., and Bäckström, G. (1988) *Rev. Sci. Instrum.* **59**, 2269.
40. Andersson, O., Sundqvist, B., and Bäckström, G. (1992) *High Pressure Res.* **10**, 599.
41. Carslaw, H. S. and Jaeger, J. C. (1959) *Conduction of Heat in Solids*, 2nd ed. (Clarendon, Oxford, 1959), p. 341.
42. Johari, G. P. and Whalley, E. (1976) *J. Chem. Phys.* **64**, 4484.
43. Johari, G. P. and Whalley, E., (2001) *J. Chem. Phys.* **115**, 3274.
44. Johari, G. P. and Whalley, E. (1979) *J. Chem. Phys.* **70**, 2094.
45. Kawada, S. (1988) *J. Phys. Soc. Jpn* **57**, 3694.
46. Johari, G. P. and Pyke, D. (2000) *Phys. Chem. Chem. Phys.* **2**, 5479.
47. Ross, R. G. and Andersson, P. (1982) *Can. J. Chem.* **60**, 881–892.
48. Bosio, L., Johari, G. P., and Teixeira, J. (1986) *Phys. Rev. Lett.* **56**, 460.
49. Andersson, O. and Johari, G. P. (2004) *J. Chem. Phys.* **121**, 3936.
50. Andersson , O. and Inaba, A. (2005) *J. Chem. Phys.* **122**, 124710.
51. Stal'gorova, O. V., Gromnitskaya, E. L., and Brazhkin, V. V. (1995) *JETP Lett.* **62**, 356.
52. Gromnitskaya, E. L., Stal'gorova, O. V., and Brazhkin, V. V. (1997) *JETP Lett.* **85**, 109.
53. Stal'gorova, O. V., Gromnitskaya, E. L., Brazhkin, V. V., and Lyapin, A. G. (1999) *JETP Lett.* **69**, 694.
54. Mishima, O. (1994) *J. Chem. Phys.* **100**, 5910.
55. Mishima, O. (1996) *Nature* (*London*) **384**, 546.
56. Majumdar, C. K. (1971) *Solid State Comm.* **9**, 1987.
57. Bendler, J. T. and Schlesinger, M. F. (1985) *Macromolecules* **18**, 591.
58. Scherrer, G. W. (1986) *Relaxations in Glasses and Composites* (John Wiley, N.Y. 1986), 1986), Chapter 12.
59. Schmidt-Rohr, K. and Spiess, H. W. (1991) *Phys. Rev Lett.* **66**, 3020.
60. Schiener, B., Bohmer, R., Loidl, A., and Chamberlin, R. V. (1996) *Science* **274**, 752.
61. Cicerone, M., Blackburn, F. R., and Ediger, M. F. (1995) *J. Chem. Phys.* **102**, 471.
62. Plonka, A. (2001) *Dispersive Kinetics* (Kluwer, Dordrecht, 2001).
63. Hemley, R. J., Chen, L. C., and Mao, H. K. (1989) *Nature* **338**, 638.
64. Salzmann, C. G., Kohl, I., Loerting, T., Mayer, E., and Hallbrucker, A. (2003) *Phys. Chem. Chem. Phys.* **5**, 3507.
65. Floriano, M. A., Handa, Y. P., Klug, D. D., and Whalley, E. (1989) *J. Chem. Phys.* **91**, 7187.
66. Johari, G. P., Pascheto, W., and Jones, S. J. (1994) *J. Chem. Phys.* **100**, 4548 and references therein.

67. Hobbs, P. V. (1074) *Ice Physics* (Clarendon, Oxford, 1974).
68. Salvetti, G., Tombari, E., and Johari, G. P. (1995) *J. Chem. Phys.* **102**, 4987.
69. Born, M. and Huang, K. (1954) *Dynamic Theory of Crystal Lattices* (Clarendon, Oxford, 1954).
70. Tse, J. S., Klug, D. D., Tulk, C. A., Swainson, I., Svensson, E. C., Loong, C. K., Shpakov, V., Belosludov, V. R., Belosludov, R. V., and Kawazoe, Y. (1999) *Nature* **400**, 647.
71. Perez, J., Mai, C., and Vassoile, R. (1977) *J. Glaciology* **21**, 361.
72. Coulson, C. A. and Eisenberg, D. (1966) *Proc. Roy. Soc.* (*London*) **A291**, 454.
73. Johari, G. P. and Jones, S. J. (1985) *Jour. de Chim. Phys. Phys.-Chim. Biol.* (Fr.) **82**, 1019.
74. Sikka, S. K. (2004) *J. Phys. Conden. Matter* **16**, S1033.
75. Johari, G. P. (2004) *J. Chem. Phys.* **121**, 8428.
76. Tulk, C. A., Benmore, C. J., Urquidi, J., Klug, D. D., Neuefeind, J., Tomberli, B., and Egelstaff, P. A. (2002) *Science* **297**, 1320.
77. Koza, M. M., Schober, H., Fischer, H. E., Hansen, T., and Fujara, F. (2003) *J. Phys. Condens. Matter* **15**, 321.
78. Guthrie, M., Urquidi, J., Tulk, C. A., Benmore, C. J., Klug, D. D., and Neuefeind, J. (2003) *Phys. Rev. B* **68**, 184110.
79. Andersson, O. (2005) *Phys. Rev. Lett.* **95**, 205503.
80. Klotz, S., Hamel, G., Loveday, J. S., Nelmes, R. J., and Guthrie, M. (2003) *Z. Kristallogr.* **218**, 117.
81. Cole, K. S. and Cole, R. H. (1941) *J. Chem. Phys.* **9**, **341**.
82. Hasted, J. B. (1973) *Aqueous Dielectrics* (Chapman and Hall, London, 1973).
83. Okada, K., Yao, M., Hiejima, Y., Kohno, H., and Kajihara, Y. (1999) *J. Chem. Phys.* **110**, 3026.
84. Johari, G. P. (2005) *Phys. Chem. Chem. Phys.* **7**, 1091.
85. Johari, G. P. (2005) *J. Chem. Phys.* **122**, 144508.
86. Johari, G. P. (2003) *J. Phys. Chem.* **107**, 9063.
87. Johari, G. P., Hallbrucker, A., and Mayer, E. (1987) *Nature* **330**, 552.
88. Richards, T. G. and Johari, G. P. (1988) *Phil. Mag. B* **58**, 445.
89. Salzmann, C. G., Mayer, E., and Hallbrucker, A. (2004) *Phys. Chem. Chem. Phys.* **6**, 5156.
90. Loerting, T., Kohl, I., Salzmann, C., Mayer, E., and Hallbrucker, A. (2002) *J. Chem. Phys.* **116**, 3171.
91. Koza, M. M., Schober, H., Geil, B., Lorenzen, M., and Requardt, H. (2004) *Phys. Rev. B* **69**, 024204.
92. Strassle, T., Saitta, A. M., Klotz, S., and Braden, M. (2004) *Phys. Rev. Lett.* **93**, 225901.
93. Johari, G. P. and Whalley, E. (1981) *J. Chem. Phys.* **75**, 1333.

COUPLED ORDERING IN SOFT MATTER: COMPETITION OF MESOSCALES AND DYNAMICS OF COUPLED FLUCTUATIONS

MIKHAIL A. ANISIMOV
Department of Chemical and Biomolecular Engineering and Institute for Physical Science and Technology University of Maryland at College Park, College Park, MD 20742, USA

Abstract. Coupling between order parameters is ubiquitous in soft matter. Such a coupling produces a variety of complex phases with unique properties that can be used in new emerging technologies. As examples, I discuss coupled ordering in thermotropic and lyotropic liquid crystals, in near-critical polymer solutions, and in an amphiphilic system (building artificial membranes). At meso- and nano-scales, coupling between different order parameter is associated with competition between different correlation lengths and results in coupled dynamics of fluctuations. These phenomena are expected in a broad range of soft-matter materials such as polymer solutions, microemulsions, and lyotropic liquid crystals, wherever phase separation couples with a mesoscopic structure.

Keywords: Soft matter, mesoscale, nanoscale, fluctuations, phase transitions, critical phenomena, liquid crystals, polymer solutions, lamellar stricture.

1. Introduction

The concept of the order parameter was introduced by Landau about 70 years ago.[1] The order parameter, φ, is a "density-like" thermodynamic variable that characterizes a degree of ordering in condensed matter. It is absent in the disordered phase and emerges in the ordered phase at the phase-transition point, breaking the breaking the symmetry of the disordered phase.

A starting point to consider a phase transition is the Landau expansion (also known as the Landau–Ginzburg functional) of local free-energy density in powers of the order parameter and its gradients:[2]

S. J. Rzoska and V. A. Mazur (eds.), Soft Matter under Exogenic Impacts, 75–90.

$$f(\varphi) = A\varphi^2 - B\varphi^3 + C\varphi^4 + \cdots + c_0 \nabla \varphi^2 + \cdots \tag{1}$$

The order parameter can be a scalar (density, concentration, or entropy) – to describe phase separation in fluids and solids as well as order–disorder transitions in binary alloys.[3] It can be a tensor – to characterize liquid-crystalline nematic ordering[4] – or an n-component vector ($n=1$, 2, or 3) for ferromagnetics and ferroelectrics, superfluids, and superconductors.[5] Like superconductors, liquid–crystalline smectic-A and smectic-C phases are described by a two-component vector order parameter (n=2);[4] while a long polymer chain in good solvent can be formally described by a "zero-component" vector order parameter ($n=0$).[6] In the Landau theory of phase transitions, which is essentially a mean-field approximation, the difference between scalar, tensor, or vector order parameters manifests itself only in the existence or absence of odd (asymmetric) terms and in a particular form of gradient terms (for the sake of simplicity, the order parameter is represented in expansion (1) by its amplitude and the gradient term corresponds to an isotropic case).

Various phase-transition scenarios are possible, depending on behavior of the coefficients in expansion (1). The coefficient A is proportional to the inverse susceptibility, $A = a(T - T^*) \propto \chi^{-1} = \partial^2 f / \partial\varphi^2$. Vanishing A at $T=T^*$ defines spinodal (T^* is the temperature of stability limit for the disordered phase). A critical point (second-order transition) is defined by the conditions $A=0$ and $B=0$ ($B \equiv 0$ if φ is a vector). When $B>0$ or/and $C<0$, the transition is first order. The conditions $A=0$, $B=0$, and $C=0$ define a tricritical point that separates second-order transitions and first-order transitions. In all these cases the coefficient c_0 is assumed to be positive. Physically, it means that inhomogeneities in the order parameter are not favorable. If c_0 becomes negative, such inhomogeneities may become thermodynamically stable. The conditions $A=0$, $B=0$, and $c_0=0$ define a so-called "the Lifshitz point" that separates homogeneous and locally heterogeneous states.

In soft-matter materials, such as complex fluids, there are always more than one order parameters. These different order parameters may couple (interact). Such coupling can

1. Change stability limit, enhancing phase separation, and changing the temperature dependence of the inverse susceptibility A.
2. Produce a cubic term ($B>0$), thus making the transition to be first order.

3. Change the sign of C, thus producing tricriticality and also driving the transition to first order.
4. Change the sign of c_0 and produce a modulated (microheterogeneous) phase.

It must be noted that the Landau theory is a mean-field approximation, which assumes the order-parameter fluctuations, represented by the gradient term in expansion (1), do not affect macroscopic thermodynamic properties. Dimensionality of the system plays no role either. However, it is well known, that in many systems the order-parameter fluctuations strongly affect the macroscopic properties in the critical region.[3] The effects of fluctuations strongly depend on the space dimension and on the nature of the order parameter. Even in the framework of Landau theory the correlation length of fluctuations ξ diverges when $A = 0$, being defined as $\xi = (c_0/A)^{1/2}$, while the macroscopic properties remain unaffected. For more accurate description, the analytical Landau model of phase transitions is to be replaced by non-analytical scaling theory.[7] A microscopic foundation of the scaling phenomenology is renormalization-group theory of critical fluctuations.[7,8] Nevertheless, the classical Landau treatment of phase transitions is a useful first approximation as it always gives a physically transparent guidance on a variety of phase behavior in complex fluids. Below I present a few characteristic examples of classical coupling between different order parameters, qualitatively explained by Landau theory.

2. Coupling between orientational and translational ordering in liquid crystals

In thermotropic liquid crystals the free energy depends on coupling between orientational (nematic) tensor order parameter Q and translational smectic-A vector order parameter ψ (one-dimensional density wave in the three-dimensional space). In lowest approximation

$$f(Q,\psi) = f_{\mathrm{N}}(Q) + f_{\mathrm{S}}(\psi) + \lambda Q\psi^2 + \cdots \tag{2}$$

The terms $f_{\mathrm{N}}(Q)$ and $f_{\mathrm{S}}(\psi)$ represent the contributions from the nematic and smectic subsystems, respectively, while the term $\lambda Q\psi^2$ represents the coupling between these two subsystems. The terms $f_{\mathrm{N}}(Q)$ and $f_{\mathrm{S}}(Q)$ can then be expanded in powers of the order parameters similar to Eq. (1).

The coupling constant λ is proportional to the slope of the nematic–smectic-A (NA) transition line in T–Q coordinates and thus can be obtain experimentally. The coupling term can be eliminated by a Legendre transformation and renormalization of the coefficients in Eq. (2) so that an effective free-energy functional, $\hat{f}(\psi)$, will depend only on the smectic order parameter:

$$\hat{f}(\psi) = \hat{A}\psi^2 + \hat{C}\psi^4 + E\psi^6 \cdots \tag{3}$$

with

$$\hat{A} = a_{\mathrm{NA}}\left(T - T_{\mathrm{NA}}^*(\lambda)\right), \quad \hat{C} = C - \frac{2\lambda^2}{a_{\mathrm{NI}}\left(T - T_{\mathrm{NI}}^*\right)}, \tag{4}$$

where T_{NA}^* and T_{NI}^* are the temperatures of stability limit of the isotropic (when N is stable) and nematic (when A is stable) phases, respectively. As one can see from Eqs. (4), the coefficient $\hat{C}$ can change its sign upon increase of the orientational susceptibility, $\propto\left(T - T_{\mathrm{NI}}^*\right)^{-1}$, when the gap between smectic A and isotropic phases becomes small enough. When $\hat{C} = 0$, a tricritical point emerges on the NA transition line, separating second-order and first-order NA transitions. This effect was first predicted by McMillan[9] and de Gennes[10] and later confirmed experimentally (see Ref. 3).

Another, more subtle, feature of NA transition is a coupling between fluctuations of nematic director and smectic order parameter. Such a coupling generates a non-trivial cubic term in the free-energy expansion and makes the transition always first order. This effect, known as the Halperin–Lubensky–Ma (HLM) effect,[11] is usually very small and cannot be easily detected in experiments. However, in the vicinity of the Landau–McMillan–de Gennes tricritical point, defined by the condition $\hat{C} = 0$ in expansion (3), the HLM effect becomes detectable and has been confirmed by accurate measurements of the transition latent heat and of the speed of phase front propagation.[12]

In the isotropic phase of liquid–crystalline materials with a narrow nematic gap, close to the weakly first-order nematic–isotropic (NI) transition, a coupling between strong orientational fluctuations and translational smectic fluctuations causes an enhancement in the pretransitional orientational susceptibility. This effect has been detected by light scattering on fluctuations of anisotropy in the isotropic phase.[3] The inverse light-scattering intensity, which, in this particular case, is proportional to the inverse susceptibility $A = A_{\mathrm{NI}}$ in Eq. (1), shows increasing deviations from the "Curie-Weiss" linear law ($\chi^{-1} \propto A_{\mathrm{NI}} = a_{\mathrm{NI}}\left(T - T_{\mathrm{NI}}^{*}\right)$) in the isotropic phase, when the nematic range decreases and the smectic phase decreases and the smectic phase becomes closer to NI transition.

A simple example of coupling between a scalar order parameter (molar concentration x) and a nematic tensor order parameter Q is liquid–liquid phase separation in non-ideal solutions of non-mesogenic substances in thermotropic nematic liquid crystals.[3]

The lowest-order coupling term in the free energy is $\lambda Q^2 x$, where the coupling constant $\lambda \propto dT_{\mathrm{NA}}/dx$. This coupling renormalizes the constant $C = C_{\mathrm{NI}}$ in the nematic–isotropic free energy expansion such that the effective coefficient

$$\hat{C}_{\mathrm{NI}} = C_{\mathrm{NI}} - \frac{2\lambda^2}{a\left(T - T_{\mathrm{c}}\right)}, \tag{5}$$

where T_{c} is the "bare" critical temperature of liquid–liquid separation without account for the coupling. Upon cooling, the sign of $\hat{C}_{\mathrm{NI}}$ can change. Large negative values of $\hat{C}_{\mathrm{NI}}$ mean enhancing the phase separation, since the transition becomes strongly first-order, despite a small value of B, and the two-phase region becomes wider. Phase diagrams for solution of a thermotropic liquid crystal 4-*n*-butyl-4′-methoxyazoxybenzene (BMOAB) in various alcohols are shown in Fig. 1.

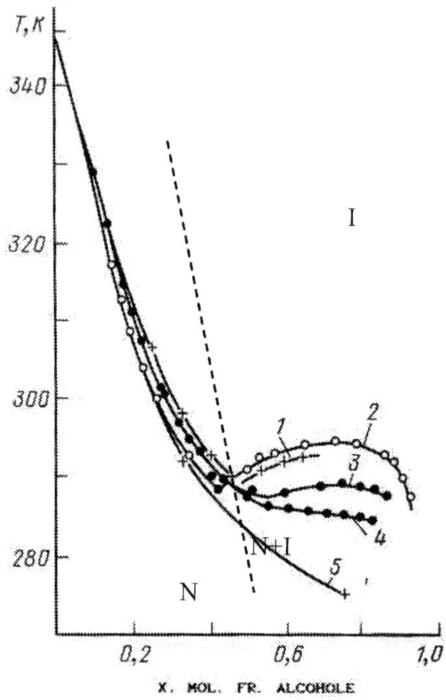

Figure 1. Phase diagrams of BMOAB-alcohol solutions.[3] **I** indicates the isotropic region, **N** the nematic region, **N+I** the two-phase region. Clearing temperatures (the last droplet of the nematic phase disappears): 1 propanol, 2 isopentanol, 3 *n*-butanol, 4 *n*-pentanol, 5 *n*-decanol. Dashed line: the last droplet of the isotropic phase disappears (shown for BMOAB-*n*-decanol solutions). In solutions of BMOAB in propanol, isopentanol, and *n*-butanol two isotropic phases coexist. Note, the clearing-temperature curve for BMOAB-*n*-pentanol solution is very flat, implying a close critical phase separation.

3. Self-assembling in aqueous solutions of a two-tail amphiphile

It is well known that amphiphilic molecules (with polar "heads" and non-polar "tails") in aqueous solutions usually are self-assembling, forming micelles. There is an interesting phenomenon in aqueous solutions of didodecyldimethylammonium bromide (DDAB).[13] Each amphiphilic molecule of DDAB has two hydrophobic tails, so that DDAB molecules spontaneously form inverted micelles in hydrocarbons and nano-size spherical vesicles in dilute aqueous solutions. However, in aqueous solutions at the DDAB concentration of about 10^{-3}mol/liter the spherical vesicles are self-assembling into a multilayer "onion-like" structure shown in Fig. 2.

Figure 2. Micrographs of a replica from the $8{\cdot}10^{-4}$ mol/liter DDAB aqueous solution obtained by freeze-fracturing electron microscopy.[13] The distance between the layers is about 100 nm. The layer thickness and the size of individual vesicle (seen in the upper left corner of the micrograph) is about 20 nm.

It is also remarkable that at the same DDAB concentration, the interfacial tension between water and a hydrocarbon (*n*-octane) almost vanishes, becoming less that 0.1 mN/m, suggesting that the formation of the multilayer structure may be accompanied by water–oil interface instability. One way to approach this problem theoretically is to apply the mesoscopic Landau–Ginzburg functional given by Eq. (1) and consider a Lifshitz-point mechanism.[14] At the presence of amphiphilic molecules the coefficient c_0 that characterizes a response of the free energy to the emergence of inhomogeneities may vanish, then changing its sign, and thus next-order gradient terms are to be added. For simplicity, consider a one-dimensional (*z*-coordinate) case with $B = 0$.[15] In lowest approximation

$$f(\varphi) = A\varphi^2 + C\varphi^4 + c_0\left(\frac{d\varphi}{dz}\right)^2 + \frac{d_0}{2}\left(\frac{d^2\varphi}{dz^2}\right)^2, \qquad (6)$$

where the order parameter φ is the difference between amphiphile concentration in the aqueous and "oil" phases. In the range of $c_0 > 0,\ A > 0,\ \varphi = 0$, or, when $A < 0,\ \varphi = \pm(-A/2C)^{1/2}$. The condition $A = 0$ at $c_0 > 0$ defines a critical line of oil–water phase separation softened by amphiphile. The Lifshitz point is defined by two conditions, $A = 0$ and $c_0 = 0$. In the range of negative c_0, a modulated phase becomes stable. If one approximates the modulation of the order parameter by a sinusoid, [15]

$$\varphi(z) = \varphi_0 \cos qz, \tag{7}$$

where q is the wave number of modulation, and takes into account that $\langle \cos^2 qz \rangle = 1/2,\ \langle \cos^4 qz \rangle = 3/8$, the free energy obtained by integration of Eq. (6) over volume V becomes

$$\frac{F}{k_B T} = \frac{1}{V} \int f dV = \frac{1}{2} \hat{A}(q) \varphi_0^2 + \frac{3}{8} C \varphi_0^4, \tag{8}$$

where $\hat{A}(q) = A + c_0 q^2 + (d_0/2) q^4$. At negative c_0, $\hat{A}(q)$ has minimum at $q = q_0 \equiv (-c_0/d_0)^{1/2}$, the equilibrium wave number of modulation. At the Lifshitz point $q_0 = 0$, meaning that the distance between lamellae becomes infinite. By minimizing Eq. (8) with respect to φ_0, one obtains the equilibrium amplitude of modulation $\varphi_0 = \pm\left(-2\hat{A}(q_0)/3C\right)^{1/2}$, where $\hat{A}(q_0) = A - (c_0^2/2d_0) = 0$ defines the line of second-order transitions between the homogeneous and modulated phases. Finally, the equation for the line of first-order transitions between the two-phase region and the modulated phase is found by equalizing the free energies of the two-phase region, $F/k_B T = -A^2/4C$, and the energy of the modulated phase, $F/k_B T = -\hat{A}^2/3C$:

$$A = -\frac{1}{\sqrt{6} - 2} \frac{c_0^2}{d_0}. \tag{9}$$

The phase diagram with a Lifshitz point is schematically shown in Fig. 3.

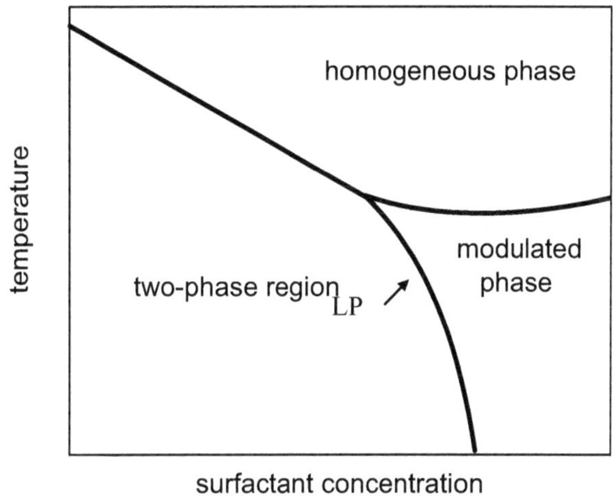

Figure 3. Schematic phase diagram of surfactant solution forming a modulated phase. The Lifshitz point (LP) separates the critical line of liquid–liquid phase separation (on the left) and the first order phase separation (lower curve on the right). The transition between the homogeneous disordered phase and the modulated phase is second order.

4. Competition of mesoscales in near-critical polymer solutions

What happens if a solution of long monodisperse polymer chains, characterized by the degree of polymerization N, radius of gyration R_g, and polymer volume fraction ϕ, is cooled down to the critical point of liquid–liquid demixing? A schematic phase diagram of such a system is shown in Fig. 4.

The critical locus of liquid–liquid separation, shown in Fig. 4 by the thick curve, is terminated at the *theta* point ($T = \Theta$, $\phi = 0$, experimentally defined as the limiting $N \rightarrow \infty$ point of phase separation in infinite dilution). There are two approaches to describe the nature of the *theta* point. The first approach, commonly known to physical chemists, based on the definition given by Flory[17]: two leading virial coefficients in a virial expansion of the osmotic pressure Π, the first A_1 and the second A_2, simultaneously vanish at the *theta* point. The virial expansion reads:

$$\frac{\Pi v_0}{k_B T} = A_1\phi + A_2\phi^2 + A_3\phi^3 + \cdots \tag{10}$$

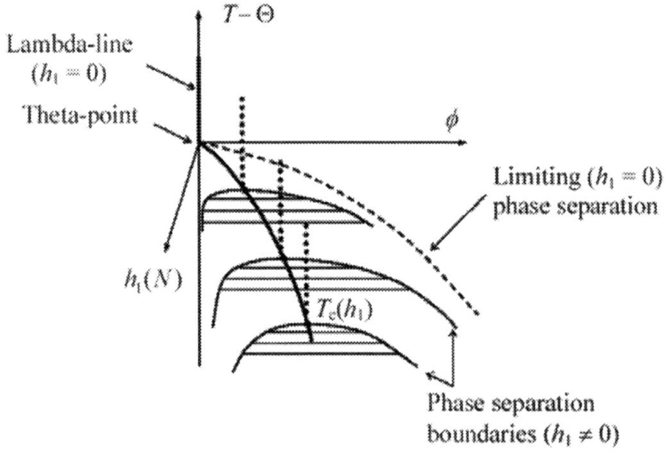

Figure 4. Schematic *theta*-point phase diagram.[16] The dotted vertical lines indicate the paths of our light-scattering experiments.

with $A_1 = 1/N$ and $A_2 = (1/2) - (w/k_B T)$, where w is the Flory interaction parameter and v_0 is the solvent molecular volume. While vanishing A_1 means the possibility of phase separation at infinite dilution, A_2 sign change separates a "good-solvent" state (above *theta* temperature, $A_2 > 0$) and a "poor-solvent" state (below theta temperature, $A_2 < 0$). Thus the *theta* temperature is defined as $T = \Theta \equiv 2w/k_B$ ("*theta*-solvent" state).

Another approach is based on the renormalization-group theory of phase transitions. It was pioneered by de Gennes[18] and further developed by Duplantier.[19] According to the theory, the *theta* point is a tricritical point, which separates the states of critical-like self-avoiding-walk singularities at $T > \Theta$ (vertical "λ-line" at $\phi = 0$ in Fig. 4) and first-order phase separation at $T < \Theta$. At $T = \Theta$ an "infinite" polymer chain exhibits random-walk behavior. Below the *theta* temperature the system is phase separated and the polymer chain entirely collapses into one phase (the dashed curve in Fig. 4). This collapse of the infinite polymer chain is accompanied by the emergence of a finite polymer volume fraction in this phase and thus can be regarded as first-order transition. A "vector-like" ordering scaling field h_1, conjugated to a "vector-like" order parameter ψ, is zero along the *lambda* line but becomes non-zero below the *theta* temperature and everywhere else at finite degrees of polymerization *N*. Neither the scaling field h_1 nor the order parameter ψ are experimentally observable. The amplitude of the polymer order parameter is proportional to the concentration of polymer-chain endpoints, so that the polymer volume fraction $\phi \propto \psi^2$, and thus vanishes along the *lambda* line. The correlation length associated with the polymer order parameter is proportional to the radius of gyration which diverges in the limit of infinite degree of polymerization. The locus of critical phase-separation points is a non-zero-h_1 critical (wing) line originating from the tricritical point. The order parameter for the fluid–fluid phase separation, associated with the polymer volume fraction ϕ, and the polymer-chain order parameter ψ belong to different classes of universality. Tricriticality emerges as a result of a coupling between these two order parameters and in three dimensions exhibits mean-field behavior with small logarithmic corrections.[19]

The appropriate field-dependent thermodynamic potential to describe theta- -point tricriticality is the osmotic pressure, which can be represented by a Landau expansion in powers of the polymer order parameter[16]:

$$\frac{\Pi v_0}{k_B T} = h_1 \psi - \left(\frac{1}{N} \psi^2 + C\psi^4 + D\psi^6 + \cdots \right). \tag{11}$$

The field h_1 is found by minimization of Π with respect to ψ and then is to be substituted into Eq. (11). By comparing the result and Eq. (10), one can conclude that $C \propto A_2$ vanishes at the theta-point, as required by tricriticality.

The next question to be answered is how to describe crossover from fluctuation-dominated critical behavior of liquid–liquid separation at finite degrees of polymerization (in the vicinity of the "wing" critical line) to the theta-point mean-field tricritical behavior in the limit of infinite degree of polymerization. In other words, "the thing is to successfully marry the second virial coefficient and scaling theory".[20] The criticality of fluid phase separation and tricriticality at the theta point are described by different order parameters belonging to different classes of universality. Correspondingly, there exist two different mesoscopic correlation lengths, characterizing the "size" of the order-parameter fluctuations. Upon approaching the critical point of phase separation, the correlation length of concentration fluctuations eventually becomes so large that it exceeds the radius of gyration making the polymer chain almost irrelevant. However, upon increase in degree of polymerization, the range of phase-separation criticality shrinks, yielding the *theta*-point tricriticality. Accurate light-scattering studies[16] performed on polystyrene-cyclohexane solutions unambiguously show that the crossover line, which separates criticality and tricriticality, is defined by the condition $\xi \cong R_g / \sqrt{3}$ (Fig. 5).

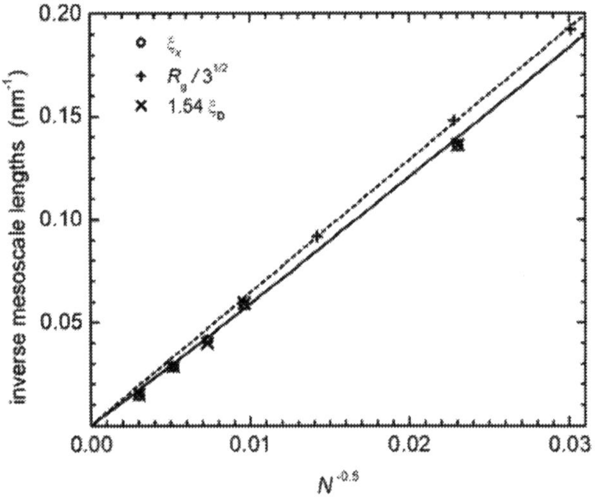

Figure 5. Inverse mesoscale lengths as a function of $N^{-1/2}$.[16] Circles are inverse correlation lengths ξ_x at the crossover temperature; × are theoretically calculated inverse screening length ξ_D multiplied by 1.54; + are inverse normalized radii of gyration $R_g/\sqrt{3}$ from Ref. 21. The solid curve represents a theory that includes logarithmic corrections.

Interestingly, the crossover absolute temperatures appear to be always very close to temperatures of limiting $N \to \infty$ phase separation in infinite dilution shown in Fig. 4 by the dashed line.

5. Dynamics of coupled fluctuations and "avoided crossing" in soft matter

Recently, a strong coupling between diffusive relaxation of concentration fluctuations and viscoelastic relaxation has been found in a near-critical high-molecular-weight polymer solution.[22,23] This coupling results in the emergence of two experimentally observable (effective) modes (Fig. 6). Two original soft modes (diffusive and viscoelastic) are associated with relaxation of two order parameters, each belonging to a different universality class in dynamics. The coupling occurs when these two original modes should cross but they do not. Near the point, where they should cross, the modes mix and, while avoiding crossing, they change their character. Neither of these observable modes is the original diffusion mode nor the original viscoelastic mode. From the behavior of coupled dynamic modes we are able to deduce a molecular-weight dependence of the viscoelastic parameters. One may call this procedure "fluctuation microrheology" as an alternative to traditional microrheology: instead of tracing Brownian particles in a viscoelastic medium, one can couple slow critical fluctuations, spontaneously generated in the system, with viscoelasticity. "Avoided crossing" can also be observed in the isotropic phase of an aqueous solution of cromolyn.[24] Cromolyn is a representative of lyotropic chromonic liquid crystals with elongated rods formed by self-assembled aggregates of disc-like molecules. Upon cooling, the cromolyn solution forms a nematic phase with a preferable orientation of the anisotropic aggregates. Close to the isotropic–nematic transition, the near-critical fluctuations of anisotropy are coupled with concentration fluctuations. "Avoided crossing" is observed even in simple binary fluids near the vapor–liquid critical point,[25] where the diffusive relaxation of concentration fluctuations is coupled with the diffusive relaxation of entropy fluctuations. A similar effect has been observed in polymer solutions subjected to shear.[26] Ubiquity of "avoided crossing" extends beyond condensed matter: the conceptually similar "Landau–Zener effect" describes coupled quantum oscillators.[27] We expect the "avoided crossing" to be found in a broad range of soft-matter materials such as polyelectrolytes, microemulsions, and lyotropic liquid crystals, wherever concentration fluctuations couple with self-assembling and mesoscopic structure.

The avoided crossing in cromolyn solutions was not as conclusive as that in polymer solutions, since large polydispersity of the elongated aggregates makes the dynamic modes too broad. To obtain a clear picture of coupling between the diffusion mode and orientational relaxation, a study of nematic thermotropic liquid crystals in molecular solvents where the emergence of the nematic ordering is accompanied by near-critical phase separation, is needed. No such studies have been performed so far. A good candidate for such an experiment would be solutions BMOAB, discussed in Section 3, in *n*-pentanol and pentanol isomers. BMOB is a very well studied nematic liquid crystal, chemically stable and pure, with well established molecular anisotropy.

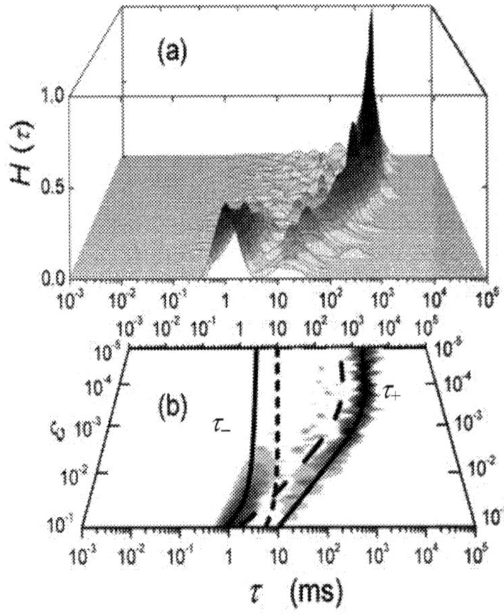

Figure 6. Avoided crossing of two dynamic modes in a critical polystyrene-cyclohexane solution (molecular weight 11.4 million) as a function of the reduced temperature distance to the critical temperature of phase separation, $\varepsilon = (T - T_c)/T$, detected by dynamic light scattering[23] at the scattering angle 30°. (a) Decay-time (τ) probability distribution $H(\tau)$ extracted from the experimental dynamic correlation functions. (b) Projection of the decay-time distribution on an $\varepsilon-\tau$ plane. The solid curves represent the theoretical relaxation times of the effective "slow" (τ_+) and "fast" (τ_-) modes. The long-dashed curve represents the uncoupled critical diffusion time. The short-dashed curve represents the uncoupled viscoelastic relaxation time.

6. Conclusion

Coupling between order parameter at mesoscales generates a rich variety of new phenomena both in statics and dynamics. Mesoscopically, such a coupling is associated with competition of characteristic correlation lengths which belong to different interacting order parameters. New experiments, especially on liquid crystalline and biomolecular systems, are highly desirable.

References

1. Landau, L. D. (1937) *Phys. Zs. Sowjet.* **8**, 113 (1935); *Phys. Zs. Sowjet.* **11**, 26; *Phys. Zs. Sowjet.* **11**, 545.
2. Landau, L. D. and Lifshitz, E. M. (1980) *Statistical Physics* (Pergamon Press, Oxford.
3. Anisimov, M. A. (1991) *Critical Phenomena in Liquids and Liquid Crystals* (Gordon & Breach Science Publishers, New York).
4. de Gennes, P. G., and Prost, J. (1993) *The Physics of Liquid Crystals*, 2nd Edition (Oxford Univ. Press, Oxford).
5. Kleman, M. and Lavrentovich, O. D. (2003) *Soft Matter Physics: An Introduction* (Springer, New York).
6. de Gennes, P. G. (1979) *Scaling Concepts in Polymer Physics* (Cornell University, Ithaca, NY).
7. Fisher, M. E. (1982) "Scaling, universality, and renormalization group theory" in *Lecture Notes in Physics*, F. J. W. Hahne, ed. (Springer, Berlin), Vol. **186**, p.1.
8. Zinn-Justin, J. (1996) *Quantum Field Theory and Critical Phenomena* (Clarendon, Oxford).
9. McMillan, W. L. (1971) *Phys. Rev. A* **4**, 1238.
10. de Gennes, P. G. (1972) *Solid State Comm.* **10**, 753.
11. Halperin, B. I., Lubensky, T. C., and Ma, S. K. (1974) *Phys. Rev. Lett.* **32**, 282.
12. Anisimov, M. A., Cladis, P. E., Gorodetskii, E. E., Huse, D. A., Podneks V. E., Taratuta, V. G,. van Saarloos, W., and Voronov, V. P. (1990) *Phys. Rev. A* **41**, 6749.
13. Zhuravleva, E. V., Anisimov, M. A., Svitova, T. F., and Lobanova, V. Yu. (1992) "New Kind of Water/Oil Interface Instability and the Nature of Ultra-low Interfacial Tension", in *Physical Chemistry of Colloids and Interfaces in Oil Production*, Proc. of the 6th IFP Exploration and Production Research Conference, 1991, Saint-Raphael. Editions Technip, Paris, 133–139.
14. Anisimov, M. A., Gorodetsky, E. E., Davydov, A. J., and Kurliandsky, A. S. (1992) *Liq. Cryst.* **11**, 941.
15. Michelson, A. (1977) *Phys. Rev. B* **16**, 577.
16. Anisimov, M. A., Kostko, A. F., Sengers, J. V., and Yudin, I. K. (2005) *J. Chem. Phys.* **123**, 16490/1–17.

17. Flory, J. (1953) *Principles of Polymer Chemistry* (Cornell University, Ithaca, NY), p. 613.
18. de Gennes, P. G. (1975) *J. Physique Lett.* (*Paris*) **36**, L-55.
19. Duplantier, B. (1982) *J. Physique* **43**, 991.
20. Stell G. (personal communication).
21. Melnichenko Y. B. and Wignall, G. D. (1997) *Phys. Rev. Lett.* **78**, 686.
22. Tanaka, H., Nakanishi, Y., and Takubo, N. (2002) *Phys. Rev. E* **65**, 021802/1-4.
23. Kostko, A. F., Anisimov, M. A., and Sengers, J. V. (2002) *Phys. Rev. E* **66**, 020803 (R)/1-4.
24. Nastishin, Yu. A., Liu, H., Shiyanovskii, S. V., Lavrentovich, O. D., Kostko, A. F., and Anisimov, M. A. (2004) *Phys. Rev. E* **70**, 051706/1-9.
25. Anisimov, M. A., Agayan, V. A., Povodyrev, A. A., Sengers, J. V., and Gorodetskii, E. E. (1998) *Phys. Rev. E* **57**, 1946.
26. Dixon, K., Pine, D. J., and Wu, X. -I. (1992) *Phys. Rev. Lett.* **68**, 2239.
27. Landau L. D. and Lifshitz, E. M. (1977) *Quantum Mechanics: Non-Relativistic Theory* (Pergamon Press, New York).

ALL STANDARD THEORIES AND MODELS OF GLASS TRANSITION APPEAR TO BE INADEQUATE: MISSING SOME ESSENTIAL PHYSICS

K.L. NGAI
Naval Research Laboratory,
Washington, DC 20375-5320 USA

Abstract. Many-molecule relaxation dynamics manifest themselves as general properties of the structural α-relaxation of glass-forming liquids independent of chemical composition and physical structure. An example is the Kohlrausch stretched exponential correlation function, $\phi(t) = \exp[-(t/\tau_\alpha)^{1-n}]$, where $0 \leq n < 1$ is a convenient measure of the dispersion of the α-relaxation and the extent of the many-molecule relaxation dynamics. Plenty of the other general properties are shown to be either governed by n or correlated with n, and these empirical facts collectively indicate the importance of many-molecule relaxation dynamics. Practically, all standard theories and models of glass transition have not taken many-molecule relaxation dynamics adequately into consideration and it is unsurprising that they cannot explain some of the general properties. Furthermore, they only address the structural α-relaxation without considering the universal Johari–Goldstein secondary relaxation and the role it plays as the precursor of the many-molecule relaxation dynamics. Experiments have shown that the Johari–Goldstein relaxation has properties that mimic the α-relaxation, Its relaxation time τ_{JG} bears a general relation to τ_α, which is governed by n. These general properties of the Johari–Goldstein relaxation indicate its fundamental importance, which is not recognized or considered in all standard theories and models. Thus, we conclude that all standard theories and models of glass transition are inadequate.

Keywords: Glass transition, glasses, structural relaxation, Johari–Goldstein relaxation, coupling model.

1. Introduction

The glass transition is a general phenomenon found in many different kinds of materials including inorganic substances with widely different chemical compositions, organic small molecules, synthetic polymers, side chain liquid crystalline polymers, metallic compounds, plastic crystals, pharmaceuticals, and

S. J. Rzoska and V. A. Mazur (eds.), Soft Matter under Exogenic Impacts, 91–111.

some biomaterials. Upon decreasing temperature T or increasing pressure P, the structural α-relaxation time, τ_α, of a noncrystallizing liquid or the associated transport coefficients can change by many orders of magnitude. Eventually the material will vitrify; that is τ_α becomes so long that the system cannot attain an equilibrium configuration in the available time. This marked decrease in mobility has led others to construct theories of glass transition based concomitant changes of conjugate thermodynamic variables such as free volume and configurational entropy. Each of these theories was considered by the original author or authors as an adequate solution of the glass transition problem, and this claim was upheld by others who apply such theories. For example, free volume model is often used and considered adequate to address glass transition of amorphous polymers by workers in polymer viscoelasticity as evidenced by the standard textbooks on this subject.[1] The model of Adam and Gibbs[2] also advocate that configurational entropy and its change with temperature are sufficient to explain the glass transition problem. The main objective of these theories is to explain the rapid change of τ_α with temperature, but not the many other properties of τ_α. This is no fault of the originators of these theories because the many other properties of τ_α were found in subsequent years by novel or improved experimental techniques.[3,4] Nevertheless, this is obviously an inadequacy of these standard theories.

In spite of this optimism and even allowing the possibility that both free volume and configurational entropy contribute, these theories have not completely solved the glass transition problem. Firstly, these theories address only the α-relaxation but not the processes that have transpired before it, which include caging of the molecules[5] and the universal Johari–Goldstein secondary relaxation.[5-13] These processes are universal and have properties indicating that they bear some relations to the α-relaxation and likely play some fundamental role in glass transition. Before jumping into explaining the change of τ_α with temperature, like the free volume and configurational entropy models, one should recognize that a more fundamental problem is to explain the evolution of dynamics from short times to the long time α-relaxation at *constant* temperature. Secondly, the conventional theories do not address many well established and general properties of the α-relaxation and its relaxation time τ_α.[3,4,14] Several of these general properties either is correlated with or governed by the width of the dispersion of the α-relaxation, or equivalently the fractional exponent n in the Kohlrausch function,[15]

$$\phi(t) = \exp[-(t/\tau_\alpha)^{1-n}], \quad 0 \le n < 1, \tag{1}$$

used to describe the correlation function of the α-relaxation.[3,4] The standard theories in their original forms did not consider the dispersion of the α-

relaxation, and occasionally was considered in some elaborations of the theories by others. However, the α-dispersion was derived as a separate consequence independent of τ_α, usually with further assumptions. The purported α-dispersion bears no fundamental relationship to the value of τ_α or its properties. Thus, the conventional theories and their embellished versions cannot explain many key experimental facts about the α-relaxation. Such theories cannot be viable or considered as satisfactory solution of the glass transition problem. Nevertheless, their legitimacy is maintained to present days. Others continue to provide additional experimental data to further substantiate their limited predictions, even though glaring contradictions to other experimental facts remain.

The conventional theories explain the temperature dependence of τ_α above T_g in terms of the Vogel–Fulcher–Tammann–Hesse (VFTH)[2] equation or the equivalent Williams–Landel–Ferry (WLF)[1] equation. Even this explanation is inadequate. In most glass-formers, τ_α has an Arrhenius temperature dependence at high temperatures $T > T_A$ when τ_α becomes short, a VFTH dependence in the range $T_A > T > T_B$, and a different VFTH dependence at lower temperatures within the range $T_B > T > T_g$.[16,17] The three temperatures regimes are marked by some characteristics involving the width of the dispersion of the α-relaxation or n.[18,19]

One may suspect that the cause of the inadequacy of conventional theories is the neglect to include some fundamental physics. There is no doubt that (free) volume and/or (configurational) entropy do determine molecular mobility and are essential. However, these thermodynamic quantities by themselves are not sufficient, and as we shall show that they were applied in the wrong place in the standard theories. The purpose of this paper is to use experimental facts as guides to identify the missing physics necessary for a fully successful and rigorous theory to be constructed in the future. Meanwhile, a schematic approach capturing the essential physics is proposed to pave the way for someone to finally solve the glass transition problem.

2. Many-molecule relaxation dynamics

The entire problem of glass transition is composed of at least three parts: (1) the evolution of dynamics from processes at short times to the terminal structural α-relaxation at constant T and P; (2) the many-molecule relaxation dynamics of the structural α-relaxation at constant T and P; and (3) the changes of all processes including the α-relaxation and its relaxation time τ_α with changes of T and P. From this breakdown of the problem into three parts, it is clear that the conventional theories only attack the third part of the problem and just the α-relaxation and τ_α. Actually, the first two parts belong to a broader problem of relaxation and diffusion in many systems with nontrivial interactions between the relaxing units. The relaxing or diffusing units are not necessarily molecules. Glass-forming liquids are just special cases. No wonder that some dynamic properties are shared by all these systems.[20–23] Thus, I submit that "many-molecule relaxation dynamics" is a suitable nomenclature for the first two parts of the problem. The broader problem of many-body relaxation dynamics is difficult to solve because of the nontrivial interactions between the relaxing units, many-body considerations are necessary in an irreversible process. It is interesting to relate this problem to Einstein's solution of Brownian motion in 1905, particularly because 2005 marks the centenary of the three important papers that Einstein published in 1905, one of which is on Brownian motion[24]. While the particles undergoing Brownian motion in Einstein's problem have no interactions with each other, this is not true in the complex systems of interest nowadays including the glass-forming substances. The introduction of interactions into Einstein's problem has made the many-body diffusion problem difficult to solve for the last 100 years. The same can be said of Debye's solution of rotational diffusion of a dipolar molecule in a solvent. The presence of intermolecular interaction again makes the reorientation relaxation of molecules in glass-forming substances an unsolved problem.

One may be aware that the many-molecule relaxation caused by intermolecular interaction is important in considering the α-relaxation and the glass transition. However, the dilemma is the absence of an established method to treat many-body relaxation. Naturally, this important factor of dynamics was put aside or deemphasized by most, if not all, architects of the conventional theories of the glass transition. Conventional treatments have oversimplified the many-body problem by descriptions based on the thermodynamic quantities such as free volume and configurational entropy. It is not surprising that the thermodynamic descriptions cannot fully capture the dynamics of diffusion and relaxation of glass-formers and many-body interacting systems in general. Even density fluctuation treated as mean field in the mode coupling theory is not a full treatment of many-body relaxation as evidenced by the limitations in the description of the dynamics beyond caging or below its critical temperature T_c. Considerations of energy landscape in configurational space also have oversimplified the problem. Certainly, the only approach to relaxation in interacting system that guarantees the essential many-body physics of the relaxation dynamics will not be lost is one that starts from the Hamiltonian level, with the interaction potential taken into account on a fundamental basis. This is an enormously difficult task because it is already difficult to treat a irreversible process at a Hamiltonian level, and now on top of it many-body interactions have to be included into the theoretical treatment. This task seems unavoidable if a rigorous solution of the problems is required. The best substitute for a solution at this time is molecular dynamics simulation, which is however only a computer experiment. Interactions in most amorphous materials such as van der Waals glass-forming liquids, vitreous ionic conductors, polymers, etc. involve potentials (e.g., Lennard–Jones and screened Coulomb) that are highly anharmonic or "nonintegrable". In classical mechanics, such anharmonic interactions cause *chaos* (i.e., nonlinear Hamiltonian dynamics)[25] in the phase space of the system. Unfortunately, at present no one knows how to apply nonlinear Hamiltonian dynamics to solve the many-body relaxation problem in real materials. Only simple analogous models have yielded results.[26–30]

We have indicated that many-molecule relaxations are central in considering the structural α-relaxation and the glass transition. This is intuitively obvious because of the presence of intermolecular interactions between the molecular units. There is much experimental evidence for them and some examples are given here. Many-molecule relaxation is responsible for the observed dynamically heterogeneous character of the relaxation (i.e., existence of faster and slower relaxing molecular units which exchange roles with times of the order of τ_α)[31–33] and the correlation function often observed to have the Kohlrausch stretched exponential form (Eq. 1). Here the nonexponentiality parameter, n, a positive fraction of unity, is a convenient measure of the degree of the many-molecule relaxation. Fourier transform of $\phi(t)$ gives the dispersion of the α-relaxation in the frequency domain, and n prescribes the shape and the width of the α-dispersion in the frequency domain. Nuclear magnetic resonance experiments have shown that n correlates with the length scale of the dynamic heterogeneity of the α-relaxation.[34] Also, broadband dielectric relaxation measurements of many glass-formers from different classes have shown that n determines the properties of fundamental processes seen at shorter times than the α-relaxation, including the universal Johari–Goldstein secondary relaxation[5,10,11,13,35] and the nearly constant loss when the molecules are still caged.[5,35,36] Hence, n is legitimately the parameter that characterizes many-molecule relaxation. Even without solving the many-molecule relaxation problem, n can be used as a surrogate of the many-molecule relaxation. Evidence of the importance of many-molecule relaxation on glass transition can be gleaned from the control that n (or equivalently the dispersion of the α-relaxation) has on various experimental properties of relaxation related to glass transition.[3,4, 5,10,11,13,35–37]

There is a good deal of experimental evidence that the dynamic properties of the α-relaxation and its relaxation time τ_α in bulk glass-formers are governed by or correlated with the dispersion of the α-relaxation. The most recent and definitive experimental evidence comes from broadband dielectric relaxation measurements of many polymeric and nonpolymeric glass-formers at various temperatures and pressures.[14] A general experimental fact has been discovered in glass-formers: for a given material at a fixed value of τ_α, the dispersion of the

α-relaxation is constant, independent of thermodynamic conditions (T and P).[14] In other words, the shape of the structural α-relaxation function depends only on the relaxation time τ_α. Some representative data of small molecule and polymeric glass-formers are shown in Figs. 1 and 2 respectively. This general property is seen also by other experimental techniques including photon correlation spectroscopy and dynamic specific heat spectroscopy, although dielectric relaxation is widely used because of its advantage of broad frequency range and ease in applying pressure up to a few GPa.

Conventional theories of the glass transition, including the well-known free volume and configurational entropy models, explain the glass transition by invoking some quantity (but not n or the α-dispersion) that governs the structural relaxation time. The α-dispersion is either not addressed or derived as a separate consequence independent of τ_α. The α-dispersion bears no fundamental relationship to the value of τ_α or its properties. Thus, the conventional theories are inconsistent with experiments.

Not only does τ_α uniquely define the dispersion, as shown herein, but also many properties of τ_α are governed by the dispersion or the fractional exponent n. Examples of these properties are described in several papers and reviews.[4,20,21] Here we mention a few in the following:

(i) In quasielastic neutron scattering experiments, at constant T, the dependence of τ_α on the scattering vector Q is given by $Q^{-2/(1-n)}$. This is to be contrasted with the Q^{-2}-dependence of Brownian motion of noninteracting bodies.

(ii) The separation, $[\log(\tau_\alpha)-\log(\tau_{JG})]$, of τ_α from the Johari–Goldstein secondary relaxation time τ_{JG} at a fixed value of τ_α correlates with n. Not only is the Johari–Goldstein relaxation universal, but also this correlation is obeyed in small molecular, polymeric,[38] plastic crystalline,[39] and metallic[40,41] glass-formers.

(iii) The observed crossover of the temperature dependence of τ_α from one Vogel–Fulcher–Tammann–Hesse (VFTH) equation to another,[16,17] coincident with the apparent onset of bifurcation of τ_{JG} from τ_α, and the onset of decoupling of translation diffusion from viscosity, at some temperature T_B above T_g, are all related to the small values of n at temperatures above T_B and the latter's more rapid increase with decreasing temperature below T_B.[18,42]

(iv) The ratio of the two values of τ_α at T_g given by the two VFTH equations used to fit τ_α at temperatures above and below T_B increases with increasing n.[42]

(v) The steepness or "fragility" index m correlates with n for glass-formers belonging to the same family.[4,43]

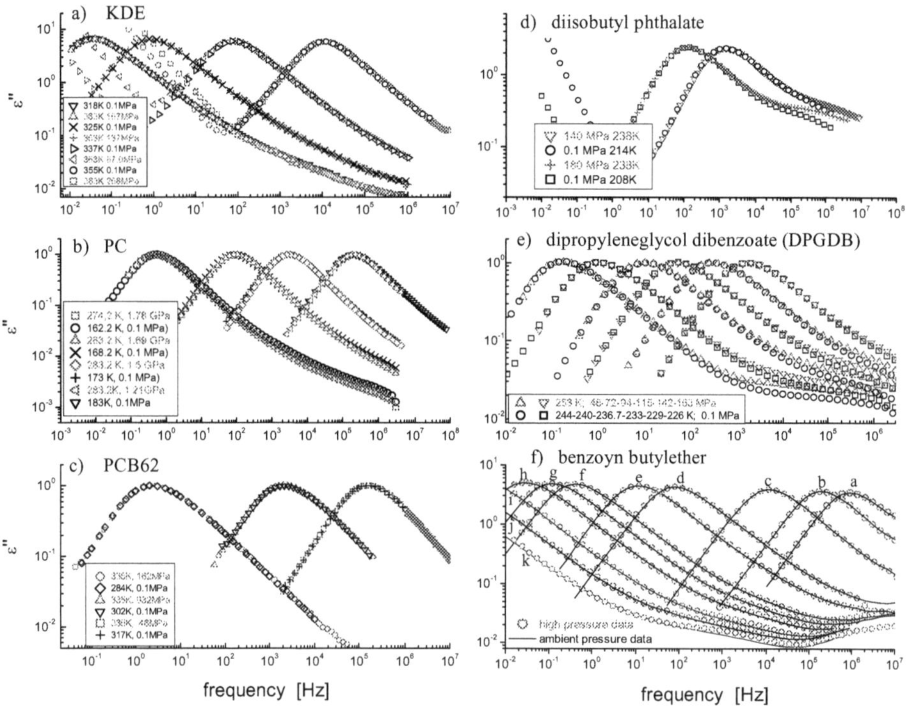

Figure 1. Dielectric loss data at various combinations of temperature and pressure as indicated to demonstrate the invariance of the dispersion of the α-relaxation at constant α-loss peak frequency ν_α or equivalently at constant α-relaxation time τ_α for:

a) Cresolphthalein–dimethylether (KDE)

b) Propylene carbonate (PC) (loss normalized to the value of the maximum of the α-loss peak).

c) Chlorinated biphenyl (PCB62)

d) Diisobutyl phthalate (DiBP)

e) Dielectric loss of dipropyleneglycol dibenzoate (DPGDB). Loss normalized to the value of the maximum of the α-loss peak. The d.c. conductivity contribution has been subtracted. Red triangles are isothermal measurements at $T=253$ K and $P=48$, 72, 94, 115, 142, 163 MPa (from right to left). Black symbols are isobaric measurements done at $P=0.1$ MPa and $T=244$, 240, 236.7, 233, 229, 226 K (from right to left). The spectrum at $T=226$ K has been shifted along x-axis by multiplying frequency by a factor 1.3.

f) Dielectric loss of benzoyn isobutylether (BIBE) at different T and P. The d.c. conductivity contribution has been subtracted. Spectra obtained at higher P are normalized to the value of the maximum of the loss peak obtained at the same frequency at atmospheric pressure. From right to left: black lines are atmospheric pressure data at $T=271$ (a), 263 (b), 253 (c), 240 (d), 236 (e), 230 (f), 228 (g), 226 (h), 223 (i), 220.5 (j), 218 (k) K. Symbols are high pressure data: $T=278.5$ K and $P=32$ (a), 65 (b), 118 (c), 204 (d), 225 (e), 320 (h), 370 (j), 396 (k) MPa, $T=288.2$ K and $P=350$ (f), 370 (g), 423 (i), 450 (j) MPa, $T=298$ K and $P=330$ (d), 467 (h) MPa.

(vi) The systematic increase of n with change in chemical structure that increase the intermolecular interaction or constraints.[44]

(vii) For polymers, there is also the correlation of the degree of thermo-rheological complexity of the viscoelastic spectrum with n, as exemplified by polystyrene and polyisobutylene.[45,46]

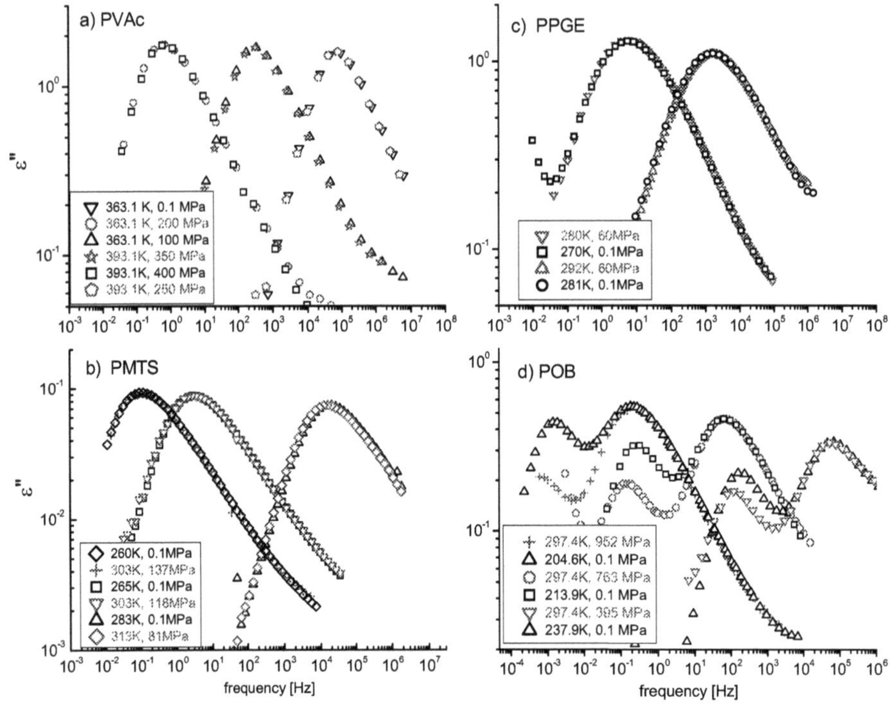

Figure 2. Dielectric loss data at various combinations of temperature and pressure as indicated to demonstrate the invariance of the dispersion of the α-relaxation at constant α-loss peak frequency ν_α or equivalently at constant α-relaxation time τ_α for: a) poly(vinylacetate) (PVAc); b) poly(methyltolylsiloxane) (PMTS); c) poly(phenyl glycidyl ether)-*co*-formaldehyde (PPGE); d) poly(oxybutylene) (POB). In all cases spectra obtained at higher P are normalized to the value of the maximum of the loss peak obtained at the same frequency at atmospheric pressure.

From the discussion above, it is clear that there is a plethora of experimental evidence showing the importance of the α-relaxation dispersion or n in determining the properties of various dynamics of glass transition. In stark contrast, there is hardly any well-known conventional theory of the glass transition that makes the α-relaxation dispersion play such a decisive role. Since the α-relaxation dispersion (or n) is just the manifestation of the many-molecule relaxation, this unfortunate situation is due to the inability of anyone to solve the many-molecule relaxation problem. Nevertheless, this important aspect of glass transition should be universally recognized and taken into consideration in future work, particularly of a theoretical nature. Otherwise the research will go on and on in circles in the future, and closure of the problem will never be reached. This comment applies as well to other fields of relaxation in interacting systems. An example is ionic conductivity relaxation in glasses, crystals and melts, where experiments have shown that the dispersion or n governs the ion dynamics[35,47,48] but not often taken into account in theories proposed.

3. The Johari–Goldstein secondary relaxation

A complete theory of glass transition must be able to describe all the processes starting from short times when all molecules are caged and the evolution of the many-molecule relaxation with time culminating in the α-relaxation at a constant temperature. Moreover, the dependences of all these processes on temperature and pressure have to be accounted for in order to consider the glass transition. Although the structural α-relaxation defines the glass transition and is the most prominent feature in the relaxation spectrum, it originates from relaxation of individual molecules. At sufficiently short times, the many-molecule relaxation is reduced to isolated local motions independent of each other, which correspond to the primitive relaxation of the coupling model.[5,10,11,13,14,26–31,48–50] On increasing time, more molecules participate and their motion can no longer be independent because of intermolecular interaction/constraints. Many-molecule heterogeneous dynamics take over and the number of participating molecules increases with time. The final stage is the terminal α-relaxation having correlation function given by Eq. (1) and the maximum number of participating molecules (or length-scale) determined by the intermolecular interaction/constraints at the temperature and pressure of the glass-former. The primitive relaxation, acting as the initiator of the evolving many-molecule dynamics, should be seen as part of the faster processes in the relaxation spectrum, which is traditionally called the secondary relaxation. For the purpose of correctly identifying with the primitive relaxation, the secondary relaxations coming from an isolated part of the molecule are excluded, because the primitive relaxation by definition involves essentially the entire molecule[11]. Those secondary relaxation which involves the motion essentially of the entire molecule, or the entire repeat unit in the case of a polymer, is called the

Johari–Goldstein (JG) secondary relaxation to honor their important discovery of the existence of secondary relaxation in rigid molecular glass-formers having no intramolecular degree of freedom.[6–9] More will be said in Section 5 on the remarkable good correspondence between the primitive relaxation time and the JG relaxation time found in many glass-formers.

The relaxation time of the JG secondary relaxations, τ_{JG}, also have dynamic properties similar to or correlated with that of the α-relaxation time τ_{α}.[5,8,11–13,51–59] Examples include the dependence of τ_{JG} on pressure and physical aging, non-Arrhenius temperature dependence of τ_{β} at temperatures above T_g, and the correlation of the ratio τ_{α}/τ_{JG} with the width of the α-relaxation or n for different glass-formers at constant any τ_{α}.[5,10,11] The last correlation continues to hold when mixing with another glass-former,[54–59] or increasing the number of covalent bonds by polymerization or crosslinking.[58] The relaxation strength, $\Delta\varepsilon_{\beta}$, of the JG relaxation is found to change on heating through the glass transition temperature in a similar manner as the changes observed in the enthalpy H, entropy S, and volume V. The derivative of $\Delta\varepsilon_{\beta}$ with respect to temperature, $d\Delta\varepsilon_{\beta}/dT$, is raised from lower values at temperatures below T_g to higher values at temperatures above T_g, mimicking the same behavior of the specific heat C_p and the expansion coefficient, which are the derivatives dH/dT and dV/dT, respectively.[8] For some glass-formers that have weak JG relaxation in the glassy state such as salol, its strength $\Delta\varepsilon_{\beta}$ can be enhanced by rapid quenching, and this indicates that the JG relaxation is sensitive to the thermodynamic state of the glass. These characteristics are also used to distinguish them from other secondary relaxations of lesser importance.[11] In the last few decades, experiments have repeatedly found the existence of the JG relaxation in all kinds of glass-formers, small molecular,[5–11] polymeric,[38] plastic crystalline,[39] metallic,[40,41] inorganic oxide[60] glass-formers, and ionic conducting salts.[61] Thus, from their universal occurrence and general properties, the JG secondary relaxation is a fundamental process of the many-molecule relaxation leading to the α-relaxation, and cannot be ignored in any serious theoretical effort to solve the glass transition problem. Remarkably, in all standard theories or models of the glass transition, the JG relaxation either does not enter into the consideration or, if considered, it bears no relation or has no correlation with the α-relaxation. If this practice continues in the future, there is not much hope that the problem of many-molecule relaxation and the glass transition problem will have any chance of being solved.

4. More complicated systems

Glass-forming materials are useful and applications often go beyond the neat material in the bulk state. For examples, mixtures of two glass-formers can give more desirable properties, and size reduced to nanometer scale has technological applications such as ultrathin polymer films in the fabrication of electronic devices. These systems are more complicated than a neat glass-forming material in the bulk state. Nevertheless, progress in research on glass transition in these more complicated systems necessarily has to be based on a theory of neat and bulk glass-formers, and problems will arise if the theory is a conventional one that has not taken into account the many-molecule dynamics properly. Two examples are given here.

The first example is the problem of the glass transition of each component in a binary mixture of two glass-former, polymeric or nonpolymeric. The presence of two different kinds of molecules makes the consideration of the dynamics of each component leading to glass transition more complicated. Most theories of mixtures/blends are constructed by identifying some new feature of mixtures, such as concentration fluctuations[62,63] and "self-concentration" if the component is a polymer,[64] and incorporating this feature into a conventional theory of the glass transition of a neat glass-former (e.g., the free volume model in the case of polymers). We have shown that such the conventional theories have limited validity and are already unacceptable as solution of the problem of glass transition in neat glass-formers. Thus, there is no reason to expect that these theories of binary mixtures based on it[62–64] will fare well as shown.[65] If the concentration of one component is small, the problem is reduced to the dynamics of a probe molecule in a host. Concentration fluctuation is absent and the theory of mixtures, based solely on this feature, has nothing to say on the probe dynamics. "Self-concentration" is irrelevant for binary mixture of non-polymeric glass-formers, and the theory based solely on this feature[64] becomes inapplicable.[65] However, the component dynamics of binary mixtures of non-polymeric glass-formers as seen by experiments are no different than if the components are polymers, as far as glass transition is concerned.

The glass transition of materials in nanoconfinement including polymer thin films is another problem more complicated than that of the bulk materials due to finite size effect and the presence of surfaces and interfaces. Again, in this field, the tendency of researchers is to employ conventional theories of glass transition of bulk materials, incorporate some new elements into it, and generate a theory of glass transition for nanoconfined materials. We can expect limited success of such theories. In fact, in a comprehensive review of experiments and theory,[66] it was concluded that "... the existing theories of T_g are unable to explain the range of behaviours seen at the nanometer size scale, in part because the glass transition phenomenon itself is not fully understood."

5. A Scheme to guide the search for a solution

We have pointed out that many-molecule relaxation is the problem impeding progress. Experiments have shown that the dispersion of the α-relaxation (or n) and the JG relaxation are the vestige of the many-molecule relaxations. They should play a prominent role in any successful approach even if it is short of being the final solution. This requirement is fulfilled in the coupling model (CM).[5,10–13,20,26–31,49,50] The CM is neither a solution of the full many-molecule relaxation problem nor is a complete theory of glass transition. Nevertheless, the CM recognizes the importance of many-molecule relaxation, and the role of its surrogates, i.e., the primitive or the JG relaxation and n or the dispersion of the α-relaxation. In the CM, there is a relation between the structural α-relaxation time τ_α and the primitive relaxation time τ_0 via the parameter n that characterizes the α-dispersion, e.g.,

$$\tau_\alpha = [t_c^{-n}\tau_0]^{1/(1-n)} \tag{2}$$

By means of this CM relation, the α-dispersion (or n) governs the properties of the structural α-relaxation and hence helps to explain them. This feat of the CM has been demonstrated in the past in connection with the dynamic properties of bulk glass-formers[49] and the viscoelasticity of bulk polymers.[46,49] Good correspondence between the theoretically calculated τ_0 and the experimental τ_{JG} has been found in many glass-formers, $\tau_{JG}(T,P) \approx \tau_0(T,P)$, and hence JG relaxation has been taken into account in the CM.[5,10–13,53–59] The features of the CM mentioned above can be summarized by one equation:

$$\tau_\alpha(T,P) = [t_c^{-n}\tau_0(T,P)]^{1/(1-n)} \approx [t_c^{-n}\tau_{JG}(T,P)]^{1/(1-n)} \tag{3a}$$

Although the dependences of τ_α and τ_0 (or τ_{JG}) on T and P are not given or derived in the CM, this relation between τ_α and τ_{JG} is obeyed in all glass-formers tested so far. This fact lead us to conclude[13,59] that the T and P dependences of τ_α are not original but originate from that of τ_{JG} or τ_0. The stronger T- and P-dependences of τ_α compared with that of τ_0 or τ_{JG} is merely the consequence of the many-molecule dynamics in α-relaxation, which amplifies the dependences of τ_0 or τ_{JG} by the superlinear power, $1/(1-n)$, in the equation, corresponding to a longer length scale. A schematic theory of the glass transition that includes many-molecule dynamics and the influence of the thermodynamic variables T and P has been constructed based on this observation.[59] This scheme should be useful as guide for others in constructing a rigorous theory of many-molecule relaxation and glass transition hopefully in the near future.

The concepts that free volume and configurational entropy determines molecular mobility would be better served if applied to the primitive or the JG relaxation. The Adam–Gibbs model[2] was originally designed to describe the α-relaxation. Several assumptions were made in the AG theory. The first one is that the transitions of the cooperative regions involve the z^* molecules surmounting simultaneously the individual potential energy barriers, $\Delta\mu$, hindering their cooperative rearrangement, which is a temperature independent constant. This assumption enabled AG to write $\tau_\alpha(T)$ as $\tau_\alpha(T)=\tau_\infty \exp(z^*\Delta\mu/kT)$. AG further assumed that the configurational entropy associated with the cooperative rearrangement of the z^* molecules, s_c^*, is temperature independent. With the second assumption, AG expressed z^* in terms of the molar configurational entropy S_c of the macroscopic sample as $z^*(T)=N_A s_c^*/S_c(T)$ where N_A is Avogadro's number. The experimentally deduced $z^*(T)$ increases monotonically with decreasing temperature from a value of 2 at $T=T_B$ to $z^*(T_g)$ at $T=T_g$.[67,68] However, $z^*(T_g)$ is not large. For example, it is less than 5 for ortho-terphenyl,[67] for glycerol it is 3.2,[68] and for tri-α-naphthal benzene it is 2.4,[68] which contradict the much larger number of molecules involved in the α-relaxation near and above T_g as deduced by NMR measurement.[69] On the other hand, the small $z^*(T)$ deduced is more like that of local motions that are forerunners of the α-relaxation and closely related to the primitive relaxation or the JG relaxation. Thus, it seems that the configurational entropy theory of AG, if valid, is actually more appropriate for the processes transpired at shorter times than the α-relaxation with the Kohlrausch correlation function. There is another piece of evidence to support this statement from the expression for the α-relaxation time of the AG theory,

$$\tau_\alpha(T) = \tau_\infty \exp(\Delta\mu s_c^*/kTS_c) = \tau_\infty \exp(C/TS_c) \tag{3b}$$

which predicts a linear relation between log τ_α and $(TS)_c^{-1}$ at temperatures above T_g. However, deviation from the linear relation between log τ_{exp} and $(TS_c)^{-1}$ was found in many glass-formers at higher temperatures starting somewhere in the vicinity of T_B. The extent of the breakdown is larger for liquids that have larger value of n at T_g. For details and references see Ref.. Recall that T_B is where the temperature dependence of τ_α crosses over from on VFTH relation to another, and T_B separates the regime of small n (for $T>T_B$) and regime of larger n (for $T>T_B$). It has been shown that τ_0 calculated from the Eq. (3) can be described by a single VFTH equation over the entire temperature range,[19] indicating the crossover of temperature dependence of τ_α at T_B is due to the change of n across T_B. More importantly for the present discussion, there is now a linear relation between log τ_0 and $(TS_c)^{-1}$ at all temperatures above T_g. This is another indication that the AG theory is more appropriate for the primitive processes acting as forerunners of the terminal Kohlrausch α-relaxation. The expression on the right-hand-side of Eq. (3) stands for τ_0 or τ_{JG}. Then the CM equation (2) takes it to τ_α. The proposed new configuration entropy model gives the following expression for τ_α:

$$\tau_\alpha(T) = [t_c^{-n}\tau_\infty \exp(C/TS_c)]^{1/(1-n)} \tag{4}$$

In free volume are defined differently by different authors, but invariably all free volume theories relate free volume to the α-relaxation. In the free volume theory of molecular transport as derived by Cohen and Turnbull,[71] τ_α is given by

$$\tau_\alpha = C\exp(\gamma V_f^*/V_f) \tag{5}$$

where $V_f = (V - V_{occ})$ is the mean specific free volume, and V_f* is the minimum specific free volume required for the occurrence of the α-process. Another relation,

$$\tau_\alpha = C\exp(\gamma V_{fh}^*/V_{fh}) \tag{6}$$

is obtained[72] by replacing the specific free volume quantities in Eq. (5) by the specific hole free volume quantities, V_{fh} and V_{fh}*, calculated from the Simha–Somcynsky equation of state.[73] Eq. (6) predicts a linear relation between log τ_α and $1/V_{fh}$. However, this is not accord with experimental data. On the other hand, log τ_{JG} varies linearly with $1/V_{fh}$, indicating that the specific hole free volume is a better description of the mobility of the JG or the primitive relaxation. The viscosity η was found not following Eq. (4) either.[74] However, as is frequently observed, Eq. (5) well describes the transport of small molecules[75] in polymers. This is understandable because small molecule motion in a polymer is similar to the local JG relaxation of the host polymer.

Free volume deduced from positronium annihilation lifetime spectroscopy measurements have been successfully reinterpreted as for the primitive processes or the JG relaxation. The results will be published elsewhere.[76]

Dielectric relaxation data taken at different combinations of T and P have shown that $\tau_{JG}(T, P)$ of unresolved JG relaxation (manifested as the excess wing, see Figs.1a–1d) or resolved JG relaxation (see Figs.1e and 1f) is also constant as long as $\tau_\alpha(T, P)$ is kept constant. This fact means that the quantities $\tau_\alpha(T, P)$

and $\tau_{JG}(T,P)$ are simultaneous constants, and t_c is independent of T and P. Hence the remaining quantity in Eq. (2), n, also has to be independent of T and P. Since n characterizes the α-dispersion, the required constancy of n explains why the α-dispersion is invariant to changes of T and P as seen in Figs. 1a–1e and Figs. 2a–2d.

The CM has been applied to the more complicated systems including the component dynamics of mixtures, and glass-formers in nanoconfinement discussed here before. For mixtures, it has explained the component dynamics of nonpolymeric and polymeric mixtures, and the limiting case of probe dynamics.[65,77] As mentioned before, the ratio τ_α/τ_{JG} of a component was observed to increase with its most probable n in mixtures,[54–57] as predicted by the CM. When the concentration of that component is low such that broadening of the α-relaxation by concentration fluctuation is negligible, n can be determined from the fit to the α-relaxation by Eq. (1), and this prediction of the ratio τ_α/τ_{JG} becomes exact,[54] and is verified by experiment.[78] This is one of the features of component dynamics that other theories[62–64] cannot address. For dynamics of glass-formers in nanoconfinement or ultrathin polymer films, the CM can explain the range of behaviors seen at the nanometer size scale.[79–81]

6. Epilogue

I have used principally experimental data and presumably logical deductions to show that all standard theories of glass transition in their present forms are inadequate because some essential pieces of physics is either totally missing or not adequately taken into consideration. The missing physics are the many-molecule relaxation dynamics of the structural α-relaxation and the forerunning processes that includes the universal Johari–Goldstein relaxation. It is the evolution of the forerunners with time that eventually leads to the terminal α-relaxation with a correlation function described by the Kohlrausch stretched exponential function. The width of the dispersion of the Kohlrausch function or the exponent, n, is a measure of the extent of the many-molecule involvement in the α-relaxation. The best indication of many-molecule relaxation is the fact that properties of the α-relaxation and its relaxation time τ_α are either correlated with or governed by n. The plethora of experimental evidence also points to the fundamental role played by the Johari–Goldstein relaxation as the precursor of the many-molecule relaxation dynamics, and is the process where free volume and configurational entropy first enter into affecting molecular mobility. A general relation exists between the Johari–Goldstein relaxation time τ_{JG} and τ_α, and it is also governed by n. This empirical relation further indicates the connection between the Johari–Goldstein relaxation and the α-relaxation. Without revision to incorporate these essential elements, the standard theories will continue to be incomplete, mislead the uninformed, and hinder progress in the solution of glass transition as a basic research problem. This current situation also has an ill effect on progress and development in many other areas of science and technology where the glass transition is fundamental in the understanding of even more complex processes involved in these disciplines. These areas include current research in pharmaceuticals, nanometer technology, food science, biomolecular materials, and life sciences, as well as traditional areas such as polymer viscoelasticity. Without a correct theory of glass transition to describe the local segmental relaxation, the basic problems of polymer viscoelasticity[45] cannot be resolved.

The coupling model (CM) does not solve the many-molecule relaxation dynamics problem either. Nevertheless, it recognizes at the outset[50] the importance of the many-molecule interactions and correlations in the α-relaxation. It has the primitive relaxation as the precursor of the many-molecule α-relaxation processes, and a relation that shows that its primitive relaxation time τ_0 in conjunction with n determine the α-relaxation time τ_α. Thus, in particular, n or the width of the dispersion of the α-relaxation time either correlates with or governs τ_α and its properties, as indicated by experiments. The relation clearly shows that τ_0 is where temperature, pressure, free volume, and configurational entropy first enter. The dependences of τ_0 on these thermodynamic variables are magnified in τ_α by the participation in α-relaxation of many more molecules. The relation also allows τ_0 to be calculated from the parameters n and τ_α of the experimentally observed α-relaxation. One of the remarkable recent achievements of the CM is that, for many glass-formers of various kinds, the calculated τ_0 is approximately the same as the characteristic Johari–Goldstein relaxation time τ_{JG} obtained from experiment. Thus, the CM is consistent in all respects with the experimental facts. A schematic solution of the glass transition problem based on it is offered to pave the way for someone to generate a rigorous theory hopefully in the near future.

Acknowledgment

The author thanks many colleagues for collaborations in research. In writing this article, he is benefited particularly by works in collaboration with S. Capaccioli, R. Casalini, M. Paluch, C.M. Roland, S. Hensel-Bielowka, K. Kessairi, S. Pawlus, D. Prevosto, M. Lucchesi, P.A. Rolla, and J. Habasaki. He thanks Don J. Plazek for a careful reading of the manuscript. This work was supported by Office of Naval Research.

References

1. Ferry, J.D. (1980) *Viscoelastic Properties of Polymers,* 3rd ed. (Wiley, New York).
2. Adam, G., and Gibbs, J.H. (1965) On the temperature dependence of cooperative relaxation properties in glass-forming liquids. *J. Chem. Phys.* **43**, 139–146.
3. Angell, C.A., Ngai, K.L., McKenna, G.B., McMillan, P.F., and Martin, S.W. (2000) Relaxation in glassforming liquids and amorphous solids. *J. Appl. Phys.* **88**, 3113–3157.
4. Ngai, K.L. (2000) Dynamic and thermodynamic properties of glass-forming substances. *J. Non-Cryst. Solids* **275**, 7–51.
5. Ngai, K.L. (2003) *J. Phys.:Condens. Matter* **15**, S1607–S1632.
6. Johari G.P., and Goldstein, M. (1970) Viscous liquids and the glass transition. II. Secondary relaxations in glasses of rigid molecules. *J. Chem. Phys.* **53**, 2372–2388.
7. Johari, G.P. (1973) Intrinsic mobility of molecular glasses. *J. Chem. Phys.* **58**, 1766–1770.
8. Johari, G.P., Power, G., and Vij, J.K. (2002) Localized relaxation's strength and its mimicry of glasssoftening thermodynamics. *J. Chem. Phys.* **116**, 5908–5909.
9. Johari, G.P. (2002) Localized molecular motions of β-relaxation and its energy landscape. *J. Non-Cryst. Solids* **307–310**, 313–325.
10. Ngai, K.L. (1998) Relation between secondary relaxations and the α-relaxations in glass-forming materials according to the coupling model. *J. Chem. Phys.* **109**, 6982–6994.
11. Ngai, K.L., and Paluch, M. (2004) Classification of secondary relaxation in glass-formers based on dynamic properties. *J. Chem. Phys.* **120**, 857–873.
12. Nemilov, S., and Johari, G.P. (2004) A mechanism for spontaneous relaxation of glass at room temperature. *Philos. Mag.* **83**, 3117–3132; **84**, 845–852 (2003).
13. Ngai, K.L. (2004) Johari–Goldstein or primitive relaxation: Terminator of caged dynamics and precursor of *α*-relaxation. *AIP Conf. Proceedings* **708**, 515–522. Third International Symposium on "Slow Dynamics in Complex Systems". November 3–8. 2003. Sendai. Japan. M. Tokuyama and I. Oppenheim, Eds. (Am. Institute of Physics, New York).
14. Ngai, K.L., Casalini, R., Capaciolli, S., Paluch, M., and Roland, C.M. (2005) Do theories of the glass transition, in which the structural relaxation time does not define the dispersion of the structural relaxation, need revision? *J. Phys. Chem. B* **109**, 17356–17360.
15. Kohlrausch, R. (1847) Nachtrag uber die elastische Nachwirkung beim Cocon und Glasfaden, etc. *Pogg. Ann. Phys. (III)* **12**, 393–399. Kohlrausch R. Theorie des electrischen Ruckstandes in der Leidener Flasche, *Pogg. Ann. Phys.* (1854) *(IV)* **1**, 79–86.
16. Ngai, K.L., Magill, J.H., and Plazek, D.J. (2000) Flow, diffusion and crystallization of supercooled liquids: Revisited. *J. Chem. Phys.* **112**, 1887–1892.
17. Stickel, F., Fischer, E.W., and Richert, R. (1995) Dynamics of glass forming liquids.: I. Temperature derivative analysis of dielectric relaxation times. *J. Chem. Phys.* **102**, 6251–6258.
18. Ngai, K.L. (1999) Synergy of entropy and intermolecular coupling in supercooling liquids, *J. Chem. Phys.* **111**, 3639–3645.

19. Casalini, R., Ngai, K.L., and Roland, C.M. (2003) Connection between the high-frequency crossover of the temperature dependence of the relaxation time and the change of intermolecular coupling in glass-forming liquids. *Phys. Rev. B* **68**, 014201, 1–5.
20. Ngai, K.L., and Rendell, R.W. (1997) *Supercooled Liquids, Advances and Novel Applications,* edited by Fourkas, J.T., Kivelson, D. Mohanty, U., and Nelson, K. eds., ACS Symposium Series Vol. **676**, Chapter 4, 45–66 (Am. Chem. Soc., Washington, DC).
21. Ngai, K.L. (1994) *Disorder Effects on Relaxational Properties*, edited by R. Richert and A. Blumen, Chapter 3 (Springer Verlag, Berlin).
22. Ngai, K.L. (1996) A review of critical experimental facts in electrical relaxation and and ionic diffusion in ionically conducting glasses and melts. *J. Non-Cryst. Solids* **203**, 232–245.
23. Ngai, K.L., and Phillies, G.D.J. (1996) Polymer dynamics in solutions: Probe diffusion and viscosity. *J. Chem. Phys.* **105**, 8385–8395.
24. Einstein, A. (1905) *Ann. Phys.* **17**, 549.
25. *Hamiltonian Dynamical Systems* (1987) edited by R.S. MacKay and J.D. Meiss, (Adam Hilger, Bristol UK).
26. Ngai, K.L., and Rendell, R.W. (1991) Toward a theory of relaxation in correlated systems: diffusion in the phase space of a chaotic Hamiltonian. *J. Non-Cryst. Solids* **131–133**, 233–237.
27. Ngai, K.L., Peng, S.L., and Tsang, K.Y. (1992) Fractal phase space transport dynamics and relaxations in complex correlated systems. *Physica A* **191**, 523–531.
28. Tsang, K.Y., and Ngai, K.L. (1997) Relaxation in interacting arrays of oscillators. *Phys. Rev. E* **54**, 3067–3071.
29. Tsang, K.Y., and Ngai, K.L. (1997) Dynamics of relaxing systems subjected to nonlinear interactions. *Phys. Rev. E* **56**, R17–R20.
30. Ngai, K.L., and Tsang, K.Y. (1999) Similarity of relaxation in supercooled liquids and interacting arrays of oscillators. *Phys. Rev. E* **60**, 4511–4517.
31. Ngai, K.L., and Rendell, R.W. (1990) The symmetric and fully distributed solution to a generalized dining philosophers problem: An analogue of the coupling theory of relaxations in complex correlated systems in relaxation, *Complex systems and Related Topics*, NATO ASI Series. Series B: Physics Vol. **222**, edited by Ian A. Campbell and C. Giovannella, pp. 309–316 (Plenum, New York and London).
32. Schmidt-Rohr, K., and Spiess, H.W. (1991) Nature of nonexponential loss of correlation above the glass transition investigated by multidimensional NMR, *Phys. Rev. Lett.*, **66**, 3020–3023.
33. For reviews of other works, see Böhmer, R., Chamberlin, R.V., Diezemann, G., Geil, B., Heuer, A., Hinze, G., Kuebler, S.C., Richert, R., Schiener, B., Sillescu, H., Spiess, H.W., Tracht, U., and Wilhelm, M. (1998) Nature of the non-exponential primary relaxation in structural glass-formers probed by dynamically selective experiments. *J. Non-Cryst. Solids* **235–237**, 1–9.
34. Reinsberg, S.A., Heuer, A., Doliwa, B., Zimmermann, H., and Spiess, H.W. (2002) Comparaitive study of the NMR length scale of dynamic heterogeneities of three different glassformers. *J. Non-Cryst. Solids* **307–310**, 208–214.

35. Ngai, K.L., Habasaki, J., León, C., and Rivera, A. (2005) Comparison of dynamics of ions in ionically conducting materials and dynamics of glass-forming substances: Remarkable similarities. *Z. Phys. Chem.* **219,** 47–70.
36. Ngai, K.L. (2004) Why the fast relaxation in the picosecond to nanosecond time range can sense the glass transition. *Philos. Mag.* **84**, 1341–1353.
37. Ngai, K.L. (2005) Do theories of glass transition that address only the α-relaxation need a new paradigm? *J. Non-Cryst. Solids* **351**, 2635–2643.
38. Roland, C.M., Schroeder, M.J., Fontanella, J.J., and Ngai, K.L. (2004) Evolution of the dynamics in 1,4-polyisoprene from a nearly constant loss to a Johari–Goldstein β-relaxation to the α-relaxation. *Macromolecules* **37**, 2630–2635.
39. Brand, R., Lunkenheimer, P., and Loidl, A. (2002) Relaxation dynamics in plastic crystals. *J. Chem. Phys.* **116**, 10386–10401.
40. Pelletier, J.M., Van de Moortele, B., and Lu, I.R. (2002) Viscoelasticity and viscosity of Pd–Ni–Cu–P bulk metallic glasses. *Mater. Sci. Eng. A* **336**, 190–195.
41. Rosner, P., Samwer, K., and Lunkenheimer, P. (2004) Indications for an "excess wing" in metallic glasses from the mechanical loss modulus in $Zr_{65}Al_{7.5}Cu_{27.5}$. *Europhys. Lett.* **68**, 226.
42. León, C., and Ngai, K.L. (1999) Rapidity of the change of the Kohlrausch exponent of the α-relaxation of glass-forming liquids at T_B or T_β and consequences. *J. Phys. Chem. B* **103**, 4045–4051.
43. Böhmer, R., Ngai, K.L., Angell, C.A., and Plazek. D.J. (1993) Nonexponential relaxations in strong and fragile glass formers. *J. Chem. Phys.* **99**, 4201–4209.
44. Ngai, K.L., and Roland, C.M. (1993) Chemical structure and intermolecular cooperativity: Dielectric relaxation results. *Macromolecules* **26**, 6824–6830.
45. Ngai, K.L., and Plazek, D.J. (1995) Identification of different modes of molecular motion in polymers that cause thermorheological complexity. *Rubber Chem. Technol. Rubber Review* **68**, 376–434.
46. Ngai, K.L., Plazek, D.J., and Rendell, R.W. (1997) Some examples of possible descriptions of dynamic properties of polymers by means of the coupling midel. *Rheol. Acta* **36**, 307–319.
47. Ngai, K.L. (1996) A review of critical experimental facts in electrical relaxation and ionic diffusion in ionically conducting glasses and melts. *J. Non-Cryst. Solids* **203**, 232–245.
48. Ngai, K.L., Greaves, G.N., and Moynihan, C.T. (1998) Correlation between the activation energies for ionic conductivity for short and long time scales and the Kohlrausch stretching parameter β for ionically conducting solids and melts. *Phys. Rev. Lett.* **80**, 1018–1021.
49. Ngai, K.L. (2001) Coupling model explanation of salient dynamic properties of glass-forming substances. *IEEE Trans. Dielectrics Electrical Insulation* **8**, 329–344.
50. Ngai, K.L. (1979) Universality of low-frequency fluctuation, dissipation and relaxation properties of condensed matter, 1. *Comments Solid State Phys.* **9**, 127–140.
51. Schneider, U., Brand, R., Lunkenheimer, P., and Loidl, A. (2000) Excess wing in the dielectric loss of glass formers: A Johari–Goldstein β relaxation? *Phys. Rev. Lett.* **84**, 5560–5563.

52. Paluch, M., Roland, C.M., Pawlus, S., Zioło, J., and Ngai, K.L. (2003) Does the Arrhenius temperature dependence of the Johari–Goldstein relaxation persist above T_g? *Phys. Rev. Lett.* **91**, 115701, 1–4.
53. Prevosto, D., Capaccioli, S., Lucchesi, M., Rolla, P.A., and Ngai, K.L. (2004) Dynamics of supercooled and glassy dipropyleneglycol dibenzoate as functions of temperature and aging: Interpretation within the coupling model framework. *J. Chem. Phys*. **120**, 4808–4815.
54. Capaccioli, S., and Ngai, K.L. (2005) Relation between the α-relaxation and Johari–Goldstein β-relaxation of a component in binary miscible mixtures of glass-formers. *J. Phys. Chem. B* **109**, 9727–9735.
55. Ngai, K.L., and Capaccioli, S. (2004) Changes of the primary and secondary relaxation of sorbitol in mixtures with glycerol. *J. Phys. Chem. B* **108**, 11118–11124.
56. Psurek, T., Maslanka, S., Paluch, M., Nozaki, R., and Ngai, K.L. (2004) Effects of water on the primary and secondary relaxation of xylitol and sorbitol: Implication on the origin of the Johari–Goldstein relaxation, *Phys. Rev. E* **70**, 011503, 1–6.
57. Ngai, K.L. (1999) Correlation between β-relaxation and α-relaxation in the family of poly(*n*-butyl methacrylate-*stat*-styrene) random copolymers. *Macromolecules* **32**, 7140–7146.
58. Beiner, M., and Ngai, K. L. (2005) Interrelation between primary and secondary relaxations in polymerizing systems based on epoxy resins. *Macromolecules* **38**, 7033–7042.
59. Capaccioli, S., Prevosto, D., Lucchesi, M., Rolla, P.A., Casalini, R., and Ngai, K.L. (2005) Identifying the genuine Johari–Goldstein β-relaxation by cooling, compressing, and aging small molecular glass-formers. *J. Non-Cryst. Solids* **351**, 2643–2651.
60. Mai, C., Etienne, S., Perez, J., and Johari, G.P. (1985) Modulus and internal friction in phosphate–silicate bioactive glass. *J. Non-Cryst. Solids* **74**, 119–127.
61. Mai, C., Etienne, S., Perez, J., and Johari, G.P. (1985) Mechanical relaxation of CKN. *Philos. Mag. B* **50**, 657–663.
62. Zetsche, A., and Fischer, E.W. (1994) Dielectric studies of the α-relation in miscible polymer blends and its relation to concentration flucutations. *Acta Polym.* **45**, 168–175.
63. Kamath, S., Colby, R.H., and Kumar, S.K. (2003) Dynamic heterogeneity in miscible polymer blends with stiffness disparity: Computer simulations using the bond fluctuation model. *Macromolecules* **36**, 8567–8573.
64. Lodge, T., and McLeish, T.C.B. (2000) Self-concentrations and effective glass transition temperatures in polymer blends. *Macromolecules* **33**, 5278–5284.
65. For a review of mixtures and polymer blends, see Ngai, K.L., and Roland, C.M. (2004) Models for the component dynamics in blends and mixures. *Rubber Chem. Technol.* **77**, 579–590.
66. Alcoutlabi, M., and McKenna, G.B. (2005) Effects of confinement on material behaviour at the nanometre size scale. *J. Phys.: Condens. Matter* **17**, R461–R524.
67. Yamamuro, O., Tsukushi, I., Lindqvist, A., Takahara, S., Ishikawa, M., and Matsuo, T. (1998) Calorimetric study of glassy and liquid toluene and ethylbenzene: Thermodynamic approach to spatial heterogeneity in glass-forming molecular liquids. *J. Phys. Chem. B* **102**, 1605–1609.

68. Johari, G.P. (2000) Contributions to the entropy of a glass and liquid and the dielectric relaxation time. *J. Chem. Phys.* **112**, 7518–7523; Johari, G.P. (2000) A resolution for the enigma of a liquid's configurational entropy-molecular kinetics relation. *J. Chem. Phys.* **112**, 8958–8969.
69. Reinsberg, S.A., Heuer, A., Doliwa, B., Zimmermann, H., and Spiess, H.W. (2002) Comparative study of the NMR length scale of dynamic heterogeneities of three different glass formers. *J. Non-Cryst. Solids* **307–310**, 208.
70. Ngai, K.L. (1999) A modification of the Adam–Gibbs model of glass transition for for consistency with experimental data. *J. Chem. Phys.* **103**, 5895–5902.
71. Cohen, M.H., and Turnbull, D. (1959) Molecular transport in liquids and glasses. *J. Chem. Phys.* **31**, 1164-1169; Turnbull, D., and Cohen, M.H. (1970) On the free-volume model of the liquid-glass transition. *J. Chem. Phys.* **52**, 3038–3041.
72. Dlubek, G., Kilburn, D., and Alam, M.A. (2005) Temperature and pressure dependence of α-relaxation and free volume in poly(vinyl acetate). *Macromol. Chem. Phys.* **206**, 818–826.
73. Simha, R., and Somcynsky, T. (1969) On statitical thermodynamics of spherical and chain molecule fluids. *Macromolecules* **2**, 342–350.
74. Utracki, A. (1983) On free volume. *Polym. Eng. Sci.* **23**, 446.
75. Dlubek, G., Kilburn, D., and Alam, M.A. (2004) Comments to the paper "The need to reconsider tradional free volume theory for polymer electrolytes." *Electrochem. Acta* **49**, 5241–5247; Okamoto, K., Tanaka, K., Katsube, M., Kida, H., Sueoka, O., and Ito, Y. (1993) *Polym. J.* **25**, 275; Tanaka, K., Kawai, T., Kida, H., Okamoto, K., and Ito, Y. (2000) *Macromolecules* **33**, 5513; Bamford, D., Reiche, A., Dlubek, G., Alloin, F., Sanchez, J.-Y., and Alam, M.A. (2003) Ionic conductivity, glass transition, and local free volume in poly(ethylene oxide) electrolytes: Single and mixed ion conductors. *J. Chem. Phys.* **118**, 9420–9432.
76. Bartos, J., and Ngai, K.L. to be published.
77. The "self-concentration" factor for polymer blends can be easily incorporated into the CM without distracting from the main point, which is the necessity of treating the many-molecule dynamics in both neat glass-formers and their mixtures.
78. Capaccioli, S., Kessairi, K., Prevosto, D., Lucchesi, M., and Ngai, K.L., (2005) Genuine Johari–Goldstein β-relaxations in glass-forming binary mixtures. *J. Non-Cryst. Solids.* submitted.
79. Ngai, K.L. (1999) Removal of cooperativity in glass-forming materials to reveal the primitive α-relaxation of the coupling model. *J. Phys.: Condens. Matter* **11**, A119–A130.
80. Ngai, K.L. (2002) Mobility in thin polymer films ranging from local segmental motion, Rouse modes to whole chain motion: A coupling model consideration. *Eur. Phys. J. E* **8**, 225–232.
81. Ngai, K.L. (2005) Interpreting the dynamics of nano-confined glass-formers and thin polymer films: Importance of starting from a viable theory for the bulk. *J. Polym. Sci. B: Polym. Phys.* **55**, 2980–2995.

PART II: GLASS FORMING LIQUIDS

POSITRON ANNIHILATION LIFETIME SPECTROSCOPY AND ATOMISTIC MODELING – EFFECTIVE TOOLS FOR THE DISORDERED CONDENSED SYSTEM CHARACTERIZATION

J. BARTOŠ*, D. RAČKO
Polymer Institute of SAS, Dúbravská cesta 9, 842 36 Bratislava Slovakia

O. ŠAUŠA, J. KRIŠTIAK
Institute of Physics of SAS, Dúbravská cesta 9, 842 28 Bratislava, Slovakia

Abstract. The complex structure–property relationships in the disordered systems under normal and exogenic conditions can be understood after characterizing the spatial arrangement of constituents. Here, an integral approach including the relevant experimental technique, phenomenological, and theoretical analyses as well as atomistic modeling is presented. Application of such a combined approach is demonstrated for the cases of glycerol and propylene glycol.

Keywords: Free volume, Positron annihilation, free volume models, atomistic modeling.

1. Introduction

Many important condensed materials are characterized by the irregular spatial arrangements of the particles' constituents. Understanding the complex structural–property relationships in such systems under normal and exogenic conditions needs as complete as possible a characterization of the irregular sites of the matrix. At present, the relevant experimental technique is *positron annihilation lifetime spectroscopy* (*PALS*). In order to utilize fully the characterization potential of this method there is required a complete understanding of the universal as well as the specific features of the PALS

* To whom correspondence should be addressed.

S. J. Rzoska and V. A. Mazur (eds.), Soft Matter under Exogenic Impacts, 113–131.

response of a given material. To solve this complex task requires a combined approach including phenomenological analyses of the PALS response using the thermodynamic and dynamic data, its appropriate theoretical treatments, i.e., *free volume analyses* (*FVA*) and theoretical and atomistic modeling approaches consisting in appropriate *free volume models of liquid state* or *molecular dynamics* (*MD*) simulation followed by the *cavity analysis* (*CAVA*), respectively. The individual aspects of this combined approach will be presented together with application on two model systems.

2. Positron annihilation lifetime spectroscopy

Positron annihilation lifetime spectroscopy (PALS) is based on the unique annihilation behavior of ortho-positronium (o-Ps)[1–5] being very sensitive indicator of the local regions of lowered electron density in condensed matrix such as vacancies in the ordered (crystal) systems or the so-called free volume holes in the disordered (amorphous or semicrystalline) ones.

When a positron, e^+, from the positron source ^{22}Na enters the condensed system it loses its kinetic energy very quickly under forming a positron radiation track which contains the ionized entities such as electrons, e^-, and cations of constituents, M^+, and the excited species. The large majority of these ionized species are immediately neutralized by their mutual recombinations within the positron track but some of electrons can escape it and be preferentially trapped in the local positively charged regions of the matrix. The electrons from track and the trapped electrons can capture the original or next entering positron at forming a neutral bound positron–electron particle, the so-called positronium (Ps). Depending on the mutual spin orientation of positron and electron in Ps, two distinct states differing in the respective lifetimes, τ, can arise. Singlet with antiparallel spins: para-positronium (p-Ps) has a typical lifetime, $\tau_1 \cong 0.12$ ns being almost independent on temperature. On the other hand, a triplet state with parallel spins forms the ortho-positronium (o-Ps) which is more stable with a typical lifetime, τ_3, ranging from ~0.6–5 ns in normal organic substances. This specific form of Ps annihilates by an interaction of the positron with some electron from the surrounding medium with antiparallel spin in the so-called pick-off decay mechanism being temperature dependent. Finally, in addition to the above-mentioned bounded forms of positron, the unbounded “free” positron can survive in the matrix with a typical lifetime, $\tau_2 \cong 0.3$–0.4 ns which annihilate by an interaction with some electron in the matrix.

2.1. TYPICAL PALS RESPONSE, ITS PHENOMENOLOGICAL ANALYSIS AND CORRELATIONS

Figure 1 shows the typical o-Ps responses of two small molecular glass-formers: *glycerol* (*GL*) HO–CH_2–CH(OH)–CH_2–OH[6] and *1,2-propylene glycol* (*PG*) HO–CH_2–CH(OH)–CH_3.[7] In both the chemical compounds, the o-Ps lifetime, τ_3, as a function of temperature exhibits four regions of different thermal behavior depicted as regions I, II, III and IV. Linear analyses of these regions define the characteristic PALS temperatures in the liquid state: T_g^{PALS}, T_{b1}^L and T_{b2}^L.[8] In the case of GL: $T_g^{PALS} = 189$ K, $T_{b1}^L = 241$ K and $T_{b2}^L \cong 290$ K, while for PG: $T_g^{PALS} = 172$ K, $T_{b1}^L = 220$ K and $T_{b2}^L \cong 265$ K.

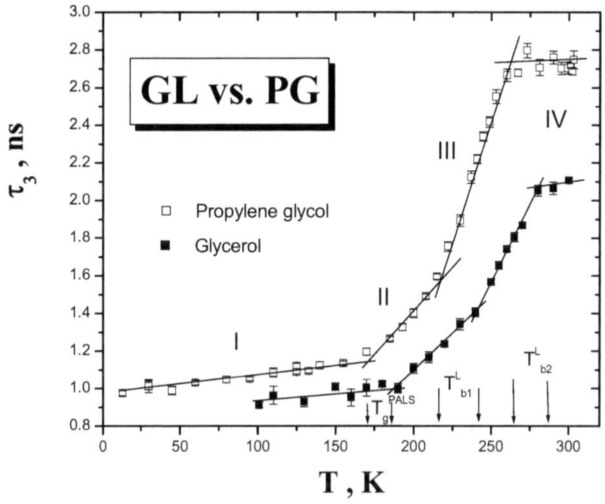

Figure 1. The o-Ps lifetime, τ_3, as a function of temperature for glycerol (GL) and propylene glycol (PG).

The first pronounced change in slope on τ_3–T plot takes place near the corresponding quasi-static glass to liquid transition temperatures, T_g^{THERM} as obtained by classical macroscopic thermodynamic techniques such as dilatometry[9,10] as well as to the dynamic glass temperatures, T_g^{DYN}, being commonly defined as the temperature at which the mean relaxation time of the primary α relaxation reaches some arbitrary value, e.g., $\tau_\alpha = 100$ s.[11,12] These findings mean that the expansion behavior of the localized free volume regions detectable via the annihilation of the o-Ps probe is closely related to the expansion behavior of the macroscopic volume as well as to the cooperative dynamics which is believed to be responsible for the most characteristic dynamic process, i.e., the primary α relaxation.

In the glassy state region I up to T_g^{PALS}, GL and PG are characterized by the relatively low o-Ps lifetimes, whereby the values of τ_3 for GL are systematically smaller than those for PG over the whole temperature range investigated. By comparing GL and PG with small molecular systems of van der Waals type as studied by us such as *meta-tricresyl phosphate* (*m-TCP*)[5] or *diethyl phthalate* (*DEP*)[13] it indicates that both the systems are relatively effectively packed ones. This appears to be consistent with the hydrogen-bonding nature of both the systems, whereby the lower values for GL may be attributed to the higher density of hydroxyl groups capable to form more intermolecular hydrogen bonds compared to PG. Anyway, the presence of H-bonding leads to the more compact and stiffer microstructure of GL matrix with respect to the PG one as well as to some typical van der Waals substances such as *m*-TCP and DEP.[5,13]

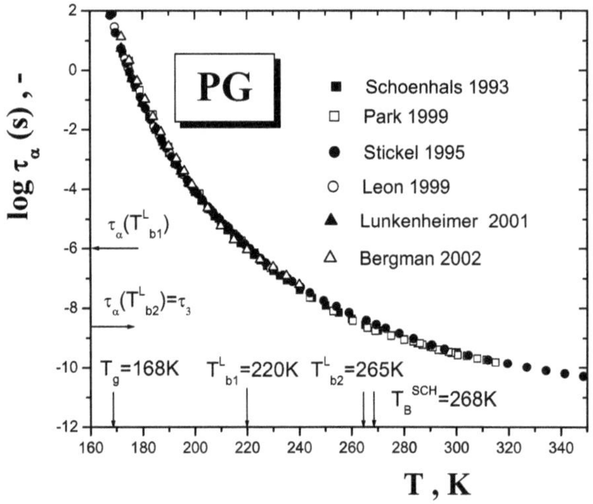

Figure 2. Compilation of the primary α relaxation times for propylene glycol (PG) together with the characteristic PALS temperatures (T_g, T_{b1}^L, and T_{b2}^L) and the DS temperature T_B^{SCH} showing empirical correlation between T_{b2}^L and T_B^{SCH}.

In the liquid state above T_g^{PALS}, the two further bend effects are evident in Fig. 1: i) a slight change in slope in strongly cold liquid at $T_{b1}^L = 1.275\, T_g^{PALS}$ for GL and $T_{b1}^L = 1.28 T_g^{PALS}$ for *PG* which defines mild crossover between regions II and III and finally ii) a large change in slope in weakly cold liquid at higher temperatures to a quasi-plateau level at $T_{b2}^L = 1.53 T_g^{PALS}$ for GL and $T_{b2}^L = 1.54\, T_g^{PALS}$ for PG defining drastic crossover between regions III and IV.

First, we analyze the former of the mentioned effects. Figure 2 represents a compilation of the mean relaxation time of the primary α process, τ_α, as a function of temperature for PG as obtained from several dielectric relaxation studies in the literature. As can be seen, at T^{L}_{b1} the mean dielectric relaxation time of the primary α process reaches just the value of 10^{-6} s, now being depicted as $T_\alpha(-6)$. Similar situation can be found also for further glass-forming systems. Figure 3 demonstrates such an empirical correlation between the T^{L}_{b1}'s values and the characteristic DS temperature $T_\alpha(-6)$ for a series of small molecular as well as polymer glass-formers. This finding suggests that a change in the expansion behavior of the localized free volume regions at around 1.2–1.3T_g^{PALS} seems to reflect some universal feature of glass-forming liquids which occurs when the mean α relaxation time reaches just a microsecond level.

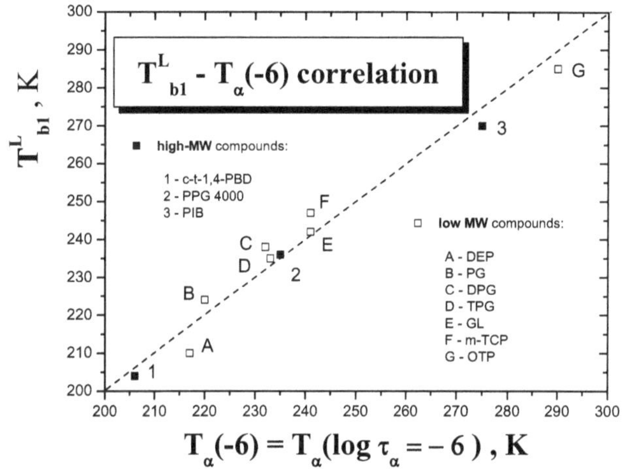

Figure 3. The empirical correlation between the first characteristic PALS temperature, T^{L}_{b1}, and the temperature, $T_\alpha(-6)$, at which the primary α relaxation time, τ_a, reaches just 10^{-6} s for a series of small molecular and polymer glass-formers.

Explanation of this feature in the o-Ps response is the subject of a continuing interest. The two possible-not necessary excluding-explanations based on either structural or dynamical hypothesis have recently been suggested by us.[5] Here, we discuss the latter possibility consisting in the influence of some more rapid local mobility, generally named as secondary relaxations, on a redistribution of the free volume hole population. Figure 4 presents the temperature dependences of relaxation time for both the primary α relaxation and the secondary β process for *glycerol.* Their values were obtained by a spectral deconvolution of the dielectric spectra in terms of the Cole–Davidson (CD) function for the α process and the Cole–Cole (CC) function for the β one.[14] In particular, the closeness between the o-Ps lifetime, τ_3, and the mean secondary β relaxation time, τ_β, at temperature 253 K quite close to $T_{b1}^{L} = 241$ K suggests that the local rapid β relaxators might contribute to the effective free volume distribution accessible for o-Ps probe. According to the general length–frequency principle of the dynamic events,[15] presumably smaller free volume holes requesting the smaller amplitude local motion might be eliminated from detection by an o-Ps probe. This elimination may result into the dominance of the larger free volume holes above T_{b1}^{L}.[16]

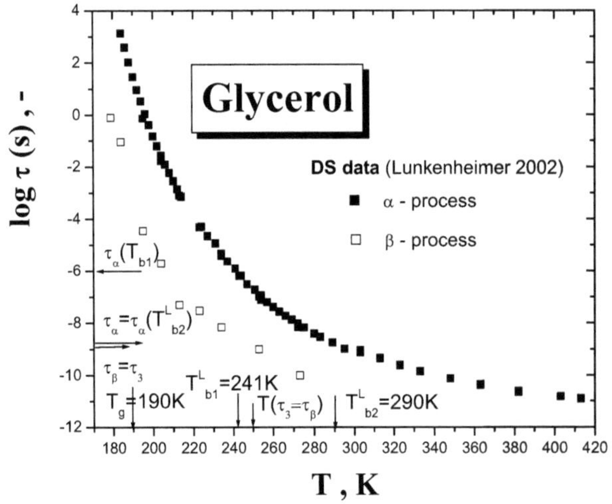

Figure 4. Relationship between the o-Ps lifetime, τ_3, and the secondary β_{JG} relaxation time, for τ_β for GL.

On increasing further the temperature, a dramatic discontinuity on τ_3–T plot occurs, at $1.53T_g^{PALS}$ in GL and 1.54 T_g^{PALS} in PG, characterized by a crossover from the steeply increasing linear trend in the weakly cold liquid to a quasi-plateau level in the warm liquid. The onset of the plateau effect correlates with several dynamic features of the primary α process as observed in dielectric spectroscopy studies.[17,18] First, Fig. 5 presents the so-called Stickel's plot for DS data on GL taken from Ref. 17. Here, the two linear regions indicate a crossover between the two different motional regimes at the so-called Stickel temperature, $T_B^{ST} = 287$ K. Similar type of analysis for PG gives $T_B^{ST} = 270$ K.[17] In both the cases, good coincidence between the onset of plateau region at T_{b2}^{L} and this characteristic DS temperature can be found for GL and PG. Secondly, Fig. 6 shows the so-called Schönhals' plot for the relaxation strength of the α process for PG, $\Delta\varepsilon_\alpha$, vs. logarithm of the α relaxation time, log τ_α.[18] Evidently, a dramatic change at the characteristic DS temperature, T_B^{SCH} = 268 K can be found being in the vicinity of the T_{b2}^L = 265 K. Good

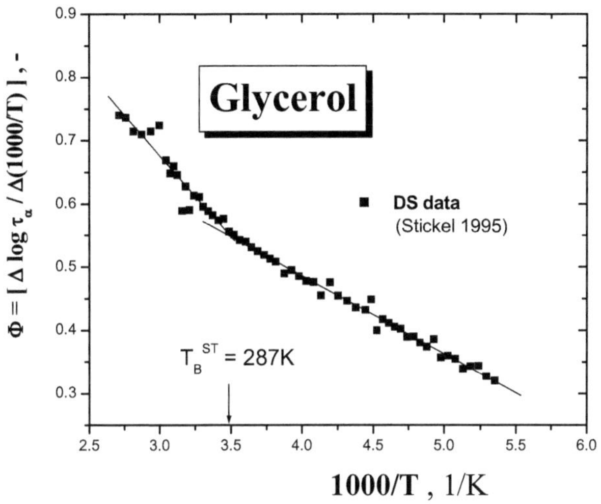

Figure 5. Stickel's plot for the primary α relaxation time, τ_α for GL demonstrating the change in the dynamical regime at the characteristic DS (Stickel) temperature, T_B^{ST} correlating with the characteristic PALS temperature T_{b2}^L.

agreement between T_B^{ST} and T_B^{SCH} for GL and PG and its correlation with the dramatic change in the PALS response at T_{b2}^L indicates the common origin of these significant changes in the relaxation time, τ_α, the relaxation strength, $\Delta\varepsilon_\alpha$, and the o-Ps lifetime, τ_3. This feature will be addressed further below.

Finally, in addition to the above-mentioned coincidence between the characteristic PALS and DS temperatures reflecting *the changes* in both the annihilation and relaxation quantities, more direct relationship between the corresponding time scales of both the methods exists. In other words, a comparison of the τ_3 value at the onset of the quasi-plateau region in Fig. 1 with the mean relaxation times of the α process, τ_α, from Fig. 2 for PG confirms our so far findings of the equality between both the time parameters for other small molecular and polymer systems.[8] The same has already been found to be valid for GL.[6] This fact indicates very close relationship between the progressively fluctuating matrix via the less cooperative α dynamics and the mean lifetime of the o-Ps probe[8] at the relatively higher temperatures, and at higher macroscopic volumes, $V(T)$, and then, at higher static empty free volumes $V_f^e(T) = V(T) - V^W$, where V^W is the van der Waals volume of the

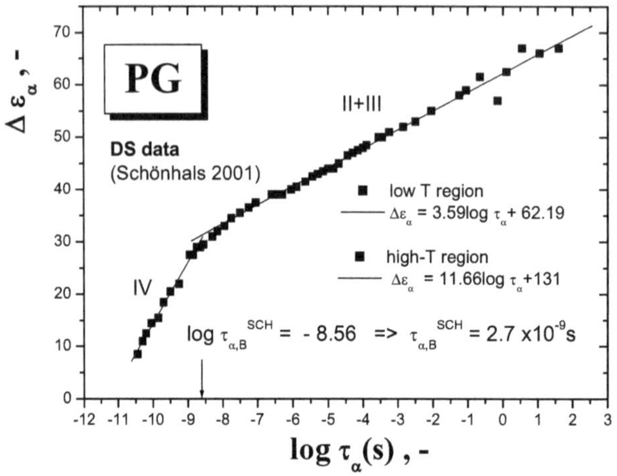

Figure 6. Schönhals' plot for the primary α relaxation time, τ_α, for PG showing the change in the α relaxation strength at the characteristic DS (Schönhals) temperature, T_B^{SCH} correlating with the characteristic PALS temperature T_{b2}^{L}.

molecules. The continuously increasing macroscopic volume of the system is accompanied by a continuously increasing empty free volume and the still less co-operative α process characterized by the more intense cage reorganization. Both these aspects together result into a quasi-plateau feature in the PALS response above T_{b2}^{L}, where o-Ps seems to be formed and localized in the sufficiently long living dynamic free volume holes only.

2.2. FREE VOLUME INTERPRETATION OF THE O-PS RESPONSE USING SEMIEMPIRICAL MODELS

The o-Ps response can be transformed into free volume information by means of existing physically plausible models which address two basic aspects of free volume characterization of any disordered material, i.e., free volume hole size, V_h, and free volume hole fraction, f_h. The former quantity is obtained using a standard quantum-mechanical model of o-Ps in a spherical hole which provides the following relationship between the o-Ps lifetime, τ_3, and the free volume hole radius R_h:[19]

$$\tau_3 = \tau_{3,0}\left[1-\frac{R_h}{R_h+\Delta R}+\frac{1}{2\pi}\sin\left(\frac{R_h}{R_h+\Delta R}\right)\right]^{-1}, \tag{1}$$

where $\tau_{3,0}$ is 0.5 ns and ΔR is the thickness of electron layer about hole obtained as a calibration parameter from fitting the observed o-Ps lifetimes to known vacancy or free volume hole sizes in molecular crystals or zeolites. In fact, the form of free volume entities is not strictly spherical because of the non-spherical form of constituent's particles so that the standard model is used usually in the sense of a volume equivalent spherical hole size $V_h = (4\pi/3)R_h^3$. Other geometrical forms of holes such as ellipsoid, cuboid, and cylinder have been considered.[20] However, as it was recently shown some parameters derived from PALS such as the free volume fraction are not strongly influenced by the choice of the free volume hole geometry.[21]

The mean free volume hole sizes in GL and PG at the above-mentioned characteristic temperatures are summarized in Table 1. In the glassy state as well as in the supercooled liquid one below the second liquid PALS temperature, T_{b2}^{L}, the mean hole sizes represent a part of the van der Waals volume of the corresponding molecules, while above T_{b2}^{L}, the corresponding mean hole sizes overcome the own molecular size apparently in consistency with the low viscosity normal liquid character of the systems.

TABLE 1. The mean free volume hole characterization of GL and PG according to Eq. (1) at the characteristic PALS temperatures.

System	T_g^{PALS}	$\overline{V_h}$	$\overline{V_h}/V^W$	T_{b1}^{L}	$\overline{V_h}$	$\overline{V_h}/V^W$	T_{b2}^{L}	$\overline{V_h}$	$\overline{V_h}/V^W$
	K	Å	–	K	Å	–	K	Å	–
GL	189	20	0.23	241	47	0.54	290	105	1.2
PG	172	30	0.38	220	61	0.78	265	165	2.1

3. Extended free volume model and its relationship to the free volume data from PALS

Some microscopic models based on free volume concept can be tested by using the above-mentioned free volume information. Here, we present an application of the extended free volume (EFV) model formulated by Cohen–Grest.[22] Briefly, this microscopic model of the liquid state accounts for a liquid as being formed by solid-like and liquid-like cells, only the latter ones contain a free volume. Using the percolation ideas, the specific mean free volume within the liquid-like clusters is given by the four-parameter equation:

$$v_f = A'\left\{\left(T - T_0^{CG}\right) + \left[\left(T - T_0^{CG}\right)^2 + C'T\right]^{1/2}\right\}, \tag{2}$$

where $A' = (v_m \ln e/B)$, and B, C' and T_0^{CG} are material-dependent coefficients; here v_m is the molecular volume and T_0^{CG} represents the temperature at which percolated liquid cluster should occur.

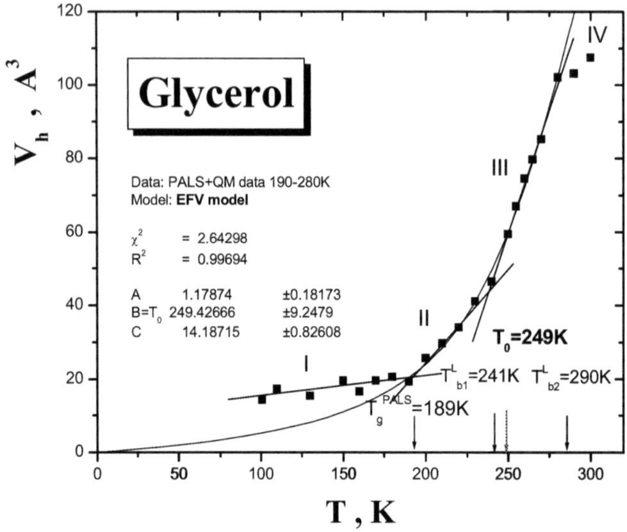

Figure 7. Application of the EFV model on the free volume data for glycerol.

Figure 7 shows the result of fitting Eq. (2) to the mean free volume hole sizes being identified with v_f from the EFV model. As can be seen, the EFV model is able to describe the free volume hole data over the whole range of the equilibrated supercooled liquid from the high temperature limit at T_{b2}^{L} down to T_g^{PALS}. Interestingly, the characteristic temperature of the EFV model, T_0^{CG}, is quite close to the first characteristic PALS temperature. In particular, for GL we find T_0^{CG} (PALS) = 249 ± 9 K close to T_b^L = 241 K and for PG T_0^{CG} (PALS) = 237 ± 7 K not too distant from T_{b1}^{L} = 220 K. According to the EFV model, this coincidence of the temperature parameters could be interpreted as the occurrence of the free volume percolation at around T_{b1}^{L} being responsible for the steeper slope in slightly supercooled liquid. However, the situation is not so simple as it will be discussed below.

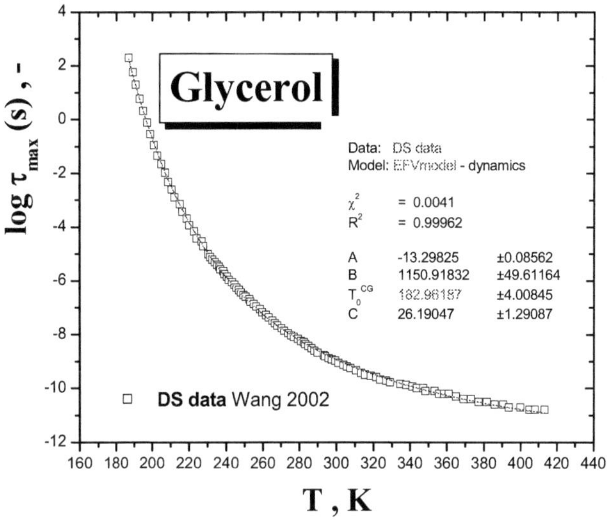

Figure 8. Application of the EFV model on the primary α relaxation data for glycerol.

4. Atomistic modeling of free volume microstructure of supercooled liquids

The significance of free volume information, as obtained from the FVA of the o-Ps response can be evaluated by means of *atomistic modeling*. In general, this approach consists in a combination of the generation of disordered systems by using, e.g., *molecular dynamics* (*MD*) *simulation* followed by an appropriate free volume analysis of the generated microstructures. Here, we demonstrate such an atomistic modeling approach for both the model systems. The first step of modeling has been performed by detailed MD simulations by means of the ORAC program.[23] The next step consists in sampling of the free space between the molecular bodies in the simulated microstructures by inserting the rigid probe of a given geometry and size and the volume integration of the inserted probes. In the case of our cavity analysis (CAVA) program,[24] the o-Ps probe is modeled by the hard sphere of diameter 1.06 Å and the molecular bodies, formed by atoms with van der Waals radii employed, represent the matrix particles. The CAVA code provides a set of the cavity parameters which can be compared with the free volume hole characteristics derived from the above-mentioned free volume analysis (FVA).

The results of our MD plus CAVAinvestigations for GL are presented in Ref. 24 and for PG here. Figure 9 compares the calculated volume from our very extended MD simulations of GL and PG with the experimental macroscopic volume from dilatometry (DIL)[9,10] in terms of the volume of simulation box containing the sufficient amount of molecules. In general, satisfactory agreement between both the macroscopic volume quantities can be found, especially at higher temperatures.

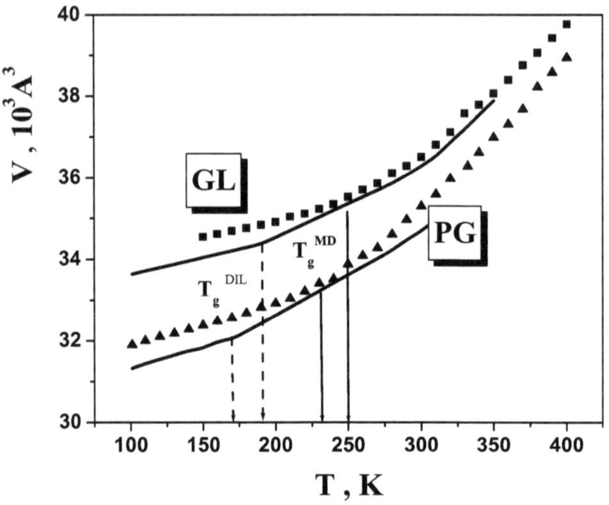

Figure 9. The simulated (points) and experimental (lines) box volumes of 300 molecules for glycerol (GL) and propylene glycol (PG).

In particular, for *GL* the relative difference between the simulated volume data and the measured ones reaches 0.3% at higher temperature regions and 1.4% at lower ones. Evidently, the larger deviations for the lower temperature regions are connected with the effectively very high cooling rate in MD simulation[25] and the associated easier onset of the glass transition phenomenon. The onset of deviations defines the simulated glass transition temperature, T_g^{MD}, which can be related to the experimental glass transition temperature, T_g, from dilatometry via fragility, m_g, by the following Donth relation:[15]

$$\frac{T_g}{m_g} = -\frac{T_g - T_g^{DIL}}{\Delta \log \Phi_c}, \tag{3}$$

where $\Delta \log \Phi_c$ is the difference between the simulated and experimental cooling rates. Table 2 summarizes the dilatometric, T_g^{DIL}, the estimated, T_g, and the simulated, T_g^{MD}, glass transition temperatures for both the systems. Good agreement between the estimated and simulated values indicates that a linear approximation for cooling rate dependence of glass temperature is valid even for the 14 orders of magnitude change in Φ_c, so that the simulated microstructures should correspond to the real physical ones as obtained with very rapid cooling.

TABLE 2. The experimental, T_g^{DIL}, estimated, T_g^*, and simulated, T_g^{MD}, glass temperatures for GL and PG according to Eq. (3) for $\Delta \log \Phi_c = 14$. The mg values from Beiner, M. (2001) *J. Non-Cryst. Solids* **279**, 126.

System	T_g^{DIL}	m_g	T_g^*	T_g^{DIL}
	K	-	K	K
GL	189	49	240	250
PG	172	48	220	230

The obtained static microstructures of GL and PG have been analyzed by using the CAVA program.[24] On the basis of determination of the sizes and the number of the individual cavities, Figs.10a and b show for the case of GL and PG a comparison between the mean free volume hole, V_h, from Eq. 1 and the mean cavity volume, V_{cav}, as calculated from the following equation:[26]

$$\overline{V}_{cav} = \sum_{V_{cav,min}}^{V_{cav,max}} V_{cav,i} \Big/ \sum_{i}^{N_{cav}} N_{cav,i} \,. \tag{4}$$

As can be seen, the mean cavity volumes are smaller than the mean free volume ones over the whole temperature range predominantly due to a large amount of the small cavities sampled by very small o-Ps probe ($V_{o\text{-}Ps}$ = 0.64 Å^3). Having on mind that the simulated microstructures are the static ones while PALS is performed on the fluctuating matrix, we introduce within the framework of general length scale–frequency principle[15] the dynamic aspect via a concept of the limiting cavity size.[24] This is expressed by a set of curves for different $V_{cav,lim's}$. As the result, a good agreement for the equilibrated liquid above T_{b1}^{L} between both the mean free volume quantities is achieved with $V_{cav,lim} \approx$ 20–30 Å^3. The physical background for this concept accounts for a possible blocking effect of some small scale mobility on a part of the total free volume distribution. Experimental support for this concept in terms of the secondary β relaxation has been reported for GL elsewhere.[16] In fact, from Fig. 4 it follows that the time scale of the β process above 250 K is shorter than the mean o-Ps lifetime so that a part of the free volume hole population is fluctuating too rapidly to be accessible for the o-Ps probe localization. The finding about the size of $V_{cav,lim}$ is consistent with other estimations as obtained by other ways.[27]

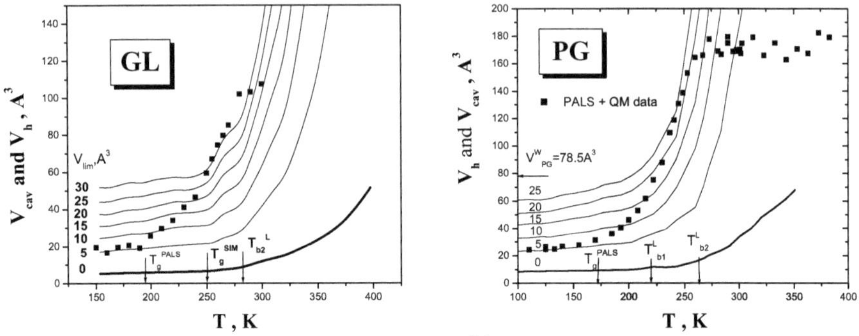

(a) (b)

Figure 10. Mean cavity volumes, calculated from Eq. (4) for several cases of V_{lim}, together with the experimental mean hole volumes (points) from PALS for GL (a) and PG (b).

Figure 11 presents the integral volume distributions of all the cavities at several temperatures in PG. Two qualitatively different temperature regions can be distinguished. At lower temperatures below 260 K, a continuously increasing integral distribution is observed which changes into a discontinuous bimodal one in the higher temperature region above 270 K. The detailed visual inspection of the *static* cavity configurations reveals an increasing contribution of the larger cavities which results into the fully percolated cavities over the whole simulation box. This is evident from the temperature evolution of three-dimensional cavity microstructures in GL (Fig. 12). Interestingly, the presence of percolated cavity phenomenon from our MD + CAVA approach in between 260 and 270 K is commensurable with the characteristic PALS temperature of the onset of the plateau region at T_{b2}^{L} = 265 K as well as with the characteristic DS temperatures T_B^{ST} and T_B^{SCH} as mentioned above. Moreover, the equality between $\tau_3 \equiv \tau_\alpha$ at $T_{b2}^{L} \cong T_B$'s is observed – Fig. 2 – so that for $T < T_{b2}^{L}$ the $\tau_3 < \tau_\alpha$ relationship is valid, while for $T > T_{b2}^{L}$ the $\tau_3 > \tau_\alpha$ relationship is found. All these facts allow us to formulate a concept of the dynamic free volume facility, i.e., the presence of the *static* fully percolated cavity volume facilitates for a qualitative change in the primary α relaxation. Thus, according to our proposal, the crossover effects in PALS and DS data result from the mutual interplay between the increasing empty free space between particles with the progressively increasing appearance of percolating cavities and the associated intense cooperative rearrangements of particle constituents of the matrix.

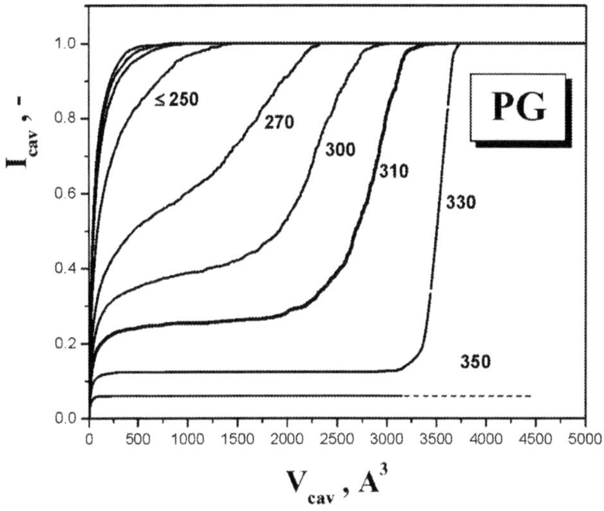

Figure 11. Integral cavity distribution in PG for a series of temperatures exhibiting the change to the bimodal cavity distribution in between 250 K and 270 K coinciding with $T^{\mathrm{L}}_{\mathrm{b2}}$.

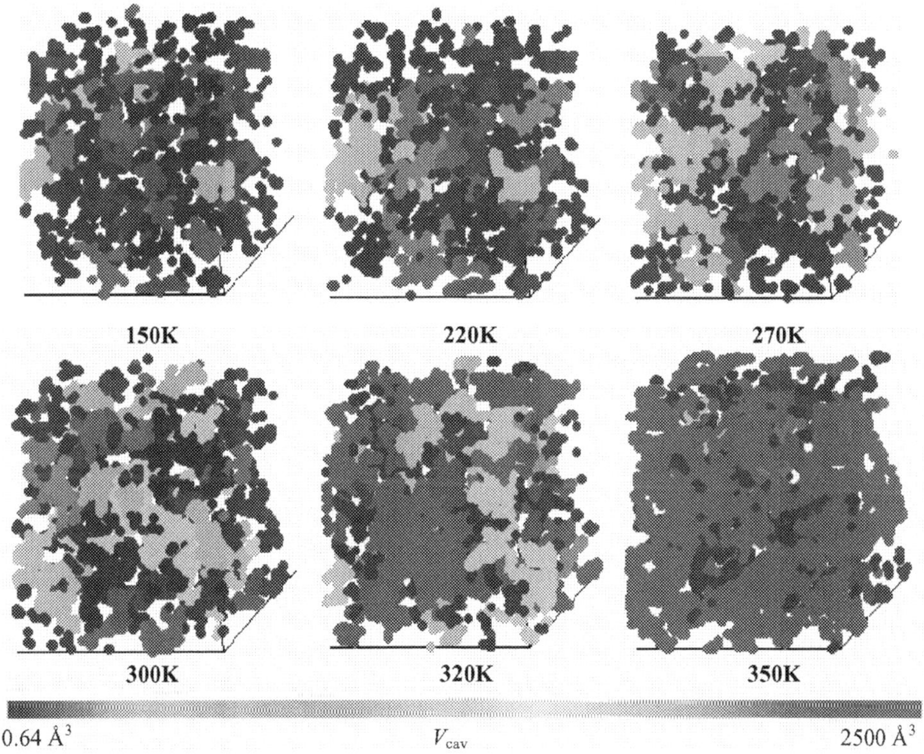

Figure 12. Temperature evolution of the cavity microstructure in 3D space for GL. Yellow color indicates the presence of the percolated cavity volume over the whole simulation box.

5. Crossover phenomenon from a viewpoint of the present microscopic models and atomistic modeling

Finally, it is of interest to compare the explanations of the crossover phenomenon in the primary α relaxation properties at T_B proposed so far with the results of our atomistic modeling studies on GL and PG. Several interpretations of this important aspect of the supercooled liquid dynamics in terms of various microscopic models have been offered in the literature. They are based either on a coincidence between the crossover temperature, T_B, and a certain characteristic temperature corresponding to the specific microscopic model or on the ability to describe the cooperative primary α relaxation via the preceding simpler process the so-called primitive, i.e., independent relaxation, after extraction of many-body aspect of the cooperative α dynamics. The former group is represented by two cases: i) the characteristic free volume temperature, T_0^{CG}, from the extended free volume (EFV) model[22] at which the onset of the free volume percolation should take place or ii) the critical temperature, T_c^{MCT}, of the mode coupling theory (MCT)[28] where a critical slowing down of the α process should occur.

As for the EFV model, the relaxation time is expressed in the following form:[22]

$$\log \tau(T) = \mathrm{A} + \mathrm{B} \Big/ \left\{ \left(T - T_0^{CG}\right) + \left[\left(T - T_0^{CG}\right)^2 + \mathrm{C}T\right]^{1/2} \right\}, \tag{5}$$

where A, B, and C are material coefficients and T_0^{CG} is the temperature at which a continuity of the liquid-like cells is attained. The closeness between the crossover temperature at $T_B \cong 1.2T_g$ from DS data and the characteristic temperature T_0^{CG}(DS) from application of Eq. (5) on the DS data has been observed for a series of six van der Waals systems.[29] Consequently, the crossover at T_B is interpreted as the result of free volume percolation leading to the change in the primary or α dynamics. Very recently, one of us has demonstrated for the case of DEP a good coincidence between T_B^{ST} from DS and T_{b1}^{L} from PALS with the T_0^{CG} value determined not only from DS data alone T_0^{CG}(DS) but also from the free volume ones from PALS T_0^{CG}(PALS).[13] On the other hand, as mentioned above, the EFV model can describe the PALS data of GL and PG with a reasonable coincidence between T_0^{CG}(PALS) and the first bend temperature T_{b1}^{L}, but not between T_0^{CG}(PALS) and the second characteristic PALS temperature $T_{b2}^{L} \cong T_B$ as indicated by atomistic modeling. Moreover, for our hydrogen-bonded liquids GL and PG Eqs. 2 and 5 are not able to account for both the free volume and the relaxation data simultaneously, with the same T_0^{CG}(PALS) = T_0^{CG}(DS) parameter as predicted by the EFV model.[22] In fact, the T_0^{CG}(DS) values for GL (Fig. 8) and PG (not shown) are 178.5 ± 7 K or 162 ± 6 K, being close to the respective T_g's, are not consistent with the above-mentioned T_0^{CG}(PALS) ones: 249 ± 9 K and 237 ± 7 K, respectively. Thus, the inability to describe both PALS and DS data simultaneously as well as the discrepancy with the results of atomistic modeling allow us to conclude that the EFV model appears to be irrelevant for strongly hydrogen-bonded glass-forming liquids such as GL and PG, at least. On the other hand, the validity of the EFV model for slightly hydrogen-bonded and van der Waals' systems remains to be further tested.

Regarding the interpretation within the mode coupling theory (I-MCT), although $T_B \approx T_c$ is often observed,[30,31] the predicted divergence of the α relaxation time is not actually observed so that the ideal version of MCT needs some modification. The extended mode coupling theory (E-MCT) when applied on a few van der Waals systems such as ortho-terphenyl gives T_c(I-MCT) ≅ T_c(E-MCT)[32] so that $T_B \approx T_c$(I-MCT) should imply an onset of the thermally activated dynamics. Very recent I-MCT analysis of GL provided $T_c = 288$ K[31] in agreement with the above-mentioned coincidence: $T_B^{ST} \cong T_B^{SCH} \cong T_{b2}^{L}$ as well as close to the onset temperature of the percolated free volume from our MD and CAVA approach. Unfortunately, the E-MCT has not been applied so far on our more complicated hydrogen-bonded model systems so that the range of its validity remains unknown.

Finally, the latter class is represented by the coupling model (CM) of Ngai et al.[33] For a series of five van der Waals liquids, the crossover at T_B^{ST} is explained in terms of the large change in the heterogeneous character of the primary dynamics via the change in the strength of the intermolecular cooperativity on the preceding primitive relaxation process from strongly cooperative process below T_B to less cooperative one at higher temperatures.[34] In the last years, a growing amount of the evidences that the primitive process may be identified with the secondary Johari–Goldstein β process has appeared.[35,36] The original set of the five compounds does not include strongly intermolecularly hydrogen-bonded systems.[34] In the case of GL, the closeness of the β relaxation time, τ_β, as obtained from a deconvolution of the DS spectra, with the primitive relaxation time, τ_0, calculated from the primary α relaxation time, τ_α, via $\tau_0 = t_c^n . \tau_\alpha^{1-n}$, where n is a measure of cooperativity connected with the Kohlrausch–Williams–Watts (KWW) exponent, β_{KWW}, by relation $n = 1-\beta_{KWW}$ as well as the mutual merging of α and β processes at around 300 K, close to $T_{b2}^{L} = 290$ K[37], seem to suggest some important change in the microstructure of the GL matrix, which may change the character of the α dynamics. This quasi-microscopic CM explanation together with the fact of the equality of the α relaxation time, τ_α, and the mean o-Ps lifetime, τ_3, appears to be qualitatively consistent with our MD and CAVA results. Thus, the onset of the fully percolated static free volume in the MD microstructures close to T_{b2}^{L} might imply some significant influence on the character of the cooperative α dynamics. As mentioned above, one possibility may be that the transition to the percolated free volume facilitates the crossover from a strongly cooperative mobility below T_B^{ST} to a slightly cooperative one above T_B^{ST} together with a full diminishing the more local β process just above T_B^{ST}.

Acknowledgment

The authors would like to express their thanks to the VEGA Agency, Slovakia for the financial support by grants 2/3026/23 (JB) and 2/4103/24 (JK) as well as the APVT Agency, Slovakia for grant APVT 51-045302 (JK&JB).

References

1. Brandt, W. and Dupasquier, A. (eds.) (1983) *Positron Solid State Physics*, North-Holland, Amsterdam.
2. Jean, Y.C. (1995) Characterizing free volume and holes in polymers by positron annihilation spectroscopy in Dupasquer, A. (ed.) *Positron Spectroscopy of Solids*, IOS, Ohmsha, Amsterdam, p. 563.
3. Bartoš, J. (2000) Positron annihilation spectroscopy of polymers and rubbers in R.A. Meyers (eds.) *Encyclopedia of Analytical Chemistry*, Wiley & Sons, Chichester, p. 7968.
4. Jean, Y.C., Mallon, P.E., and Schrader, D.M. (eds.) (2003) *Principles and Applications of Positron & Positronium Chemistry*, World Scientific Publ., New Jersey.
5. Bartoš, J., Šauša, O., and Krištiak, J. Annihilation response of the ortho-positronium probe from positron annihilation lifetime spectroscopy and its relationships to the free volume and dynamics of glass-forming systems *in ARW NATO Series: Nonlinear Dielectric Phenomena in Complex Liquids*, Kluwer, Acad. Publ., Dordrecht, The Netherlands, p. 289.
6. Bartoš, J., Šauša, O., Krištiak, J., Blochowicz, T., and Rössler, E. (2001) *J. Phys.-Cond. Matter* **13**, 11473.
7. Šauša, O., Bartoš, J., and Krištiak, J. to be published.
8. Bartoš, J., Šauša, O., Bandžuch, P., Zrubcová, J., and Krištiak, J. (2002) *J. Non-Cryst. Solids* **307–310**, 417.
9. Kovacs, A. (1963) *Adv. Polym. Sci.* **3**, 394.
10. Parks, G.S. and Huffman, H.M. (1927) *J. Phys. Chem.* **11**, 1842.

11. Angell, C.A. (1985) Strong and fragile liquids in Ngai, K., Wright, G.S. (eds.) *Relaxations in Complex Systems*, NTIS, Springfield, p. 1.
12. Leon, C., Ngai, K., and Roland, C.M. (1999) *J. Chem. Phys.* **110**, 11585.
13. Pawlus, S., Bartoš, J., Šauša, O., Krištiak, J., and Paluch, M. (2005) *J. Chem. Phys.* **124**, 104505.
14. Lunkenheimer, P. and Loidl, A. (2002) *Chem. Phys.* **284**, 205.
15. Donth, E., *The Glass Transition,* Springer, Berlin, 2001.
16. Bartoš, J., Šauša, O., Račko, D., Krištiak, J., and Fontanella, J.J. (2005) *J. Non-Cryst. Solids* **351**, 2599.
17. Stickel , F.J. (1995) Ph.D. Thesis, Shaker–Verlag, Aachen.
18. Schönhals, A. (2001) *Europhys. Letts.* **56**, 815.
19. Tao, J. (1972) *J. Chem. Phys.* **56**, 5499; Eldrup, M., Lightbody, D., Sherwood, J.N. (1981) *Chem. Phys.* **63**, 51; Nakanishi, H., Wang, S.J., and Jean, Y.C. (1988) in S.C. Sharma (eds.), *Positron Annihilation Studies of Fluids*, World Science, Singapore, p. 292.
20. Jean, Y.C. and Shi, H. (1994) *J. Non-Cryst. Solids* **172**, 806; Jasinska, B., Koziol, A.E., and Goworek, B. (1999) *Acta Phys. Polonica A* **95**, 557; Olson, B.G., Prodpran, T., and Jamieson, A.M., Nazarenko, S. (2002) *Polymer* **43**, 6775.
21. Consolati, G. (2002) *J. Chem. Phys.* **117**, 7279.
22. Cohen, M.H. and Grest, G.S. (1979) *Phys. Rev.* **B20**, 1077; Grest, G.S., and Cohen, M.H. (1980) *Phys. Rev.* **B21**, 4113; Grest, G.S. and Cohen, M.H. (1981) *Adv. Chem. Phys.* **48**, 455.
23. Procacci, P., Darden, T.A., Paci, E., and Marchi, M. (1996) *J. Phys. Chem.* **100**, 10464.
24. Račko, D., Chelli, R., Bartoš, J., Cardini, G., and Califano, S. (2005) *Eur. Phys. J.* **E32**, 289.
25. Angell, C.A., and Wang, L.M. (2003) *Biophys. Chem.* **105**, 621.
26. Sastry, S., Truskett, T.M., Debenedetti, P.G., Torquato, S., and Stillinger, F.H. (1998) *Mol. Phys.* **95**, 289.
27. McCullagh, C.M., Yu, Z., Jamiesson, A.M., Blackwell, J. and McGervey, J.D. (1995) *Macromolecules* **28**, 6100; Yu, Z. (1995) Ph.D. thesis, Case Western Reserve University, Cleveland, OH.
28. Götze, W., and Sjörgen, L., (1992) *Rep. Progr. Phys.* **55**, 241.
29. Paluch, M., Casalini, R., and Roland, C.M. (2003) *Phys. Rev.* **E67**, 021508.
30. Novikov, V.N., and Sokolov, A.P. (2003) *Phys. Rev.* **E67**, 031507.
31. Adichtchev, S., Blochowicz, T., Tschirwitz, Ch., Novikov, V.N., and Rössler, E.A. (2003) *Phys. Rev.* **E68**, 011504.
32. Götze, W. (1999) *J. Phys.-Cond. Matt.* **11**, A1.
33. Ngai, L.K. (1979) *Comment Solid State Phys.* **9**, 127; Ngai, K.L., Rendell, R.W. (1997) in Fourkas, J.T., Kivelson, D., Mohanty, U., and Nelson, K. (eds.) *Super-cooled Liquids. Advances and Novel Applications*, ACS Symposium Series, Vol. 676 *Amer. Chem. Soc.* Washington, DC p. 45.
34. Casalini, R., Ngai, K.L., and Roland, C.M. (2003) *Phys. Rev.* **B68**, 014201.
35. Ngai, K.L., and Paluch, M. (2004) *J. Chem. Phys.* **120**, 857.
36. Ngai, K.L. (2005) *J. Non-Cryst. Solids* **351**, 2635.
37. Ngai, K.L., Lunkenheimer, P., Leon, C., Scheneider, U., Brand, R., and Loidl, A. (2001) *J. Chem. Phys.* **115**, 1405.

SEGMENTAL AND CHAIN DYNAMICS IN POLYMERS

C.M. ROLAND[1,*] and R. CASALINI[1,2]
[1]*Naval Research Laboratory, Chemistry Division, Code 6120 Washington, DC 20375-5342, USA*
[2]*George Mason Univ., Chemistry Department, Fairfax, V22030 USA*

Abstract. All polymers exhibit a breakdown of time–temperature superpositioning in the softening zone, where both local and global chain modes contribute to the measured dynamic response. This differing behavior of the two modes is also seen in the volume and pressure dependences of the respective relaxation times. Nevertheless, the superposition of local segmental and global chain modes can be achieved by expressing the relaxation times as a function of the product of temperature, T, times specific volume, V, with the latter raised to a constant γ. Moreover, the value of γ is the same for the two modes. These experimental facts reveal that the relative contribution of temperature and volume is the same for the two modes, although their response to either variable differs; that is, while having the same functional form, the combined effect of T and V on the respective modes is different.

Keywords: Segmental relaxation, structural relaxation, chain modes, global motion.

1. Breakdown of time–temperature–pressure superpositioning

It has been known for at least 40 years that polymers are thermorheologically complex.[1,2] This breakdown of time–temperature superpositioning transpires not only in mechanical measurements, but is also apparent in dielectric spectra[3] and

*To whom correspondence should be addressed.

S. J. Rzoska and V. A. Mazur (eds.), Soft Matter under Exogenic Impacts, 133–140.

NMR.[4] Ironically, one of the most dramatic manifestations of the phenomenon is observed in a polyisobutylene NBS standard.[5] Measurements in the 1950s on this very sample in laboratories around the world were instrumental in the development and acceptance of the idea of time–temperature superpositioning in polymers.[6] The thermorheological complexity of polymers is most evident in the transition zone, where the local modes change more strongly with temperature than do the chain modes. At temperatures sufficiently above T_g, the respective T-dependences become parallel. This is seen in data on atactic-polypropylene in Fig. 1, obtained using several different spectroscopies.

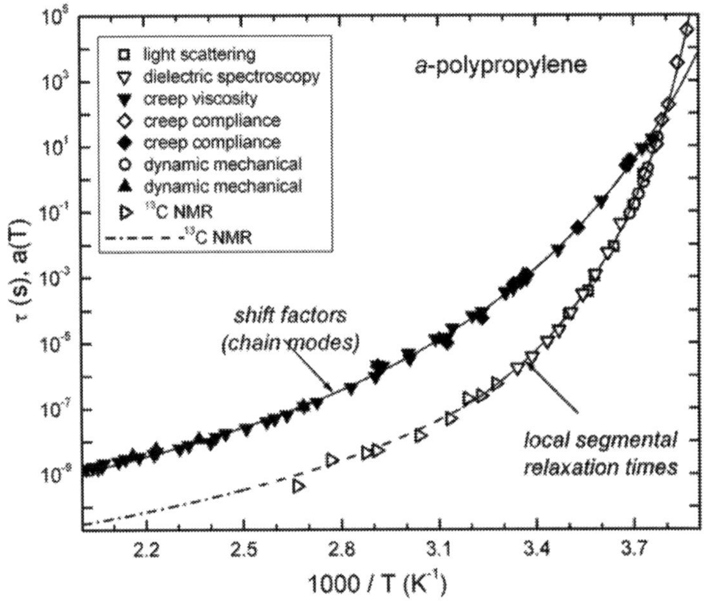

Figure 1. Segmental relaxation times (hollow symbols) and shift factors for the terminal (chain) relaxation times (solid symbols) for atactic-polypropylene measured by various spectroscopies.

The different response to pressure of the two modes is seen in dielectric measurements at elevated pressure, as first shown by Floudas and coworkers,[7,8] wherein the local segmental motion has a larger activation volume. When combined T- and P-dependences are converted to V-dependences, the results are the same – a stronger variation of the segmental mode in comparison to the chain modes. This is illustrated in Fig. 2 for polyoxybutylene.[9]

One interesting aspect of this universal behavior of polymers is the absence of its recognition in most textbooks of polymers science. Commonly, ostensibly satisfactory master curves are shown for polystyrene or another polymer, along with a description of the time–temperature superposition principle. However, invariably there is no mention of the failure of *t–T* superpositioning whenever local and global modes both contribute (e.g., in the glass transition zone). This error likely arises from expectations drawn from theory. Both the Rouse[10] and reptation[11] models of polymer dynamics express the chain relaxation times with expressions whose only *T*-dependence is the ratio ζ/T. ζ is the local, monomeric friction coefficient, which is identified with the friction coefficient describing local segmental motions.

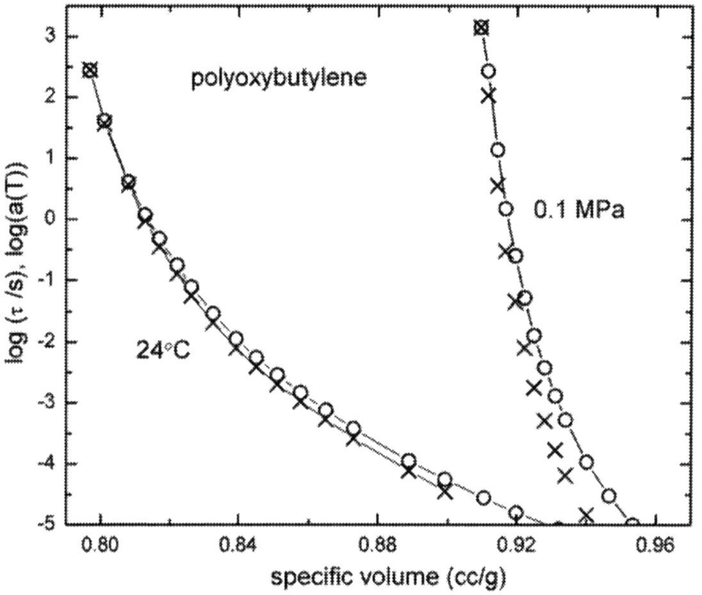

Figure 2. Segmental (crosses) and normal mode (circles) dielectric relaxation times for polyoxybutylene versus the specific volume. The τ for the segmental process have been shifted vertically to superpose at long times with the global relaxation times.

The fallacy of master curves that span the viscoelastic spectrum from the glassy zone to the terminal flow regime is demonstrated in Fig. 3, showing an ostensibly accurate master curve for high molecular weight 1,4-polyisoprene. Due to the paucity of data points in the transition zone, the various isotherms appear to superpose. However, as seen in the inset, there is a marked change in the shape of the loss tangent in the transition zone, revealing the thermorheological complexity.

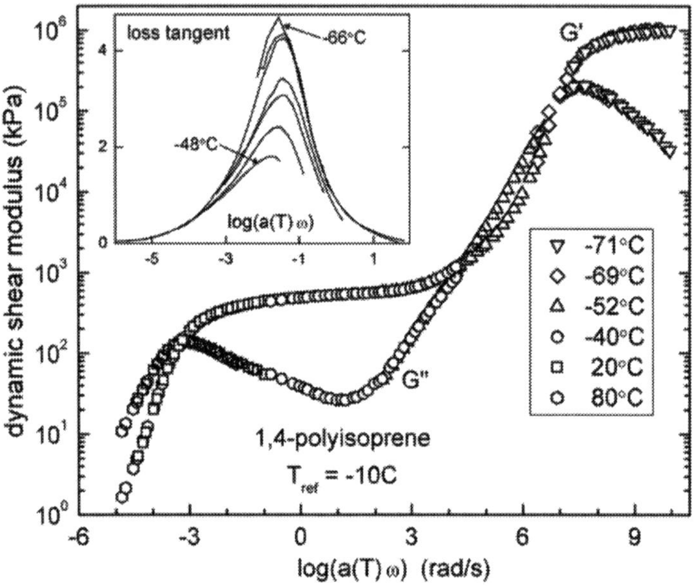

Figure 3. Master curve of the mechanical storage and loss moduli for 1,4-polyisoprene (M_w=500 kg/mol). The inset shows the loss tangent in the transition zone, where thermorheological complexity is evident.

2. Scaling of relaxation times

The local segmental relaxation times of polymers (and correspondingly the structural or α-relaxation times of molecular liquids) superpose when plotted versus TV^{γ}, where γ is a material constant.[12–14] The scaling exponent γ is usually determined empirically by shifting τ measured under isothermal and/or isobaric conditions. It can also be calculated from *PVT* measurements using the relation $\gamma = -T/\alpha_\tau$, where α_τ is the volume expansion coefficient at constant τ. With some assumptions $\gamma \sim \gamma_G$,[15,16] where γ_G is the Grüneisen parameter

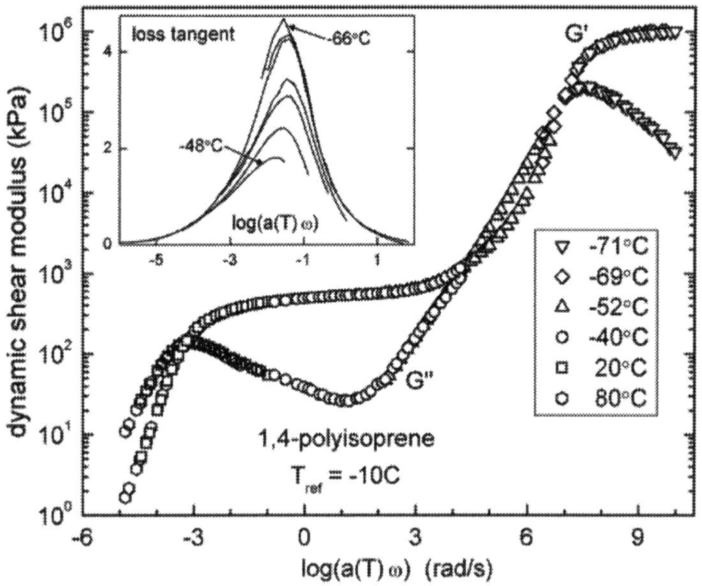

Figure 4. Dielectric relaxation times for the local segmental (squares) and normal mode (circles processes in polyoxybutylene as a function of the product TV^{γ} with $\gamma = 2.65$. Vertical shifting o the former to superpose at long times evidences the differing dependences on the scaled data.

quantifying the anharmonicity of the vibrational motions. This TV^{γ} scaling is illustrated in Fig. 4 for polyoxybutylene.[17]

The successful TV^{γ} scaling of local segmental relaxation times raises the question of the behavior of the global motions: do the relaxation times for the chain modes scale in the same fashion, and if so, what is the value of the scaling exponent? For polymers having a dipole moment parallel to the chain, these global dynamics are reflected in the dielectric normal mode, which is sensitive to the chain end-to-end vector. Included in Fig. 4 are the normal mode relaxation times measured for the polyoxybutylene. Note that these do indeed superimpose when plotted versus TV^{γ}; moreover, the value of γ is the same for both relaxation modes. This surprising result appears to be generally valid, having been found as well for polypropylene glycol and 1,4-polyisoprene.[14,18]

3. Reconciling the dynamics of the local and global modes

The fact that the segmental and chain modes, which have different *T*, *P*, and *V* dependences, both superimpose when expressed as a function of the TV^{γ} using the same value of γ deserves further consideration. In Fig. 4 the segmental relaxations times have been shifted vertically to superimpose at longer τ with the normal mode relaxation times. It is seen for polyoxybutylene (and also for polypropylene glycol and 1,4-polyisoprene[14,18]) that there is a larger change in the segmental dynamics with change in TV^{γ} than in the chain dynamics. This greater TV^{γ}-sensitivity mirrors the former's stronger *T*, *P*, and *V*-dependences.

Since the exponent γ reflects the contribution of volume relative to that from temperature to the dynamics, we can conclude that the relative effect of these two variables is the same for the two relaxation modes. However, this does not mean that the combined effect of *V* and *T* are the same; clearly, changes in *T* and *V* exert a stronger influence on the segmental dynamics. It is only that the apportionment of τ variations to *V* and *T* is the same for segmental relaxation as for the chain motions.

While the relaxation times for the local and the chain modes are functions of the same variable TV^{γ} with the same γ, the functions *per se* are not the same. What can be said about the nature of the respective functions? According to the coupling model,[19–21] measured properties such as the relaxation times for local segmental and chain motions are related to a primitive friction coefficient, ζ_0 (primitive in that it is strictly noncooperative intermolecularly). The characteristics of this primitive process, which serves as a precursor to the glass transition, suggests identification with the Johari–Goldstein secondary relaxation.[22] The friction factor underlying the motions in dense phase are related to ζ_0 by the fundamental equation of the coupling model, $\zeta \sim \zeta_0^{1/\beta}$, where β is the Kohlrausch exponent describing the breadth of the relaxation function. This breadth is larger for local segmental motion than for the chain dynamics (since intermolecular cooperativity is stronger for the former); thus, any *T*, *P*, or *V* dependences of ζ_0 are amplified nonlinearly to a greater degree for the local segmental modes. This accounts for the stronger dependences on these variables of the local segmental relaxation times.

Implicit in this discussion is that ζ_0 must be a function of TV^{γ}. This follows from the experimental fact that τ is a function of TV^{γ}, provided β is constant over the range of τ under consideration.

4. Summary

The global dynamics is less sensitive to thermodynamic variables such as temperature, pressure, and volume than is the local segmental dynamics. This is a consequence of their different friction factors. Nevertheless, relaxation times for both processes can be superimposed when plotted as a function of TV^{γ}, where the material constant γ is the same two modes. These seemingly contradictory results can be understood through consideration of the underlying, noncooperative dynamics, which is expected to conform to the same TV^{γ} scaling. This noncooperative motion, which can be identified with the Johari-Goldstein relaxation, can be quantitatively related to the measured dynamic properties through application of the coupling model.

Acknowledgment

This work was supported by the Office of Naval Research.

References

1. Plazek, D.J. (1965) Temperature Dependence of the Viscoelastic Behavior of Polystyrene, *J. Phys. Chem.* **69**, 3480–3487.
2. Ngai, K.L. and Plazek, D.J. (2005) Identification of Different Modes of Molecular Motion in Polymers that Cause Thermorheological Complexity, *Rubber Chem. Tech.* **68**, 376–434.
3. Santangelo, P.G. and Roland, C.M. (1998) Temperature Dependence of Mechanical and Dielectric Relaxation in cis-1,4-Polyisoprene, *Macromolecules* **31**, 3715–3719.
4. Roland, C.M., Ngai, K.L., Santangelo, P.G., Qiu, X.H., Ediger, M.D., and Plazek, D.J. (2001) Temperature Dependence of Segmental and Terminal Relaxation in Atactic Polypropylene Melts, *Macromolecules*, **34**, 6159–6160.
5. Plazek, D.J., Chay, I.-C., Ngai, K.L., and Roland, C.M. (1995) Viscoelastic Properties of Polymers. 4. Thermorheological Complexity of the Softening Dispersion in Polyisobutylene, *Macromolecules* **28**, 6432–6436.
6. Ferry, J.D. (1980) *Viscoelastic Properties of Polymers*, 3rd edition, Wiley, New York, p. 606.
7. Floudas, G. and Reisinger, T. (1999) Pressure Dependence of the Local and Global Dynamics of Polyisoprene, *J. Chem. Phys.* **111**, 5201–5204.
8. Floudas, G., Gravalides, C., Reisinger, T., and Wegner, G. (1999) Effect of Pressure on the Segmental and Chain Dynamics of Polyisoprene. Molecular Weight Dependence, *J. Chem. Phys.* **111**, 9847–9852.
9. Casalini, R. and Roland, C.M. (2005) Temperature and Density Effects on the Local Segmental and Global Chain Dynamics of Poly(oxybutylene), *Macromolecules* **38**, 1779–1788.

10. Yamakawa, H. (1971) *Modern Theory of Polymer Solutions*, Harper and Row, New York, chapter 6.
11. Doi, M. and Edwards, S.F. (1986) *The Theory of Polymer Dynamics*, Clarendon Press, Oxford, p. 226.
12. Casalini, R. and Roland, C.M. (2004) Thermodynamical Scaling of the Glass Transition Dynamics, *Phys. Rev. E* **69**, 062501.
13. Roland, C.M. and Casalini, R. (2004) Comment on Ddisentangling Density and Temperature Effects in the Viscous Slowing Down of Glass-forming Liquids, *J. Chem. Phys.* **121**, 11503–11504.
14. Roland, C.M., Hensel-Bielowka, S., Paluch, M., and Casalini, R. (2005) Supercooled Dynamics of Glass-forming Liquids and Polymers under Hydrostatic Pressure, *Rep. Prog. Phys.* **68**, 1405–1478.
15. Casalini, R. and Roland, C.M. (2005) A Thermodynamic Interpretation of the Scaling of the Dynamics of Supercooled Liquids, *J. Chem. Phys.*, submitted.
16. Roland, C.M., Feldman, J.L., and Casalini, R. (2005) Scaling of the Local Dynamics and the Intermolecular Potential, *J. Non-Cryst. Solids*, **352**, 4895–4899.
17. Casalini, R and Roland, C.M. (2005) Temperature and Density Effects on the Local Segmental and Global Chain Dynamics of Poly(oxybutylene), *Macromolecules* **38** 1779–1788.
18. Roland, C.M., Casalini, R., and Paluch, M. (2004) Effect of Volume and Temperature on the Global and Segmental Dynamics in Polypropylene Glycol and 1,4-polyisoprene, *J. Polym. Sci. Polym. Phys. Ed.* **42**, 4313–4319.
19. Ngai, K.L. (1979) *Disorder Effects on Relaxation Processes* (R. Richert and A. Blumern, eds.), Springer-Verlag, Berlin, pp. 89–150.
20. Ngai, K.L., Casalini, R., and Roland, C.M. (2005) Volume and Temperature Dependences of the Global and Segmental Dynamics in Polymers: Functional Forms and Implication on the Glass Transition, *Macromolecules* **38**, 4363–4370.
21. Ngai, K.L., Casalini, R., Capacciolli, S., Paluch, M., and Roland, C.M. (2005) Dispersion of the Structural Relaxation and the Vitrification of Liquids, *Adv. Chem. Phys.*, submitted.
22. Ngai, K.L. and Paluch, M. (2004) Classification of Secondary Relaxation in Glass-formers Based on Dynamic Properties, *J. Chem. Phys.* **120**, 857–873.

ISOBARIC AND ISOCHORIC PROPERTIES OF GLASS-FORMERS

R. CASALINI[1,2,*] and C. M. ROLAND[1]
[1]*Naval Research Laboratory*
Chemistry Division, Code 6120
Washington, DC 20375-5342
USA
[2]*George Mason Univ., Chemistry Department*
Fairfax, VA 22030
USA

Abstract. From new results for the equation of state of dibutylphthalate in combination with previously reported dielectric relaxation data, an analysis of the volume contribution to the dynamics was carried out for this fragile glass-former ($m_P = 84$). We find that the ratio of the isochoric and isobaric fragilities equals 0.75, the largest found to date for any molecular van der Waals liquids, reflecting a strong temperature effect, consistent with the large fragility. The relaxation times superpose when plotted versus temperature times the specific volume raised to the 3.2 power. This is a somewhat larger value of the exponent than expected based on the fragility of DBP. The implication is that for molecular glass-formers, the classical Lennard-Jones repulsive potential may represent the limit for fragile liquids.

Keywords: glass transition, pressure dependence, structural relaxation.

1. Introduction

Although the art of glass-making dates to ancient times and glasses are part of everyday life, efforts to understand the phenomenon at the molecular level continue unabated. The most spectacular change, and thus a primary focus of research, is the dramatic slowing down (by more than 14 decades) of molecular motions in a liquid during vitrification, which transpires without change in molecular arrangement and only modest (a few percent) changes in thermodynamic parameters such as T and V. The lack of any signature of the glass transition means that without a temporal reference, the

*To whom correspondence should be addressed.

S. J. Rzoska and V. A. Mazur (eds.), Soft Matter under Exogenic Impacts, 141–147.

molecular motions of a glass cannot be differentiated from those of the corresponding supercooled liquid. In light of this, much of the effort to understand the glass transition is devoted to finding correlations of the dynamics with other physical properties.

Here we report some new data for dibutylphthalate (DBP) and briefly review some results from the study of small molecular glass formers under high pressure.

2. Fragility and scaling

A popular metric for classifying the structural dynamics is the parameter m, referred to as the steepness index or the fragility[1,2]

$$m = \left.\frac{\partial \log(x)}{\partial (T_g/T)}\right|_{T=T_g} . \tag{1}$$

Here T_g is the glass transition temperature, often defined as $\tau(T_g) = 100\text{s}$, and x can be the viscosity or a relaxation time τ, with the latter measured by dielectric spectroscopy, photon correlation, neutron scattering, Brillouin scattering, or other experimental techniques. The fragility varies among small molecules and polymers in the range $35 \leq m \leq 214$,[3–7] indicating the drastic differences in dynamics among different materials. The fragility has been shown to depend on pressure, generally decreasing with increasing pressure,[8] apart for cases in which hydrogen bonds are present. Using the equation of state (EOS) it is possible to calculate the fragility for isochoric conditions (m_V). The ratio of m_V with the (more usual) isobaric fragility determined at atmospheric pressure (m_P) gives a direct measure of the relative importance of volume to the dynamics at atmospheric pressure. If τ is a unique function of volume, then $m_V = 0$ (no change of τ at constant V); conversely, $m_V = m_P$ if τ depends only on T. Of course, it follows that $0 \leq m_V \leq m_P$.

The isochoric fragility m_V can be determined from m_P using the relation[9,10]

$$m_P = m_V\left(1 + \gamma\alpha_P T_g\right), \tag{2}$$

where α_P is the isobaric volume expansion coefficient and γ is a material constant, obtained from the scaling relation

$$\tau(T,V) = \Im\left(TV^{\gamma}\right). \tag{3}$$

The superposition of relaxation times when expressed as a function of TV^{γ} has been verified for many glass-formers from dielectric relaxation and other measurements.[9–17]

3. Results for DBP

For the case of DBP using published dielectric relaxation times[18] and new data on the EOS it is possible to determine the volume dependence of the relaxation time (Fig. 1)

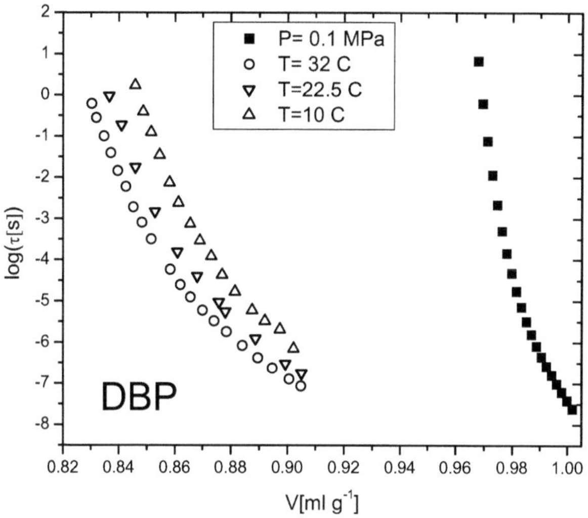

Figure 1. Dielectric relaxation time for DBP versus specific volume. The τ data are from Ref. 18.

and the relative importance of temperature and volume for the dynamics. From Fig. 1 it is evident that the dynamics of DBP is far from being a unique function of V. The relative dependence on T and V is also in this case well represented by the scaling (Eq. (3)). In Fig. 1, the $\tau(T,V)$ data fall all in a master curve when plotted versus the scaling variable TV^{γ} for $\gamma = 3.2$. The value of γ is the smallest value found so far for molecular glass formers without hydrogen bonds.

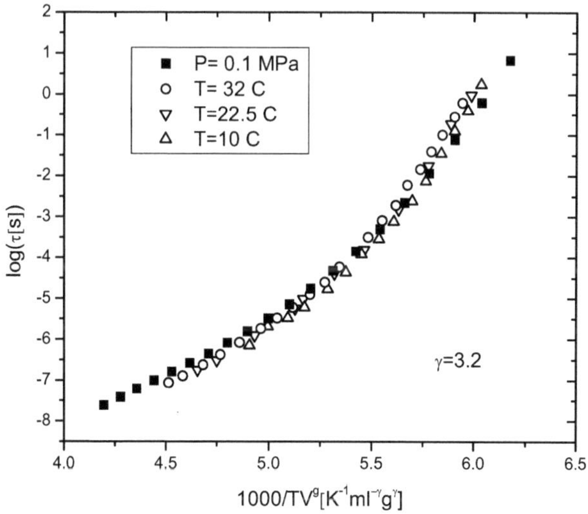

Figure 2. Dielectric relaxation times at $P = 0.1$ MPa (atmospheric pressure) and three fixed temperatures for DBP versus the product TV^{γ}.

From the data at atmospheric pressure, we determined the isochoric fragility $m_P = 83.8$ and $T_g = 177.4$ K. Knowing the isobaric fragility and the isobaric expansion coefficient $\alpha_P = 5.78 \times 10^{-4} C^{-1}$, using equation (2) it is possible to calculate the ratio of isobaric and isochoric fragility $m_V/m_P = 0.75$ and the isochoric fragility, $m_V = 63.1$. The large ratio m_V/m_P indicates that temperature plays a more important role than volume in the dynamics. For small molecule glass formers without hydrogen bonds the values of m_V/m_P have been found in the range $0.38 \le m_V/m_P \le 0.64$;[19] therefore, the value found for DBP is larger than any other found so far. This unusually large value of the ratio m_V/m_P correlates well with the rather small value of γ.

From data on various materials, including molecular and polymeric glass-formers, a correlation was found between m_V and m_P. $m_P = (33 \pm 4) + (0.92 \pm 0.05) m_V$.[20,21] An important consequence of this correlation is that large m_P is associated with large ratios of m_V/m_P; thus, temperature exerts a stronger relative effect on more fragile materials.

In Fig. 3 we report the isobaric versus isochoric fragility for 34 materials including DBP. The results or DBP seems to agree with the general correlation found previously however, the value of m_P is smaller than that observed for other materials having comparable m_V (or larger m_V for comparable m_P).

Finally we tested the correlation between the parameter γ and the isochoric fragility, by plotting (Fig. 4) γ versus the inverse isochoric fragility for DBP and comparing with the values found previously for other molecular glass formers. While in general m_V decreases with increasing γ, DHIQ and PC are outliers.

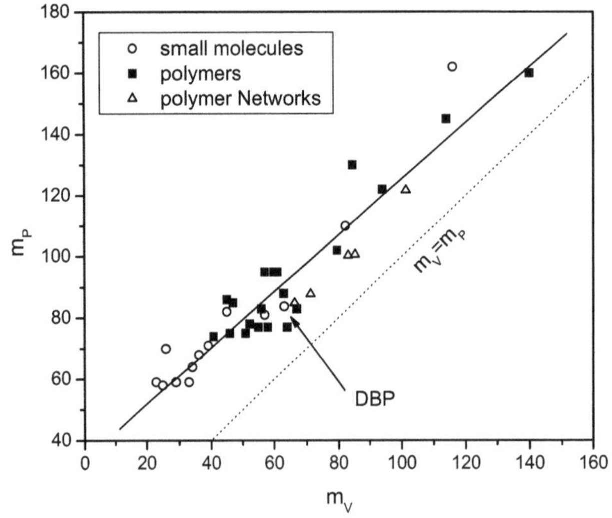

Figure 3. Isobaric fragility m_P (at atmospheric pressure) versus isochoric fragility m_V for 33 materials (in order of increasing m_V).[20]

The implication is that for molecular glass-formers, $\gamma \sim 4$ is a lower limit, rather than $\gamma = 0$. This result deserves further investigation on other fragile liquids.

The dependence of the relaxation time for DBP on volume and temperature shows some interesting features distinguishing it from other materials. The high isobaric fragility is associated with a high isochoric fragility; however; the value of m_V is somewhat higher than expected from the correlation between m_V and m_P established for most glass-formers (solid line in Fig. 4). This implies that volume has a less important role than expected for this fragile material. The importance of volume is also reflected in the value of γ, which is large in comparison to the scaling exponents found for comparably fragile materials (i.e., polymers and associated liquids[19]). But there is an important difference: for polymers, the small values of γ is due to the large fraction of intramolecular bonds, which are insensitive to pressure, thus minimizing the effect of V.[22] On the other hand, for a simple molecule like DBP, the small magnitude of γ reflects directly the nature of the intermolecular interactions.

The idea of using Eq. (3) to analyze relaxation times arose from the notion that for local dynamics, the intermolecular potential could be approximated by an inverse repulsive power law.[13,23] Such a potential is most appropriate for spherically symmetric molecules with a dominant repulsive part, with the scaling exponent γ then interpreted in terms of the exponent 3γ of the inverse power law potential. Certainly, molecular glass-formers lacking hydrogen bonds most closely approximate this ideal case. We find in Fig. 5 that the limiting value of γ approaches the value for the classical Lennard-Jones potential, $3\gamma = 12$.

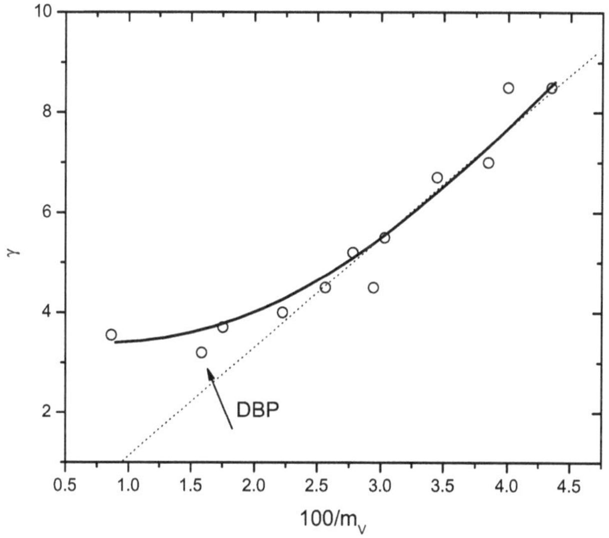

Figure 4. Parameter γ versus the inverse isochoric fragility for 12 amorphous materials. The small-molecule glass-formers (in order of increasing m_V) are: PCB62, BMMPC, BMPC, PCB54, PCB42, KDE, salol, PDE, OTP, PC, DBP, and DHIQ. The dashed line is a linear fit of all data including polymers [Ref. 20], while the solid line is a guide to the eye for the behavior of molecular glass-formers.

4. Summary

From PVT measurements on DBP in combination with previously published dielectric relaxation times, the relative contributions of temperature and volume (density) to the T-dependence of the structural relaxation dynamics were quantified. The ratio of the isochoric and isobaric activation enthalpies = 0.75, which is the highest value of m_V/m_P found to date for molecular glass-formers absent hydrogen bonding. Since DBP is not the most fragile molecular glass-former, this large value is somewhat at odds with the observation that a large m_V/m_P ratio corresponds to large m_P (or large m_V is associated with large m_P). The value of m_V, however, is somewhat larger than expected from the general correlation between m_V and m_P. This means that volume plays a less significant role than expected from correlations of these quantities.

The relaxation times are found to superpose onto a master curve when plotted as a function of $TV^{3.2}$. This exponent is larger than anticipated from the assumption that γ is inversely proportional to m_V (dotted line in Fig. 5). The suggestion is that for molecular glass-formers, γ tends to a limiting value ~4 for large m_V; further work is needed to confirm this. Assuming the exponent γ reflects the nature of the intermolecular repulsive potential, this limiting value of the exponent corresponds to the repulsive term in the classical Lennard-Jones 6–12 potential. Smaller values of γ have been found but only for polymers and hydrogen-bonded liquids.

Acknowledgments

This work was supported by the Office of Naval Research.

References

1. Angell, C. A. (1985) Strong and fragile liquids, in *Relaxation in Complex Systems*, eds. K. L. Ngai and G. B. Wright, Washington DC.
2. Angell, C. A. (1995) Formation of glasses from liquids and biopolymers, *Science* **267**, 1924.
3. The values of m_P for inorganic glass-formers or plastic crystal are not considered since measurements under high pressure are unavailable.
4. Plazek, D. J. and Ngai, K. L. (1991) Correlation of polymer segmental chain dynamics with temperature-dependent time-scale shifts, *Macromolecules* **24**, 1222.
5. Böhmer, R., Ngai, K. L., Angell, C. A., and Plazek, D. J. (1993) Nonexponential relaxations in strong and fragile glass formers, *J. Chem. Phys.* **99**, 4201.
6. Huang, D. and McKenna, G. B. (2001) New insights into the fragility dilemma in liquids, *J. Chem. Phys.* **114**, 5621.
7. The value for PEO ($m_P = 23$) reported previously by Huang and McKenna was calculated from data far from T_g ($\tau < 10$-4s); m is well known to decrease for shorter τ.

8. Casalini, R. and Roland, C. M. (2005) Scaling of the supercooled dynamics and its relation to the pressure dependences of the dynamic crossover and the fragility of glass formers, *Phys. Rev. B* **71**, 041210.
9. Casalini, R. and Roland, C. M. (2004) Thermodynamical scaling of the glass transition dynamics, *Phys. Rev. E* **69**, 062501
10. Casalini, R. and Roland, C. M. (2004) Scaling of the segmental relaxation times of polymers and its relation to the thermal expansivity, *Colloid Polym. Sci.* **283**, 107.
11. Roland, C. M., Casalini, R., and Paluch, M. (2004) Effects of the volume and temperature on the global and segmental dynamics in poly(propylene glycol) and 1,4-polyisoprene, *J. Polym. Sci. Polym. Phys. Ed.* **42**, 4313.
12. Roland, C. M. and Casalini, R. (2005) Density scaling of the dynamics of vitrifying liquids and its relationship to the dynamic crossover, *J. Non-Cryst. Solids* **351**, 2581.
13. Tölle, A. (2001) Neutron scattering studies of the model glass former ortho-terphenyl, *Rep. Prog. Phys.* **64**, 1473.
14. Dreyfus, C., Aouadi, A., Gapinski, J., Matos-Lopes, M., Steffen, W., Patkowski, A., and Pick, R. M. (2003) Scaling the alpha-relaxation time of supercooled fragile organic liquids, *Physical Review E* **68**, 011204.
15. Dreyfus, C., Le Grand, A., Gapinski, J., Steffen, W., and Patkowski, A., (2004) Temperature and pressure study of Brillouin transverse modes in the organic glass-forming liquid orthoterphenyl, *Eur. Phys. J.B.* **42**, 309.
16. Alba-Simionesco, C., Calliaux, A., Alegria, A., and Tarjus, G. (2004) Scaling out the density dependence of the relaxation in glass-forming polymers, *Europhys. Lett.* **68**, 58.
17. Budzien, J., McCoy, J. D., and Adolf, D. B. (2004) General relationships between the mobility of a chain fluid and various computed scalar metrics, *J. Chem. Phys.* **121**, 10291.
18. Sekula, M., Pawlus, S., Hensel-Bielowka, S., Ziolo, J., Paluch M., and Roland, C. M. (2004) Structural and secondary relaxations in supercooled di-*n*-butylphthalate and di-isobutylphthalate at elevated pressure, *J. Phys. Chem. B*, **108**, 4997.
19. Roland, C. M., Hensel-Bielowka, S., Paluch M., and Casalini, R. (2005) Supercooled dynamics of glass-forming liquids and polymers under hydrostatic pressure, *Rep. Prog. Phys.* **68**, 1405–1478.
20. Casalini, R. and Roland, C. M. (2005) Why liquids are fragile, *Phys. Rev. E* **72**, 031503.
21. Casalini, R., Mc Grath, K. J., and Roland, C. M. (2005) Isobaric and isochoric properties of an extremely fragile glass-former, *J. Non Crystalline Solids*, submitted.
22. Casalini, R. and Roland, C. M. (2005) Temperature and density effects on the local segmental and global chain dynamics of poly(oxybutylene), *Macromolecules* **38**, 1779.
23. Hoover, W. H. and Ross, M. (1971) Statistical theories of melting, *Contemp. Phys.* **12**, 339.

INFLUENCE OF DIFFERENCES IN MOLECULAR STRUCTURE ON BEHAVIOR OF α AND β RELAXATION PROCESSES IN DIISOOCTYL PHTHALATE AND DIISOOCTYL MALEATE

[1]S. PAWLUS, [1]M. PALUCH, [1]M. MIERZWA, [1]S. HENSEL-BIELOWKA, [1]E. KAMIŃSKA, [1]K. KAMIŃSKI, [1]S. J. RZOSKA, [2]M. MAŚLANKA

[1]Institute of Physics, Silesian University, ul. Uniwersytecka 4, 40-007 Katowice, Poland ; [2]Institute of Chemistry, Silesian Universitym ul. Szkolna 9, 40-007 Katowice, Poland

Abstract. Dielectric relaxation measurements have been performed to study effect of the molecular structure on the primary and secondary relaxations in two materials with very similar structure: diisooctyl phthalate (DIOP) and diisooctyl maleate (DIOM). The dielectric spectra were measured over ten decades of frequency and a broad range of temperature. The results show that the temperature dependence of the α-relaxation times cannot be described by a single Vogel–Fulcher–Tammann equation and the temperature dependence of the β-relaxation times in the glassy state obey the Arrhenius law with different activation energies, i.e., $E_A/k = 4231$ K and $E_A/k = 2917$ K for DIOP and DIOM, respectively. Moreover, third relaxation process visible as an excess wing of the high frequency part of the α loss peak was identified in dielectric spectra of DIOM.

Keywords: Glass transition, dielectric spectroscopy, high pressures.

1. Introduction

It is well known fact that almost every liquid can be supercooled or suppressed below melting point avoiding crystallization. However, rate of cooling or pressurizing required to form glassy state depend strongly on the molecular structure of a material. Even within family of substances with small differences in the structure very pronounced differences in relaxation and transport properties are often observed. Consequently, it means that not only thermodynamic variables but also structure and molecular weight are important in understanding molecular dynamics of the liquids. This situation is clearly visible in many different classes of materials: polyalcohols,[1,2] polymers,[3–5] liquid crystals.[6]

S. J. Rzoska and V. A. Mazur (eds.), Soft Matter under Exogenic Impacts, 149–159.

Very recently, the influence of the molecular structure difference on molecular dynamics was also examined in phthalates. This class of materials is particularly interesting because it includes a large number of representatives exhibiting various relaxation processes. Moreover, phthalates are usually good glass formers, i.e., they can be easily supercooled or superpressed.[7] Due to the fact that these materials posses strong permanent dipole moment it makes them very convenient systems for broadband dielectric measurements.[8]

The slowest and the most prominent relaxation process observed in the dielectric spectra of supercooled phthalates is the α process. This process is associated with the relaxation of the liquid structure and is observed as asymmetric and broad peak in relaxation spectra, moving with decreasing temperature or increasing pressure toward lower frequencies. Temperature dependence of the α-relaxation times is often described by the Vogel–Fucher–Tamman (VFT) relation:

$$\tau = \tau_\infty \exp\frac{DT_0}{T - T_0}, \tag{1}$$

where D is strength parameter and T_0 is identified with the temperature of the ideal glass transition.

The shape of the α process is usually described by the Kohlrausch–Williams–Watts (KWW) fractional exponential function, $\phi(t)=\exp[-(t/\tau_\alpha)^{1-n}]$, used to fit the dielectric loss peak by the one sided Fourier transform,

$$\varepsilon''(\omega) = \Delta\varepsilon \int_0^\infty dt \sin(\omega t)(-d\phi/dt). \tag{2}$$

As the main relaxation process moves to the lower frequencies, next relaxation process emerges few decades above the maximum of the α relaxation. This relaxation is called secondary or β process. It appears in the vicinity of the glass transition (about $1.2T_g$). The relaxation peak of the β process has broad and symmetric shape, and its relaxation strength is relatively small in comparison to the α-process. The temperature dependence of the β-relaxation times under T_g obeys the Arrhenius law:

$$\tau_\beta = \tau_0 \exp\left(E/k_B T\right). \tag{3}$$

However, very recently Dyre and Olsen claimed that the temperature dependence of the β–relaxation times of tri-propylene glycol exhibit the minimum in the vicinity of the glass transition. Their observation was confirmed by measurements under high pressure.[9,10]

Although temperature dependence of the secondary relaxation times in various glass formers obey the same pattern of behavior below T_g (i.e., the Arrhenius law) the molecular mechanisms underlying this process can be different for various glasses. In some materials the specific motion of some part of the molecules contributes to the secondary relaxation, thus, it has intramolecular character. In other, the β process is due to the local motion of essentially the entire molecule.[11] Such type of secondary relaxation has intermolecular character and was recently classified as the Johari–Goldstein (JG) relaxation. In addition it was shown that there is also quite large number of materials which have more than one secondary process. In some cases they can be identified in dielectric spectra only under specific thermodynamic conditions (e.g., at elevated pressure or after aging).[12]

Above mentioned types of secondary processes exist in phthalates and the physical properties strongly depend on the molecular structure. For example, both JG and the non-JG type of β relaxation are observed in dioctyl phthalate (DOP) whereas only non-JG relaxation can be identified in isomer of DOP, i.e., in diisooctyl phthalate (DIOP). On the other hand, in DOP the minimum of the temperature dependence of the non-JG relaxation times is observed and in DIOP minimum does not exist.[13] Activation energy and strength of the non-JG relaxation process is also markedly different in these two materials although temperature behavior of the primary relaxations is almost the same. In order to deeper our knowledge about the vitrification process it is also important to study more thoroughly connection between changes in structure of the molecules and behavior of relaxation processes.

The main goal of this paper is to compare some features of the α and β relaxation in two structurally similar materials Diisooctyl Phthalate and Diisooctyl Maleate (DIOM). They have the same alkyl chains whereas there is no phenyl ring in DIOM (see structure in Fig. 2).

2. Experimental

Diisooctyl phthalate (99.5%) was supplied by Aldrich Chemicals and used as purchased. DIOM was synthesized in Institute of Chemistry, University of Silesia. The molecular structures of both samples are displayed in inset to Fig. 2. From temperature dependence of the α-relaxation we have determined the $T_g(\tau_\alpha=100$ s) which is equal to 191 K and 171 K for DIOP and DIOM, respectively.

The temperature-dependent dielectric measurements were carried out using the experimental set-up made by Novocontrol GmbH. This system was equipped with a Novo-Control GMBH Alpha dielectric spectrometer (10^{-2}–10^{7} Hz) and Agilent 4291B impedance analyzer (10^{6}–1.8×10^{9} Hz) for measurement of the complex dielectric permittivity $\varepsilon^*(\omega)=\varepsilon'(\omega)-i\varepsilon''(\omega)$. The sample was placed in a parallel plate cell (diameter 15 mm, gap 0.05 mm – Alpha, diameter 5 mm, gap 0.1 mm – Agilent). The temperature was controlled using a nitrogen-gas cryostat, with temperature stabilization better than 0.1 K.

3. Results and discussion

Representative dielectric loss spectra of DIOM at various temperatures from 123 up to 312 K are shown in Fig. 1. At higher temperatures, only the α relaxation peak and dc conductivity are observed. As temperature is decreased, the β relaxation peak starts to emerge in dielectric spectra.

Figure 2 shows the dielectric loss for DIOP measured at T = 223, 208, and 196 K as well as spectra for DIOM at T = 206, 191, and 179 K. In Fig. 2 we compare spectra with almost the same α-relaxation time. In addition the data have been scaled with the α relaxation peak amplitude of DIOM in order to facilitate comparisons of the loss processes. Although shapes of the α processes are the same for both liquids it is clearly visible that there is a large variation of the relative position and strength of the secondary processes. In DIOP relative distance between the two relaxations is smaller than in DIOM; however, amplitude of β peak is much greater. It is similar behavior as in the case of DOP. In addition we also observe marked difference in temperature behavior of the α-relaxation times between DIOM and DIOP.

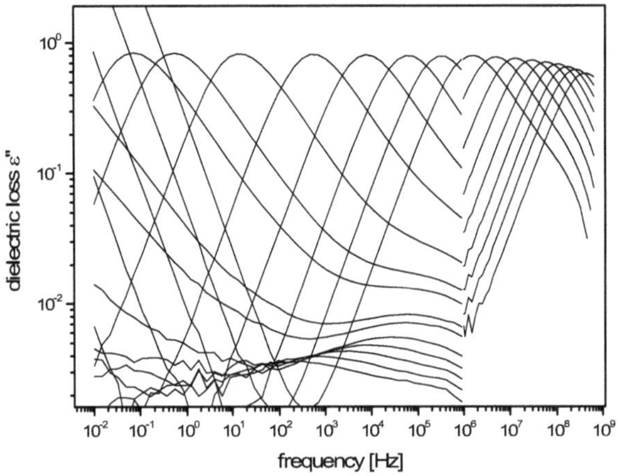

Figure 1. The selected dielectric loss spectra for DIOM at different temperatures.

To determine the temperature dependence of the relaxation time for each processes, we analyzed the spectra using a Cole–Cole[14] relaxation function to describe the secondary peak, and a Havriliak–Negami[15] function for the primary relaxation. At lower temperatures, where only the high-frequency side of the α peak is present in dielectric spectra, a power law function ($\varepsilon''_\alpha \propto f^{-\gamma}$) was used to describe the α relaxation. The relaxation times for both processes, determined from the frequency of the peak maximum ($\tau = 1/2\pi f_{max}$), are presented in Fig. 3. At first sight it is visible that the temperature dependence of the secondary relaxation times starts to deviate from simple Arrhenius behavior in the vicinity of T_g for DIOP and DIOM.

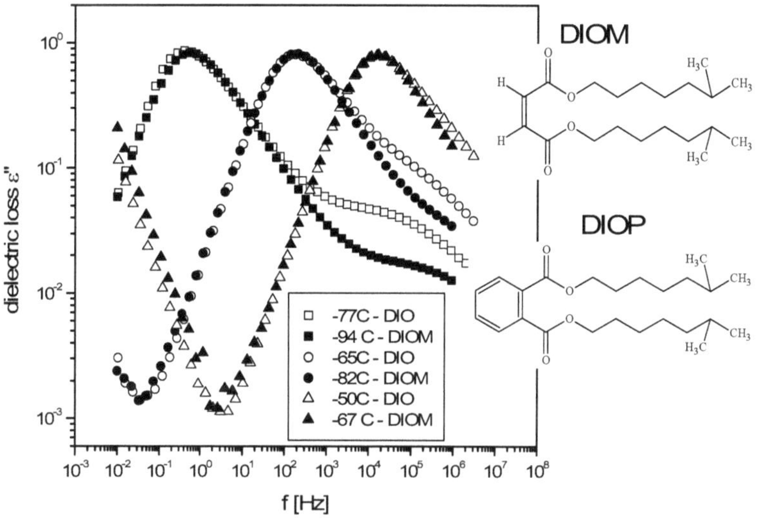

Figure 2. A comparison of the dielectric loss spectra obtained from isobaric measurements for DIOP and DIOM. The data for DIOP was slightly vertically shifted to achieve superimposition. The chemical structures of studied substances are depicted.

Since both glass temperatures and temperature dependences of the relaxation times are different for both materials (see Table 1) we present data from main part of Fig. 3 in the lower inset as T_g-normalized Arrhenius plot. It is a common approach to classify the temperature behavior of α-relaxation times, especially to quantifying the degree of departure from the Arrhenius dependence. This classification scheme has been proposed by Angel,[16–18] who also proposed term "fragility plot". According to "fragility" classification, strong materials are those having a nearly Arrhenius dependence of relaxation times on T_g/T plot, whereas for fragile materials relaxation times strongly depend on the normalized temperature variable. Quantitatively, fragility is expressed by steepness index $m = d\log\tau(T_g)/d(T_g/T)_p$. In the case of examined materials fragility of DIOM is smaller than DIOP ($m_{DIOM} < m_{DIOP}$) although both liquids can be classified as the "fragile materials". It is also visible that temperature dependences of the secondary relaxation times below T_g are slightly different.

They were parameterized by Arrhenius law (3). Estimated activation energy E_A for DIOP is, greater than for DIOM (see Table 1).

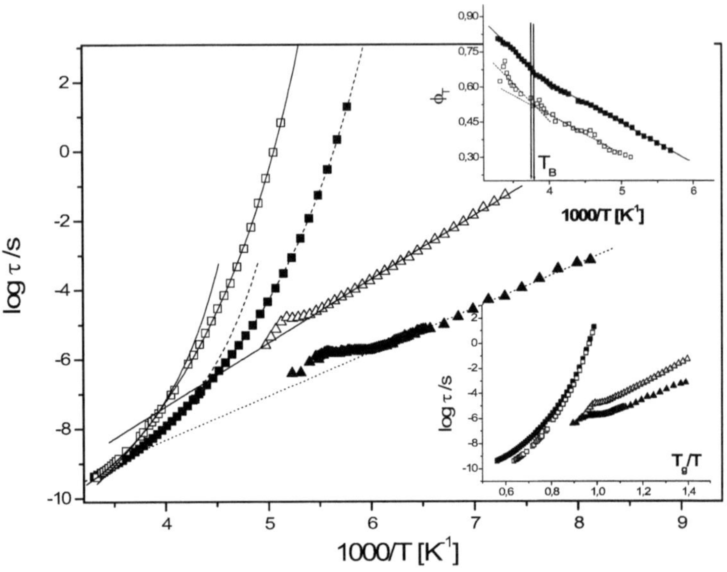

Figure 3. Arrhenius plot of the α- and β-relaxation times for DIOM (solid points) and DIOP (empty points) (solid and dashed lines and curves represent fit by the VFT and activation law). Lower inset present the same data rescaled to glass temperature T_g. Upper inset show Stickel plot of the data.

Analyzing the behavior of the α-relaxation times it turned out that the temperature dependence of τ_α cannot be described by means of only one VFT law (1) in the whole temperature range. Systematic deviations of the temperature dependence of relaxation times from the VFT behavior are observed at $T > 1.37T_g$ and $T > 1.51T_g$ for DIOP and DIOM, respectively. By use of a model-independent derivative operator introduced by Stickel and coworkers[19] $\phi_T = [d(\log(\tau))/d(1000/T)]^{-\frac{1}{2}}$ we find that the high-temperature region region of the temperature dependence of the α-relaxation times is described by a second VFT law (see upper inset in Fig. 3). The transition between these different temperatures dependencies (point of intersection of two straight lines with finite slope on the Stickel plot, obtained from the VFT_1 and VFT_2) defines a crossover temperature $T_B > T_g$. From analysis of the relaxation data for different glassforming liquids Novikov and Sokolov[20] noted that relaxation time at T_B is almost a universal constant, equal $10^{-6.5\pm0.5}$ s. However, there are a number of materials, for which the relaxation time corresponding to the change of dynamics is significantly different from $10^{-6.5\pm0.5}$ s.[21] It is worth to emphasize that in propylene carbonate temperature and pressure behavior of α-relaxation times, in combination with *PVT* data, reveal that the change in dynamics occurring above T_g arises when the liquid acquire a given degree of mobility (reflected in the given value of relaxation time), independent of the particular combination of T, P, and V.[22] In the case of studied herein materials we found that the temperatures T_B are nearly the same.

Due to a better separation of α and β processes as well as lower amplitude of the secondary relaxation in DIOM we also analyze dielectric loss spectra to check if the excess wing (EW observed previously in DOP) can be revealed in this material. Excess wing is manifested as an extra contribution on the high-frequency side of the α relaxation peak. The origin of the EW is one of the most disputed aspects of the glass transition dynamics. Early works considered the EW to be an inherent part of the main relaxation. On the other hand, in some liquids the EW can be transformed into a flank of a nascent peak by physical aging[23,24] or by high pressure.[8] Such results indicate that the EW may be a submerged relaxation process.

A theoretical tool for prediction the JG relaxation times τ_{JG} comes from the the coupling model (CM) by Ngai. According to this model the JG relaxation time can be calculated as

$$\tau_{JG} = t_c^{\,n} \tau_\alpha^{1-n}, \tag{4}$$

where t_c (≈2ps for molecular liquids) is a crossover time from primitive to cooperative relaxation and ($1-n=\beta_{KWW}$) is the KWW stretch exponent. The CM provides particularly interesting results in the case of liquids revealing the EW in the dielectric loss spectra. It turns out that the value of predicted τ_{JG} correspond to the frequency range of the EW. This finding supports the notion that the EW is a JG relaxation but hardly to observe as separate peak due to overlap of the close laying α peak.

Four representative dielectric loss spectra are shown in inset of the Fig 4, one measured at T_g and other three at lower temperatures. In the Fig. 4 the KWW fit, with $\beta_{KWW} = 0.62$ is also depicted (solid line). From the spectra depicted in the inset no EW is clearly evident. However, some deviation from the KWW power law can be noted in the high frequency flank of the α process. In the main part of the figure we show a master curve, constructed by shifting along the abscissa the three lower temperature spectra. In this representation the data provides a strong suggestion of the presence of an excess wing. Moreover, the JG relaxation time (indicated by arrow) which was calculated by Eq. (4) for the estimated value of β_{KWW} lies in the area of the EW.

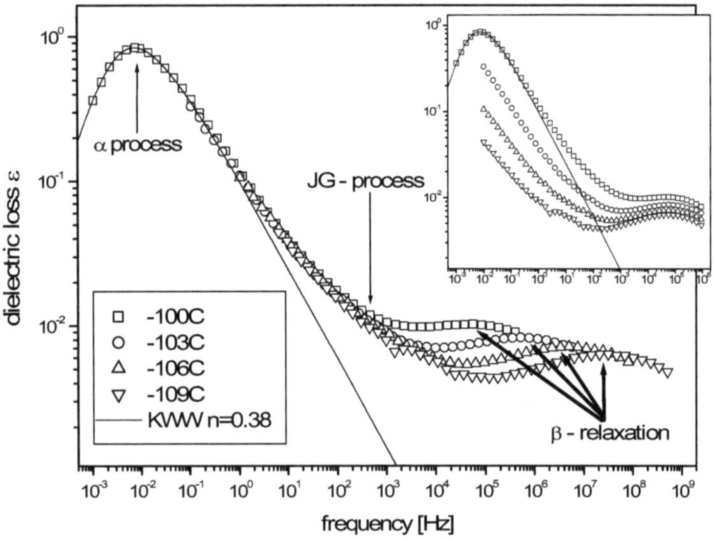

Figure 4. Dielectric loss spectra for DIOM at the four indicated temperatures in the vicinity of T_g after shifting in frequency to superimpose the high frequency flank of the main relaxation. The solid curve is a KWW function with $1-\beta=0.38$. In the inset are showed the same spectra before superpositioning.

From these results, we conclude that the EW in DIOM could be a sub-merged JG peak. On the other hand, the fact that the EW cannot be identified in the dielectric loss spectra of DIOP can be a consequence of the proximity between α and β peaks.

TABLE 1. The values of the characteristic parameters for studied compounds.

Name of material	T_g [K]	T_B [K]	$\log \tau(T_B)$	m	E_A/k [K]
DIOP	191	261	−7.91	69	$4231_{\pm 25}$
DIOM	171	257.8	−8.26	65	$2917_{\pm 50}$

4. Conclusions

We analyzed the relaxation dynamics in materials with similar molecular structure. They have the same alkyl chains whereas there is no phenyl ring in DIOM. The difference in the structure influences on glass transition temperature, fragility, activation energy of the β-relaxation times, the position, and the amplitude of the β peak. On the other hand, the differencein the in the structure of studied molecules does not influence the shape parameters of the primary relaxation process. Absence of the phenyl ring in DIOM molecule causes better separation between α and β peaks and lowering of amplitude of the secondary peak. Moreover, changes into the molecular structure enables the observation of the excess wing on the high frequency shoulder of the primary relaxation peak in the vicinity of T_g. Based on the CM we estimated position of the JG relaxation times which lies in the area of the EW. It suggests that the EW could be a submerged JG-relaxation peak.

Acknowledgments

The support of the Polish State Committee for Scientific Research (Project No. 1P03B 075 28) are gratefully acknowledged. One of us (S. P.) also acknowledges a domestic scholarship from Foundation for Polish Science (FNP'05). S. J. Rzoska was supported by Polish State Committee for Scientific Research (KBN, project no 2PO 3B 034 25, resp: S. J. Rzoska).

References

1. Döß, A., Paluch, M., Sillescu, H., and Hinze, G. (2002) From strong to fragile glass formers: secondary relaxation in polyalcohols, *Phys. Rev. Lett.* **88**, 095701; Döß, A., Paluch, M., Sillescu, H., and Hinze, G. (2002) Dynamics in supercooled polyalkohols: primary and secondary relaxation, *J. Chem. Phys.* **117**, 6582.
2. Paluch, M., Roland, C. M., Pawlus, S., Ngai, K. L., and Zioło, J. (2003) Does the Arrhenius temperature dependence of the Johari–Goldstein relaxation persist above T_g? *Phys. Rev. Lett.* **91**, 115701-4.
3. Paluch, M., Roland, C. M., and Pawlus, S. (2002) Temperature and pressure dependence of the α-relaxation in polymethylphenylsiloxane, *J. Chem. Phys.* **116**, 10932; Paluch, M., Pawlus, S., and Roland, C. M. (2002) Pressure and temperature dependence of α-relaxation in poly(methyltolylsiloxane), *Macromolecules* **35**, 7338; Pawlus, S., Roland, C. M., Rzoska, S. J., Zioło, J. and Paluch, M. (2003) Effect of temperature and pressure on segmental relaxation in polymethylphenylsiloxane, *Rubber Chem. Techno.* **76**, 1106.
4. Leon, C., Ngai, K. L., and Roland C. M. (1999) Relationship between the primary and secondary dielectric relaxation processes in propylene glycol and its oligomers, *J. Chem. Phys.* **110**, 11585.
5. Mattsson, J., Bergman, R., Jacobsson, P., and Börjesson, L. (2003) Chain-length-dependent relaxation scenerios in an oligomeric glass-forming system: From merged to well-separated α and β loss peaks, *Phys. Rev. Lett.* **90**, 075702-4.
6. Rzoska, S. J., Paluch, M., Pawlus, S., Drozd-Rzoska, A., Zioło, J., Jadżyn, J., Czupryński, K., and Dąbrowski, R. (2003) Complex dielectric relaxation in supercooling and superpressing liquid crystalline chiral isopentylcyanobiphenyl, *Phys. Rev. E* **68**, 031705; Drozd-Rzoska, A., Rzoska, S. J., Paluch, M., Pawlus, S., Zioło, J., Santangelo, P. G., Roland, C. M., Czuprynski, K., and Dabrowski, R. (2005) Mode

coupling behavior in glass-forming liquid crystalline isopentylcyanobiphenyl, *Phys. Rev. E* **71**, 011508.

7. Cook, R. L., King, H. Ee., Jr., Herbst, Ch. A., and Herschbach, D. R. (1994) Pressure and temperature pependent viscosity of two glass forming liquids: Glycerol and dibutyl phthalate, *J. Chem. Phys.* **100**, 5178.
8. Paluch, M., Ziolo, J., Rzoska, S. J., and Habdas, P. (1996) High-pressure and temperature dependence of dielectric relaxation in supercooled di-isobutyl phthalate, *Phys. Rev. E* **54**, 4008; Paluch, M., Ziolo, J., Rzoska, S. J., and Habdas, P. (1997) The influence of pressure on dielectric relaxation for phthalate derivatives in the supercooled state, *J. Phys.: Condens. Matter* **9**, 5485; Pawlus, S., Paluch, M., Sekuła, M., Ngai, K. L., Rzoska, S. J., and Ziolo, J. (2003) Changes in dynamic crossover with temperature and pressure in glass-forming diethyl phthalate, *Phys. Rev. E* **68**, 021503; Paluch, M., Sekuła, M., Pawlus, S., Rzoska, S. J., Roland, C. M., and Ziolo, J. (2003) Test of the Einstein-Debye relation in supercooled dibutyl phthalate at pressures up to 1.4 GPa, *Phys. Rev. Lett.* **90**, 175702; Hensel-Bielówka, S., Sekuła, M., Pawlus, S., Psurek, T., and Paluch, M. (2004) Influence of molecular structure on dynamics of secondary relaxation in phthalates, NATO Science Series Kluwer Academic Publishers; Sekula, M., Pawlus, S., Hensel-Bielowka, S., Ziolo, J., Paluch, M., and Roland, C. M. (2004) Structural and secondary relaxations in supercooled di-*n*-butylphthalate and di-isobutylphthalate at elevated pressure, *J. Phys. Chem. B* **108**, 4997; Paluch, M., Pawlus, S., Hensel-Bielowka, S., Kaminski, K., Psurek, T., Rzoska, S. J., Ziolo, J., and Roland, C. M. (2005) Effect of glass structure on the dynamics of the secondary relaxation in diisobutyl and diisooctyl phthalates, *Phys. Rev. E* **72**, 224205.
9. Dyre, J. C. and Olsen, N. B. (2003) Minimal model for beta relaxation in viscous liquids, *Phys. Rev. Lett.* **91**, 155703.
10. Pawlus, S., Hensel-Bielowka, S., Grzybowska, K., Ziolo, J., and Paluch M. (2005) Temperature behavior of secondary relaxation dynamics in tripropylene glycol, *Phys. Rev. E* **71**, 174107; Grzybowska, K., Grzybowski, A., Pawlus, S., Hensel-Bielowka, S., and Paluch, M. (2005) Dielectric relaxation processes in water mixtures of tripropylene glycol, *J. Chem. Phys.* **123**, 204506.
11. Ngai, K. L. and Paluch, M. (2004) Classification of secondary relaxation in glass-formers based on dynamic properties, *J. Chem. Phys.* **120**, 857.
12. Casalini, R. and Roland, C. M. (2003) Pressure Evolution of the Excess Wing in Type-B Glass Former, *Phys. Rev. Lett.* **91**, 015702; Casalini, R. and Roland, C. M. (2004) Excess wing in the dielectric loss spectra of propylene glycol oligomers at elevated pressure, *Phys. Rev. B* **69**, 094202; Pawlus, S., Hensel-Bielowka, S., Paluch, M., Casalini R., and Roland, C. M. (2005) Hydrogen bounding and secondary relaxations in propylene glycol trimer, *Phys. Rev. B* **72**, 064201.
13. Kaminska, E., Kaminski, K., Hensel-Bielowka, S., Paluch, M., and Ngai, K. L. – unpublished.
14. Cole, K. S. and Cole, R. H. (1941) Dispersion and absorption in Dielectrics I. Alternating Current Characteristics, *J. Chem. Phys.* **9**, 341.
15. Havriliak, S. and Negami, S. (1966) A Complex plane analysis of alpha-dispersions in some polymer systems, *J. Polym. Sci., Part C: Polym. Symp.* **14**, 89.

16. Angell, C. A. (1991) Relaxation in liquids, polymers and plastic crystals—strong/fragile patterns and problems, *J. Non-Cryst. Solids* **131–133**, 13.
17. Stickel, F., Fischer, E. W., and Richert, R. (1995) Dynamic of glass-forming liquids. I. Temperature-derivative analysis of dielectric relaxation data, *J. Chem. Phys.* **103**, 6251.
18. Novikov, V. N. and Sokolv, A. P. (2003) Universality of the dynamic crossover in glass-forming liquids: A "magic" relaxation time, *Phys. Rev. E* **67**, 031507.
19. Casalini, R., Paluch, M., and Roland, C. M. (2003) Dynamic crossover in supercooled liquids induced by high pressure, *J. Chem. Phys.* **118**, 5701.
20. Pawlus, S., Casalini, R., Roland, C. M., Paluch, M., Rzoska, S. J., and Ziolo, J. (2004) Temperature and volume effects on the change of dynamics in propylene carbonate, *Phys. Rev. E* **70**, 061501.
21. Schneider, U., Brand, R., Lunkenheimer, P., and Loidl, A. (2000) Excess wing in the dielectric loss of glass formers: A Johari–Goldstein β relaxation? *Phys. Rev. Lett. Lett.* **84**, 5560.
22. Lunkenheimer, P., When, R., Riegger, Th., and Loidl, A. (2002) Excess wing in the dielectric loss of glass formers: further evidence for a β-relaxation, *J. Non-Cryst. Solids* **307–310**, 336.
23. Roland, C. M., Casalini, R., and Paluch, M. (2003) Isochronal temperature-pressure superpositioning of the α-relaxation in type-A glass formers, *Chem. Phys. Lett.* **367**, 259.
24. Hensel-Bielowka, S., Pawlus, S., Roland, C. M., Ziolo, J., and Paluch, M. (2004) Effect of large hydrostatic pressure on the dielectric loss spectrum of type-A glass formers, *Phys. Rev. E* **69**, 050501.

ORIENTATIONALLY DISORDERED GLASSY PHASES

JOSEP Ll. TAMARIT*
Department of Physics and Nuclear Engineering, ETSEIB, Laboratory of Characterization of Materials, Diagonal, 647 08028 Barcelona, Catalonia, Spain

SEBASTIAN PAWLUS, ALEKSANDRA DROZD-RZOSKA and SYLWESTER J. RZOSKA
Institute of Physics, Silesian University, ul. Uniwersytecka 4 40-007 Katowice, Poland

Abstract. Broadband dielectric spectroscopy (BDS) studies for a remarkable temperature range (230 K) in a supercooled orientationally disordered phase given rise to an orientational glass have been performed in mixed crystal formed between NPG ($(CH_3)_2C(CH_2OH)_2$) and NPA ($(CH_3)_3C(CH_2OH)$) compounds. The α-relaxation slows down and the width of the dielectric spectra becomes broader with decreasing temperature without any additional relaxation process, making easy the fitting procedure of the relaxation loss spectra. The existence of three dynamic domains is made evident: For the high-temperature region the relaxation time obeys an Arrhenius behavior, while when decreasing temperature two crossover temperatures define the temperature intervals for which two different VFT regimes are clearly visible. The shape parameters α and $\alpha \cdot \beta$ according to the fit of HN function have been determined. Approaching T_g, the common feature of many structural glass-forming materials (decrease of the shape parameters with temperature) is found, indicating the strong deviation from the Debye behavior. Both parameters seem to reach the claimed universal value of 0.5.

According to MCT theory, the temperature dependence of the relaxation time seems to validate the general trend given by the MCT equation. But a fine-tuning reveals the existence of three different regimes governed by three different exponents γ.

Keywords: Orientational disorder, plastic crystal, glass transition, orientational glass, dielectric spectroscopy, dynamic crossovers.

---

* To whom correspondence should be addressed. Josep Ll. Tamarit, jose.luis.tamarit@upc.edu

S. J. Rzoska and V. A. Mazur (eds.), Soft Matter under Exogenic Impacts, 161–187.

1. Introduction

1.1. GLASS TRANSITION AND STRUCTURAL GLASSES

By cooling a liquid fast enough to prevent crystallization; it could result in a supercooled metastable liquid in which the material retains its local structure. Subsequent cooling ultimately forces the system to freeze-in a disordered state to become a nonequilibrium state in a thermodynamic sense which corresponds to the nonergodic state of the ergodic liquid; such a nonergodic state is related to a system for which some property is time-dependent and is often referred to as *structural glass* (*SG*) or *liquid glass*. This state and, more particularly, the microscopic description of the precursor cooperative mechanisms which originate at the high-temperature domain in the low time-scale (picoseconds to nanoseconds), i.e., far above the glass transition temperature T_g, for which the time-scale is assumed to be around 100 s, is an old and unresolved question in the field of condensed matter of physics.[1,2]

The dynamics describing the liquid, the supercooled liquid and the structural glass involves generally a characteristic bimodal distribution of relaxation time, as shown by means of dielectric spectroscopy. The low-frequency peak originates from the collective or correlated dynamics of the α-process involving many molecules. The high-frequency peak is attributed to localized diffusion and is commonly referred as the Johari-Goldstein relaxation[3] or β-process and it has been shown that these secondary relaxation exists even in many rigid molecular glassformers without any intramolecular degree of freedom. It is now commonly suspected, mainly from many theoretical and numerical results, the potential role of the Johari-Goldstein relaxation (non-cooperative process) on the cooperative α-process, responsible of the vitrification.[4–18]

Nevertheless, a different point of view has been also proposed in which the β-process is the noncooperative motion and the α process the cooperative motion of a group of atoms.[19–23]

1.2. ORIENTATIONAL GLASSES

An open question is the extent to which the loss of periodicity when going from an ideal crystal to a true structural glass, controls the microscopic dynamics and gives rise to the variety of phenomena characteristic of glasses. To shed some light onto these aspects, particularly concerning the issue of how orientational and translational disorder can control the dynamics of the vitrification process, the study of systems displaying exclusively one type of disorder should be considered as the simplest way.

There is a kind of molecular materials for which reorientational motions of the molecules in solid crystalline state could occur. The molecules forming such materials are usually *globular* in shape, a fact that provides little steric hindrance for reorientationl freedom. These compounds are known as *plastic crystals* as well as Orientational Disordered Crystals (ODIC for short) and the solid state or phase which presents this orientational disorder can be denoted as an OD-state. For such a state, the molecular centres of mass display long-range translational order, but the molecules are dynamically disordered in orientation.

The OD-state can be supercooled fast enough to prevent phase transformations to the low-temperature ordered crystalline state, resulting in a supercooled metastable OD-state in which the material retains the characteristic orientational disorder and the regular lattice of the stable OD-state. In the same way as for structural glasses, subsequent cooling ultimately forces the system to freeze-in an orientational disordered state as well as crystalline state which was referred to as a *glassy state*[24] or as an *orientational glass* (*OG*)[25] since only the orientational degrees of freedom are frozen-in. ODIC are often considered as model systems for structural glass formers and much simpler to treat in theoretical and simulation approaches to the glassy state.

From a dynamic point of view, the dielectric spectra are dominated by the α-relaxation process like in structural glasses. Because of the presence of a long-range translational order, such orientationally disordered crystals offer the possibility of a fully analytical treatment for a number of static and dynamical properties also considered as being characteristic of the glassy state, such as the softening of the elastic constants due to orientational disorder, the frequency dependence of the sound velocity and the well-known anomalies in the dielectric behavior seen in most glasses.[26]

The ideal material to be studied would appear when for the same molecular system different phases can show glassy behavior, in a close temperature domain, one phase retaining the long-range translational order and the other not. Unfortunately, there is only a limited number of experimental studies in orientational glasses which can be supercooled without special experimental conditions giving rise to different glassy states emerging from the supercooled liquid (structural glass) or from the supercooled orientationally disordered phase (orientational glass), the ethanol being one of the best examples.[27–39]

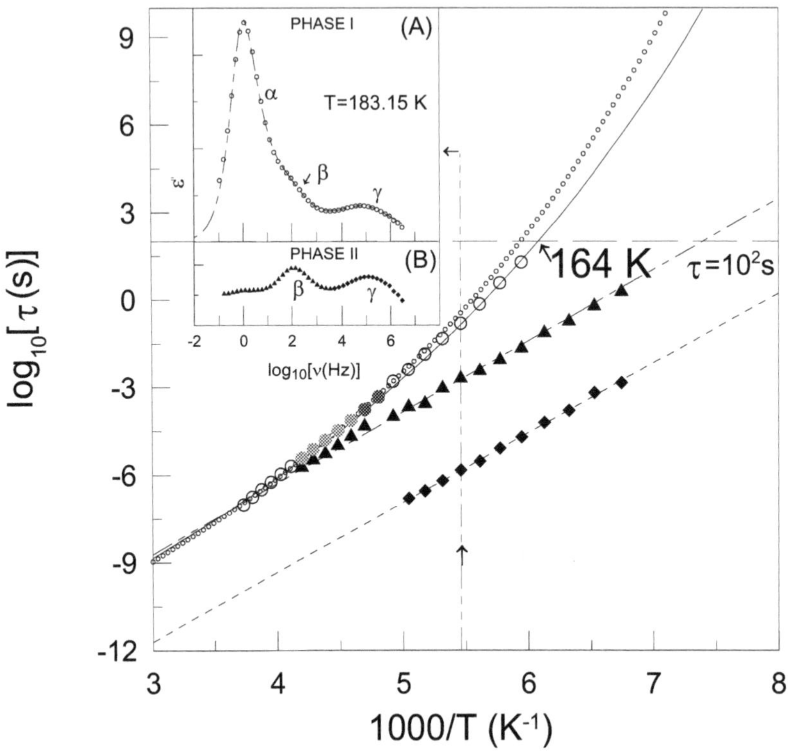

Figure 1. Relaxation time of α-, β- and γ-processes in an Arrhenius plot. The solid line corresponds to the VFT-fitting of the α-relaxation data in phase I (open circles and solid-grey circles) and the dotted-line corresponds to the Brands' VFT-fitting according to Brand et al.. [44] The dot–dashed and dashed lines are fits of β-relaxation data (solid triangles) and γ-relaxation data (solid diamonds), respectively, both in phase II, according to Arrhenius behavior. The inset shows the typical dielectric loss spectra measured in supercooled phase I (A) and in phase II (B) at 183.15 K (from Puertas et al.[41]).

Unfortunately, there are only a few experimental papers until now that report results concerning the α and β processes in ODICs. Phase I of Cyclooctanol[40–46] is a typical representative of orientationally disordered phases showing the occurrence of both processes, in addition to a γ process. These secondary relaxation phenomena have been attributed to the different conformations of the molecular *three-dimensional-shapes of the C-octagon* and to the equatorial and axial conformations of the OH-group, much faster than the conformational changes of the C-octagon. Moreover, cyclooctanol displays a low-temperature phase (phase II) for which β and γ processes are also present, providing then a unique possibility to "follow" such processes for a large temperature domain (see Fig. 1). In comparison with earlier works on cyclooctanol,[42–46] the recently published by Puertas et al.[41] provides additional data in the temperature gap 205–240 K for the supercooled phase I and, the most important, new data relating to the primary β-like process in phase II, which is followed for the whole temperature range of existence as a stable or possibly supercooled state.

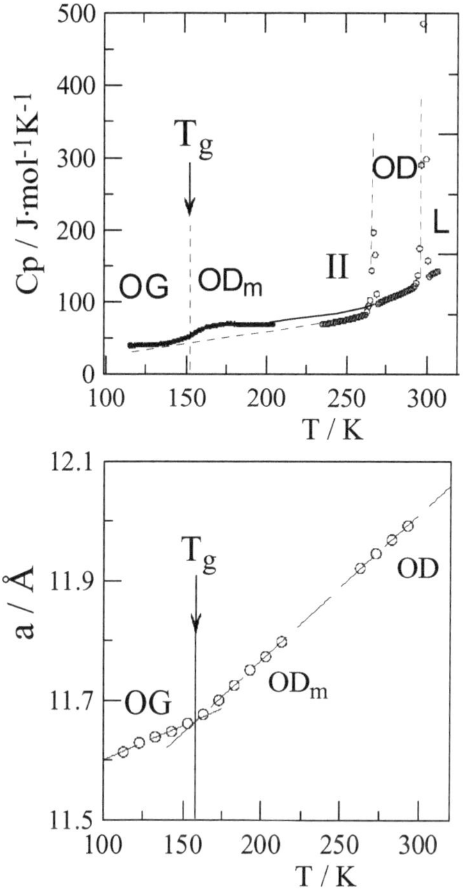

Figure 2. (a) Specific-heat as a function of temperature for the orientational glass phase (OG), orientationally disordered supercooled phase (OD_m), orientationally disordered phase (OD) and liquid phase. (b) Breakdown of the slope of the simple cubic lattice parameter as a function of temperature at the glass transition. Both figures are for cyclooctanol.

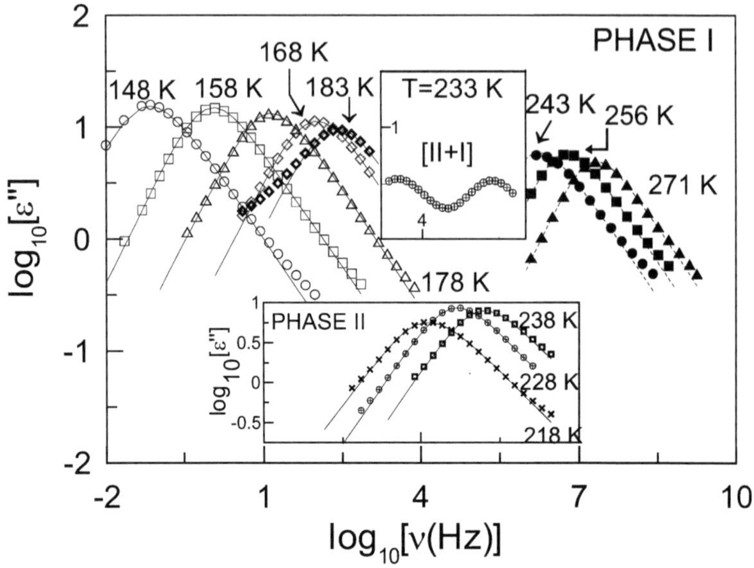

Figure 3. α-Relaxation dielectric loss peak of phase I of cycloheptanol for several temperatures. The lower inset shows the dielectric loss of phase II after annealing at 233 K of the spectra shown in the inset at the top. The lines are fits using CD-function (from Puertas et al.[48]).

The use of different experimental techniques in the analysis of the glass transition, as calorimetry and X-ray powder diffraction could provide close glass transition temperatures, but in both cases, it was not possible to bypass the concomitant irreversible transition from the supercooled OD phase to the low-temperature ordered phase. Figure 2 reveals such a trasition by using reveals such a transition by using the aforementioned techniques.

Another well-known representative of a molecular compound displaying orientationally disordered phases is cycloheptanol as firstly demonstrated by Adachi et al.[45] This compound shows two orientationally disordered phases (I and II) and the dielectric relaxation phenomena have been well analysed.[46–48] Figure 3 shows some characteristic dielectric loss spectra of the complex dielectric permittivity corresponding to the α-relaxation peaks in phases I and II. For phase II, dielectric spectra were obtained after annealing the sample (in which coexistence of phase I and II was present) for 24 h.

The relaxation time as a function of temperature for both phases I and II (empty circles and full diamonds, respectively) obtained from CD-fits, are shown as an Arrhenius plot in Fig. 4. The behavior of the α-relaxation time of phase I against temperature (Fig. 4) seems to be well described by means of the phenomenological Vogel–Fulcher–Tamann (VFT) law

$$\tau = \tau_o \exp\left(\frac{A}{T - T_o}\right), \tag{1}$$

where τ_o is an attempt time and T_o ($T_o < T_g$) is the divergence temperature (Vogel temperature) for the relaxation time. For phase II shows an Arrhenius dependence,

$$\tau = \tau_o \exp(E/RT). \tag{2}$$

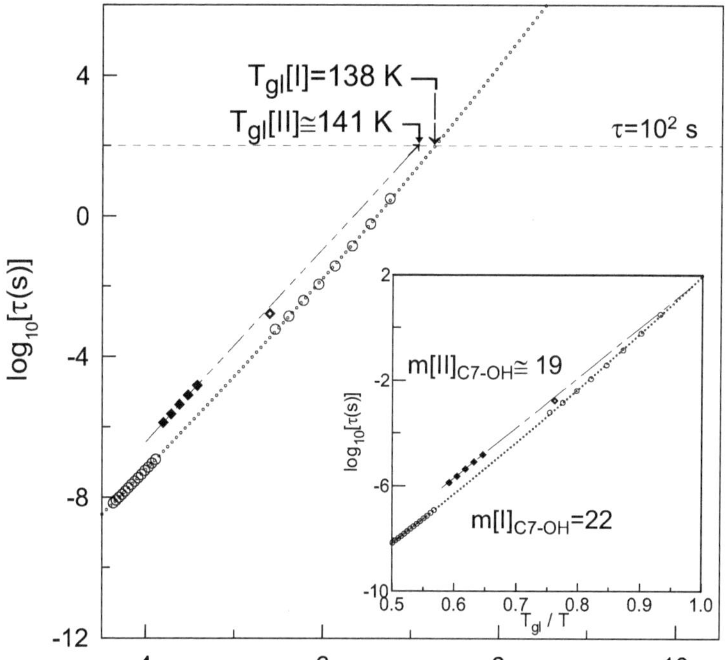

Figure 4. α-Relaxation time for phases I (empty circles) and II (full diamonds) of cycloheptanol in an Arrhenius plot. The lines correspond to the VFT-fitting of the α-relaxation data in phase I and to the Arrhenius fitting for phase II. The inset shows the typical Angell's plot for the α-relaxation in phases I and II with the fragility index (from Puertas et al.[48]).

1.3. MIXED CRYSTALS

In spite of the studies performed on these cyclic molecules two experimental problems, concomitant with the way to obtain the orientational glass appear. On the one hand, a fast cooling should be performed in order to reach the non-equilibrium glass state from the ergodic orientationally disordered phase. On the second hand, after heating from the orientationally disordered supercooled phase once the glass transition has taken place, the irreversible phase transition to the stable phase (the low-temperature phase) appears, making impossible to study, with continuity the dynamical properties, as the relaxation time associated to the different processes, or related thermodynamic properties, as specific-heat or volume. These problems, which many times do not appear on the studies of the structural glasses, can be bypassed by the use of mixed crystals. Hence, it has been shown for cyclootanol–cycloheptanol, which are known to form continuous solid solutions in the orientationally disordered phase that mixed crystals can be supercooleed with any cooling rate and that the supercooled OD phase does not transform to the low-temperature ordered phase.[49]

Some other mixed crystals giving rise to orientational glasses have studied till now. This is the case of $(C_2F_6)_{1-X}(CClF_3)_X$, although in this case the T_g/T range was relatively narrow (0.85-0-95).[50,51] Related mixed crystals are $(CF_4)_{0.46}$ $(CClF_3)_{0.54}$.[52] Cyano-adamantane, one of the most typical molecular compounds providing orientational glass mixed with adamantane also gives mixed crystals which have been the focus of many experimental studies.[43,53–61] In particular, these molecular crystals are really appreciated because good quality single crystals can be obtained and quenched without damage, enable then the analysis of many structural properties.[62–65]

Among the examples of OD mixed crystals, those formed between the tetrahedral molecules neopentylalcohol (NPA), $(CH_3)_3C(CH_2OH)$, and neopentylglycol (NPG), $(CH_3)_2C(CH_2OH)_2$, represent a continuous series for which the properties can be studied easily from an experimental point of view.

The structural and thermodynamic properties of pure compounds NPA and NPG are well known and the readers are addressed to references.[66–69]

Mixed crystals between any of these pure compounds and related tetrahedral molecules have also been the focus of many works.[70–76] For the special case of mixed crystals $([(CH_3)_3C(CH_2OH)]_{1-X}$ $[(CH_3)_2C(CH_2OH)_2]_X$, the existence of a isomorphism relationship between the OD phase of pure compounds and the appearance of the orientational glass state for mixed crystals with X at least up to 0.5, has been also reported in previous works.[77–79]

This chapter is devoted to the analysis of $(CH_3)_3C(CH_2OH)]_{0.70}$ $[(CH_3)_2C(CH_2OH)_2]_{0.30}$ mixed crystal, with special attention to the dynamics of the relaxation phenomenon appearing as a consequence of the molecular orientational disorder. Although the detail of many of the experimental results will be published elsewhere, we will focus our interest on the enormous possibilities of these kinds of materials for the study of the dynamics of orientationally disordered phase which can be easily obtained with almost any experimental restriction.

2. Experimental

Some previous results obtained by means of Modulated Differential Scanning Calorimetry (MDSC) as well as X-ray powder diffraction will be recalled in order to introduce the different experimental ways the orientational glass corresponding to the $NPA_{0.70}$ $NPG_{0.30}$ mixed crystal could be obtained. Readers are addressed to references () to find additional details of the techniques as well as for the results.

The pure compounds were supplied by Aldrich Chemicals (with 99% purity) and were submitted to a purification process consisting of growing crystals from vacuum sublimation at 313 and 343 K for NPA and NPG, respectively. Mixed crystals were prepared from the melt of the pure materials in the selected molar composition by slow cooling to room temperature.

The main results described here concern dielectric relaxation technique. Broad-band dielectric spectroscopy measurements at normal pressure were performed by means of an *Alpha impedance analyzer* (10^{-3}–10^{7} Hz) from *Novocontrol* and with a coaxial line reflectometer (Agilent 4291B impedance analyzer) from 10^{6}–$1.8{\cdot}10^{9}$ Hz on the same planar cell consisting of two gold-plated brass electrodes –diameter 50 mm– separated by two 50 mm silica spacers making a plane capacitor. The sample was held in a cryostat, the temperature being controlled by a *System Quatro* from *Novocontrol* using a heated nitrogen gas stream with temperature stabilization better than ±0.1 K.

3. Results

3.1. PURE COMPOUNDS

The main features at ordinary pressure are sketched in this section. Figure 5 shows the lattice parameter of the OD fcc phase or NPA as a function of temperature after the sample, contained in a Lindemann capillary, was quenching into liquid nitrogen. The change of the slope is enhanced by calculation of the derivative of the volume, which accounts for the thermal expansion coefficient. The breakdown of such a coefficient evidences the glass transition at T_g. It should be noticed that soon after the transition from the orientational glass to the supercooled OD phase, the sample transforms to the low-temperature (monoclinic) phase making impossible to follow the temperature evolution.

As far as dielectric measurements, Fig. 6 displays some dielectric loss spectra at several temperatures corresponding to the stable temperature domain of the OD fcc phase. For this kind of measurements it was impossible to obtain the OG by supercooling the OD phase.

Similar dielectric measurements concerning the NPG were obtained and all the efforts to reach the OG by quenching the OD stable phase were not successful. Thus only a narrow temperature domain of the OD fcc phase could be measured.

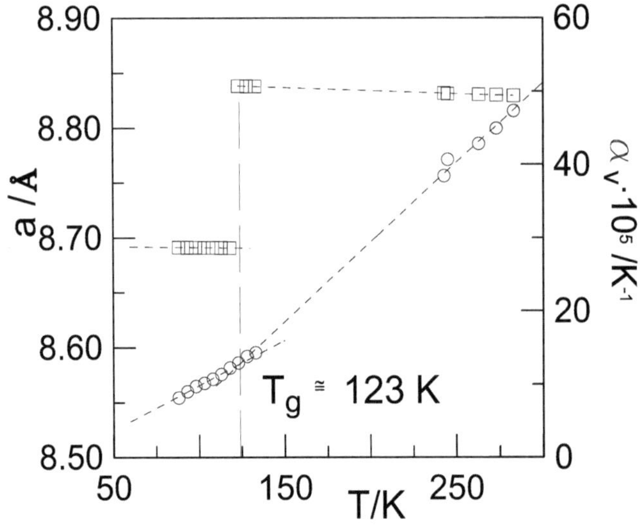

Figure 5. Lattice parameter (a(Å)) and volume expansivity (α(K^{-1})) as a function of temperature for NPA.

The relaxation time as a function of the reciprocal temperature normalised by the glass-transition temperature, Angell plot, is depicted in Fig. 7. The α-relaxation time were fitted according to the empirical Vogel–Fulcher–Tammann (VFT) law (see Eq. (1)), one of the most used.[80–82] This VFT law, which is known to fit the relaxation time as the temperature approaches T_g as a consequence of the enhanced co-operative motions, gives rise to a strong consistency with the experimental data. In the present case it should be noticed that T_g temperature for both compounds were obtained by superimposing

$\tau(T_g)$=100 s, which in the case of NPA agrees quite well with the transition temperature measured by means X-ray powder diffraction.

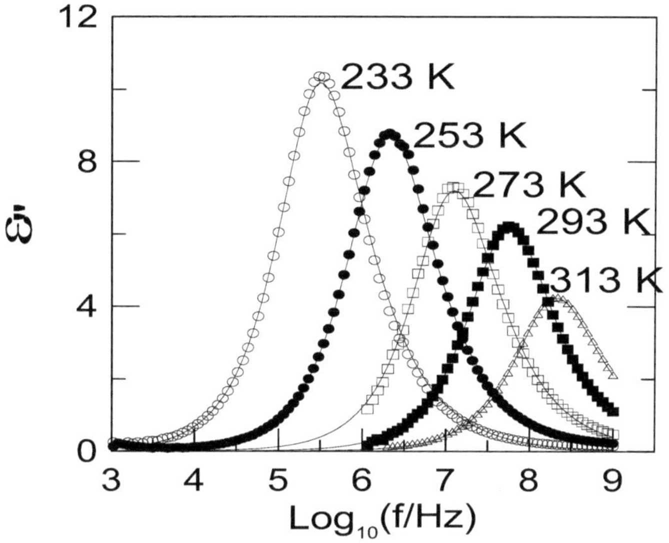

Figure 6. The frequency dependence of the dielectric loss ε′′ for NPA at different temperatures in the OD fcc phase (left panel) and relaxation time as a function of the reciprocal temperature (from López et al.[69]).

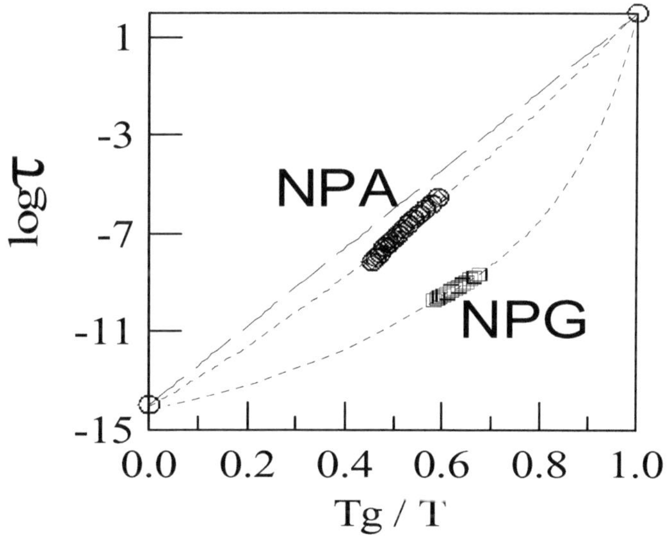

Figure 7. Angell plot for the α-relaxation of NPA and NPG pure compounds.

3.2. MIXED CRYSTALS: THERMODYNAMIC BEHAVIOR

The equilibrium phase diagram of the two-component system NPA+NPG is sketched in Fig. 8 together with the glass-transition temperatures (obtained by means MDSC) as a function of the molar composition.

As revealed by Fig. 8, glass-transition varies linearly with composition. In addition, it should be noticed that T_g for NPG, which was obtained by extrapolation of the relaxation time as a function of temperature by more than six decades is close to the linear behavior evidenced by the mixed crystals with mole composition lower than 0.5.

3.3. $NPA_{0.70}NPG_{0.30}$ MIXED CRYSTAL

In this section evidence is provided that at lowest temperatures the quenched $NPA_{1-X}NPG_X$ mixed crystals exhibit a glassy state. We focus on discussing mixed crystal with $x = 0.30$ which have been studied most thoroughly.

Figure 10 shows the α-relaxation through the liquid state at 363 K, until the orientational glass down to 130 K (well below the glass transition temperature, 161 K), which progressively slows down and the width becomes broader with decreasing temperature. It can be also seen that any additional relaxation process appears in the temperature domain studied. This fact makes easy the fitting procedure of the relaxation loss spectra.

Figure 11 shows the dependences of the α-relaxation times on temperature at ordinary pressure. The temperature dependence of the α-relaxation time is often described by more than one Vogel–Fulcher–Tammann equation as Fig. 11 evidences. Such a experimental fact is conveniently brought out by the method proposed by Stickel et al.[83–86] The method transforms the VFT relaxation time dependence into a linear dependence on the reciprocal temperature by defining a new function:

$$\phi(\tau) = \left[\frac{d \ln \tau}{d(1000/T)} \right]^{-1/2}. \tag{3}$$

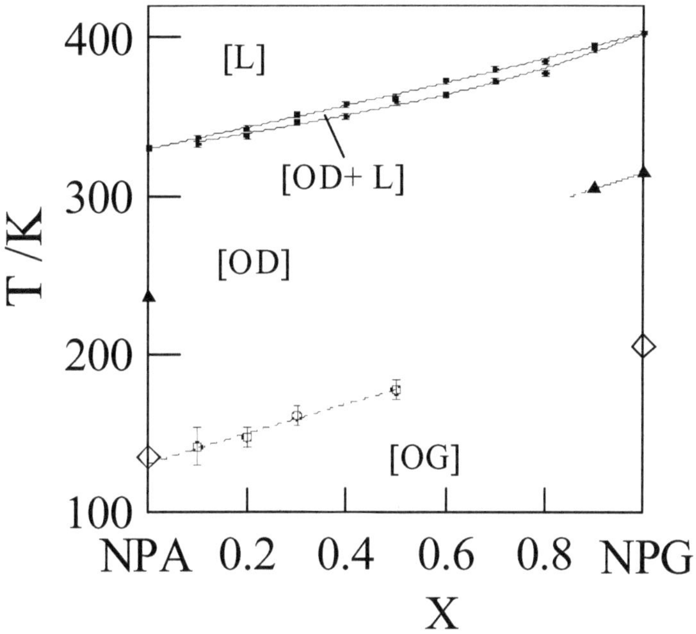

Figure 8. Equilibrium phase diagram for the NPA+NPG two-component system together with the glass-transition temperatures as determined from specific-heat measurements, except for pure compounds for which they were obtained from dielectric measurements by superimposing $\tau(T_g)$=100 s. Full symbols were obtained by scanning calorimetry.[78,79]

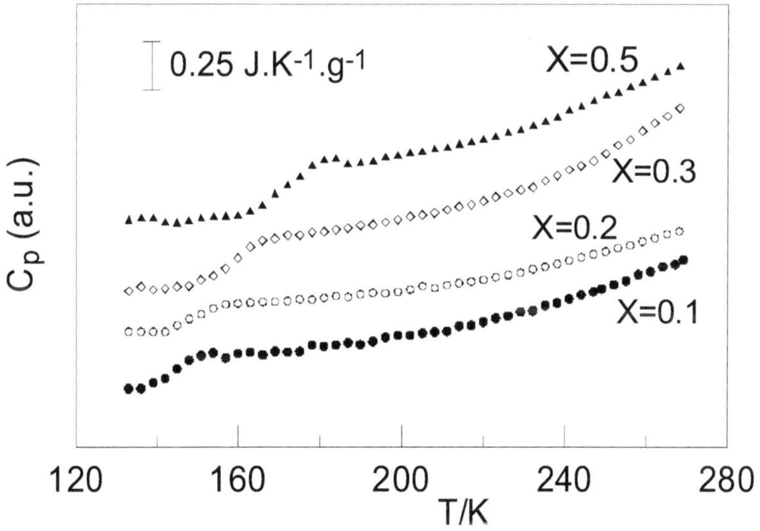

Figure 9. Specific-heat measurements as a function of temperature for some $NPA_{1-X}NPG_X$ mixed crystals (from Salud et al.[78]).

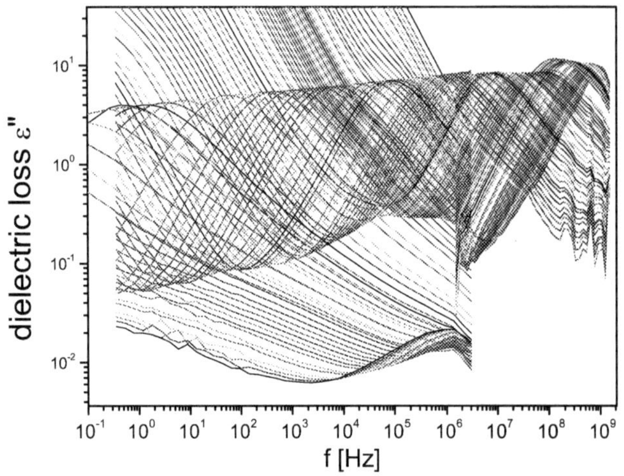

Figure 10. The frequency dependence of the dielectric loss ε″ for $NPA_{0.70}NPG_{0.30}$ mixed crystal at different temperatures from 363 K (liquid state) down to 130 K. Measurements were performed every 2 K on slow cooling.

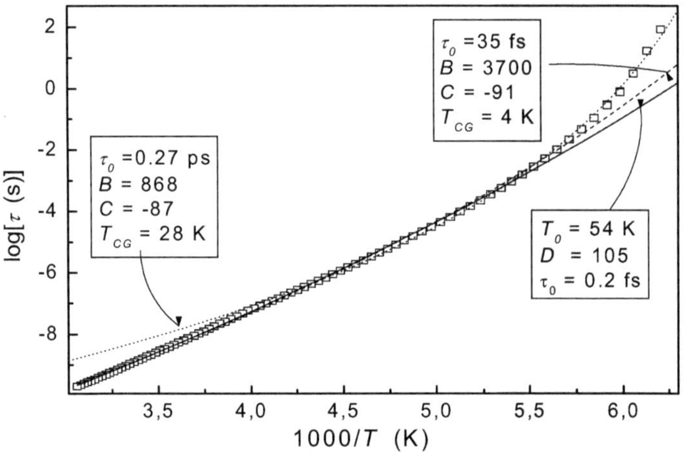

Figure 11. The Arrhenius plot of dielectric relaxation time. Black solid curve shows the single VFT fit results function. The dotted and dashed–dotted curves portray the Cohen–Grest function, based on low-and high-temperature data respectively. Fitted parameters, according to relations (1) and (4), are given in the figure.

Figure 12 clearly evidences the existence of three dynamic domains. For the high-temperature region the relaxation time obeys an Arrhenius behavior. When decreasing temperature two crossover temperatures, $T_2 = 227$ K and $T_1 = 186$ K define the temperature intervals for which two different VFT regimes are clearly visible. Although the most common equation to describe the temperature dependence of τ is the VFT aforementioned, different approaches have been considered in the past. One way is the use of the Cohen–Grest function[87–90] which treats the supercooled dynamics considering diffusion to be governed by the free-volume

$$\log_{10}\tau = \log_{10}\tau_o + \frac{B}{T - T_{CG} + \left[(T - T_{CG})^2 + CT\right]^{1/2}} \cdot \qquad (4)$$

This Eq. (4) contains four adjustable parameters and thus, two fewer adjustable parameters that it would be required by using the fitting procedure of two VFT equations. Nevertheless, Fig. 11 evidences that Cohen–Grest function cannot account for the temperature dependence of τ for the whole temperature domain and, at least, the high-temperature interval, which shows an Arrhenius behavior (see Fig. 11) should be put apart.

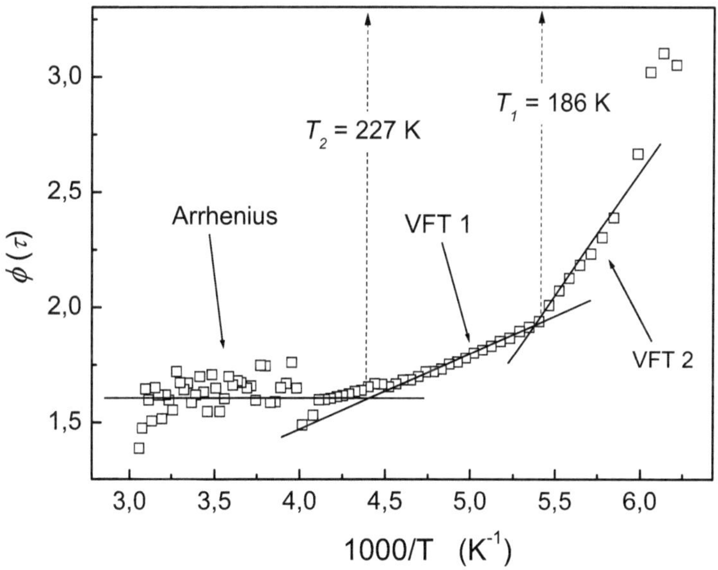

Figure 12. Stickel-plot of the temperature dependence of dielectric relaxation time. The sloped lines reveals regions described by two different VFT functions (Eq. (1)) while the non-sloped lines indicates the Arrhenius behavior (Eq. (2)).

As usual, dielectric losses were found to be asymmetrically broadened with respect to simple Debye behavior and were fitted according to the empirical Havriliak–Negami equation,[91] for which the complex dielectric permittivity is given by

$$\varepsilon^* = \varepsilon_\infty + \frac{\Delta\varepsilon}{\left[1 + (i2\pi\tau)^\alpha\right]^\beta}, \tag{5}$$

where ε_∞ stands for the dielectric high-frequency limit, $\Delta\varepsilon = \varepsilon_s - \varepsilon_\infty$ for the dielectric strength, α and $\alpha{\cdot}\beta$ parameters determine the low- and high-frequency spectral shape and τ is the relaxation time.

The relaxation strength as a function of the relaxation time as determined from the fit of the HN function is shown in Fig. 13. According to the Onsager equation,[92] the strength should be proportional to the inverse of the temperature, i.e., to the relaxation time in log scale. The deviation indicates the existence of a co-operative behavior that discards the possibility of the molecules rotating as free dipoles.

The shape parameters α and $\alpha\beta$ which characterize the low- and high-frequency wind of the dielectric loss have been also determined from the fit of HN function (the α and $\alpha\beta$ parameters are plotted in Fig. 14). The former is close to 1 in the high-temperature domain (close to T_2 crossover temperature), which means, as observed in other OD compounds, that the Cole–Davison (CD) function can accurately describe the non-Debye relaxation process. The temperature independent behavior of β in the high-temperature region is in agreement with the mode coupling theory (MCT[93]) for which the independent behavior is predicted at temperatures above crossover. At a temperature close to the onset of the mentioned co-operative effects, it is no longer possible to fit accurately the dielectric loss by using the CD function. Approaching the glass transition temperature, the common feature of many structural glass-forming materials, i.e., a decrease of the shape parameters with temperature, is found, indicating the strong deviation from the Debye behavior. In addition, at temperature close to T_g, both parameters seem to reach the claimed universal value of 0.5.

Experimental works covering very a wide relaxation time range have indicated the existence of, at least, two relaxation temperature domains of the α-relaxation process. The temperature where qualitative changes in the dynamics occur is the crossover temperature T_c.[94] This temperature has been determined through the analysis of the temperature variations of the relaxation time in spite of the non-existence of a critical behavior for the relaxation time itself.[84] Nevertheless, values of T_c around 1.15–1.2 T_g are reported mainly for glass-forming liquids that, according to the Angell classification, are generally fragile glasses. Analyses of strong network glass formers (like B_2O_3) has evidenced the existence of this dynamical transition at a temperature well above the glass-transition temperature (T_c = 1.5 T_g).[95,96] Then, it is interesting to investigate whether the T_c temperature defined in the MCT scenario[93,97] can be determined for materials displaying non-far-from-Arrhenius behavior, i.e. strong orientational glasses from the Angell classification like the molecular compound studied in this work.[98,99]

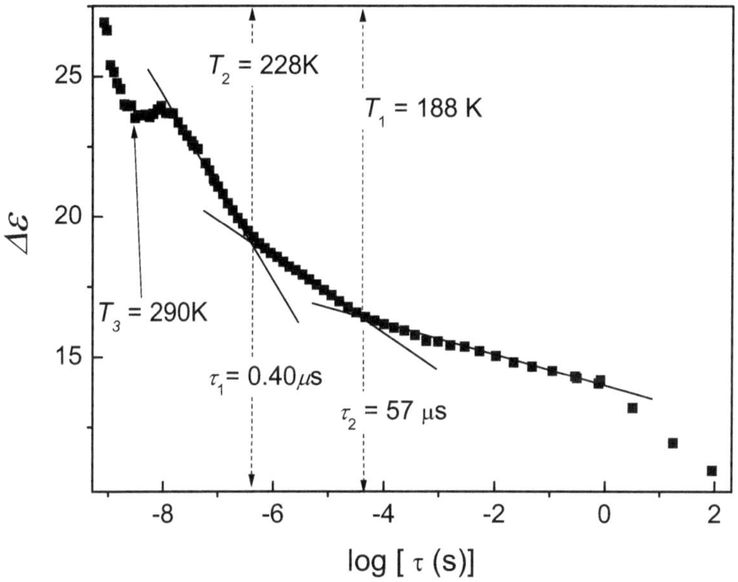

Figure 13. Dielectric strength as a function of inverse temperature as determined from the fit of the Havriliak–Negami function for the molecular mixed crystal $NPA_{0.7}NPG_{0.3}$. Vertical dashed lines reveal three dynamic regimes although crossover temperatures are slightly shifted when compared to those determined by the Stickel-plot.[84]

According to MCT theory, the temperature dependence of the relaxation time can be written by means of a power law with three parameters:

$$\tau = A(T - T_X)^{-\gamma}. \tag{6}$$

Figure 15a, which shows the α-relaxation time as a function of T-T_X, seems to validate the general trend given by the MCT equation. Nevertheless, a fine-tuning (Fig. 15b) reveals the existence of three different regimes governed by three different exponents γ. As far as the values of the γ exponents it can be observed that they are relatively high and exponents of this magnitude have no theoretical interpretation according to the MCT. In addition, it has been demonstrated that for values near to 12, the Eqs. (6) and (1) are mathematically equivalent. In spite of this possible unphysical meaning for the γ exponents it is remarkable the closeness of the temperatures at which the "γ-regime" change with those of Figs. 12 and 13.

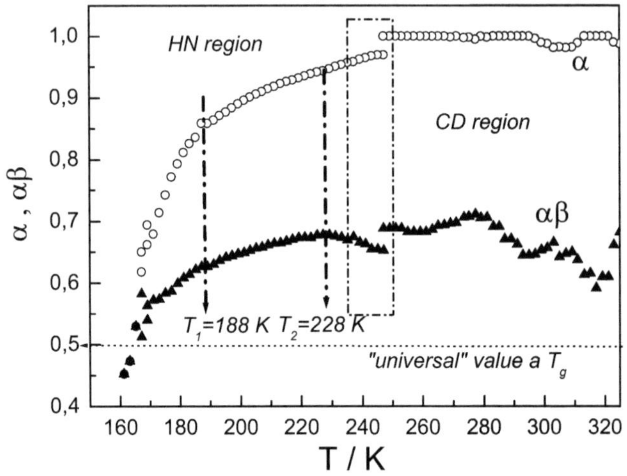

Figure 14. Evolution of the low- and high-frequency shape parameters, α and $\alpha\beta$, with temperature as determined from the fit of the Havriliak–Negami function for the molecular mixed crystal $NPA_{0.7}NPG_{0.3}$. Rectangle indicates the transition-region in which the relaxation spectra go from CD (α=1) to HN.

To end with this section, we have also investigated the validity of the Debye–Stokes–Einstein equation,[85,100] which establishes $\sigma\tau^s$=const with s=1. Figure 16 shows the relation between conductivity and the relaxation time for the temperature domain of 200–250 K. It evidences that such a relation falls down not only at temperatures lower than the crossover temperatures previously determined, but also in the high-temperature domain. In addition, the fractional DSE relation (DSE relation with s coefficients lower than 1) gives two s coefficients, the change being at a temperature around 220 K.

As found in many systems with high viscosity on approaching the glass transition, the DSE law breakdowns at temperature close to the crossover temperature. In the present case, the question arises: Why the decoupling between viscosity and conductivity, in fact, the relaxation time, appears at such relatively high temperatures? In other words, the relaxation becomes (at least) bimodal in such way that the different modes of transport have a different time-scale.

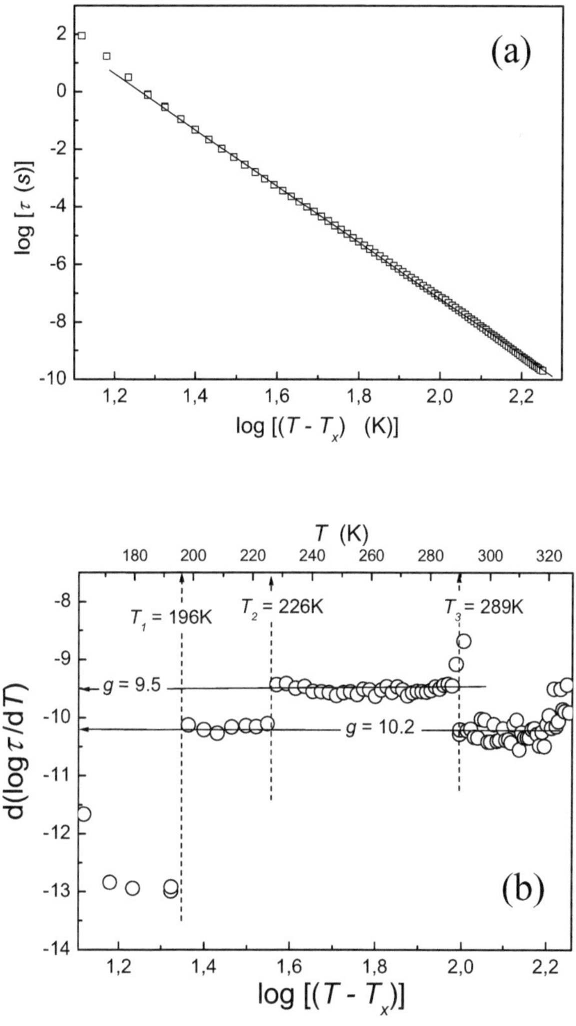

Figure 15. (a) The log–log plot showing the possible validity of the MCT "critical" like behavior $\tau \propto (T - T_X)^{-\gamma}$, with γ = 9.8 and T_X = 148 K. (b) The derivative of the log–log MCT plot given in (a). It confirms the validity of the MCT "critical-like relation" but reveals existence of three different regimes, with different values of the power exponent γ. The crossover temperatures agree well with temperatures indicated in Figs. 12 and 13.

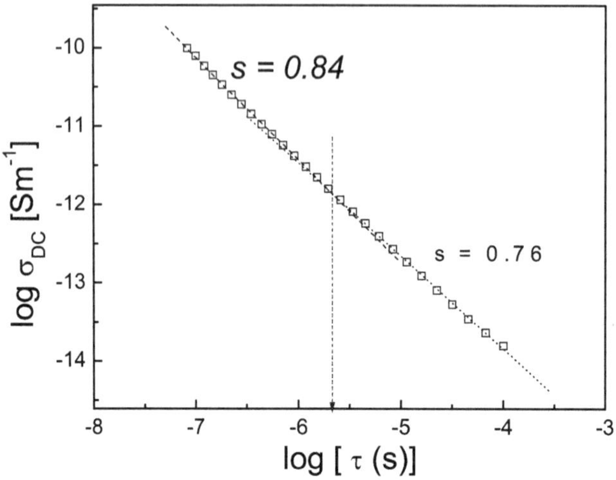

Figure 16. Conductivity (σ) vs α-relaxation times (τ) on a log–log scale. Vertical dashed line indicates the transition from one fractional Debye–Stokes–Einstein regime to another.

4. Conclusions

The relaxation dynamics of the mixed crystal $NPA_{0.7}NPG_{0.3}$ has been studied in detail in the orientationally disordered stable state as well as in its supercooled state. It is dominated by the appearance of primary α-relaxation process and indications of fast secondary relaxations were not detected in the whole temperature range. The parameters characterizing the α-relaxation in terms of temperature have shown that some crossovers appear. For the high-temperature region the relaxation time obeys an Arrhenius behavior, while when decreasing temperature two crossover temperatures define the temperature intervals for which two different Vogel–Fulcher–Tammann regimes are clearly visible.

As for as the shape parameters α and β according to the fits of Havriliak–Negami function have been determined. Approaching the glass transition temperature, the common feature of many structural glass-forming materials (decrease of the shape parameters with temperature) is found, indicating the strong deviation from the Debye behavior. Both parameters seem to reach the claimed universal value of 0.5. Cole–Davison function can accurately describe the non-Debye relaxation process. The temperature independent behavior of β in the high-temperature region is in agreement with the mode coupling theory (MCT) for which the independent behavior is predicted at temperatures above crossover.

The analysis according to the MCT scenario seems to validate the general trend given by the MCT equation. Nevertheless, a fine-tuning reveals the existence of three different regimes governed by three different exponents γ. The magnitude of the exponents is relatively high and theoretical interpretation according to the MCT seems to be difficult. In spite of this possible unphysical meaning for the γ exponents it is remarkable the closeness of the temperatures at which the "γ-regimes" change with those crossover temperatures found by means of the VFT fits.

The validity of the fractional Debye–Stokes–Einstein equation has also been investigated. Two regimes for such a relation have been found, making evident the existence of a crossover temperature for the decoupling between viscosity and conductivity.

Acknowledgments

This research was sponsored by the Spanish Ministry of Education and Science, under grant FIS2005-00975 and by the Generalitat de Catalunya, DURSI department, under grant 2005SGR-00535. S. J. Rzoska would like to ackowledge the Committee for Scientific Research (KBN, Poland) – Ministry for Higher Education and Science, via the grant 2PO3B 034 25 for years 2003–2006 (grant responsible: S. J. Rzoska).

References

1. Angell, C. A., Ngai, K. L., McKenna, G. B., McMillan, P. F., and Martin, S. W. (2000) Relaxation in glassforming liquids and amorphous solids, *J. Appl. Phys.* **88**, 3113–3157.
2. Ngai, K. L. (2000) Dynamic and thermodynamic properties of glass-forming substance, *J. Non-Cryst. Solids* **275**, 7–51.
3. Johari, G. P. and Goldstein, M. (1970) Viscous Liquids and the Glass Transition. II. Secondary relaxations in glasses of rigid molecules, *J. Chem. Phys.* **53**, 2372–2388.
4. Kudlik, A., Benkhof, S., Blochowicz, T., Tschirwitz, C., and Rössler, E. (1999) The dielectric response of simple organic glass formers, *J. Mol. Struct.* **479**, 201–218.
5. Johari, G. P., Powers, G., and Vij, J. K. (2002) Localized relaxation's strength and its mimicry of glass-softening thermodynamics, *J. Chem. Phys.* **116**, 5908–5909.
6. Corezzi, S., Beiner, M., Huth, H., Schröter, K., Capaccioli, S., Casalini, R., Fioretto, D., and Donth E. (2002) Two crossover regions in the dynamics of glass forming epoxy resins, *J. Chem. Phys.* **117**, 2435–2448.
7. Pisignano, D., Capaccioli, S., Casalini, R., Lucchesi, M., Rolla, P. A., Justl, A., and Rössler, E. (2001) Study of the relaxation behaviour of a tri-epoxy

compound in the supercooled and glassy state by broadband dielectric spectroscopy, *J. Phys.: Condens. Matter* **13**, 4405 –4419.
8. Ngai, K. L. (1998) Relation between some secondary relaxations and the α-relaxations in glass-forming materials according to the coupling model, *J. Chem. Phys.* **109**, 6982–6994.
9. Ngai, K. L. (1999) Correlation between beta-relaxation and alpha-relaxation in the family of poly(*n*-butyl methacrylate-stat-styrene) random copolymers, *Macromolecules* **32**, 7140-7146.
10. Olsen, N. B., Christensen, T., and Dyre, J. C. (2000) *Beta* relaxation of nonpolymeric liquids close to the glass transition, *Phys. Rev. E* **62**, 4435–4438.
11. Rault, J. (2000) Origin of the Vogel–Fulcher–Tammann law in glass-forming materials: the alpha-beta bifurcation, *J. Non-Cryst. Solids* **271**, 177217.
12. Richert, R. (2001) Spectral selectivity in the slow β-relaxation of a molecular glass, *Europhys. Lett.* **54**, 767–773.
13. Ngai. K. L. Lunkenheimer, P., León, C., Schneider, U., Brand, R., and Loidl, A. (2001) Nature and properties of the Johari–Goldstein β-relaxation in the equilibrium liquid state of a class of glass-formers, *J. Chem. Phys.* **115**, 1405–1413.
14. Döß, A., Paluch, M., Hinze, G., and Sillescu, H. (2002) From strong to fragile glass formers: secondary relaxation in polyalcohols, *Phys. Rev. Lett.* **88**, 095701.
15. Ngai K. L., (2003) An extended coupling model description of the evolution of dynamics with time in supercooled liquids and ionic conductors, *J. Phys: Condens. Matter* **15**, S1107–S1125.
16. Ngai K. L. and Paluch, M. (2003) Inference of the evolution from caged dynamics to cooperative relaxation in glass-formers from dielectric relaxation data, *J. Phys. Chem. B* **107**, 6865–6872.
17. Hensel-Bielowka, S. Paluch, M. Ziolo, J. and Roland, C. M. (2002) Dynamics of sorbitol at elevated pressure, *J. Phys.Chem. B* **106**, 12459–463.
18. Johari, G. P. (1982) Effect of annealing on the secondary relaxations in glasses, *J. Chem. Phys.* **77**, 4619–4626.
19. Kudlik, A., Schirwitz, C., Blochowicz, T., Benkhoff, S., and Rössler, E. (1997) Slow secondary relaxation process in supercooled liquids, *Europhys. Lett.* **40**, 649–654.
20. Hofmann, A., Kremer, F., Fisher, E. W., and Schonhals, A (1994) in *Disorder Effects on Relaxational Processes*, edited by R. Richertand and A. Blumens, Springer-Verlag, Heidelberg, Chap. 11.
21. Fisher, E. W., Donth, E., and Stiffen, W. (1992) Temperature dependence of characteristic length for glass transition, *Phys. Rev. Lett.* **68**, 2344–2346.
22. Donth, E. (1992) *Relaxation and Thermodynamics in Polymer-Glass Transitions* (Academic-Verlag, Berlin.
23. Brand, R., Lunkenheimer, P., Schneider, U., and Loidl, A. (1999) Is There an Excess Wing in the Dielectric Loss of Plastic Crystals? *Phys. Rev. Lett.* **82**, 1951–1954.
24. Suga, H. and Seki, S. (1974) Thermodynamic investigation on glassy states of pure simple compounds, *J. Non-Cryst. Solids* **16**, 171–194.

25. Höchli, U. T., Knorr, K., and Loidl, A. (1990) Orientational glasses, *Adv. Phys.* **39**, 405–615.
26. Loidl, A. and Böhmer, R. (1994) in *Disorder Effects on Relaxational Processes*, edited by R. Richert and A. Blumen (Springer-Verlag, Berlin,), p. 659.
27. Fayos, R., Bermejo, F. J., Dawidowski, J., Fischer, H. E., and González, M. A. (1996) Direct experimental evidence of the relationship between intermediate-range order in topologically disordered matter and discernible features in the static structure factor, *Phys. Rev. Lett.* **77**, 3823–3826.
28. Ramos, M. A., Vieira, S., Bermejo, F. J., Dawidowski, J., Fischer, H. E., Schober, H., González, M. A., Loong C. K., and Price, D. L. (1997) Quantitative assessment of the effects of orientational and positional disorder on glassy dynamics, *Phys. Rev. Lett.* **78**, 82–85.
29. Descamps, M., Bée, M., Derollez, P., Willart, J. F., and Carpentier, L. (1994) in *Quasielastic Neutron Scattering*, edited by J. Colmenero *et al.* (World Scientific, Singapore), p. 107.
30. Cabrillo, C., González, M. A., Cuello, G. J., Bermejo, F. J., Saboungi, M. L., and Price, D. L. (2004) Microscopic origin of the non-Gaussian behavior of dynamic structure factors of glassy matter, *Phys. Rev. B* **69**, 134202.
31. Fischer, H. E., Bermejo, F. J., Cuello, G. J., Fernandez-Diaz, M. T., Dawidowski, J., Gonzlez, M. A., Schober, H., and Jimenez-Ruiz, M. (1999) Quantitative evaluation of anharmonic and disorder effects on glassy dynamics, *Phys. Rev. Lett.* **82**, 1193–1196.
32. Gonzalez, M. A., Enciso, E., Bermejo, F.J., and Bee, M. (2000) Molecular approach to the interpretation of the dielectric relaxation spectrum of a molecular glass former, *Phys. Rev. B* **61**, 6654–6666.
33. Bermejo, F. J., Criado, A., Fayos, R., Fernandez-Perea, R., Fischer, H E., Suard, E., Guelylah, A., and Zúñiga, J. (1997) Structural correlations in disordered matter: An experimental separation of orientational and positional contributions, *Phys. Rev. B* **56**, 1136–1145.
34. Criado, A., Jiménez-Ruiz, M., Cabrillo, C., Bermejo, F. J., Grimsditch, M., Fischer, H. E., Bennington, S. M., and Eccleston, R. S. (2000) Role of low-frequency vibrations on sound propagation in glasses at intermediate temperature, *Phys. Rev. B* **61**, 8778–8783.
35. Jiménez-Ruiz, M., Gonzalez, M. A., Bermejo, F. J., Miller, M. A., Birge, N. O., Cendoya, I., and Alegria, A. (1999) Relaxational dynamics in the glassy, supercooled liquid, and orientationally disordered crystal phases of a polymorphic molecular material, *Phys. Rev. B* **59**, 9155–9166.
36. Miller, M. A., Jimenez-Ruiz, M., Bermejo, F. J., and Birge, N. O. (1998) Comparison of the structural and orientational glass-transition dynamics in ethanol, *Phys. Rev. B.* **57**, R13977–R13980.
37. Haida, O., Matsuo, T., Suga, H., and Seki, S. (1974) Calorimetric study of glassy state. 10. Enthalpy relaxation at glass-transition temperature of hexagonal ice, *J. Chem. Thermodyn.* **9**, 815–825.
38. Talón, C., Ramos, M. A., Vieira, S., Cuello, G. J., Bermejo, F. J., Criado, A., Senent, M. L., Bennington, S. M., Fischer, H. E., and Schober, H. (1998) Low-

temperature specific heat and glassy dynamics of a polymorphic molecular solid, *Phys. Rev. B* **58**, 745–755.
39. Criado, A., Jimenez-Ruiz, M., Cabrillo, C., Bermejo, F. J., Fernandez-Perea, R., Fischer, H. E., and Trouw, F. R. (2000) Rotational dynamics in the plastic-crystal phase of ethanol: Relevance for understanding the dynamics during the structural glass transition, *Phys. Rev. B* **61**, 12082–12093.
40. Leslie-Pelecky, D. L. and Birge, N. O. (1994) Universal scaling of the relaxation near a model glass transition, *Phys. Rev. Lett.* **72**, 1232–1235.
41. Puertas, R., Rute, M. A., Salud, J., López, D. O., Diez, S., van Miltenburg, J. C., Pardo, L. C., Tamarit, J. Ll., Barrio, M., Pérez-Jubindo, M. A., de la Fuente, M. R. (2004) Thermodynamic, crystallographic, and dielectric study of the nature of glass transitions in cyclo-octanol, *Phys. Rev. B* **69**, 224202.
42. Forsman, H. and Andersson, O. (1991) Dielectric relaxation and a new phase of cyclooctanol at pressures to 1 GPa, *J. Non-Cryst. Solids* **131–133**, 1145–1148.
43. Leslie-Pelecky, D. L. and Birge, N. O. (1994) Dielectric measurement of the model glass transition in orientationally disordered cyclooctanol, *Phys. Rev. B* **50**, 13250–13258.
44. Brand, R., Lunkenheimer, P., and Loidl, A. (1997) Relaxations and fast dynamics of the plastic crystal cyclooctanol investigated by broadband dielectric spectroscopy, *Phys. Rev. B* **56**, R5713–R5716.
45. Lunkenheimer, P., Brand, R., Schneider, U., and Loidl, A. (1999) Excess wing and high frequency dynamics in plastic crystals, *Philos. Mag. B* **79**, 1945–1951.
46. Tyagi, M. and Murthy, S. S. N. (2001) Study of the nature of glass transitions in the plastic crystalline phases of cyclo-octanol, cycloheptanol, cyanoadamantane and cis-1,2-dimethylcyclohexane, *J. Chem. Phys.* **114**, 3640–3652.
47. Adachi, K., Suga, H., and Seki, and S. (1972) Calorimetric study of glassy state. 7. Phase changes between crystalline phases of cycloheptanol with various degrees of stability, *Bull. Chem. Soc. Jpn.* **45**, 1960–1967.
48. Puertas, R., Salud, J., López, D. O., Rute, M. A., Diez, S., Tamarit, J. Ll., Barrio, M., Pérez-Jubindo, M. A., de la Fuente, M. R., and Pardo, L. C. (2005) Static and dynamic studies on cycloheptanol as two-orientational glass-former, *Chem. Phys. Lett.* **401**, 368–373.
49. Rute, M. A., Salud, J., Negrier, P., López, D. O., Tamarit, J. Ll., Puertas, R., Barrio, M., and Mondieig, D. (2003) Two-component system Cycloheptanol (C7) + Cyclooctanol (C8): An extraordinary system, *J. Phys. Chem. B* **107**, 5914–5921.
50. Böhmer, R. and Loidl, A. (1991) Relaxation dynamics in molecular alloys. Annealed $(C_2F_6)_{1-X}(CClF_3)_X$ mixtures, *J. Chem. Phys.* **94**, 2143–2148.
51. Böhmer, R. and Loidl, A. (1991) Relaxation dynamics in molecular alloys. II. Supercooled $(C_2F_6)_{1-X}(CClF_3)_X$ plastic crystals, *J. Chem. Phys.* **94**, 7397–7401.
52. Böhmer, R. and Loidl, A. (1990) in *Basic Features of the Glassy State*, edited by J. Colmenero and A. Alegria (World Scientific, Singapore), pp. 215–219.
53. Denicourt, T., Hédoux, A., Guinet, Y., Willart, J.-F., and Descamps, M. (2003) Raman scattering investigations of the stable and metastable phases of Cyanoadamantane glassy crystal, *J. Phys. Chem. B* **107,** 8629–8636.

54. Brand, R., Lunkenheimer, P., and Loidl, A. (2002) Relaxation dynamics in plastic crystals, *J. Chem. Phys.* **116**, 10386–10401.
55. Affouard, F., Cochin, E., Decressain, R., and Descamps, M. (2001) Experimental and numerical signatures of dynamical crossover in orientationally disordered crystals, *Europhys. Lett.* **53**, 611–617.
56. Descamps, M., Bee, M., Derollez, P., Willart, J. F., and Carpentier, L. (1993) *Quasielastic Neutron Scattering* (World Scientific, Singapore).
57. Amoureux, J. P., Noyel, G., Foulon, M., Bee, M., and Jorat, L. (1984) Low-frequency dielectric-properties of 1-cyanoadamantane $C_{10}H_{15}CN$, *Mol. Phys.* **41**, 161–171.
58. Pathmanathan, K. and Johari, G. P. (1985) Molecular relaxations in a rigid molecular glassy crystal, *J. Phys. C* **18**, 6535–6545.
59. Amoureux, J. P., Decressain, R., Sahour, M., and Cochon, E. (1992),Molecular motions in glassy crystal cyanoadamantane – a proton spin-lattice relaxation study, *J. Phys. II* **2**, 249–259.
60. Lusceac, S. A., Roggatz, I., Medick, P., Gmeiner, J., and Rössler, E. (2004) ^2H nuclear magnetic resonance study of the molecular motion in cyanoadamantane. I. Supercooled plastically crystalline phase, *J. Chem. Phys.* **121**, 4770–4780.
61. Sauvajol, J. L., Lefebvre, J., Amoureux, J. P. Muller, M. (1984) Low-frequency dielectric-properties of 1-cyanoadamantane $C_{10}H_{15}CN$, *J. Phys. C* **17**, 2257–2269.
62. Willart, J. F., Descamps, M., Bertault, M., and Benzakour, N. (1992) Ordering and orientational glass-transition of $(\text{cyanoadamantane})_{1-x}$ $(\text{chloroadamantane})_x$ mixed compounds, *J. Phys.: Condens. Matter* **4**, 9509–9516.
63. Willart, J. F., Descamps, M., and Benzakour, N. (1996) Polymorphism of a glass forming plastic crystal: A kinetic investigation, *J. Chem. Phys.* **104**, 2508–2517.
64. Delcourt, O., Descamps, M., Even, J., Bertault, M., and Willart, J. F. (1997) Peculiarities of the enthalpy relaxation of a glassy crystal, *Chem. Phys.* **215**, 51–57.
65. Decressain, R., Carpentier, L., Cochin, E., and Descamps, M. (2005) Nuclear magnetic resonance and dielectric investigations of molecular motions in a glassy crystal: The mixed compound $(\text{CN-adm})_{0.75}(\text{Cl-adm})_{0.25}$, *J. Chem. Phys.* **122**, 034507.
66. Tamarit, J. Ll., Legendre, B., and Buisine, J. M. (1994) Thermodynamic study of some Neopentane derivated by Thermobarometric analysis, *Mol. Cryst. Liq. Cryst.* **50**, 347–358.
67. Tamarit, J. Ll., Pérez-Jubindo, M. A., and de la Fuente, M. R. (1997) Dielectric studies on orientationally disordered phases of neopentylglycol ($(CH_3)_2C(CH_2OH)_2$) and tris(hydroxymethyl) aminomethane ($NH_2C(CH_2OH)_3$), *J. Phys.: Condens. Matter* **9**, 5469–5478.
68. Reuter, J., Büsing, D., Tamarit, J. Ll., and Würflinger, A. (1997) High Pressure DTA study on the phase behaviour in some tert-butyl compounds: pivalic acid, t-butylthiol and t-butylamine *J. Mater. Chem.* **7(1)**, 41–46.
69. López, D. O., Tamarit, J. Ll., de la Fuente, M. R., Pérez-Jubindo, M. A., Salud, J., and Barrio, M. (2000) Dielectric Spectroscopy on two orientationally disordered crystals: NPA ($(CH_3)_3CCH_2OH$) and TCE (Cl_3CCH_2OH), *J.Phys.: Condens. Matter* **12**, 3871–3882.

70. Barrio, M., Font, J., López, D. O., Muntasell, J., Tamarit, J. Ll., Negrier, P., and Haget, Y. (1994) Miscibility in plastic phases: Binary system NPG (neopentylglycol)/AMP (2-amino,2-methyl-1,3-propanediol), *J. Phys. Chem. Solids*, **55**(11), 1295–1302.
71. López, D. O., van Braak, J., Tamarit, J. Ll., and Oonk, H. A. J. (1994) Thermodynamic phase diagram analysis of three binary systems shared by five neopentane derivatives, *Calphad* **18**(4), 387–396.
72. López, D. O., van Braak, J., Tamarit, J. Ll., and Oonk, H. A. J. (1995) Molecular mixed crystals of neopentane derivatives. A comparative analysis of three binary systems showing crossed isodimorphism, *Calphad* **19**(1), 37–47.
73. Barrio, M., López, D. O., Tamarit J. Ll., Negrier, P., and Haget Y. (1995) Miscibility degree between non-isomorphous plastic phases: Binary system NPG/TRIS, *J. Mater. Chem.* **5**(3), 431–439.
74. Barrio, M., López, D. O., Tamarit, J. Ll., Negrier, P., and Haget, Y. (1996) Molecular interactions and packing in molecular alloys between non-isomorphous plastic phases, *J. Solid State Chem.* **124**, 29–38.
75. Salud, J., López, D. O., Barrio, M., Tamarit, J. Ll., Oonk, H. A. J., Haget, Y., and Negrier, P. (1997) On the crystallography and thermodynamics in orientationally disordered phases, *J. Solid State Chem.* **133**, 536–544.
76. Salud, J., Barrio, M., López, D. O., Alcobé, X., and Tamarit, J. Ll. (1998) Anisotropy of the intermolecular interactions from the study of the thermal-expansion tensor, *J.Appl. Crystall.* **31**, 748–757.
77. Salud, J., López, D. O., Barrio, M., Tamarit, J. Ll., and Oonk, H. A. J. (1999) Two-component systems of isomorphous orientationally disordered crystals. Part II: Thermodynamic Analysis of the mixed crystals, *J. Mater. Chem.* **9**(4), 917–922.
78. Salud, J., López, D. O., Barrio, M., and Tamarit, J. Ll. (1999) Two-component systems of isomorphous orientationally disordered crystals. Part I: Packing of the mixed crystals, *J. Mater. Chem.* **9**(4), 909–916.
79. Tamarit, J. Ll., López, D. O., de la Fuente, M. R., Pérez-Jubindo, M. A., Salud, J., and Barrio, M. (2000) Relaxation dynamics in orientationally disordered molecular mixed crystal $[(CH_3)_3CCH_2OH]_{0.7}$ $[(CH_3)_2C(CH_2OH)_2]_{0.3}$, *J. Phys. Condens. Matter* **12**, 8209–8220.
80. Drozd-Rzoska, A. and Rzoska, S. J. (2005) Complex dynamics of isotropic 4-cyano-4-*n*-pentylbiphenyl (5CB) in linear and nonlinear dielectric relaxation studies, *Phys. Rev. E* **65**, 041701.
81. Johari, G. P. (2003) Simple ratio for testing a supercooled liquid's relaxation time-entropy relation, *J. Phys. Chem. B* **107**, 5048–5051.
82. Lunkenheimer, P., Wehn, R., Schneider, U., and Loidl, A. (2005) Glassy aging dynamics, *Phys. Rev. Lett.* **95**, 055702.
83. Schonhals, A., Kremer, F., Hofman, A., Fischer, E. W., and Schlossser, E. (1993) Anomalies in the scaling of the dielectric α-relaxation, *Phys. Rev. Lett.* **70**, 3459–3462.
84. Stickel, F., Fischer, E. W., and Richert, R. (1995) Dynamics of glass-forming liquids. I. Temperature-derivative analysis of dielectric relaxation data, *J. Chem. Phys.* **102**, 6251–6257.

85. Hansen, C., Stickel, F., Berger, T., Richert, R., and Fisher, E. W. (1997) Dynamics of glass-forming liquids. III. Comparing the dielectric α- and β-relaxation of 1-propanol and *o*-terphenyl, *J. Chem. Phys.* **107**, 1086–1093.
86. Hansen, C., Stickel, F., Richert, R., and Fischer, E. W. (1998) Dynamics of glass-forming liquids. IV. True activated behavior above 2 GHz in the dielectric α-relaxation of organic liquids, *J. Chem. Phys.* **108**, 6408–6415.
87. Cohen, M. H. and Grest, G. S. (1979) Liquid-glass transition, a free-volume approach, *Phys. Rev. B* **20**, 1077–1098.
88. Cohen, M. H. and Grest, G. S. (1982) Nondispersive relaxation in supercooled liquids and glasses, *Phys. Rev. B* **26**, 2664–2665.
89. Grest, G. S. and Cohen, M. H. (1980) Liquid-glass transition: Dependence of the glass transition on heating and cooling rates, *Phys. Rev. B* **21**, 4113–4117.
90. Cohen, M. H., and Grest, G. S. (1984) The nature of the glass transition, *J. Non-Cryst. Solids* **61-62**, 749–759.
91. Havriliak, S., and Negami, S. (1967) A complex plane representation of dielectric and mechanical relaxation processes in some polymers, *Polymer* **8**, 161–210.
92. Frölich, H. (1949) *Theory of Dielectrics* (London: Oxford University Press).
93. Götze, W. and Sjögren, L. (1992) Relaxation processes in supercooled liquids, *Rep. Prog. Phys.* **55**, 241–376.
94. Sokolov, A. P. (1996) Why the glass transition is still interesting? *Science* **273**, 1675–1676.
95. Brodin, A., Börjesson, L., Engber, D. Torell, L. M., and Sokolov, A. P. (1996) Relaxational and vibrational dynamics in the glass-transition range of a strong glass former B_2O_3, *Phys. Rev.* B **53**, 11511–11520.
96. Engberg, D., Börjesson, L., Swenson, J., Torell, L. M., Howella, W. S., and Wannberg, A. (1999) The liquid-glass transition in a strong network glass former investigated by neutron scattering, *Europhys. Lett.* **47**, 213–219.
97. Götze, W. (1999) Recent tests of the mode-coupling theory for glassy dynamics, *J. Phys.: Condens. Matter* **11**, A1–A45.
98. Schilling, R. (2000) Mode-coupling theory for translational and orientational dynamics near the ideal glass transition, *J. Phys.: Condens. Matter* **12**, 6311–6322.
99. Rickert, M. and Schilling, R. (2005) Microscopic theory of glassy dynamics and glass transition for molecular crystals, *Phys. Rev. E* **72**, 011508.
100. Psurek, T., Hensel-Bielowka, S., Ziolo, J., and Paluch, M. (2002) Decoupling of the dc conductivity and (α-) structural relaxation time in a fragile glass-forming liquid under high pressure, *J. Chem. Phys.* **116**, 9882–9888.

PART III: LIQUID CRYSTALS

GLASSY DYNAMICS OF ROD-LIKE LIQUID CRYSTALS: THE INFLUENCE OF MOLECULAR STRUCTURE

ALEKSANDRA DROZD-RZOSKA, SYLWESTER J. RZOSKA, MAŁGORZATA JANIK
Institute of Physics, Silesian University, ul. Uniwersytecka 4, 40-007 Katowice, Poland

Abstract. The evolution of dielectric relaxation times in rod-like liquid crystalline *n*-octylcyanobiphenyl (8CB) and *n*-octyloxycyanobiphenyl (8OCB) is discussed. The linearized derivative-based analysis of data made it possible to identify dynamic crossovers linked to different Vogel–Fulcher–Tammann (VFT) equations. It is noteworthy that in the isotropic phase the dynamics can also be well portrayed by the critical-like relation but within the mode-coupling theory (MCT) for 8CB and the dynamical scaling model (DSM) for 8OCB. For both compounds the evolution of relaxation times in the Smectic A phase seems to follow the DSM description. Results obtained clearly indicate the dominant influence of heterogeneities on glassy dynamics in both liquid crystalline compounds.

Keywords: Glassy dynamics, complex liquid, liquid crystals, dielectric relaxation.

Despite decades of studies the complex dynamics of vitrifying liquids remains a puzzling phenomenon. Hence, studies in experimental model-systems which made it possible to reduce the complexity associated with canonical glass formers, namely low molecular supercooled liquids and polymers, may be of particular importance.[1–3] This idea has proved its significance in studies on colloidal liquids[4], enabling strong restrictions of orientational degrees of freedom or on orientationally disordered glassy crystals (ODIC)[5–8] where the translational freezing is coupled to the orientational disorder. To the category of experimental model-systems can be encountered rod-like compounds which on cooling usually firstly freeze in partially ordered liquid crystalline mesophases.[9–13] The mode-coupling theory (MCT) analysis of a fluid composed of hard ellipsoids of revolution, which mime rod-like nematic liquid crystals (NLC), pointed out several significant model-features of such systems, namely (i) the glassy dynamics in the isotropic phase, (ii) the significance of ordered heterogeneities in the fluidlike (isotropic) surrounding, (iii) the unique appearance of two mode coupling, ergodic – non-ergodic crossovers, linked to subsequent orientational and translational freezing.[9,10] These feature obtained

S. J. Rzoska and V. A. Mazur (eds.), Soft Matter under Exogenic Impacts, 189–199.

surprisingly fair confirmation in state-of-the-art broad-band-dielectric spectroscopy (BDS)[11,13] and optical heterodyne detected optical Kerr effect (OHD OKE)[12] studies on several isotropic rod-like NLC. Probably the most comprehensive investigations were carried solely out for *n*-pentylcyanobiphenyl (5CB) – the prominent member of *n*-cyanobiphenyls (*n*CB) homologous series.

This contribution presents BDS-based studies of dynamics in two liquid crystalline (LC) compounds, with slightly different structure and the same, isotropic (I) – nematic (N) – smectic A (SmA) – solid (S), polymorphism. This made it possible to discuss the hypothetically glassy dynamics in the isotropic phase and LC mesophases as well to test the possible influence of the structure. Results present were obtained using the Novocontrol BDS 80 spectrometers with the Quattro temperature unit, at present probably the best and the most standard broad frequency range impedance analyzer.[2] Studies have been carried in *n*-octylcyanobiphenyl (8CB) and *n*-octyloxycyanobiphenyl (8OCB), liquid crystalline compounds with a similar polymorphism, namely:[14]

8CB: *Solid – (288.8* K*) – Smectic A – (304.3* K*) – Nematic – (312.6* K*) – Isotropic liquid.*

8OCB: *Solid – (317* K*) – Smectic A – (340* K*) – Nematic – (353.1* K*) – Isotropic liquid.* For the S-SmA transition usually $T_{S-SmA} \approx 322$ K[14] is usually reported, hence the tested sample was slightly supercooled.

Both compounds are characterized by the same, large permanent dipole moment parallel to the long axis of the molecule. It is associated with the same "$\mathrm{C} \equiv \mathrm{N}$" group.[14] Structures of 8CB and 8OCB are shown in Figs. 1 and 4, respectively. Both compounds were obtained due to the courtesy of R. Dąbrowski and K. Czupryński from Technical Military University (Warsaw, Poland). They were carefully purified to reach the electric conductivity value as low as possible. All samples were degassed prior to measurements. This paper focuses on the evolution of dielectric relaxation times on cooling in 8CB and 8OCB. Relaxation times were determined from peak frequencies (f_{peak}) of loss curves via condition $\tau = 1/2\pi f_{\text{peak}}$.[2] The evolution of relaxation times in subsequent mesophases of 8CB is presented in Fig. 1. The clearly non-Arrhenius behavior in the isotropic phase terminates at the slightly discontinuous I–N transition. There is no manifestation of a discontinuity when passing the N–SmA. For the I–N transition the orientational freezing occurs and for the N–SmA this is supplemented by the one-dimensional translational ordering. The behavior in the nematic phase is anomalous, in comparison to the pattern occurring in the isotropic phase. This can be related to the predominantly parallel, in respect to the direction of the measurement electric field, orientation of rod-like molecules. In the SmA mesophase the artifacts linked to orientation of rod-like molecules ceases to be significant.[14] At first sight the evolution of relaxation times on 8CB seems to be weakly non-Arrhenius in the isotropic phase and almost Arrhenius in the SmA phase. However, recently it was pointed out that the unequivocal characterization of the underlying evolution of relaxation times can be obtained using the supplemented derivative-based, linearized transformation of data.[15] For the Vogel–Fulcher–Tammann relation,[1, 2] this yields:[15]

$$\tau(T) = \tau_0^{VFT} \exp\left[\frac{D_T T_0}{T - T_0}\right], \tag{1}$$

$$\left[\frac{d\ln\tau}{d(1/T)}\right]^{-1/2} = \left[\frac{H_a}{R}\right]^{-1/2} = \left[H_a^{'}\right]^{-1/2} = \left[(D_T T_0)^{-1/2}\right] - \frac{\left[T_0 (D_T T_0)^{-1/2}\right]}{T} = A - \frac{B}{T_0}. \tag{2}$$

where the temperature T_0 is associated with the estimation of the ideal glass transition and the strength coefficient D_T can be considered as the measure of fragility for the given temperature domain, H_a is the apparent, temperature dependent, activation enthalpy.

Consequently, the plot $(H_a^{'})^{-1/2}$ vs. $1/T$ should create a linear dependence in the region of validity of the VFT dependence with parameters $T_0 = |B/A|$ and $D_T = 1/|AB|$.[15] It is noteworthy that similar analysis was introduced by Stickel et al.[16,17] a decade ago via the formal plot $d\log_{10}\tau/d(1/T)$ vs. $1/T$, although it was solely used for detecting the temperature or pressure of dynamic crossovers between dynamical domains described by the different VFT equations.[2, 3, 15]

For the high temperature, isotropic phase one also may consider the validity of the basic prediction of the MCT, namely the critical-like description of the primary relaxation time. For canonical supercooled liquids such behavior remains valid up to ca. 20 K above the singular temperature. Including the derivative-based transformation of data this yields:[15]

$$\tau(T) = \tau_0^{MCT}\left(\frac{T - T_X}{T_X}\right)^{-g}, \tag{3}$$

$$H_a^{'}(T) = \frac{H_a(T)}{R} = R\left(\frac{d\ln\tau}{d(1/T)}\right) = \frac{gT^2}{T - T_X}. \tag{4}$$

Consequently, the plot $T^2/H_a^{'}(T) = T^2/[d(\ln\tau(T))/d(1/T)] = A + BT$ should give a linear dependence in a region of validity of relation (3) or any similar critical-like relation. A simply linear regression yields parameters $T_X = |A/B|$ and $g = |B^{-1}|$, i.e., they are determined from $T^2/H_a^{'}(T) = 0$ condition and the slope of a line.[15] The advantage of the analysis employing the derivative-based linearized transformation of data via relations (2) or (4) results from the fact that they precisely indicate domains of validity of relations (1) or (3). Next the linear regression fits yield the unequivocal set of relevant parameters for relations (1) or (2). The final fit via relation (1) or (3) is then reduced to prefactors. Results for such an analysis is presented in Figs. 2 and 3.

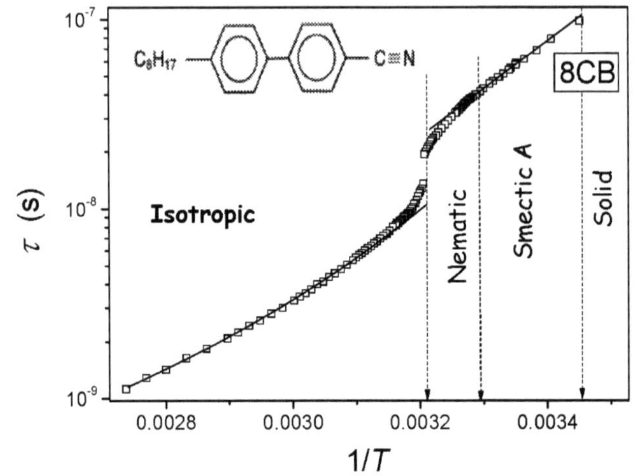

Figure 1. The Arrhenius plot of the evolution of dielectric relaxation times in isotropic and liquid crystalline 8CB. The solid curve is plotted via the VFT relation (1) with parameters taken from the linearized, derivative-based analysis given in Fig. 2: $D_T = 2.6$ and $T_0 = 218$ K. The fit yielded $\tau_0^{VFT} = 24.7$ ps for the prefactor. The glass temperature from the condition $\tau(T_g) = 100$ s yields $T_g = 238$ K.

It is clear visible that experimental data are well portrayed via the VFT relation in the isotropic phase. The obtained value of the fragility strength coefficient D_T is typical for the so-called fragile glass formers.[2,3] For $T < T_{IN} + 5$ K the fragility becomes enormous and the value of T_0 approaches the I–N transition As indicated in Ref. 13, for the isotropic 5CB this unusual behavior can be associated with the appearance of bond ordering, prenematic fluctuations appearing due to the weakly discontinuous character of the I–N transition. Values $D_T = 2.6$ and $T_0 = 218$ K were next substituted into the VFT relation to create the solid curve portraying $\tau(1/T)$ data in the isotropic phase in Fig. 1. Noteworthy is the validity of the Arrhenius type description in the SmA phase. In the derivative plot in Fig. 2 such behavior is manifested as the horizontal line. Results presented in Fig. 3 show that the same data in the isotropic phase can be also well portrayed by the critical-like relation (3). The obtained value of the power exponent φ is in fair agreement with values usually obtained for the MCT high temperature behavior in canonical, glass-forming liquids. However, for canonical glass formers the derivative-based analysis has been only recently applied.[15] Noteworthy is the relatively strong distortion, from this behavior for $T < T_{IN} + 2\ \mathrm{K}$.

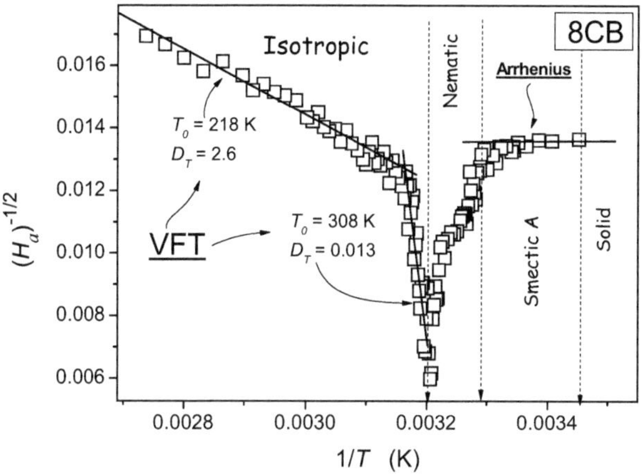

Figure 2. The linearized, derivative-based transformation of dielectric relaxation data from focused on the validity of the VFT relation in the description of dynamics in 8CB (see relations (1) and (2)). All parameters given in the figure were determined from the linear regression analysis as discussed below relation (2) and derived in Ref. 15.

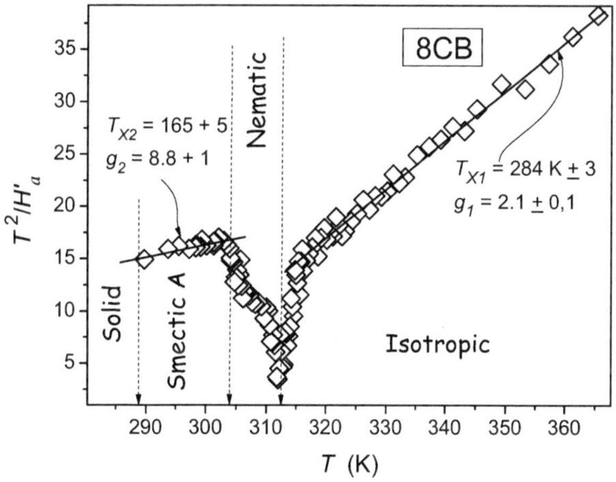

Figure 3. The linearized, derivative-based transformation of dielectric relaxation data from focused on the validity of the critical-like relation (3) in the description of dynamics in 8CB (see relations (3) and (4)). All parameters given in the figure were determined from the linear regression analysis as discussed below relation (4) and derived in Ref. 15. Note that $\tau_{VFT}(T_{X1}) \approx 1.3 \times 10^{-7}$, in fair agreement with the "magic" universal time-scale (see Refs. 20 and 21).

Surprisingly, there is a clear manifestation of the critical-like dependence (3) for the SmA phase in Fig. 3, but with the value of the exponent g and the temperature T_X definitively beyond predictions of the MCT. It is noteworthy that the latter is located well below the hypothetical glass transition temperature, determined from the VFT dependence determined for the isotropic phase via the condition $\tau_{VFT}(T_g) = 100s$.[1,2] In the opinion of the authors the set parameters obtained for the SmA phase may indicate the possibility of the validity of the dynamical scaling model (DSM) developed to portray dynamics of vitrifying liquids.[17,18] The DSM assumes the appearance of cooperative domains, despite the lack of the clear phase transition in canonical vitrifying liquids. These domains of the cooperative motion are characterized by the correlation length $\xi(T) = \xi_0[(T - T_C)/T_C]^{-\nu}$ with $\nu = 3/2$ and the resulted cooperative relaxation time:[17,18]

$$\tau(T) = \tau_0 \left(\frac{T - T_C}{T_C} \right)^{-\nu z}, \quad (5)$$

where $\phi = \nu z = 9$ and $z = 6$ is the dynamic exponent.

The critical temperature of divergence related to cooperative regions is suggested to be hidden below T_g. The DSM critical-like description was claimed to be valid below the caging temperature $1.3T_C < T_A < 1.8T_C$.[17,18] The analysis of $\tau(T)$ data in various glass-forming polymers indicated that relation (6) fairly well describe experimental data with $\nu z = 8.5 - 12$.[17,18] For supercooled liquids $\nu z = 11 - 20$ empiric exponent was suggested. This non-universality was explained as the influence of the low temperature activation processes, which introduces $\exp(E_{\mathrm{low}}/kT)$ multiplicator.[17, 18]

Figures 4 through 6 parallel the above discussion but for 8OCB, an NLC compound with the structure from slightly different than 8CB. The main difference between Fig. 1 (8CB) and Fig. 4 (8OCB) seems to be the validity of the Arrhenius description already in the isotropic phase for 8OCB. This conclusion clearly confirms the linearized derivative-based transformation presented in Fig. 5. However, distortions from such a description start already at $T \approx T_{IN} + 10$ K, what can be hardly visible at the basic plot of experimental data in Fig. 4. In the immediate vicinity of T_{IN} the dynamics is definitively non-Arrhenius. In this domain it can be portrayed by the VFT dependence but with the enormously small values of the fragility strength coefficient D_T, i.e., extremely large fragility, and the value of T_0 approaching T_{IN}.

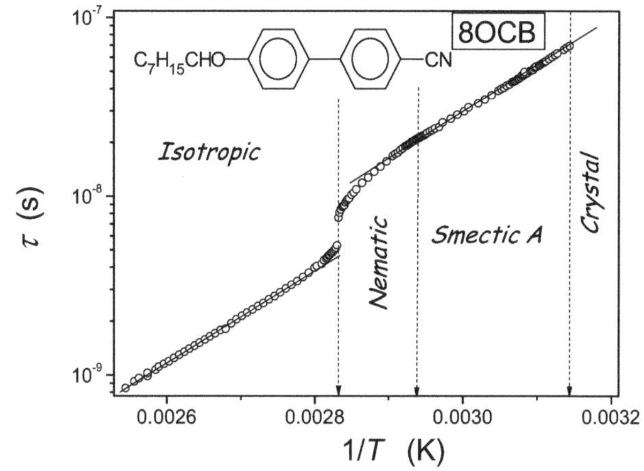

Figure 4. The Arrhenius plot of the evolution of dielectric relaxation times in isotropic and liquid crystalline 8OCB. The solid curves indicate the Arrhenius temperature behavior.

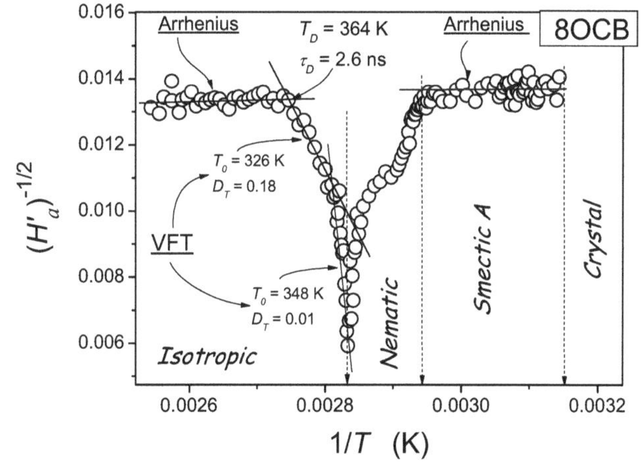

Figure 5. The linearized, derivative-based transformation of dielectric relaxation data from focused on the validity of the VFT relation in the description of dynamics in 8OCB (see relations (1) and (2)). All parameters given in the figure were determined from the linear regression analysis as discussed below relation (2) and derived in Ref. 15. The horizontal line points out Arrhenius domains.

For SmA phase the derivative plot in Fig. 5 shows the horizontal line, pointing out the validity of the Arrhenius description, similarly as for 8CB. The surprising behavior revealed the linearized derivative analysis presented in Fig. 6. It is visible that both for the Isotropic and the SmA phase the evolution of relaxation times can also be well portrayed by the critical-like dependence, with

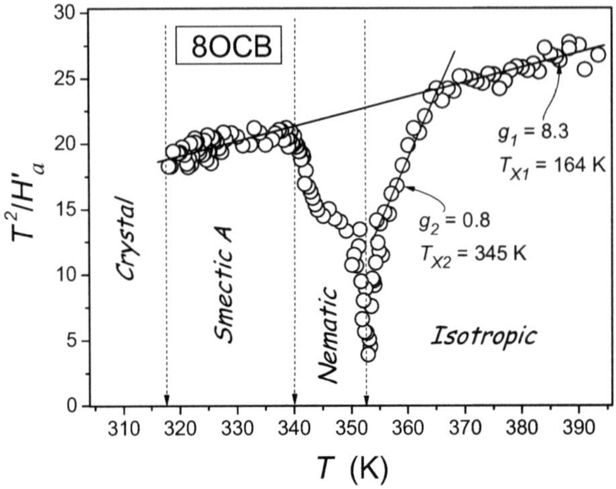

Figure 6. The linearized, derivative-based transformation of dielectric relaxation data from focused on the validity of the critical-like relation (3) in the description of dynamics in 8OCB (see relations (3) and (4)). All parameters given in the figure were determined from the linear regression analysis as discussed below relation (4) and derived in Ref. 15.

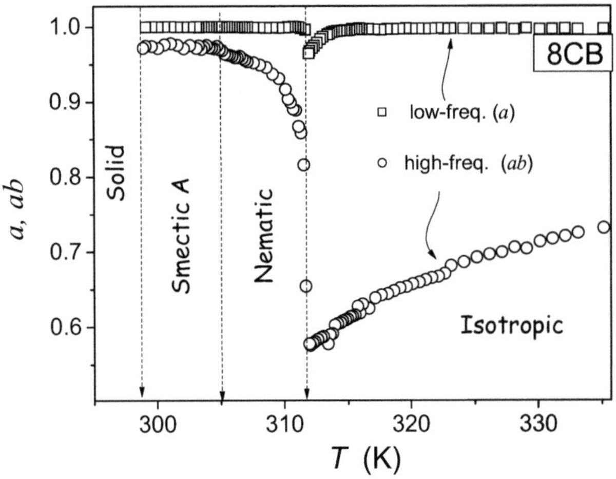

Figure 7. The evolution of Havriliak–Negami[1,2] form parameters describing the low- and the high- frequency wings of dielectric loss curves in 8CB.

the same value of the exponent $g = 8.3 \pm 1$ and the singular temperature $T_X = 164\ \text{K} \pm 2$. It is noteworthy that for the isotropic phase such behavior is valid only for $T > T_{IN} + 10$ K. This may suggest that in 8CB and 8OCB the DSM type description may be applied for both domains.

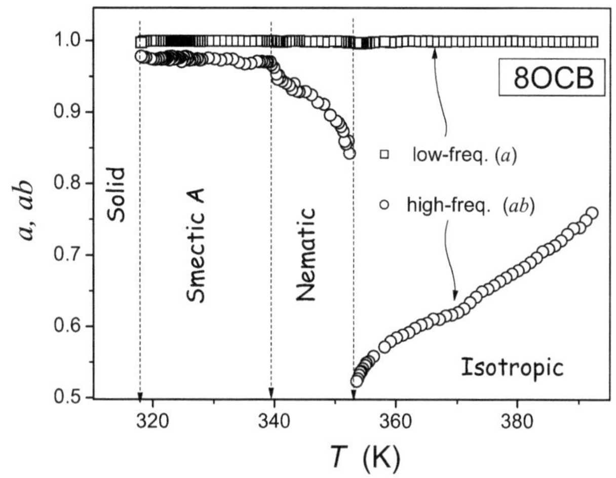

Figure 8. The evolution of Havriliak–Negami[1,2] form parameters describing the low- and the high-frequency wings of dielectric loss curves in 8OCB.

Figures 7 and 8 supplement the above discussion by the temperature evolution of the distribution of relaxation times. Both for 8CB and 8OCB the permanent broadening of loss dielectric on cooling in the isotropic liquid phase occurs, although this process seems to be dominated by the high-frequency part. In 8CB a small broadening for the low-frequency wing in the immediate vicinity of T_{IN} also takes place. In liquid crystalline mesophases the distribution of relaxation times continuously narrows, i.e., it approaches the single relaxation time Debye pattern on cooling.

Concluding, results presented above made the comparison of the evolution of dielectric relaxation times in all mesophases of 8CB and 8OCB possible. The support from the linearized derivative-based analysis revealed subtle features of the dynamics, in practice impossible to detect by the direct fit of $\tau(T)$ data. The behavior observed for the isotropic 8CB seems to be similar to the one noted in isotropic 5CB.[13] Hence, one may expect that such pattern may be common for all members of the *n*-cyanobiphenyls (*n*CB) homologous series. Both compounds tested in this paper exhibit a similar polymorphism. In both compounds the evolution of relaxation times in the SmA phase seems to follow the critical-like DSM description. The behavior in the isotropic phase can be well portrayed by the critical like dependence (3). However, the obtained set of parameters indicate the validity of the MCT description for 8CB and to the DSM description for 8OCB. Surprisingly, for 8OCB the DSM-based critical-like behavior in the isotropic phase continuous in the SmA mesophase. The distribution of relaxation times seems to follow a similar pattern in 8CB and 8OCB. Noteworthy is the fact that the form coefficient $-n = ab \rightarrow 1/2$ on $T \rightarrow T_{IN}$. Such value was recommended as the universal one for glass-forming liquid near T_g.[19] The isotropic phase in 8CB and 8OCB is definitively well above T_g. However, on approaching T_{IN} the parameterization via the VFT dependence indicates the abnormal decrease of D_T and the shift of the fictive "ideal glass" or alternatively VFT temperature T_0 towards T_{IN}. Consequently, one may define a set of fictive T_g temperatures, located near T_{IN}. Results presented above show that the dynamics of isotropic and mesophasic rod-like liquid crystalline compounds can be portrayed using the same dependences as glass-forming supercooled liquids or polymers. However, they also show a set of specific features which locates LC rod-like compounds beyond the canonical "glassy pattern", despite some theoretical expectations. Undoubtedly, results presented above point out the important role of heterogeneities on dynamics. The worth recalling feature of results presented above is the influence of a relatively small change in structure of compounds on dynamics. The comparison of 8CB and 8OCB structures indicates that the length of both rod-like compounds is different. However, this factor cannot be directly responsible for differences in dynamics noted above. This conclusion supports similar patterns of dynamics in 8CB (see above) and 5CB (see Ref. 13), compounds with different length of the molecule. In the opinion of the authors one of possible reasons for differences in dynamics between 8CB and 8OCB may be the different degree of intermolecular coupling[14] for both compounds.

Acknowledgments

Results presented above were supported by the Committee for Scientific Research (KBN, Poland) – Ministry for Higher Education and Science, via the grant 2PO3B 034 25 for years 2003–2006 (grant responsible: S. J. Rzoska).

References

1. Donth, E. (2001) *The Glass Transition: Relaxation Dynamics in Liquids and Disordered Materials* (Springer Verlag, Berlin).
2. Kremer, F. and Shoenhals, A. (eds.) (2003) *Broad Band Dielectric Spectroscopy* (Springer, Berlin).
3. Rzoska, S. J. and Zhelezny, V. (eds.) (2004) *Nonlinear Dielectric Phenomena in Complex Liquids NATO Sci. Series II*, Vol. **157** (Kluwer, Brussels).

4. Haertl, C. and Colloidal glasses, W. (2001) *Curr. Opin. Colloid Interface Sci.* **6**, 479–483.
5. Tamarit, J. Ll., Pérez-Jubindo, M. A., and de la Fuente, M. R. (1997) Dielectric studies on orientationally disordered phases of neopentylglycol ($(CH_3)_2C(CH_2OH)_2$) and tris(hydroxymethyl) aminomethane ($NH_2C(CH_2OH)_3$), *J. Phys.: Condens. Matter* **9**, 5469–5478.
6. Affouard, F. and Descamps, M. (2001) Is there something of mode coupling theory in orientationally disordered crystals?, *Phys. Rev. Lett.* **87**, 035501.
7. Puertas, R., Rute, M. A., Salud, J., López, D. O., Diez, S., van Miltenburg, J. C., Pardo, L. C., Tamarit, J. Ll., Barrio, M., Pérez-Jubindo, M. A., and de la Fuente, M. R. (2004) Thermodynamic, crystallographic, and dielectric study of the nature of glass transitions in cyclo-octanol, *Phys. Rev. B* **69**, 224202.
8. Drozd-Rzoska, A., Rzoska, S. J., Pawlus, S., and Tamarit, J. L. (2006) Dynamics crossovers and the dynamic scaling description in vitrifying orientationally disordered crystal, *Phys. Rev. B* **73**, 224205.
9. Letz, M. and Schilling, R. (1999) Mode coupling theory for molecular liquids: what can we learn from a system of hard ellipsoids, *Phil. Mag. B* **79**, 1815–1820.
10. Letz, M., Schilling, R., and Latz, A. (2000) Ideal glass transition for hard ellipsoids, *Phys. Rev. E* **62**, 5173.
11. Drozd-Rzoska, A. and Rzoska, S. J. (2002) Complex relaxation in the isotropic phase of *n*-pentylcyanobiphenyl in linear and nonlinear dielectric studies, *Phys. Rev. E* **65**, 041701.
12. Cang, H., Li, J., Novikov, V. N., and Fayer, M. D. (2003) Dynamical signature of two "ideal glass transitions" in nematic liquid crystals, *J. Chem. Phys.* **119**, 10421–10427.
13. Drozd-Rzoska, A. (2006) Heterogeneity-related dynamics in isotropic *n*-pentylcyanobiphenyl, *Phys. Rev. E* **73**, 022501.
14. Demus, D., Goodby, J., Gray, G. W., Spiess, H. W., and Vill, V. (eds). (1998) *Handbook of Liquid Crystals, Vol. 1: Fundamentals* (Springer, Berlin).
15. Drozd-Rzoska, A. and Rzoska, S. J. (2006) Derivative-based analysis for temperature and pressure evolution of dielectric relaxation times in vitrifying liquids, *Phys. Rev. E* **73**, 041502.
16. Hansen, F., Stickel, F. Richert, R., and Fischer, E. W. (1997) dynamics of glass-forming liquids. III: Comparing the dielectric alpha- and ß-Relaxation of 1-propanol and o-terphenyl, *J. Chem. Phys.* **107**, 1086–1093.
17. Erwin, B. M. E. and Colby, R. H. (2002) Temperature dependence of relaxation times and the length of cooperative motion for glass-forming liquids, *J. Non-Cryst. Solids* **307–310**, 225–231.
18. Colby, R. H. (2000) Dynamic scaling approach to glass formation, *Phys. Rev. E* **61**, 1783–1792.
19. Olsen, N. B., Christensen, T., and Dyre, J. C. (2001) Time-temperature superposition in viscous liquids, *Phys. Rev. Lett.* **86**,1271–1276.
20. Novikov, V. N. and Sokolov, A. P. (2003) Universality of the dynamic crossover in glass-forming liquids: a "magic" relaxation time, *Phys. Rev. E* **67**, 031507.
21. Casalini, R., Paluch, M., and Roland, C. M. (2003) Dynamic crossover in supercooled liquids induced by high pressure, *J. Chem. Phys.* **118**, 5701–5703.

ORDERING EFECT ON DYNAMICS IN GLASS-FORMING MIXTURE OF LIQUID CRYSTALS

M. MIERZWA*, M. PALUCH, S.J. RZOSKA, J. ZIOŁO,
Institute of Physics, University of Silesia, Uniwersytecka 4, 40-007 Katowice, Poland
U. MASCHKE
Laboratoire de Chimie Macromoléculaire, UPRESA CNRS N°8009, Université des Sciences et Technologies de Lille, France

Abstract. Relaxation dynamics in glass-forming mixture of liquid crystals (E7) is studied by means of dielectric spectroscopy in the frequency range from 10^{-2} up to 10^{9} Hz and over the temperature range from 200 to 373 K. We analyzed both the shape of relaxation spectra and temperature dependencies of relaxation times in isotropic, nematic, and glassy state. Striking relaxation behavior is observed in nematic phase. The dominant relaxation process exhibits nearly Debye form that depends on temperature in clearly the non-Arrhenius fashion. The dielectric spectra also show second distingue relaxation process which is related to α-relaxation. Additionally, we carried out the measurements of mechanical relaxation. As a result it was found that mechanical relaxation times correspond to dielectric α-relaxation times.

Keywords: Glass transition, dielectric relaxation spectroscopy, liquid crystals, molecular dynamics.

1. Introduction

Dielectric spectroscopy is one of the most efficient methods for exploring the basic properties of glass-forming liquids. This method allows one to carry out very accurate measurements of dipolar relaxation in extremely broad frequency range. Dipolar relaxation of glass-formers in the frequency domain at given temperature manifests in the non-Debye spectral shape of the absorption peak. Regarding the temperature evolution of relaxation times, usually

* To whom correspondence should be addressed. e-mail: mmierzwa@us.edu.pl

S. J. Rzoska and V. A. Mazur (eds.), Soft Matter under Exogenic Impacts, 201–214.

non-Arrhenius, slowing-down of dynamics with decreasing temperature or increasing pressure is observed. Despite great theoretical and experimental efforts there are still no final conclusions concerning molecular mechanism responsible for these phenomena but the state-of-the-art applications of broadband dielectric spectroscopy and other modern techniques causes that glass-forming liquids are probably the best learn complex liquids/soft matter system. Hence, the question may be put forward of the range of universality of the result obtained in glass-forming liquids for other complex liquids. This paper deals with dielectric absorption studies in a nematic liquid crystalline material starting from the isotropic liquid down, in a very broad range of temperatures. There is large number of dielectric spectroscopy investigations in nematogens, but none of them is carried out consequently as in glass-forming liquids, to the best of authors' knowledge.[1,2] Existing results suggest that one should expect a Debye absorption curve in the isotropic liquid and Debye or almost Debye curve in the nematic phase.[1–6] The appearance of the single-time description of relaxation of rod like molecules in the nematic and in the isotropic phase agrees with generally accepted mean field description of the isotropic (Landau-de Gennes model)[1,2,7] and the nematic (Maier-Saupe model) phases.[1,2,7–9] Nevertheless, recently a significant break in this simple picture was made. Actually, in the isotropic phase, the limited validity of the mean field description and essential advantages of recently proposed models seems to be doubtless.[10–18] Considering the temperature evolution of relaxation times situation is even more puzzling. The majority of experimental results point to the validity of the Arrhenius behavior, in agreement with mean field description.[1–6] These results give a simple possibility for determining the activation energy, a significant parameter both from the basic (theoretical) and practical point of view.[4–6] However, the experimental situation is far from clearness because there are also results giving the evidence for the Vogel–Fulcher–Tammann (VFT) type behavior.[1,19–23]

This paper presents the results of broadband dielectric studies in four component mixture E7 composed of rod-like nematogens, with permanent dipole moments positioned along the long axis of a molecule. This is a well-known mixture of great practical importance due the very large extent of the nematic phase, a feature essential also in our studies. Moreover, this is the eutectic mixture with a single, well defined, clearing temperature T_{I-N}=334 K. Our studies cover range of almost 200 K, revealing the glass-forming properties of E7 and the existence of the new low-temperature phase below 230 K. The application of "glassy" methodology showed essential similarities with "ordinary" glass-forming liquids.

2. Experimental

E7 samples were obtained from Merck and were used as received. The chemical structures of the constituent liquid crystal (E7) are depicted in Fig. 1.

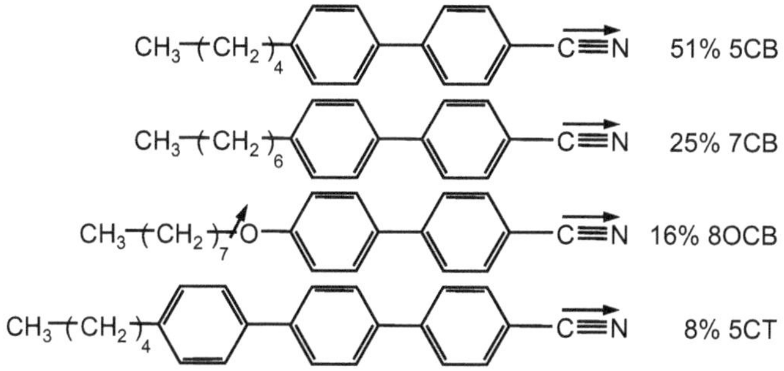

Figure 1. The chemical structure of the constituent E7. The arrows indicate the dipole moments.

The temperature-dependent dielectric measurements were performed using the experimental set-up by Novocontrol GmbH. This system was equipped with Alpha high resolution dielectric analyzer, the impedance analyzer HP4191A and temperature controller Quattro v.4.0. In the frequency range from 10^{-2} to 10^{7} Hz we measured the complex permittivity: $\varepsilon^*(\omega)=\varepsilon'(\omega)-i\varepsilon''(\omega)$ by means of Alpha high resolution dielectric analyzer. For this kind of measurements, flat parallel capacitor (diameter 10 mm, thickness 0.07 mm) was used. For measurements between 10^{6} and 10^{9} Hz a coaxial arrangement with the sample capacitor (diameter 6 mm, thickness 0.05 mm) mounted as part of the inner conductor was used. In both cases, value of AC voltage applied to the capacitor was equal to 1 V. Temperature was controlled using a nitrogen–gas cryostat and the temperature stability of the sample was better than 0.1 K.

Dynamic mechanical measurements have been performed by means of the Rheometrics RMS 800 mechanical spectrometer. Shear deformation has been applied under condition of controlled deformation amplitude, always remaining in the range of the linear viscoelastic response of studied samples. Frequency dependencies of the storage (G') and loss (G'') shear modulus have been measured at various temperatures. Geometry of parallel plates has been used with plate diameters of 6 mm and the gap between plates (sample thickness) was about 1 mm. Experiments have been performed under dry nitrogen atmosphere.

Frequency dependencies of G' and G'' measured within the frequency range of 0.1–100 rad/s at various temperatures have been used to construct master curves representing a broad range frequency dependence of these quantities. Only shifts along the frequency scale have been performed. This procedure provided a temperature dependence of shift factors ($\log(a_T)$ vs. T). The relaxation time corresponding to the transition between the Newtonian flow range at low frequencies and the glassy range at high frequencies at the reference temperature has been determined as $\tau(T_{\text{ref}}) = 1/\omega_c$, where ω_c is the frequency at which the G' and G'' dependencies cross each other. Relaxation times at other temperatures are given by $\tau(T) = \tau(T_{\text{ref}}) + \log a_T$.

The DSC measurements were performed on a Mettler 30 calorimeter. A liquid nitrogen system allows cooling experiments. The DSC cell was purged with 50 ml/min of nitrogen. Different rates (2, 4, 6, 8, and 10 K/min) were used in the temperature range from 173 to 373 K. The method consists of first cooling the sample followed by several heating and cooling ramps. At least two independently prepared samples with a weight of 10 mg were used for each rate.

The DSC curve of the eutectic mixture E7 for the rate 10 K/min (Fig. 2) shows two distinct transitions: a glass transition at T_g=211.8 K and an endothermic peak at T=334 K. The last one results from the nematic-isotropic transition. Thermograms corresponding to rates lower than 10 K/min exhibit the same two transitions. No further thermal event occurred in the temperature range between T_g and the nematic-isotropic transition. In particular, E7 does not present any crystalline state which would yield to a more complicated phase behavior.

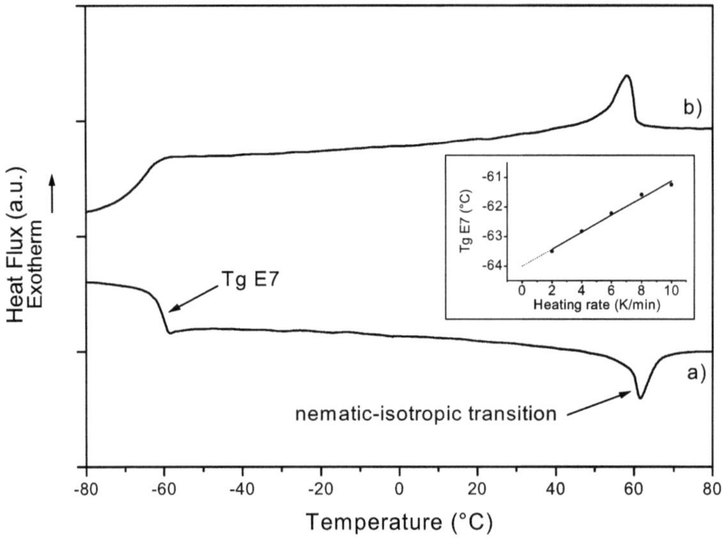

Figure 2. DSC thermogram of the eutectic mixture of the liquid crystals E7 for the second heating (a) and second cooling (b) ramp. The applied rate was 10 K/min. The inset exhibits the dependence of the glass transition as a function of the heating rate.

The inset represents the dependence of the glass transition as a function of the applied heating rate. Data analysis was carried out on the second heating ramp. In the given temperature range, the glass transition temperature increases linearly with the heating rate. The dotted line indicates the linear extrapolation to zero-heating rate yielding a T_g value of 209 K.

3. Results and discussion

The presentation of our experimental results we begin with discussion of dielectric spectra measured in temperature range from 200 to 373 K. All spectra have been collected during heating process, after rapid cooling of sample below glass transition temperature. As one can see in Fig. 3 the relaxation data cover eleven decades of frequency from 10^{-2} up to 10^9 Hz.

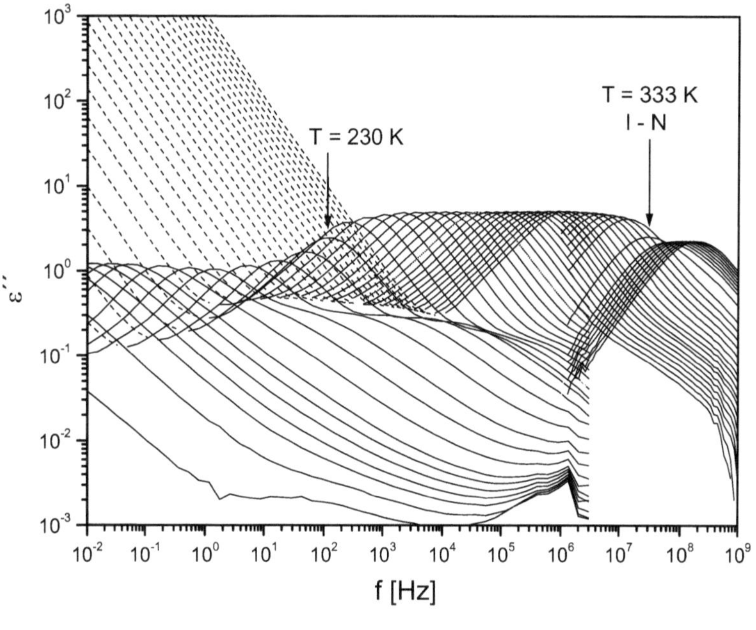

Figure 3. The dielectric loss spectra for E7 at different temperatures for a wide frequency range. At indicated temperatures the significant change of amplitude of primary relaxation occurs.

The magnitude of the main relaxation process changes enormously at two temperatures indicated in Fig. 3. At temperature $T_{I\text{-}N} = 333$ K the isotropic-nematic transition takes place. This dramatic increase of amplitude in $\varepsilon''(f)$ is related to the appearance of a high degree of order with respect to the direction of the molecular axes. In nematic phase, one can observe three relaxation modes. The most prominent relaxation process, which we will call primary relaxation, has its origin in 180° rotation of the molecules (5CB, 7CB, 8OCB) around their molecular short axes. This process is dielectrically active basically owing to a strong permanent dipole moment (5 Debye) provided by the cyano group parallel to the long molecular axis. The second process, feebly visible at lower frequencies most likely corresponds also to 180° rotation of molecular axis of 5CT. It is worth noting that Zeller[24] observed analogous low frequency relaxation arising from 5CT in two component mixture of 5CT and 5CB. Third peak occurring at higher frequencies will be assigned as α-process. In accordance to Zhong[25] α-peak is thought to be coupled to few modes with longitudinal and transverse components of the dipole moment with major contribution being from the rotation of the transverse dipole moments about their long axis. As it will be demonstrated later, this process is related to viscoelastic properties of E7. Additionally, all relaxation spectra exhibit an upgoing trend in ε'' at lowest frequencies. This part of spectrum, which has been shown in Fig. 3 as a dashed line follow a strict *1/f* dependencies. It proves that dc-conductivity process due to presence of the mobile charged carriers is also observed. In vicinity of the glass transition, approximately 20 K above the T_g, the magnitude of primary relaxation process again decrease. This change seems to be more distinct in comparison to isotropic-nematic transition. The same behavior of ε'' *(f)* at T = 230 K was also reported by Zhong.[25] Unfortunately, there is no clear explanation why such dramatic drop of amplitude of primary relaxation occurs. One can expect that this behavior is a manifestation of transition to new phase. However neither investigation by differential scanning calorimetry or structural measurements provide clear evidence for any transition in vicinity of T = 230 K. Therefore, further studies which help to resolve this problem are required.

In Fig. 4 a number of dielectric spectra measured near the T_g are shifted horizontally to the corresponding spectrum at a reference temperature. It shows that primary and the α-process overlap, forming single asymmetrical peak at glass transition temperature.

Significant differences in nematic supramolecular organization can be caused by applying a bias voltage to capacitor plates. Therefore, it is of interest to check the influence of the change in ordering extent on relaxation phenomena. To this end we carried out following experiment. The mixture was cooled from isotropic to glassy state with the same cooling rate as previously.

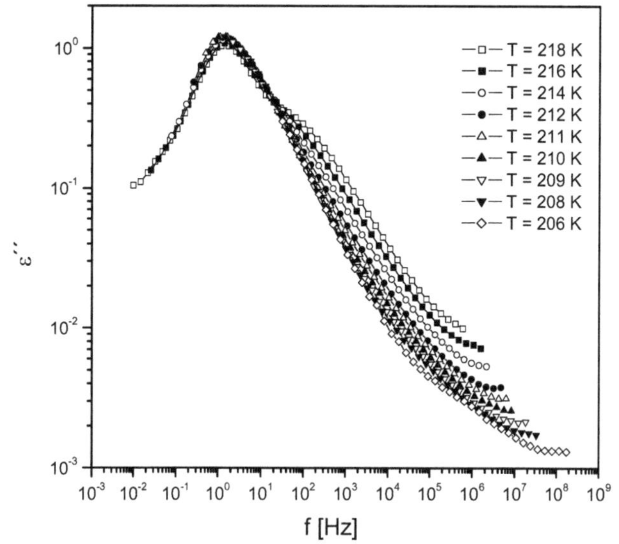

Figure 4. The normalized dielectric loss curve obtained by shifting loss curves measured at lowest temperatures.

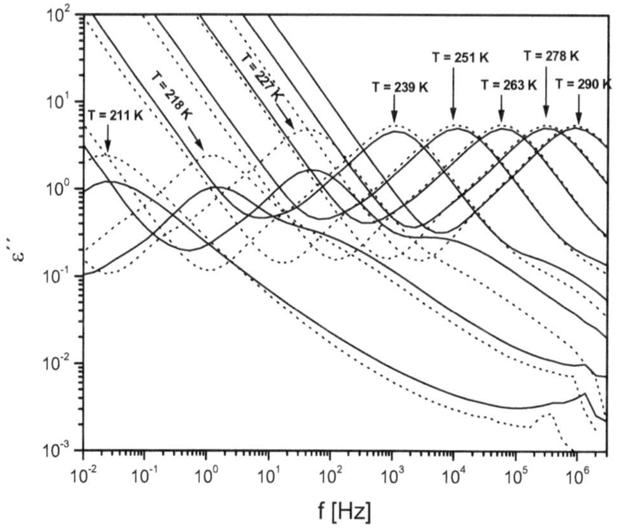

Figure 5. The effect of the bias voltage on the dielectric loss spectra (dotted lines). The solid lines represent loss spectra measured in the sample that was not treated with bias voltage.

During the cooling process, the constant value of bias voltage $U = 30$ V was applied to the capacitor plates. When the temperature of $T = 203$ K was achieved the bias voltage was switched off. Then, the dielectric spectra were recorded during the heating process with the bias voltage switched off. The results of these experiments are presented in Fig. 5.

The dielectric loss spectra of E7 obtained in the above experiment (dotted lines) are compared with the spectra (solid lines) without the bias voltage as shown in Fig. 3. Our observation regarding this comparison can be summarized as follows: firstly, intensity of peaks corresponding to the primary relaxation, which were measured close to T_g, are considerably higher for sample treated with the constant dc-voltage. At higher temperatures, above $T = 230$ K, this difference in principle vanish. As a consequence, the change of magnitude of primary relaxation in vicinity of $T = 240$ K is now less distinct. Secondly, intensity of α-relaxation process is clearly abated. Thirdly, the value of dc-conductivity, σ, decreased by about one order of magnitude. However, there is no effect of the bias voltage on relaxation times. It may mean that the change of ordering extent created by bias voltage does not influence the dynamic properties of the studied system.

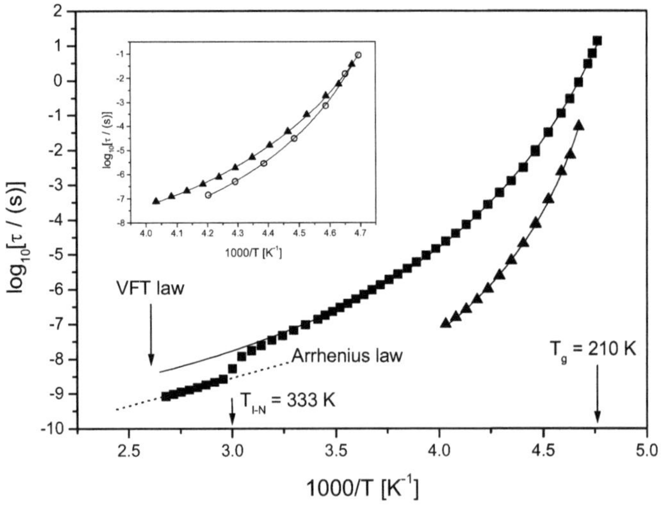

Figure 6. The temperature dependence of different relaxation processes determined from dielectric (closed symbols) and mechanical measurements (open symbols). The solid lines represent fits by means of the VFT law. The dashed line is a fit using the Arrhenius form.

In order to determine the corresponding relaxation times we analyzed the dielectric data using the Havriliak–Negami relaxation function. As three relaxation processes were generally observed in nematic phase, therefore, the experimental data are fitted by superposition of three relaxation function. In the fitting procedure, the term describing dc-conductivity: σ/ω was also included. However, analysis of dielectric spectra in isotropic phase was carried out by means of single HN function. Determined from fitting procedure the relaxation times for primary and α-relaxation processes are depicted in Fig. 6.

As a measure of relaxation rate is used of inverse of frequency at which relaxation peak shows a maximum: $\tau=1/2\pi f_{max}$. The temperature dependence of relaxation time associated with the primary relaxation exhibits two regimes. In isotropic state the log τ obey the activated temperature dependence according to

$$\tau = \tau_0 \exp\left(\frac{E}{RT}\right). \tag{1}$$

Thus, the motion of molecules is determined principally by the energy required for a rotation. The apparent activation energy E and the pre-exponential factor τ_0 equal to 34 J/mol and 1.41×10^{-14} s, respectively. In the nematic state, the relaxation data refuse to follow the Arrhenius law Eq. (1). Similarly, as it is observed in glass-forming liquids, the apparent activation energy of primary relaxation tends to increase significantly while approaching the glass transition. In this case, temperature dependence of log τ can be well reproduced by Vogel–Fulcher–Tammann law:

$$\tau = \tau_0 \exp\left(\frac{DT_0}{T - T_0}\right). \tag{2}$$

By fitting Eq. (2) to the experimental data the following values of parameters were found: $D = 6.51$, $T_0 = 169$, $\log(\tau_0) = -10.7$.

It is worth noting that the nematic–isotropic transition manifests itself in a discontinuity in log τ. However, "the second transition" which is visible in form decline of intensity of $\varepsilon''(f)$ at temperature T=240 K (see Fig. 3) has no at all influence on form of temperature dependence of the relaxation time. In other words, there is no clear evidence such transition in behavior of $\tau(T)$. In Fig. 6 the relaxation time associated with the α-relaxation that can also be parameterized using VFT law is also displayed. The variation in the relaxation time that occurs with variation in temperature is larger for the α-process than the primary relaxation. At the glass transition temperature, the value of relaxation times for both processes should be approximately the same. As a consequence, only one relaxation peak is observed at T_g (see Fig. 4).

It is well known that dielectric α-relaxation times in glass formers are usually closely related to viscosity. The question arises immediately which relaxation process detected in E7 from dielectric measurements is indeed related to viscosity. To this end we carried out mechanical relaxation measurements. The temperature–frequency measurements of the dynamic shear modulus G provide temperature dependence relaxation time, which are simply related to viscosity by Maxwell equation.

Comparison of the mechanical and the dielectric (α-process) relaxation data is presented in the inset in Fig. 6. The mechanical relaxation rate is found to change with temperature according to the VFT law and is closely related to dielectric α-process.

Direct comparison of the dielectric and mechanical spectra is presented in Fig. 7. The frequency dependencies of the storage (G') and loss (G'') moduli exhibit only one relaxation process which corresponds to the α-mode observed in dielectric spectrum of M''.

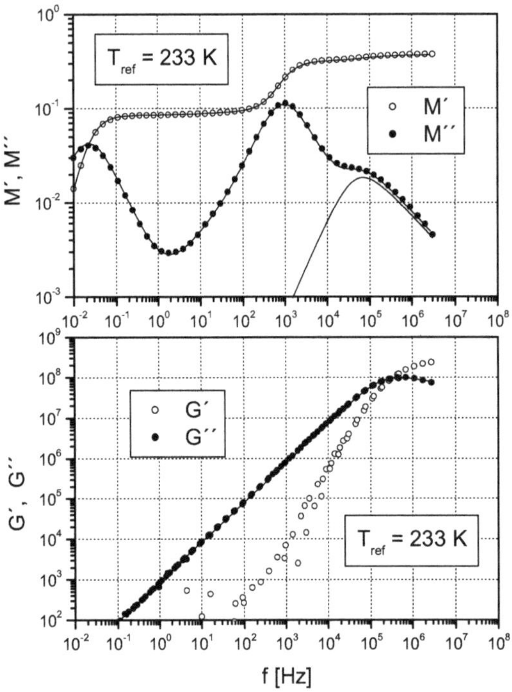

Figure 7. Top panel: the frequency dependencies of M' and M''. Bottom panel: the frequency dependencies of G' and G'' (master curve).

Note that we transformed the dielectric data to modulus representation using simple equation: $\varepsilon^*=1/M^*$. It results from fact that homogenous representation has to be chosen to get a correct result when the data obtained from different spectroscopic techniques are compared.

Finally, we also analyzed temperature behavior of dc-conductivity, σ, in nematic phase. Figure 8 shows the plot of σ against T.

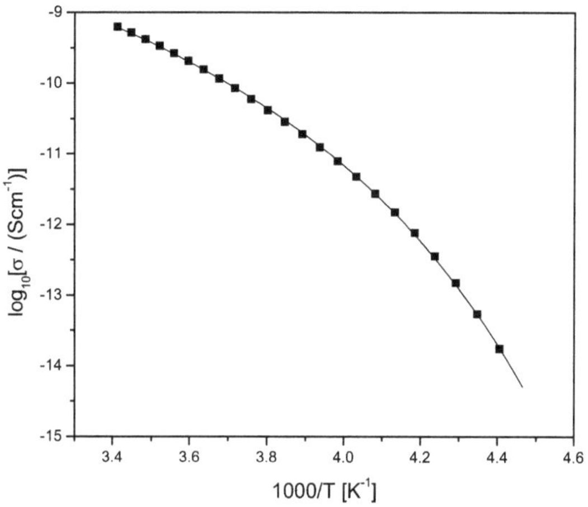

Figure 8. The temperature dependence of the relaxation times corresponding to dc-relaxation process. The solid line represents a fit to the VFT law.

One noteworthy feature is the strong curvature of $\sigma(T)$ in the activation plot. Therefore, activation form (Eq. 1) is ina dequate to describe this dependence. However, we found a satisfactory description of $\sigma(T)$ using again the VFT law. In this sense, the temperature dependence of σ behaves in analogous way as all others relaxation processes observed in E7 in the nematic phase.

Now, we return to discussion on the primary relaxation. Apart from the relaxation time, the important quantities are the shape parameters: $m = \alpha$ and $n = \alpha\beta$ which characterize the nonexponentiality of the relaxation. Additionally, these two parameters are shown to be uniquely related to distribution of relaxation times. The temperature behavior of m and n is plotted in Fig. 9. For comparison, the results for sample treated with bias voltage were also included.

In Fig. 9, one can distinguish three characteristic regions associated with peculiar behavior of shape parameters. In isotropic liquid non-Debye relaxation is observed. The value of n decrease approximately in linear way with decreasing temperature whereas m has almost constant value equals to unity. In nematic phase, the primary relaxation process exhibit quasi-Debye behavior which occurs in extraordinary broad temperature range from 333 K up to about 230 K. Below $T = 230$ K, the systematic broadening of this process is observed by approaching to glass transition. As a consequence the primary relaxation process has close to T_g clearly non-Debye character. Taking into account the correlation between non-exponentiality and deviation from Arrhenius behavior which were established for glass formers, above result seems to be obvious.

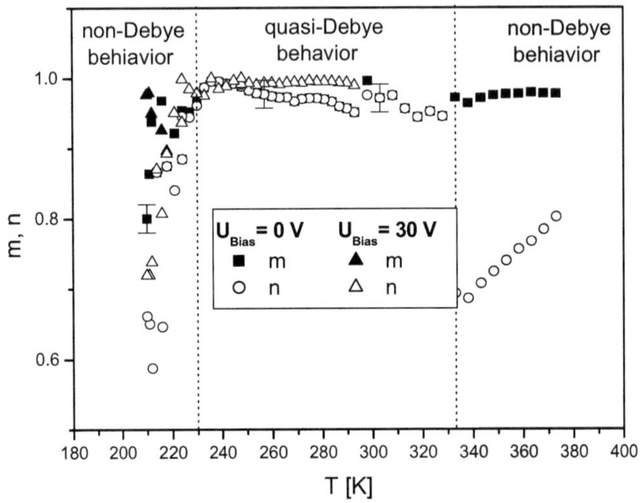

Figure 9. The shape parameters of dominated relaxation process obtained from fitting process of the HN relaxation function to the data in Fig. 1.

4. Conclusions

Using broadband dielectric spectroscopy we studied dynamics in glass-forming liquid crystal mixture E7. These studies were carried out in different phases of E7: isotropic liquid, nematic phase and glassy state. In isotropic liquid, the single non-Debye relaxation peak obeying the Arrhenius law is observed. Relaxation behavior in nematic phase is rather complex. The dominant relaxation process in nematic phase features the quasi-Debye behavior. However, the temperature dependence of relaxation time for this process is found to be well reproduced by the phenomenological Vogel–Fulcher–Tammann equation. Second relaxation process observed in nematic state is associated to the α-relaxation which takes place in all glass-forming liquids. On approaching to glassy state, both relaxation overlap to each other forming one relaxation mode at T_g.

The isotropic–nematic transition in E7 can be easily recognized basing on behavior both $\varepsilon''(f)$ and $\tau(T)$ dependencies. In case of relaxation times, this transition features by discontinuity in $\tau(T_{I-N})$ whereas $\varepsilon''(f)$ shows significant change in magnitude of $\varepsilon''(f)$ at T_{I-N}. Analogous changes in amplitude of $\varepsilon''(f)$

also takes places at temperature, approximately, 30 K above T_g. Unfortunately we did not find evidence of any phase transition at this temperature from both DSC and structural measurements.

By applying bias voltage to plates of capacitors one can change extend of ordering in E7. We found distingue effect of bias voltage on magnitude of $\varepsilon''(f)$ in vicinity glass transition temperature (below T_g+ 30 K). It was noted that bias voltage increases amplitude of $\varepsilon''(f)$ obscuring the "transition" at T_g + 30 K. Therefore, one can expect that this transition is related to change of molecules orientation.

We have also performed mechanical relaxation measurements on approaching to glass transition temperature. As a result, it was found that mechanical relaxation times correspond to dielectric α-relaxation times.

Acknowledgments

M. Mierzwa gratefully acknowledge the financial support of the Polish State Committee for Scientific Research (Project No. 1P03B 075 28). S. J. Rzoska was supported by the Polish Committee for Sci. Res. (KBN), grant no. 2PO 3B 034 25 for years 2003–2006 (grant. resp. S. J. Rzoska).

References

1. Vertogen, G. and de Jeu, W. H. (1998) *Thermotropic Liquid Crystals, Fundamentals: Springer Series in Chemical Physics 45* (Springer-Verlag, Berlin).
2. Demus, D., Goodby, J., Gray, G. W., Spiess, H.-W., and Vill ,V. (editors) (1998) *Handbook of Liquid Crystals, Fundamentals* vol. 1 (Wiley-VCH, Weinheim).
3. Jadżyn, J., Hellemans, L., Czechowski, G., Legrands, G., and Douali, R. (2000) Dielectric and viscous properties of 6CHBT in the isotropic and nematic phases, *Liq. Cryst.* **27**, 613.
4. Urban, S. and Wuerflinger, A. (1997) in *Advances in Chemical Physics*, (editors) I. Progogine and S. A. Rice (Wiley, New York), vol. XCVIII, p. 143.
5. Urban, S., Kresse, H., and Dąbrowski, L. (1997) Low frequency dielectric relaxiation process in liquid crystals with nematic and liquid-like smectic phases, *Z. Naturforsch* **52a**, 403.
6. Urban, S., Wuerflinger, A., and Gestblom, Bo. (1999) On the derivation of the nematic order parameter from the dielectric relaxation times, *PCCP* **1**, 2787.
7. de Gennes, P. G. and Prost, J. (1994) *The Physics of Liquid Crystals* (Oxford Univ. Press, Oxford).
8. Attard, G. S. and Luckhurst, G. R. (1987) Pretransitional behaviour in nematogenic mixtures. A molecular field theory, *Liq. Cryst.* **2,** 441.
9. Senbetu, L. and Chia-Wei Woo (1982) Isotropic-nematic transition: Landau-de Gennes vs. molecular theory, *Mol. Cryst. Liq. Cryst.* **84**, 101.

10. Sengupta, A. and Fayer, M. D. (1995) Theory of universal fast orientational dynamics in the isotropic phase of liquid crystals, *J. Chem. Phys.* **102**, 4193.
11. Torre, R., Tempestini, F., Bartolini, P., and Righini, R. (1998) Collective and single particle dynamics near the isotropic-nematic phase transition, *Phil. Mag. B* **77**, 645.
12. Mukherjee, P. K. (1998) The T_{NI}-T* puzzle of the nematic-isotropic phase transition, *J. Phys. Condens. Matt.* **10**, 9191.
13. Drozd-Rzoska, A., Rzoska, S. J., and Zioło, J. (1996) Critical behavior of dielectric permittivity in the isotropic phase of nematogens, *Phys. Rev. E* **54**, 6452.
14. Drozd-Rzoska, A. (1998) Influence of measurement frequency on the pretransitional behaviour of the non-linear dielectric effect in the isotropic phase of liquid crystalline materials, *Liq. Cryst.* **24**, 835.
15. Drozd-Rzoska, A. (1999) Quasicritical behavior of dielectric permittivity in the isotropic phase of *n*-hexyl-cyanobiphenyl in a large range of temperatures and pressures, *Phys. Rev. E* **59**, 5556.
16. Drozd-Rzoska, A., Rzoska, S. J., Zioło, J. (2000) Quasicritical behavior of dielectric permittivity in the isotropic phase of smectogenic *n*-cyanobiphenyls, *Phys. Rev. E* **61**, 5349.
17. Drozd-Rzoska, A., Rzoska, S. J., and Czupryński, K. (2000) Phase transitions from the isotropic liquid to liquid crystalline mesophases studied by linear and nonlinear static dielectric permittivity, *Phys. Rev. E* **61**, 5355.
18. Małecki, J. and Nowak, J. (1997) Nonlinear dielectric studies of intermolecular interactions and structure of 4-heptyl-4-cyano-biphenyl in benzene, *J. Mol. Liq.* **71**, 11.
19. Benguigui, L. (1984) Dielectric relaxation and nematic order in liquid crystals, *Phys. Rev. A* **29**, 2968.
20. Zeller, H. R. (1982) Dielectric relaxation in nematics and Doolittle's law, *Phys. Rev. A* **26,** 1785.
21. Zeller, H. R. (1982) Dielectric relaxation and the glass transition in nematic liquid crystals, *Phys. Rev. Lett.* **48**, 334.
22. Viciosa, M. T., Nunes, A. M., Fernandes, A., Almeida, P. L., Godinho M .H., and Dionisio, M. D. (2002) Dielectric studies of the nematic mixture E7 on a hydroxypropylcellulose substrate, *Liq. Cryst.* **29**, 429.
23. Drozd-Rzoska, A., Rzoska, S. J., Pawlus, S., and Zioło, J. (2005) Complex dynamics of supercooling *n*-butylcyanobiphenyl (4CB), *Phys. Rev. E* **72**, 031501.
24. Zeller, H. R. (1981) Dielectric relaxation in nematic mixtures, *Phys. Rev. A* **23**, 1434
25. Zhong, Z. Z., Schuelle, D. E., Gordon, W. L., Adamic K. J., and Akins, R. B. (1992) *J. Polymer Sci. B* **30**, 1443.

NONLINEAR DIELECTRIC SPECTROSCOPY NEAR SMECTIC A–SMECTIC C* TRANSITION IN FERROELECTRIC LIQUID CRYSTAL DOBAMBC

ALEKSANDRA DROZD-RZOSKA,
SYLWESTER J. RZOSKA, JERZY ZIOŁO
Institute of Physics, Silesian University, ul. Uniwersytecka 4, 40-007 Katowice, Poland

Abstract. Key results associated with nonlinear dielectric effect (*NDE*) studies in liquids are recalled. Particular attention is paid to advantages and disadvantages of basic experimental implementations. Subsequently, the first ever results for the novel, frequency selective, *NDE* technique are presented. They are based on measurements in *p*-(*n*-decyloxy)benzylidene-*p*-amino-(2-butyl)cinnamate (abb. DOBAMBC), a ferroelectric liquid crystalline compound, near the SmA – SmC* transition. This effect is compared with the *NDE* behavior in the isotropic phase on DOBAMBC.

Keywords: Nonlinear dielectric spectroscopy, chiral liquid crystalline compound, pretransitional behavior, strong electric field.

Nonlinear dielectric effect (*NDE*) describes changes of dielectric permittivity permittivity due to the application of a strong electric field, namely:[1–6]

$$NDE = \frac{\Delta\varepsilon(E)}{E^2} = \frac{\varepsilon(E) - \varepsilon(E \to 0)}{E^2}, \tag{1}$$

where ε and ε^E are dielectric permittivities in a strong and in a weak electric fields, respectively.

The first successful *NDE* measurements in liquids were carried out by Herweg[1] more than eight decades ago, namely in diethyl ether he noted the negative sign of *NDE*. Herweg[1] and Debye[2] associated the obtained *NDE* value with the orientation of noninteracting or weakly interacting permanent dipole moments in a strong electric field. Subsequently, basing on the extension of the Langevin–Debye series, as shown in Fig. 1, they derived the fundamental relation:[1–3]

$$NDE = -\frac{N\mu^4}{15k_B T^3} f\left(\varepsilon_S', \varepsilon_\infty'\right) R_S, \tag{2}$$

where N is the number of molecules in a unit volume, μ is for the permanent dipole moment, $f\left(\varepsilon_S', \varepsilon_\infty\right)$ denote the local field factor, $R_S \approx 1$ is

S. J. Rzoska and V. A. Mazur (eds.), Soft Matter under Exogenic Impacts, 215–227.

the nonlinear dipole–dipole correlation factor, $\varepsilon'_s, \varepsilon'_\infty$ denote the static dielectric permittivity and its infinite frequency counterpart.

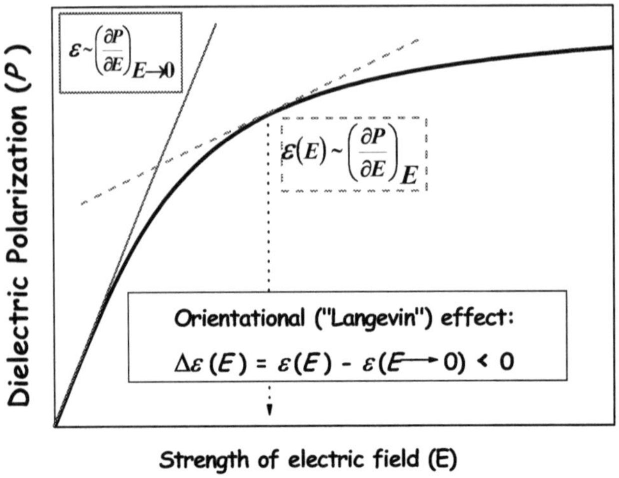

Figure 1. The behavior of dielectric polarization when increasing the intensity of a strong electric field. The value of *NDE* is determined by the difference of slopes of the dashed and the solid line.[1–3]

A decade later Piekara found that in nitrobenzene a positive sign of *NDE* may occur.[4] These results are recalled in Fig. 2, which presents measurements isothermic and concentrational behavior in a series of solutions composed of a dipolar component and a nondipolar solvent. The original studies of Piekara[3,4] carried out in nitrobenzene–carbon tetrachloride mixture are mimed by the authors measurements in nitrobenzene–hexene mixture. The structure of hexene parallels the structure of hexane what can be important when discussing *NDE* behavior in nitrobenzene–hexane mixture of limited miscibility, also presented in Fig. 2. For nitrobenzene–hexene and o-nitrotoluene–CCl_4 mixtures the negative sign of *NDE* is positive for mixtures dominated by pure dipolar compounds. For dissolved dipolar components or for nitropropane mixtures the *NDE* value is positive occurs in the whole range of concentrations. Its value follows relation (1), which predicts the linear dependence of *N*.

To explain the positive *NDE* in almost pure nitrobenzene, Piekara[3,4] suggested that for nitrobenzene the intermolecular coupling leading to the "scissor-like" arrangement of permanent dipole moments linked to nitrobenzene molecules may occur. Under a strong electric field the effective dipole moment of such bimolecular associates increases, leading to the positive sign of *NDE*.[3,4]

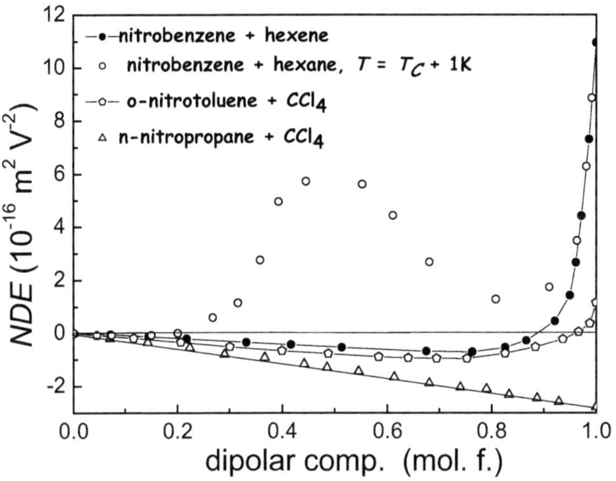

Figure 2. Isothermal behavior of *NDE* in selected solutions composed of a dipolar component and a nondipolar solvent (hexene, hexane, carbon tetrachloride). For the mixture nitrobenzene – hexane studies were carried out just above the critical consolute temperature ($T_C = 19.5^{\circ}$C). Curves in the figure are guides for eyes.

Piekara applied the same reasoning for explaining the anomalous increase towards positive values near the critical concentration of nitrobenzene–hexane mixture.[5] However, only in the early nineties it was shown that the latter effect is related solely to the anisotropic deformation of pretransitional fluctuations under strong electric field. Subsequently, the validity of the following relation was shown:[7]

$$NDE \propto \left\langle \Delta M^2 \right\rangle_V \chi_T \,, \tag{3}$$

where $\left\langle \Delta M^2 \right\rangle_V$ is the mean square of the local order parameter and χ_T is the susceptibility (compressibility).

In the homogeneous phase of binary critical mixtures $\left\langle \Delta M^2 \right\rangle_V \propto (T - T_C)^{2\beta}$, where $\beta \approx 0.325$ is the order parameter critical exponent and $\chi_T \propto (T - T_C)^{-\gamma}$, where $\gamma \approx 1.2$ is the critical exponent of susceptibility related.[7] T_C denotes here the critical consolute temperature. In subsequent years, this description was extended for the isotropic phase of nematic liquid crystals. In this case, $\left\langle \Delta M^2 \right\rangle_V = \Delta\varepsilon_f \Delta\varepsilon_0$, where $\Delta\varepsilon_f$ and $\Delta\varepsilon_0$ anisotropies of the perfectly ordered sample for the given frequency *f* and for the static limit; $\chi_T \propto \left(T - T^*\right)^{\gamma}$, where

$\gamma \approx 1$ within the mean field approximation. T^* denotes here the temperature of the extrapolated continuous phase transition.[7–9]

Although, the history of *NDE* studies in liquids is nine decades long, the number of reliable experimental results is still very limited.[3,9] This can be linked to enormous technical problems and the lack of commercial units for assembling measurements set-ups without very advanced technical background.[6,9] In the opinion of the authors important are also experimental artifacts hardly, if at all, mentioned in experimental reports so far. Probably, the latter factor is responsible for the fact that strongly different *NDE* values are reported for the same liquid, as one can conclude from data presented in Ref. 3. This fact shows that the knowledge of experimental solutions is an important factor for the critical analysis of *NDE* based results. To approach this issue a short discussion of *NDE* experimental techniques available so far is presented below. Next, the implementation of the novel technique of *NDE* measurements in ferroelectric liquid crystal DOBAMBC are shown given.

Generally, there are two principles for *NDE* measurements.[6] For the first one, a single strong electric field is used. In this case the detection of *NDE* is associated with charge-related distortions of the input signal. The spectrum of this signal should be perfectly well known and included into the output signal analysis. There are two inherent technical restrictions for the single-field *NDE* (*SF NDE*) method. Firstly, the product $frequency \times amplitude$ for the excitation (input) signal has to be limited. Secondly, the rise of the frequency of the electric field causes also the significant increase of the conductivity. This factor limits reliable results to $f_{strong} < 100\ \text{kHz}$ Consequently, the application of the *SF NDE* technique is limited to samples characterized by a very small electric conductivity and as large as possible values of $\Delta\varepsilon^E$. In practice this can be reached only for ferroelectric liquid crystals[10–14] or some polymer films.[15] The application of very high voltage input signals can significantly increase the output signal from which the value of $\Delta\varepsilon^E$ is calculated. Hence, measurements associated with a strong enough intensities of the electric field can facilitate the reliable estimation of $\Delta\varepsilon^E$. However, the high voltage input signal has to be separated from the end-module which registers the output signal. The separator also should not introduce any distortions of the signal. In practice for the *SF NDE* only voltages of the signal allowed by the manufactory of the end-module, usually $20-40$ V, are applied.[9–15] Hence, the desired high intensities of the electric field are reached by placing the tested sample in the capacitor with micrometer gap–usually $d = 5\ \mu\text{m} - 50\ \mu\text{m}$.[9–15] However, such extreme conditions may introduce a set of parasitic artifacts, namely: (i) surface-related effects: *particularly significant for anisotropic molecules* (ii) electrostriction based deformations[9,16] (iii) ionic-related heating: *always significant in liquids, particularly under the long-standing action of the strong electric field*[3] (iv) dust and gas impurities: *they can results in an enormous "false" NDE. For the SF NDE technique they are possible to detect only via the model-based analysis of the full set of data.*

All these artifacts are in practice absent for the second, dual-field *NDE* (*DF NDE*) principle of measurements. In this case changes of dielectric permittivity induced by a strong electric field are monitored via the radio-frequency weak, measuring field and the shift of the electric capacitance ΔC^E is registered. The strong electric field is applied here in the form of DC pulses, usually lasting few milliseconds. For the *DF NDE* method gas bubbles or dust impurities induce the unphysical deformation of the output signal. Similar deformation causes undesired ionic-related heating. Hence these artifacts can be detected "on line," immediately after their appearing. For the *DF NDE* method the measurement capacitor is most often placed in a resonant circuit of a generator.

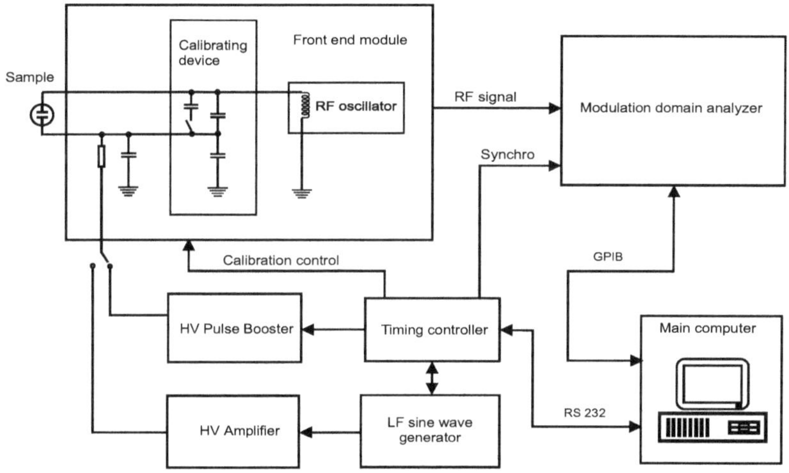

Figure 3. The scheme for the dual-field, single-generator and modulation domain nonlinear dielectric spectrometer.

In this case the shift of the resonant frequency $\Delta f^E = f(E) - f(E=0)$, proportional to ΔC^E and hence $\Delta\varepsilon^E$ is detected. The *DF NDE* technique enables measurements even extremely small (10^{-19} m^2 V^{-2})[3,9] values of the *NDE*. The desired high sensitivity and selectivity are inherent features for this implementation of *NDE* measurements. In fact for *DF NDE* much smaller intensities of the strong electric field are needed than for *SF NDE*. Moreover, there are no inherent problems associated with the separation between the high voltage source and the measurement end-module. Consequently, for *DF NDE* method, measurement capacitors with a bulk gap, ranging from 0.2 mm to 2 mm, can be applied.

Until recently, the basic idea of *DF NDE* measurements was based on the two-generators superheterodyne principle.[3,9] Unfortunately, the superheterodyne based apparatus suffered from enormous problems associated with grounding due to the inherently "scattered" design. It is worth recalling that for a typical capacitance of the measurement capacitor $C_{meas.} = 50 \div 100$ pF, the strong electric field-related changes are extremely small, namely $\Delta C^E = 5 - fF$. The fundamental problem for all *NDE* techniques was also the calibration of the output signal. In the opinion of the authors only the recently proposed reed-switch calibration unit proposed solves this problem.[17] In Ref. 6 the novel idea for the *DF NDE* apparatus, for which all problems encountered above seems to be absent, was proposed. This is the single-generator design from which the digitized signal is directly transferred the modulation domain analyzer (MDA- the registration end-module). The MDA is the equipment which appeared only recently due to its application in mobile telephones companies. The MDA extends possibilities of classical oscilloscopes, making measurements in the frequency-time domain possible.

In this contribution we present the novel extension of the single-generator *DF NDE* method for which the strong electric field can be additionally applied in the form of sine wave train (pulses). The design of such an apparatus is shown in Fig. 3. All results presented below were obtained using the apparatus presented in Fig. 3, for $f_{weak} = 3$ MHz and $U_{weak}^{peak-peak} = 1$ V. The sample was placed in flat parallel capacitor made from Invar, with 0.5 mm gap.

Studies presented were carried out for in *p*-(*n*-decyloxy)benzylidene-*p*-amino-(2-butyl)cinnamate (abb. DOBAMBC), the chiral ferroelectric liquid crystalline (FLC) compound which structure and the phase sequence is presented in Fig. 4. One may expect that for the sine wave pulses the selective *NDE* excitation associated with the frequency f_{strong} may occur. Moreover, for the AC pulse the ionic-related heating effects have less importance than for DC pulse of a strong electric field applied so far.

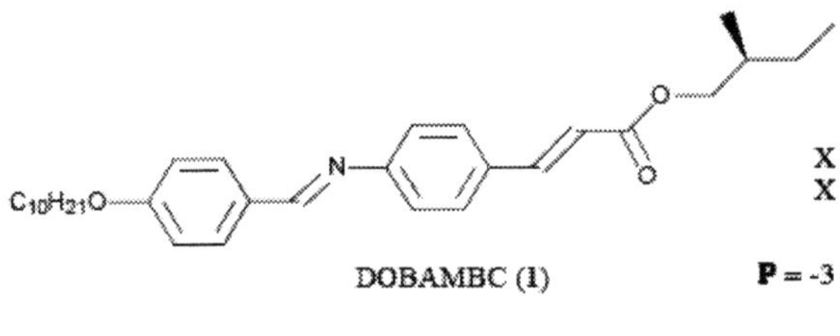

DOBAMBC (1)

X — 76 → C* ← 95 → A ← 117 → I
X ← F ← 63 — C*

P = -3 nC/cm2 ≅ -0.009 D/molecule @ 76°C

Figure 4. The ferroelectric liquid crystalline (FLC) molecule of *p*-(*n*-decyloxy) benzylidene-*p*-amino-(2-butyl)cinnamate (abb. DOBAMBC), its polymorphism, and polarity. The letter I is for the isotropic liquid, X is for the solid phase. The other letters are for Smectic mesophase. The polymorphism appearing on cooling and heating is given.[18]

Figure 5 presents the "classical" DC pulse and the novel sine wave pulse (AC) excitation.

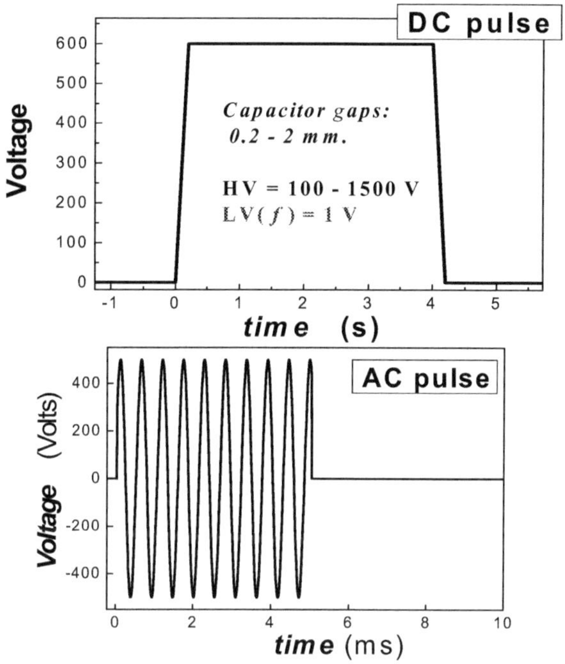

Figure 5. Forms of high voltage pulses applied in *NDE* studies in this paper.

The large nonlinear dielectric response can be expected here in the SmA phase of DOBAMC near the transition to the ferroelectric SmC* phase due to the softening of tilt fluctuations. In such a case the free energy density with the tilt angle θ and polarization P as order parameters is given by[12]

$$f = f_0 + \frac{1}{2}\left(T - T^*\right)\theta^2 + \frac{1}{4}b\theta^4 + \frac{1}{6}c\theta^6 + \frac{1}{2\chi}P^2 + \frac{1}{4}\eta P^4 - CP\theta - \frac{1}{2}\Omega P^2\theta^2 - PE, \quad (4)$$

where f_0 is the nonsingular part of f, χ is dielectric susceptibility, C and Ω are coupling constants, T_0 is related to the continuous phase transition temperature. The constant b is positive for the continuous transition and negative for the discontinuous transition. The other constants should be positive.

By minimizing the above relation in respect to θ and P, and by assuming that in a strong electric field $\theta(E) = \sum_{n=1}^{\infty} a_n E^n$ and $P(E) = \sum_{n=1}^{\infty} b_n E^n$ and finally by recalling the fact the n-th order component of dielectric permittivity

$\varepsilon_n = \partial^n D / \partial E^n$, where D denotes the dielectric displacement, one can obtain near the continuous phase transition, one can obtain[12]

$$\varepsilon_3 = -\frac{bC^4 \chi^4}{a^4 \left(T - T^*\right)^4}, \tag{5}$$

where ε_3 is the nonlinear dielectric response. When comparing relations (1) and (3) the prefactor $1/3$ should be introduced.

The validity of the above relation was shown for a variety of FLC using the *SF NDE* method.[12–14] In Fig. 6 we present the first ever results of frequency selective *DF NDE* (further abbreviated as *FS NDE*) using the double-field, single generator and modulation domain analysis principle presented above. The strong electric field was applied in the form of sine-wave pulses (up to $U_{peak-peak} = 600$ V) which induced changes of dielectric permittivity in the SmA phase, next sampled by the measuring radio-frequency field. The obtained *FS NDE* responses in the SmA phase of DOBAMBC are shown in Fig. 6.

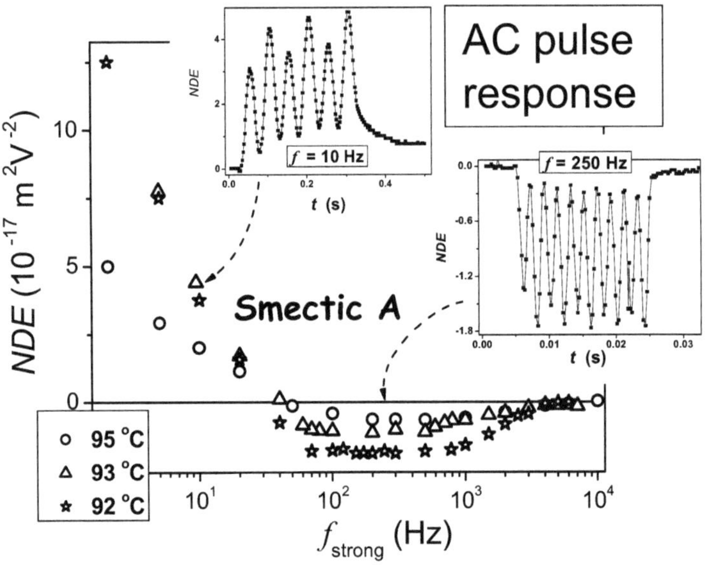

Figure 6. Basic results for the frequency selective *NDE* studies in SmA phase of DOBAMBC, on approaching SmC* ferroelectric mesophase. Insets show nonlinear dielectric responses associated with sine wave pulses of different frequencies, given in presented plots.

For lower frequencies, ca. $f_{strong} < 30$ Hz, the *FS NDE* response has a positive sign. For $150 < f_{strong} < 700$ Hz the detected values are negative and approximately constant. By the analogy to the broad band dielectric spectroscopy (BDS)[3] studies we will subsequently recall this region as the "static domain." For $f_{strong} > 10$ kHz the *FS NDE* response do not appear. Taking into account relation (3) one may conclude that *NDE* should detect mainly multimolecular *nano*- and *meso*-structures, contrary to BDS which is coupled mainly to the single molecule response. Hence, one may expect that for $f_{strong} > 10$ kHz such structures, probably pretransitional fluctuations, are not able to follow the excitation induced by a strong electric field. This factor may be considered as the reason that *NDE* relaxation time is 10^6 slower than the one obtained from BDS tests. Figure 6 also presents the form of *FS NDE* responses for different values of f_{strong}.

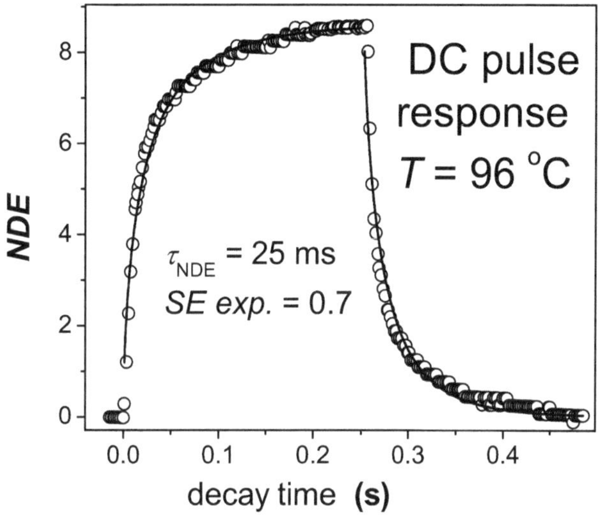

Figure 7. The NDE response induced by the DC pulse of a strong electric field in the vicinity of the SmA–SmC* transition. Solid curved are portrayed via the stretched exponential functions. $NDE(t) = NDE_{\infty}\left(1-\exp\left[-\left(t/\tau_{NL}\right)^{x}\right]\right)$ or $NDE(t) = NDE_{\infty}\exp\left[-\left(t/\tau_{NL}\right)^{x}\right]$, t denotes the elapse time and τ_{NL} is the nonlinear dielectric relaxation time; NDE_{∞} is for the saturated *NDE* value. Significant parameters are given in the figure.

The decay after switching off the strong electric field seems to present, within the limit of the experimental error, only for lower frequencies. It is noteworthy that for any tested frequency f_{strong} the decay was always described by a single exponential function, i.e., it was of the Debye-type. The DC strong electric field excitation resulted in the non-Debye decay, as shown in Fig. 7.

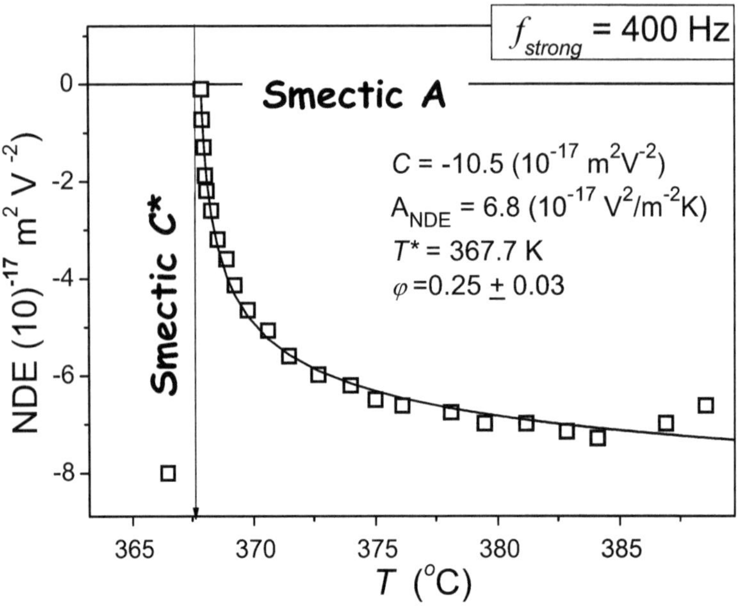

Figure 8. Results of the frequency selective *NDE* studies in SmA mesophase of DOBAMBC in the stationary nonlinear domain.

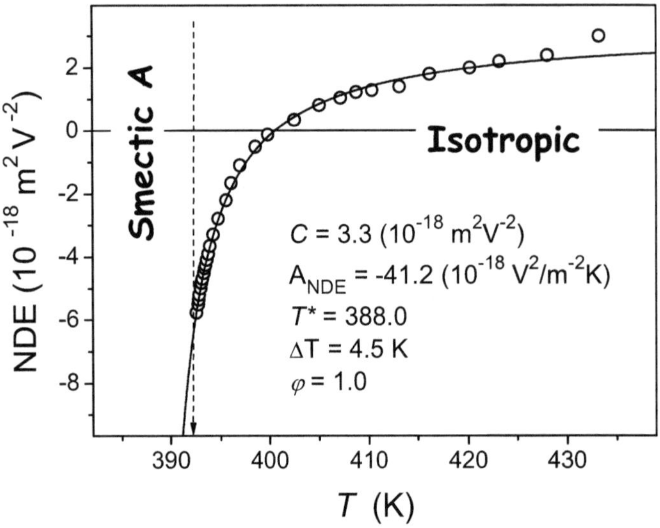

Figure 9. The stationary *NDE* response in the isotropic phase of DOBAMBC, on approaching the isotropic–smectic a transition. The response is associated with DC pulse excitation.

Figure 8 presents the temperature evolution of the *FS NDE* stationary response within the aforementioned "nonlinear static" domain. It is noteworthy, that the value of *NDE* follows the power like dependence described by the power function:

$$NDE(T) = const + \frac{A_{NDE}}{\left(T - T^*\right)^{1/4}} , \tag{6}$$

where A_{NDE} is the *NDE* amplitude and T^* within the limit of the experimental error coincides with SmA–SmC* phase transition what suggest the continuous nature of the transition.

The obtained dependence poorly coincides with relation. Moreover, $NDE\left(T^*\right) = 0$. Origins of such behavior, probably associated with the frequency selective nature of measurements and with the sampling frequency (3 MHz), are not clear for the authors. Figure 9 presents the pretransitional behavior in the isotropic phase of DOBAMBC. This result recalls preliminary measurements from Ref. 20 carried out yet by means of the superheterodyne *DF NDE* method and in more narrow range of temperatures. The obtained anomaly can be fairly well portrayed by relation (3) as shown by the solid curve in Fig. 9. Noteworthy is the first ever precise estimation of the discontinuity of the I–SmA transition for an FLC chiral compound. We would like also stress the negative sign of the *NDE* amplitude, which is beyond predictions of mean-field models for the isotropic. However, this can be explained basing on relation (3) and the structure of chiral molecules, as indicated in Ref. 19 for the case of I–N* transition in chiral isopentylcyanobiphenyl.

Concluding, this paper recalls the basic issues related to fundamental of nonlinear dielectric effect measurements, which can be considered as the natural extension of Broad Band Dielectric Spectroscopy studies. In this paper we present the first ever application of the frequency-selective *NDE*, the "nonlinear" dielectric method enabling a selective excitation of multi-molecular structures, for instance pretransitional fluctuations. This was illustrated via preliminary tests in the classical ferroelectric liquid crystalline compound DOBABMBC. The noteworthy result of this paper is the negative sign of the *NDE* pretransitional anomaly both in the isotropic liquid phase and in the SmA mesophase. It is worth recalling here that it was claimed that the only source of the negative sign of *NDE* is the orientational effect via relation (2). This paper and Ref. 19 show that this may be also the case of pretransitional anomalies in chiral liquid crystalline compounds.

Acknowledgments

Results presented above were supported by the Committee for Scientific Research (KBN, Poland) – Ministry for Higher Education and Science, via the grant 2PO3B 025 15 for years 2003–2006 (grant responsible: S. J. Rzoska).

References

1. Herweg, J. (1922) Über die Temperaturabhängigkeit der Dielektrizitäskonstanten von gasen und Dämpfen. *Z. Phys.* **3**, 36–42; Herweg, J. and Poetsch, W. (1920) Z. Phys. **29**, 105.
2. Debye, P. (1935) Die elektrischen Dipole in flüssigen Dielektricis. *Phys. Z.* **36**, 193.
3. Chełkowski, A. Dielectric Physics, *PWN-Elsevier* (Warsaw, 1980, 1990).
4. Piekara, A. and Piekara, B. (1936) *C. R. Acad Sci. Paris* **203**, 852–854
5. Piekara, A. (1936) Saturation electrique et point critique de dissolution, *C. R. Acad. Sci. Paris* **203**, 1058–1059.
6. Górny, M. and Rzoska, S. J. (2004) Experimental solutions for nonlinear dielectric studies in complex liquids, in S. J. Rzoska and V. Zhelezny (eds.), *Nonlinear Dielectric Phenomena in Complex Liquids. NATO Sci. Series II, vol. 157* (Kluwer, Brussel).
7. Rzoska, S. J. (1993) Kerr effect and nonlinear dielectric effect on approaching the critical consolute point, *Phys. Rev. E* **48**, 1136–1143.
8. Drozd-Rzoska, A. and Rzoska, S. J. (2002) Complex relaxation in the isotropic phase of n-pentylcyanobiphenyl in linear and nonlinear studies, *Phys. Rev.* E **65**, 041701.
9. Rzoska, S. J. and Zhelezny, V. (eds.) (2004) *Nonlinear Dielectric Phenomena in Complex Liquids*, *NATO Sci. Series II, vol. 157* (Kluwer, Brussels).
10. Kimura Y. and Hayakawa, R. (1993) *Jpn. J. Appl. Phys.* **32**, 4571.
11. Prasad Rao, S. K., Nair G. G., and Shankar Rao, D. S. (1999) Nonlinear dielectric response of a ferroelectric liquid crystal, *Liquid Crystals* **26**, 1187–1590.
12. Kimura, Y., Isono, H., and Hayakawa, R. (2001) Critical behavior in the SmA phase of chiral crystals, *Phys. Rev. E* **64**, 060701.
13. Kimura, Y. (2004) Frequency domain nonlinear dielectric relaxation spectroscopy, in S. J. Rzoska and V. Zhelenzny (eds.), *Nonlinear Dielctric Phenomena in Complex Liquids, NATO Sci. Series II, vol. 157,* (Kluwer, Brussels).
14. Kimura, Y. and Isono, H. (2004) Nonlinear dielectric spectroscopy of antiferroelectric liquid crystals in various smectic phases, *Ferroelectrics* **310**, 87–93.
15. Furukawa, T., Tada, M., Nakajima, K., and Seo, I. (1988) Nonlinear dielectric relaxation in vinylidene/vinyl acetate copolymer, *Jpn. J. Appl. Phys.* 27, 200 –204.
16. Roland, C. M. and Garret, J. T. (2004) Effect of constraints on electrostriction, in S. J. Rzoska and V. Zhelezny (eds.), *Nonlinear*

Dielectric Phenomena in Complex Liquids NATO Sci. Series II, vol. 157, (Kluwer, Brussels).

17. Górny, M., Zioło, J., and Rzoska, S. J. (1996) A new application of nonlinear dielectric effect for studying relaxation processes in liquids, *Rev. Sci. Instrum.* **67**, 4290–4293.
18. Demus, D., Goodby, J., Gray, G. W., Spiess, H. W., and Vill, V. (eds.) (1998) *Handbook of Liquid Crystals* (Springer, Berlin).
19. Drozd-Rzoska, A., Rzoska, S. J., Paluch, M. , Pawlus, S., Ziolo, J., Santangelo, P. G., Roland, C. M., Czupryński, K., and Dąbrowski, R. (2005) Mode coupling behavior in glass-forming liquid crystalline isopentylcyanobiphenyl, *Phys. Rev. E* **71**, 011508.
20. Zioło, J., Chrapeć, J., and Rzoska, S. J. (1989) Critical behavior in *p*-decyloxy-benzylidene-p-amino-2-methyl-butyl-cinnamate (DOBAMBC) studied by the nonlinear dielectric effect method, *Phys. Rev. A* **40**, 448–450.

CONFINED LIQUID CRYSTALINE 5CB IN 2D THERMODYNAMIC SPACE – PRELIMINARY DIELECTRIC RELAXATION STUDY

S. PAWLUS, J. OSINSKA, S. J. RZOSKA
Institute of Physics, University of Silesia, Uniwersytecka 4, 40-007 Katowice, Poland

S. KRALJ
National Centre for Scientific Research "Demokritos", 153 10 Aghia Paraskevi, Greece

G. CORDOYIANNIS
Laboratory of Physics of Complex Systems, Faculty of Education, University of Maribor, Koroška 160, 2000 Maribor, Slovenia

Abstract. Results of preliminary broadband dielectric spectroscopy studies in a wide range of temperatures and pressures range for a mixture of rod-like liquid crystalline 4-cyano-4-pentylalkylbiphenyl (5CB) and hydrophilic silica spheres (Aerosil 300) are shown. Pretransitional anomaly, observed previously in the bulk 5CB, has been found. Temperature dynamics of the mixture was investigated with via the DC conductivity σ, coupled to the reorientational relaxation. The derivative based analysis of electric conductivity showed a clear non-Arrhenius dynamics and indicated the anomalous increase of the fragility strength coefficient on approaching the isotropic–nematic transition. Pressure investigations of the solidification from the nematic phase showed the increase of the transition temperature on pressuring but with unusual increasing of dT_{NS}/dP coefficient.

Keywords: Liquid crystals, complex dynamics, random constraints, high pressures.

1. Introduction

The vast majority of studies on liquid crystals (LC) is related to bulk materials. Only recently results associated with the effect of confinement in restricted geometries appeared.[1–15] Nematic rod-like liquid crystals in restricted geometries are interesting because they are an example of an extremely complex model system. Moreover, surface interactions and orientational ordering can be investigated in details. Such studies also enable a better insight into cooperative phenomena, associated with the correlation length ξ strongly increasing with decreasing temperature. In the last years, many efforts have been

S. J. Rzoska and V. A. Mazur (eds.), Soft Matter under Exogenic Impacts, 229–238.

devoted to prove the existence of such a characteristic length scale by confining complex liquids to geometries of the order expected for ξ.[1–15] For instance, close to the glass transition temperature, ξ can be indirectly estimated to be in the range of a few nanometers.[1] The key idea of such studies is associated with the assumption that by confining a liquid in pore with a nanometer size one can reduce or remove intermolecular constraints linked to cooperativity. The dynamics of liquids in the confined geometry are found to speed up compared to the bulk sample at the same temperature.[2–11] However, for a selected matrix the relaxation dynamics showed a drastic increase of the relaxation rate. In liquid crystalline materials situation is even more interesting than in simple liquids because due to the complex sequence of phase transition coupled with the glass-like dynamics. Confined systems have a high surface-to-volume ratio, hence it is important to exploit this feature in experiments that are sensitive to physical behavior near the surface. Broadband dielectric spectroscopy (BDS) is one of fundamental research tools used to examining the influence of restricted geometry on dynamics in liquids. It is worth recalling that BDS made it possible to investigate an extraordinary wide range of frequencies. BDS studies on alkylcynobiphenyl rod-like liquid crystalline compounds (*n*CB) dispersed in porous matrices revealed slow relaxation processes absent in studies on bulk samples.[2] It was shown by Hourrin and coworkers[3] in BDS studies on mixtures of *n*CB and hydrophilic silica spheres (aerosil) that the nematic–isotropic phase transition temperature T_{IN} decreases with increasing silica concentration. The dominant relaxation process, which pertains the rotation of molecules around the short axis, is showed to follow an Arrhenius-like behavior in the nematic phase away from the T_{IN} temperature. In the vicinity of T_{IN}, the effective relaxation times decrease more gradually with increasing aerosil concentration, toward the isotropic phase value. It was suggested that this process is due to hindered rotation of the molecules located within surface layers.[3]

In this contribution, we report preliminary results BDS studies on liquid crystalline 4-cyano-4-pentylalkylbiphenyl (5CB) perturbed by the dispersion of 3% of aerosil 300. The broad applications of 5CB, its simple rod-like structure with the longitudinal permanent dipole moment and the convenient phase transition temperature, namely $T_{IN} \approx 35.2^{\circ}C$ and $T_{NS} \approx 12^{\circ}C$ ("S" is for the solid state) made from 5CB probably the most "classical" LC compound tested so far.[16,17] For the first time, studies were carried out both as a function of temperature and pressure. It is important to stress that the temperature shift influence mainly the activation energy whereas pressure is coupled to densification.

2. Experimental

The sample of 5CB was synthesized in the Military Technical University (Warsaw, Poland) by the group of R. Dąbrowski. As the randomized admixture, the hydrophilic aerosil type 300 was used. It was obtained due to courtesy of the Degussa Corp. It consists of 70-Å-diameter SiO_2 spheres with hydroxyl group covering the surface.

The temperature-dependent dielectric measurements were carried out using the Novocontrol Alfa impedance analyzer supported by the Quattro temperature control unit. The sample was placed in a parallel plate cell (diameter 20 mm, gap 0.1 mm). The temperature was controlled using a nitrogen-gas cryostat, with temperature stabilization better than 0.1 K.

For pressure measurements, the setup applied in our previous tests was used. It consists of a pressure chamber purchased from Unipress, Poland. As a pressure source, the Nova Swiss generator was used. Pressure was measured using Nova Swiss tensometric pressure meter with resolution 0.1 MPa. To stabilize the temperature, a jacket fed from Julabo HD 45S thermostat with external circulation surrounded the chamber. The sample was placed in a measurement capacitor designed in our lab. It has unique features: only 0.3 cm^3 of the tested sample is needed, the sample is in contact only with Teflon, Invar, and quartz, and the pressure can be both decreased and increased without a risk of contamination from the pressurized liquid (silicone oil). The capacitor was placed within the pressure chamber and pressure was transmitted to the sample by the deformation of 50 mm Teflon film.

3. Results and discussion

Figure 1 present real (ε') and imaginary (ε'') parts dielectric permittivity in 3% mixture of Aerosil 300 and 5CB. Measurements were conducted in the wide temperature range (80°C ··· –100°C). Noteworthy is the shift of the isotropic–nematic temperature down to T_{IN} = 25.5°C. In bulk *n*-cyanobipphenyls the anomalous decrease of the static dielectric permittivity in the isotropic phase on approaching T_{IN} occurs. It can be portrayed by the relation:[16,17]

$$\varepsilon'(T) = \varepsilon'^{*} + a(T - T^{*}) + A(T - T^{*})^{1-\alpha}, \qquad (1)$$

where $\varepsilon'(T)$ was measured for f=10 kHz (the static domain). The exponent $\alpha \approx 0.5$ is the specific heat exponent; T^{*} denotes the extrapolated temperature of hypothetical continuous transition (Fig. 2).

The analogy with the specific heat behavior is better visible when consi-sidering the derivative, namely:[16,17]

$$\frac{d\varepsilon'(T)}{dT} \propto \text{const} + (T - T^{*})^{-\alpha}, \qquad (2)$$

In fact, this dependence often enables more reliable fitting that Eq. (1), due to the significant reduction of the number of parameters. Relations (1) and (2) can be considered as the consequence of the dominant influence of prenematic fluctuations on the isotropic phase. It reflects the cancellation of permanent dipole moments ordered in an antiparallel way within prenematic fluctuations. Consequently, the static dielectric permittivity of fluctuations is smaller than the one for the isotropic surrounding.

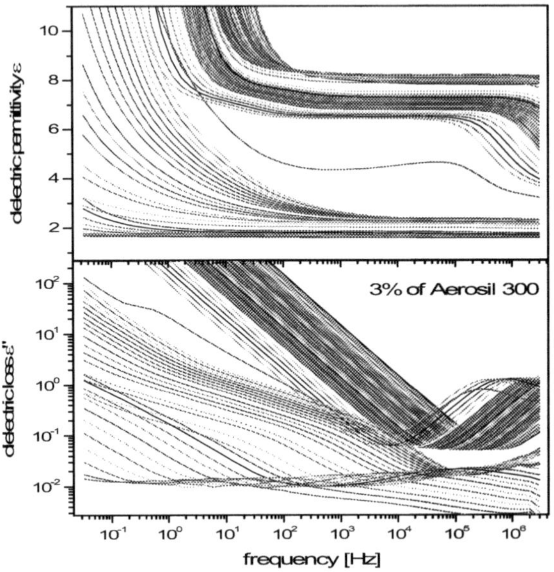

Figure 1. The frequency dependence of dielectric permittivity ε' and loss ε'' for 5CB–3% of Aerosil mixture at different temperatures.

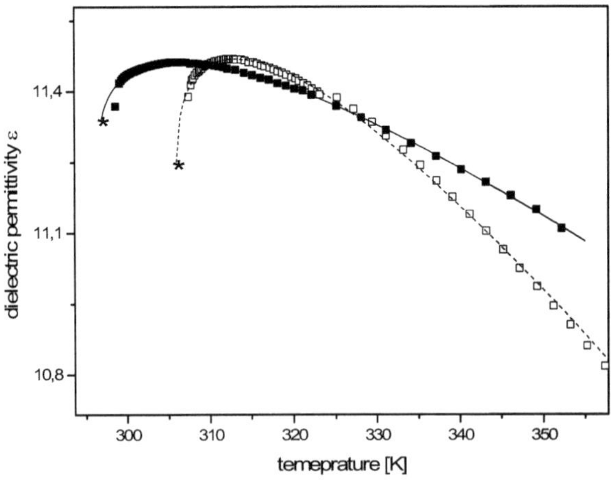

Figure 2. Temperature dependence of dielectric permittivity measured at 10 kHz for bulk 5CB (open) and mixture with Aerosil (solid). Curves are parameterized by relation (1).

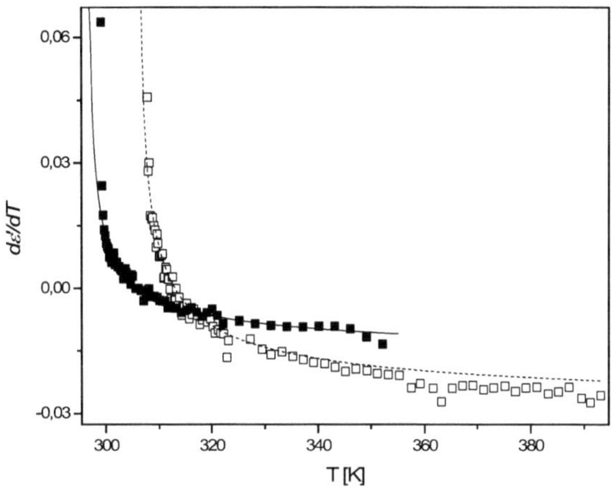

Figure 3. Derivative analysis of $\varepsilon'(T)$ data from Fig 2. The solid and the dashed curves are portrayed by eq. (2) with the same exponent $\alpha = ½$.

Hence, $\varepsilon'(T)$ has to decrease in the vicinity of $T_{I\text{-}N}$, where the volume occupied by fluctuation strongly increases. Values of T^* in bulk 5CB and for the mixture are strongly different, namely 35°C and 23.5°C, respectively. It is interesting that with addition of hydrophilic silica spheres increases the discontinuity of the phase transition, namely $\Delta T = T_{IN} - T^*$ increases up to ca. 2.1°C (Fig. 3). The increase of ΔT and the decrease of the strength of the prenematic effect in 5CB–Aerosil mixture, comparing to the bulk 5CB, can be connected with (partial) damping of prenematic fluctuations in isotropic phase by dispersed in Aerosil particles.

Because of crystallization, reorientational relaxation processes observed as peaks in dielectric loss spectra in 5CB–aerosil mixture are observed only in limited frequency range (see Fig.1). In consequence relaxation dynamics described by the relaxation times, calculated from the frequency of the peak maximum ($\tau = 1/2\pi f_{max}$) cannot be analyzed in the whole temperature range. However, it is well known that τ and and the DC conductivity (σ) are often co-related via Debye–Stokes is Einstein (DSE) relation, namely:[18–20]

$$\sigma\tau = \text{const} \tag{3}$$

Fortunately, the same situation was observed in 5CB. It means that relaxation dynamics of 5CB with addition of aerosil can be examined even in if relaxation times cannot be detected, basing on measurements of σ. Effects of such an analysis for 5CB–aerosil mixture are presented in Fig. 4.

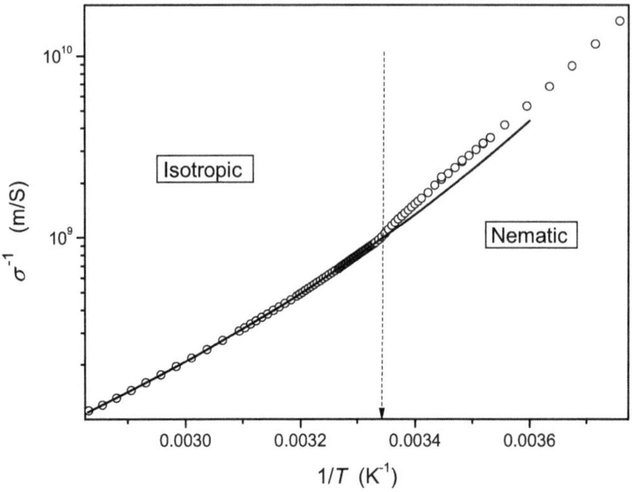

Figure 4. The Arrhenius plot of the electric conductivity in the mixture of 5CB and aerosil. The solid curve is for the VFT relation based on parameters obtained in the derivative plot in Fig. 5 for the high temperature domain.

The derivative based analysis of the same data in Fig. 5 revealed a clear non-Arrhenius behavior which can be described by a sequence of the Vogel–Fulcher–Tammann dependences, namely:[18–20]

$$\tau(T) = \tau_0^{\mathrm{VFT}} \exp\left[\frac{D_T T_0}{T - T_0}\right], \tag{4}$$

where D_T is the fragility strength coefficient and T_0 is the extrapolated glass temperature.

It was shown recently that the derivative based transformation can indicate domains of the validity of VFT relations and deliver D_T and T_0 parameters, namely:[17, 20]

$$\left[\frac{d\ln\tau}{d(1/T)}\right]^{-1/2} = \left[\frac{H_a}{R}\right]^{-1/2} = \left[H_a'\right]^{-1/2} = \left[(D_T T_o)^{-1/2}\right] - \frac{\left[T_0 (D_T T_0)^{-1/2}\right]}{T} = A - \frac{B}{T}, \tag{5}$$

where H_a is the apparent activation enthalpy and $T_0 = B/A$ and $D_T = 1/AB$.

The solid curve in Fig. 4 is based on the parameters obtained in this way. Noteworthy is the shift of T_0 towards higher temperatures and the extraordinary decrease of D_T on approaching the I–N transition. Such behavior resembles the one observed in bulk 5CB. It is noteworthy that none of dynamic crossover temperature, i.e., shifting from the one to another VFT domain, cannot be linked to the magic time scale 0.1 s suggested by Novikov and Sokolov for glassy liquids.[21]

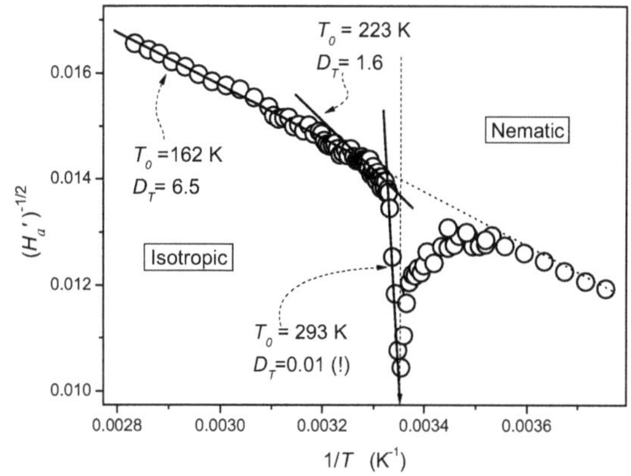

Figure 5. The derivative transformation plot of the data presented in Fig. 4. The linear regresion analysis applied for the following domains yields VFT parameters given in the figure.

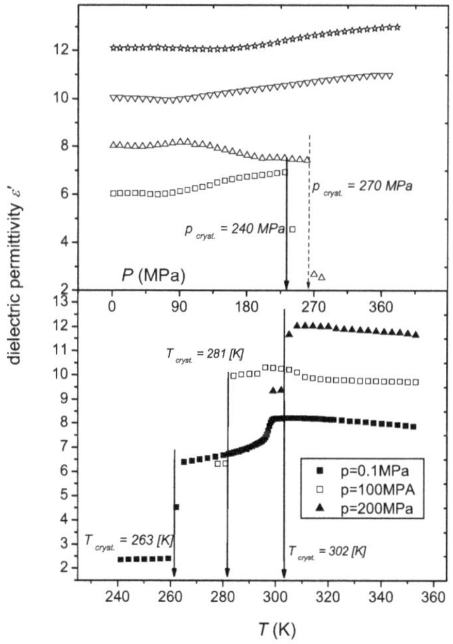

Figure 6. Temperature and pressure dependences of dielectric permittivity measured at 10 kHz for mixture of 5CB with aerosil. Arrows indicate temperatures of crystallization.

We would like to recall the similarity of the derivative analysis presented in Fig. 5 to the formal transformation of data $\phi_T = [d(\log(\tau))/d(1000/T)]^{-\frac{1}{2}}$ introducedby Stickel et al.[22] to detect the "magic" dynamic crossover.

Rough pressure measurements on the 5CB–aerosil mixture was conducted in pressure range up to 380 MPa. Isothermal and isobaric dependences the dielectric permittivities, measured at frequency 10 kHz are presented in Fig. 6. For the reasons of brevity data for different pressures and temperatures are separated by factor 2. It is visible that with increasing pressure, temperatures of crystallization T_{cryst} also increases but it is difficult to identify T_{IN} at elevated pressure because of not enough density of measured data points. Pressure dependence of T_{cryst} is showed in Fig. 7. From the figure it is obvious that pressure dependence of crystallization temperature is nonlinear. Particularly noteworthy is the fact, hardly observed in the bulk that the value of dT_{cryst}/dP increases with rising pressure.

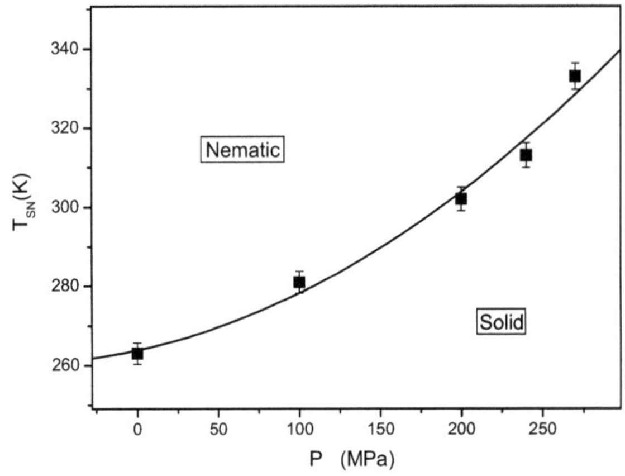

Figure 7. Pressured dependence of the solidification from the nematic phase of the 5CB–Aerosil mixture. Solid curve is a guide for the eyes.

4. Conclusions

We have investigated the effect of silica aerosil dispersion on nematic liquid crystal 5CB using broadband dielectric spectroscopy. The isotropic–nematic transition temperature is lowered by the aerosil admixture, in comparison with the bulk sample. We obtained the first ever evidence for the pretransitional anomaly of the static dielectric permittivity in a constrained nematic isotropic liquid crystalline compound. It is described by the same power exponent α, as in the bulk sample. However, the "constraint" seems to induce the increase of the discontinuity of *I-N* transition. We would like also to stress the non-Arrhenius dynamics coupled with a strong deviation from the

Debye–Stokes–Einstein dependence. The application of pressure increase phase transition temperatures both in the bulk and in the constrained samples. For the latter the makes it possible to compensate the influence of randomization. Consequently, in this was it is possible to estimate the negative pressure equivalent of the given randomized admixture.

Acknowledgments

Sylwester J. Rzoska was supported by the Polish State Committee for Scientific Research (KBN), grant 2PO3B 034 25 for years 2003–2006 (grant. resp. S.J. Rzoska). Sebastian Pawlus was supported by KBN (Poland) via grant. 5P03B 022 20) are gratefully acknowledged and a scholarship of the Foundation for Polish Science (FNP'05).

References

1. Iannacchione, G. S., Crawford, G. P., Žumer, S., Doane, J. W., and Finotello, D. (1993) Randomly constrained orientational order in porous glass, *Phys. Rev. Lett.* **71**, 2595.
2. Sinha, G. P. and Aliev, F. M. (1998) Dielectric spectroscopy of liquid crystals in smectic, nematic, and isotropic phases confined in random porous media, *Phys. Rev. E* **58,** 2001.
3. Hourri, A., Bose, T. K., and Thoen, J. (2001) Effect of silica aerosil dispersion on the dielectric properties of a nematic liquid crystal, *Phys. Rev. E* **63**, 051702.
4. Aliev, F. M., Nazario, Z., Sinha, G. P. (2002) Broadband dielectric spectroscopy of confined liquid crystals, *J. Non-Cryst. Solids* **305**, 218; Aliev, F. M. (2002) Molecular and collective relaxation in deeply supercooled confined liquid crystal, *J. Non-Cryst. Solids* **307–310**, 489.
5. Zakharov, A. V. and Thoen, J. (2004) Effect of silica aerosil on the nematic to isotropic transition: A theoretical treatment, *Phys. Rev. E* **69,** 011704.
6. Leys, J., Sinha, G., Glorieux, C., and Thoen, J. (2005) Influence on nanosized confinements on 4-*n*-decyl-4'-cyanobiphenyl (10CB): A broadband dielectric study, *Phys. Rev. E* **71**, 051709.
7. Donth, E. (1992) Relaxation and Thermodynamic in Polymers: Glass Transition, Akademie, Berlin.
8. Drozd-Rzoska, A., Rzoska, S., Pawlus, S., and Ziolo, J. (2005) Complex dynamics of supercooling *n*-butylcyanobiphenyl (4CB), *Phys. Rev. E* **72**, 031501
9. Arndt, M., Stannarius, R., Gorbatschow, W., and Kremer, F. (1996) Dielectric investigations of the dynamic glass transition in nanopores, *Phys. Rev. E* **54,** 5377; Arndt, M., Stannarius, R., Gorbatschow, W., Hempel, E., and Kremer, F. (1997) Length scale of cooperativity in the dynamic glass transition, *Phys. Rev. Lett.* **79,** 2077.
10. Barut, G., Pissis, P., Pelster, R., and Nimtz, G. (1998) Glass transition in liquids: two versus three-dimensional confinement, *Phys. Rev. Lett.* **80,** 3543.
11. Huwe, A., Kremer, F., Behrens, P., Schwieger, W. (1999) Molecular dynamics in confined space: from the single molecule to the liquid State, *Phys. Rev. Lett.* **80,**

2338; Kremer, F., Huwe, A., Arndt, M., Behrens, P., and Schwieger, W. (1999) How many molecules form a liquid? *J. Phys. Condens. Matter* **11**, A175; Kremer, F. (2002) Dielectric spectroscopy – yesterday, today and tomorrow, *J. Non-Cryst. Solids* **305**, 1.

12. Wang, L. -M., He, F., and Richert, R. (2004) Intramicellar glass transition and liquid dynamics in soft confinement, *Phys. Rev. Lett.* **92**, 095701.
13. Wübbenhorst, M., Klap, G. J., Jansen, J. C., van Bekkum, H., and van Turnhout, J. (1999) Glass transition of one-dimensional molecular chains of *p*-nitroaniline confined in $AlPO_4$-5 nanopores revealed by dielectric spectroscopy, *J. Chem. Phys.* **111**, 5637.
14. Ngai, K. L. (1999) Removal of cooperativity in glass-forming materials to reveal the primitive α-relaxation of the coupling model, *J. Phys. Condens. Matter* **11**, A119; Ngai, K. L. (2002) Relaxation in nanometer-size polymers and glass formers: application of the coupling model to some current problems, *Philosoph. Mag. B* **82**, 291.
15. Patkowski, A., Ruths, T., and Fischer, E. W. (2003) Dynamics of supercooled liquids confined to the pores of sol–gel glass: A dynamic light scattering study, *Phys. Rev. E* **67**, 021501.
16. Drozd-Rzoska, A., Rzoska, S., Pawlus S., and Ziolo, J. (2005) Complex dynamics of supercooling *n*-butylcyanobiphenyl (4CB), *Phys. Rev. E* **72**, 031501.
17. Drozd-Rzoska, A. (2006) Heterogeneity-related dynamics in isotropic *n*-pentylcyanobiphenyl, *Phys. Rev. E* **73**, 022501.
18. Casalini, R., Paluch, M., and Roland, C. M. (2003) Dynamic crossover in supercooled liquids induced by high pressure, *J. Chem. Phys.* **118**, 5701.
19. Pawlus, S., Casalini, R., Roland, C. M., Paluch, M., Rzoska, S. J., and Ziolo, J. (2004) Temperature and volume effects on the change of dynamics in propylene carbonate, *Phys. Rev. E* **70**, 061501.
20. Drozd-Rzoska, A. and Rzoska, S. J. (2006) On the derivative-based analysis for temperature and pressure evolution of dielectric relaxation times in vitrifying liquids, *Phys. Rev. E* **73**, 041502.
21. Novikov, V. N. and Sokolov, A. P. (2003) Universality of the dynamic crossover in glass-forming liquids: A "magic" relaxation time, *Phys. Rev. E* **67**, 031507.
22. Stickel, F., Fischer, E. W., and Richert, R. (1995) Dynamic of glass-forming liquids. I. Temperature-derivative analysis of dielectric relaxation data, *J. Chem. Phys.* **103**, 6251.

ANNIHILATION OF DEFECTS IN LIQUID CRYSTALS

MILAN SVETEC
Regional Development Agency MURA Ltd, Lendavska 5a, 9000 Murska Sobota, Slovenia and Laboratory of Physics of Complex Systems, Faculty of Education, University of Maribor, Koroška 160, 2000 Maribor, Slovenia

MILAN AMBROŽIČ
Jožef Stefan Institute, Jamova 39, 1000 Ljubljana, Slovenia

SAMO KRALJ*
Laboratory of Physics of Complex Systems, Faculty of Education, University of Maribor, Koroška 160, 2000 Maribor, Slovenia

Abstract. The annihilation of defect is studied theoretically in liquid crystals (LCs). We consider the annihilation of point disclinations in nematic and line edge dislocations in smectic A LC phase, respectively. We stress qualitative similarities in these processes. The whole annihilation regime is taken into account, consisting of the pre-collision, collision, and post-collision stage.

Keywords: Liquid crystals, edge dislocations, point disclinations, annihilation.

1. Introduction

The physics of topological defects[1] attracts scientists for years. They appear in phases that are reached via a continuous symmetry breaking phase transition. Because this phenomenon is very common in nature, defects are found in many (often completely different) physical systems. Note that the first theory on the origin of topological defects was made in cosmology.[2]

Due to their essentially topological basis the physics of defects exhibits many universalities.[1,3] Therefore, it is adequate to find a system in which some universal

* To whom correspondence should be addressed. e-mail: samo.kralj@uni-mb.si

S. J. Rzoska and V. A. Mazur (eds.), Soft Matter under Exogenic Impacts, 239–251.

features of defects can be studied relatively easily. For this purpose various liquid crystal (LC) phases[4,5] are very convenient by virtue of their fluidity, softness, optic anisotropy, and diversity of phases and structures. They can exhibit point and line defects in orientational ordering. In addition smectic phases exhibit also line defects in translational ordering. The defects in LCs are also called dislocations and disclinations, referring to the local non-unique description of orientational and translational hydrodynamic ordering fields, respectively. These fields are commonly represented by the nematic director field $\vec{n}$ in the nematic phase and the smectic layer phase field ϕ in the smectic phase.[4]

In general, defects are sources of relatively strong local elastic distortions. The local ordering at the defect site is consequently significantly different in comparison to the bulk ordering. This region of apparently different ordering is referred to as the *core* of defect. Its characteristic linear size is roughly given by the relevant order parameter correlation length ξ. The correct description of the core structure requires introduction of additional[6–8] order parameter fields. Their appearance mathematically removes the singularity at a defect site, which appear in the description solely using the relevant hydrodynamic field. The *core* structure of most common defects in LCs is rather well known, at least from theoretical point of view.[5–11]

Because defects are energetically costly, systems tend to minimize their number what is most often enabled by annihilation of defects. In this contribution, we make a comparative theoretically study of this process in the nematic and smectic A LC phase. We stress qualitative similarities between the annihilation of nematic point[12] and smectic A edge line[13] defects. The plan of the paper is as follows. In Section 2, we present the models we used. The nematic point defects are described in the semi-microscopic description. The dynamics is governed by the molecular Brownian dynamics. The smectic dislocations are studied with time dependent Landau–Ginzburg continuum type approach. The geometry of the problem is also presented. In Section 3, we compare simulation results obtained in the nematic and smectic phase. In the last section we summarize the results.

2. Theoretical background

In the following, we present the models that we used to describe annihilation of defects in the nematic and smectic A (SmA) phase and the geometries of the problem. We use different models in order to stress the universality of the phenomenon.

2.1. NEMATIC POINT DEFECTS

2.1.1. Model

In the nematic phase we use the semi-microscopic Lebwohl–Lasher model.[14] The pairwise interaction g_{ij} between a pair of rod-like LC *molecules* at $\vec{r}_i$ and $\vec{r}_j = \vec{r}_i + \Delta\vec{r}$, oriented along unit vectors $\vec{n}_i$ and $\vec{n}_j$, is expressed as[14]

$$g_{ij} = -\frac{J}{\Delta r^6}\left(\vec{n}_i \cdot \vec{n}_j\right)^2, \tag{1}$$

where J is a positive interaction constant. The *molecule* in our simulation corresponds in reality to a small cluster of real molecules. The average orientation of $\vec{n}_j$ over a few simulation cycles roughly corresponds to the nematic director field $\vec{n}$ in the continuum picture. The extent of fluctuation about this orientation yields nematic ordering fields. For example, the nematic uniaxial order parameter S at $\vec{r} = \vec{r}_i$ is roughly given by $S = \left\langle \frac{3}{2}\left(\vec{n} \cdot \vec{n}_i\right)^2 - \frac{1}{2} \right\rangle$, where $\langle \cdots \rangle$ describes the averaging over the simulation cycles and the first neighbors.

The interaction energy of the whole sample is calculated as the sum of all pair interactions in the sample. In the simulation we restrict the interaction volume to a sphere of radius $2a_0$, where a_0 stands for the characteristic length of the cubic lattice. The number of *molecules* in simulation ranges between 10^4 and 10^6.

A local orientation of the i-th *molecule* in the laboratory coordinate system (x,y,z) is parameterized as

$$\vec{n}_i = \left(\sin\theta\cos\phi, \sin\theta\sin\phi, \cos\theta\right). \tag{2}$$

The angles $\theta = \theta(\vec{r},t)$ and $\phi = \phi(\vec{r},t)$ are dynamic variables of the model. The rotational dynamics of the system is followed using the Brownian molecular dynamics.[15,16] At each time interval Δt (one sweep) the molecular orientation at the i-th site in the local Cartesian frame (x',y',z') is updated obeying the equations

$$\vartheta_i^{(x')} = -\frac{D\Delta t}{k_B T}\sum_{j\neq i}\frac{\partial g_{ij}}{\partial \vartheta_i^{(x')}} + \vartheta_{r,i}^{(x')},$$
$$\vartheta_i^{(y')} = -\frac{D\Delta t}{k_B T}\sum_{j\neq i}\frac{\partial g_{ij}}{\partial \vartheta_i^{(y')}} + \vartheta_{r,i}^{(y')}. \tag{3}$$

Here k_B is the Boltzmann constant, T is the temperature, and D the rotational diffusion constant. The z'-axis of the local frame is directed along the long molecular axis of the i-th molecule. The angles $\vartheta_i^{(x')}$ and $\vartheta_i^{(y')}$ correspond to small rotations about x' and y' axes. The gradient of the potential for these two rotations is calculated numerically. The quantities $\vartheta_{r,i}^{(x')}$ and $\vartheta_{r,i}^{(y')}$ are stochastic variables obeying the Gaussian distribution. The probabilities are centered at $\vartheta_{r,i}^{(x')} = \vartheta_{r,i}^{(y')} = 0$ and the widths $\Delta\vartheta_{r,i}^{(x')} = \Delta\vartheta_{r,i}^{(y')}$ of the distribution are proportional to $\sqrt{T}$.

2.1.2. Geometry

We study annihilation of nematic hedgehog and anti-hedgehog within the cylindrical tube of radius R. The pair of these defects within the geometry of the problem is schematically depicted in Fig. 1a. At the tube surface we enforce strong homeotropic anchoring (i.e., the orientation of LC molecules is set along the surface normal). At the ends of tubes, periodic boundary conditions are imposed. We enforce the defects by using the known analytic ansatz for their structures in the $\vec{n}$ representation.[17]

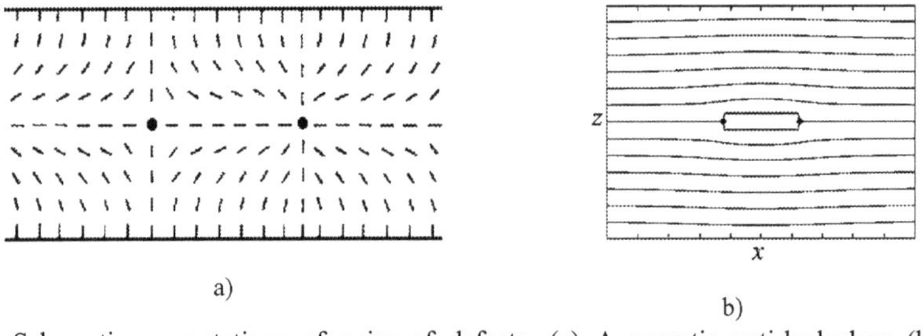

a) b)

Figure 1. Schematic presentations of pairs of defects. (a) A nematic anti-hedgehog (left) and hedgehog (right) point defect within a cylindrical confinement and (b) a pair of SmA edge dislocations. The smectic layers are stacked along the z-direction of the Cartesian coordinate system (x,y,z).

2.2. SMECTIC EDGE DISLOCATIONS

2.2.1. Model

Contrary to the nematic case, we describe the ordering in the SmA phase at the continuum level. The orientational ordering is defined by the nematic director field $\vec{n}$. The positional ordering is given by the smectic density wave $\psi = \eta^{i\phi} e$. The translational order parameter η measures the extent of smectic layering and the phase factor ϕ determines the layer position. In terms of these fields we express the smectic A free energy density contribution *f* as

$$f = \alpha|\psi|^2 + \beta|\psi|^4 + C_{||}\left|(\vec{n}\cdot\nabla\psi - q_0)^2\right| + C_\perp\left|(n\times\nabla)\psi\right|^2. \tag{4}$$

Here α and β are the material constants. The compressibility elastic constant $C_{||}$ tends to form a uniform density wave with the interlayer distance $d_0=2\pi/q_0$. The bending elastic constant $C_\perp$ tends to lock $\vec{n}$ along the smectic layer normal.

To study dynamics we use the time-dependent Ginzburg–Landau model. The corresponding dynamic equation reads

$$\gamma\frac{\partial\psi}{\partial t} = -\frac{\delta f}{\delta\psi^*}. \tag{5}$$

Here γ stands for the effective smectic viscosity constant and the symbol δ marks the functional derivative. This model[18] has been labeled by Halperin and Hohenberg as Model A. Model A usually describes systems in which there is no interaction between order parameter dynamics and classical hydrodynamics.

The scaled smectic order parameter can be expressed as $\psi = \eta e^{i\phi} = \eta e^{i(q_0 z - u)} = \psi e^{iq_0 z}$, where far from the defect u represents the conventional layer displacement field.[4] All singular contributions giving information about defect structure are contained in Ψ. In order to circumvent problems with the definition of the phase ϕ where the order parameter is small inside the defects cores, we parameterize the "bare" order parameter in terms of its real and imaginary components as

$$\Psi = \psi\, e^{-iq_0 z} = A + iB. \tag{6}$$

In the problem we limit to the case where the variational fields depend only on the x and z coordinates. We also neglect spatial variations in $\vec{n}$, which is strictly aligned along the z-axis.

2.2.2. Geometry

We monitor the annihilation of edge dislocation in the same SmA layer in the bulk sample. The layer configuration and the geometry is shown in Fig. 1b. We initially impose a pair of edge dislocations using the analytical ansatz derived in the continuum classical description[4] in terms of the displacement field u. The SmA layers are stacked along the z-axis, and the line defects run along the y-axis. Note that within this model the faced edge dislocations in the same SmA layer do not interact, what contradicts experimental observations.[19]

3. Results

Our simulations start with a pair defects. We henceforth label the nematic case with i) and the smectic with ii). The defects are well separated with respect to the i) nematic correlation length ξ_n and ii) smectic correlation length ξ_s, respectively. After a relatively short time, equal to a few i) nematic order parameter relaxation time τ_n and ii) smectic order parameter relaxation time τ_s, the defects locally attain quasi equilibrium profiles. These are presented in Fig. 2. Note that the nematic "point" defects adopt the ring-like structure, in accordance with the continuum type predictions.[8,20] In our simulation the points belonging to the *ring* (full lines in Fig. 2a) correspond to the largest connected local distortions in nematic ordering within the core of defect. On the contrary, in the case of smectic A edge dislocations the smectic ordering is melted (i.e., $\eta=0$) exactly at the defect origin, see Fig. 2b. The core structure $\eta(x,z)$ profile of well separated defects is cylindrically symmetric. For a general case we characterize the core structure with the sizes ξ_x and ξ_z. They measure the core radius along the x and z-axis, respectively. They are defined as the distances from the defect origin to the point, where the order parameter equals $\eta_b/2$. Here η_b stands for the smectic order parameter in the bulk phase.

In order to reduce elastic distortions of the system the defects approach. We refer to this stage as the *pre-collision* regime. Then the defects *collide* and annihilate. Finally, during the *post-collision* regime the equilibrium structures are gradually established. In the i) nematic phase the equilibrium structure corresponds to the so-called escaped director profile, shown in Fig. 3. In this defectless state the nematic director field structure is radial at the tube wall. On approaching the symmetry axis $\vec{n}$ gradually "escapes" along a chosen symmetry axis direction.

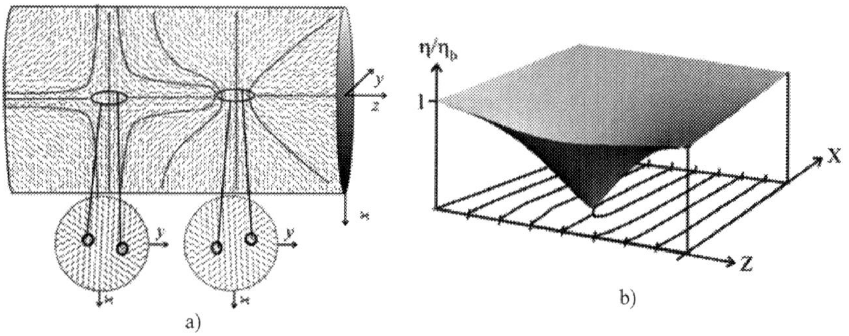

Figure 2. Core structure of defects. (a) The nematic "point" defects broaden into a ring like structures. With dotted lines we indicate the characteristic orientational ordering surrounding the anti-hedgehog (left) and hedgehog (right). Below we indicate the (x,y) cross-sections through the defects and (b) the core structure of the SmA edge dislocation. For well separated defects the $\eta(x,z)$ profile around the dislocation exhibits cylindrically symmetric profile.

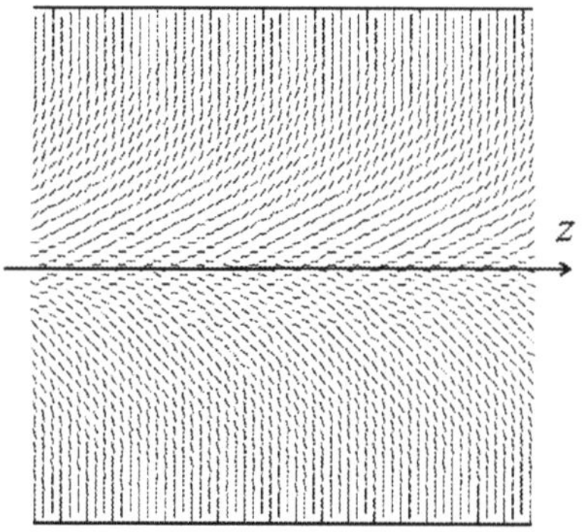

Figure 3. The escaped radial nematic structure.

In the pre-collision regime, the separation d between defects decreases with time. When $d \approx \xi$, the cores of defects are negligible influenced by the mutual attraction field. Here i) $\xi=\xi_n$ and ii) $\xi=\xi_s$ in the nematic and SmA phase, respectively. Consequently the defects can be treated to a good approximation as i) "point"-like and ii) "line"-like objects interacting via the relevant hydrodynamic field i.e., i) $\vec{n}$

and ii) ϕ. We refer to this case as the early *pre-collision regime*. The late *pre-collision regime* follows, where the cores become apparently influenced by the inter-defect attraction. The cores become prolonged along the effective attraction field direction, what is shown in i) Fig. 4a and ii) Fig. 4b. Representative hydrodynamic field structures are shown in i) Fig. 5a and ii) Fig. 6a. In the smectic case the order parameter contours are also superimposed, marking the lines with the same value of η.

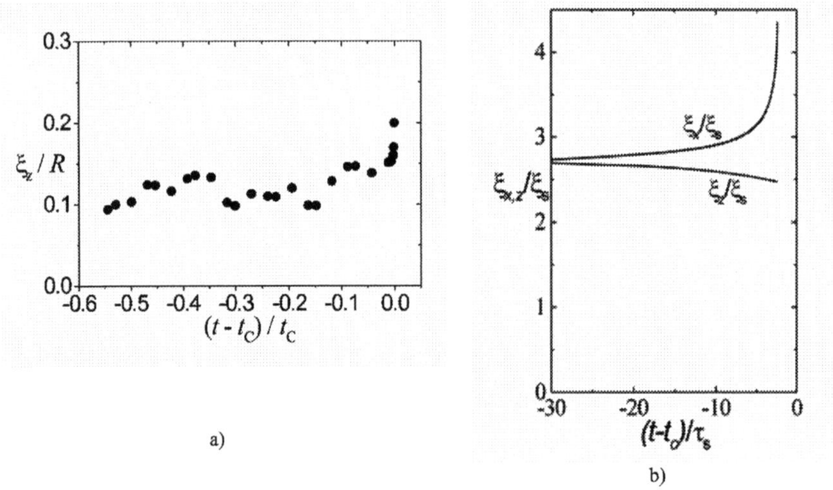

Figure 4. (a) The average ring radius as a function of time. At a given time the radii of hedgehog and anti-hedgehog are comparable and (b) the characteristic radii ξ_x and ξ_z of the edge dislocation core structure along the x and z axis. Prior the collision the value of ξ_x strongly increases. $\tau_s \approx \gamma|\alpha|^{-1}$.

The dynamics of defects in the pre-collision regime can be well fitted with the power law dependence

$$d = d_0\left(1 - \tfrac{t}{tC}\right)^{\chi}. \tag{7}$$

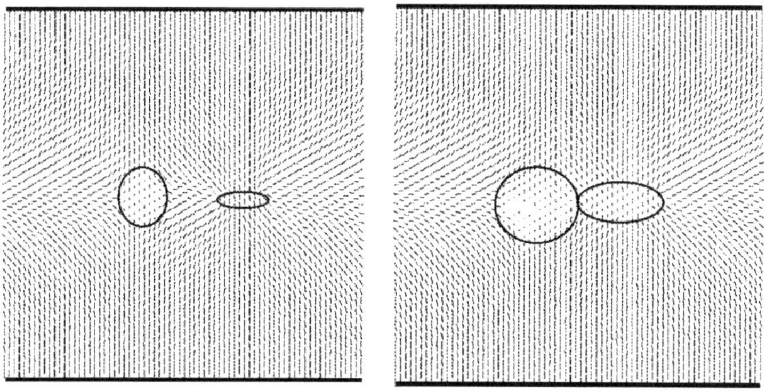

a) b)

Figure 5. Qualitatively different annihilation stages of the annihilation of nematic point defects in the (x,z) plane. (a) Early pre-collision and (b) late pre-collision regime.

Here d_0 stands for the initial separation of defects, t_c, ensures the time needed for defects to collide, and χ is the characteristic coefficient.

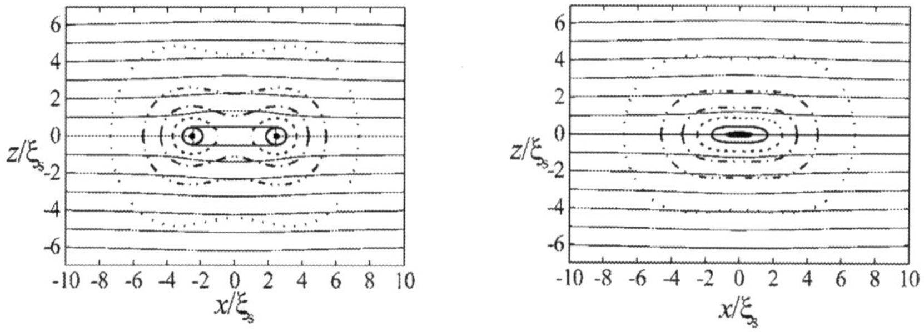

a) b)

Figure 6. Smectic layer profile with contour plot of $\eta(x,z)$ superimposed at different stages during the annihilation process. Contours correspond to η/η_b= 0.95, 0.8, 0.6, 0.4, 0.2, 0.05. (a) d=5ξ, and (b) d=0.

In the nematic case (Fig. 7a), the value of slowly changes as a function of d due to the surface induced inter-defect interaction screening. Namely, for the cylindrical confinement the inter-defect interaction becomes negligible for d>2R. In this case, the immediate environment of defects is dominantly influenced by the surface anchoring condition. For close enough defects, where the effect of confinement is negligible, we find $\chi \approx 0.5$.

In the smectic case (Fig. 7b), in which we consider the annihilation in bulk, we obtain $\chi \approx 0.5$ for all values of d. We obtain slight deviation from the behavior given by Eq. (7) at the initial simulation stage. This anomaly appears because our initially imposed structure slightly differs from the quasi-equilibrium profile that would be realized at a given value of d_0 in a "natural" way.

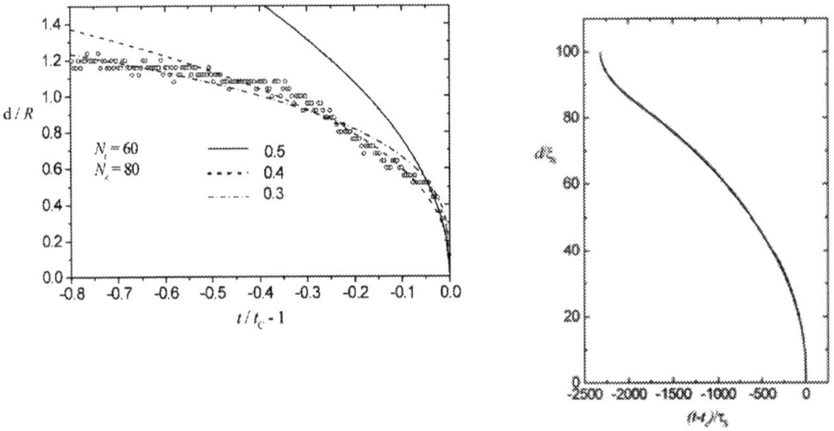

a) b)

Figure 7. $d = d(t)$ Dependence, (a) nematic point defects and (b) smectic edge dislocations.

In the collision, shown in i) Fig. 5b and ii) Fig. 6b, the defects merge to the extent that they become indistinguishable. At this event the defectless structure is formed. In the nematic case two *rings*, characterizing the defects, touch and merge into the single loop. The newly formed loop does not carry topological charge, and we therefore refer to it as the *chargeless ring*. In general this *ring* is twisted, because the relative orientation of well-separated defects is not correlated.

In the post-collision regime, the equilibrium structure is gradually established. In it the relevant order parameter attains spatially homogeneous profile. The corresponding hydrodynamic field is given by i) the escaped radial profile of $\vec{n}$ and ii) $\phi = q_0 z$. We distinguish between the early and late regimes. Early regime is characterized by apparent local variation of the relevant order parameter field. In i) the nematic case this is related to the untwisting and shrinking of the *chargeless ring*, shown in Fig. 8. In ii) the smectic case in the early post-collision regime the order parameter profile at the collision site exponentially (with the characteristic time $\approx 1.5\tau$) approaches the quasi-equilibrium $\eta(\vec{r})$ profile, fingerprinting the momentary $\phi(\vec{r})$ pattern. In the late post-collision regime the equilibrium profile is approached as $\eta_b - \eta \propto (t - t_C)^{-1}$.

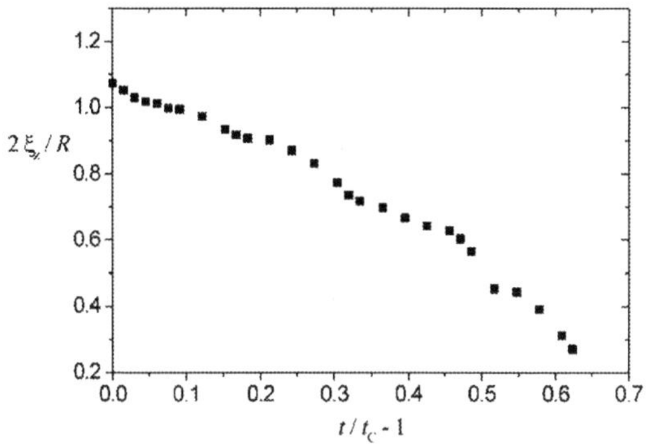

Figure 8. The time dependence of the chargeless ring size ξ_z in the post-collision regime. In the early regime the ring gradually shrinks and finally disappears.

In the late post-collision regime, the relevant order parameter is already essentially spatially homogeneous. Further relaxation toward the equilibrium is thus dominated by dynamic changes in the relevant hydrodynamic field.

4. Conclusions

We study theoretically the annihilation of defects in the nematic and SmA phase. In the nematic phase, we consider a pair consisting of a hedgehog and anti-hedgehog "point" defect within a cylindrical tube. The semi-microscopic description is used where the rod-like nematic *molecules* interact via the Lebwohl–Lasher pairwise interaction. The kinetics is studies by means of molecular Brownian dynamics, which enables us to access macroscopic time scales. In the smectic phase we study a pair of edge dislocations in the same smectic layer using the time dependent Landau–Ginzburg continuum model. Thus we study the annihilation of defects in the orientational and translational degree of ordering using completely different approaches. Despite this we observe striking similarities in the dynamic behavior.

The annihilation consists of the *pre-collision* and *post-collision stages*. In the *early pre-collision stage* the defects can be viewed as point or line-like objects. They possess inherent core structures and interact via the hydrodynamic field. In the nematic (smectic) case, this field corresponds to the nematic director (the smectic phase ϕ or, equivalently, the displacement) field, respectively. The core structure of defects is dominantly determined by the relevant order parameter. In the *late pre-collision stage* the inter-defect interaction affects the cores of defects. On approaching each other the cores become progressively prolonged along the direction of attraction. In the *collision* the defects merge and become undistinguishable. The defectless structure is formed which then gradually decays in the *post-collision stage* toward the equilibrium profile. The early and late *post-collision stages* are dominated by the relevant order parameter and hydrodynamic field dynamics, respectively.

References

1. Mermin, N. (1979) The topological theory of defects in ordered media, *Rev. Mod. Phys.* 51, 591–648.
2. Kibble, T. W. B. (1985) Topology of cosmic domains and strings, *J. Phys. A* **9**, 1387–98 (1976).
3. Zurek, W. H. (1992) Cosmological experiments in superfluid helium, *Nature* **317**, 505–508.
4. DeGennes, P. G. and Prost, J. (1993) *The Physics of Liquid Crystals* (Oxford University Press, Oxford).
5. Chuang, I., Durrer, R., Turok, N., and Yurke, B. (1991) Cosmology in the laboratory – Defect dynamics in liquid crystals, *Science* **251**, 1336–1342.
6. Schopohl, N. and Sluckin, T. J. (1987) Defect core structure in nematic liquid crystals, *Phys. Rev. Lett.* **59**, 2582–2584.
7. Schopohl, N. and Sluckin, T. J. (1988) Hedgehog structure in nematic and magnetic systems, *J. Phys. (France)* **49**, 1097–1109.
8. Kralj, S., Virga, E. G., and Žumer, S. (1999) Biaxial torus around nematic point defects, *Phys. Rev. E* **60**, 1858–1866.
9. Loginov, E. B. and Terentjev, E. M. (1987) Smectic dislocations, *Soviet Phys. Crystallogr.* **32**, 166–181.
10. Kralj, S. and Sluckin, T. J. (1995) Landau de Gennes theory of the core structure of a screw dislocation in smectic A liquid crystals, *Liq. Cryst.* **18**, 887–902.
11. Slavinec, M., Kralj, S., Žumer, S., and Sluckin, T. J. (2001) Surface depinning of smectic A edge dislocations, *Phys. Rev. E* **63**, 031705–031706.

12. Svetec, M., Bradač, Z., Kralj, S., and Žumer, S. (2004) Hedgehoge annihilation in a confined nematic liquid crystal, *Mol. Cryst Liq. Cryst.* **413**, 43–51.
13. Ambrožič, M., Kralj, S., Žumer, S., Sluckin, T. J., and Svenšek, D. (2004)Annihilation of edge dislocations in SmA liquid crystals, *Phys. Rev. E* **70**, 51704–51712.
14. Lebwohl, P. A. and Lasher, G. (1972) Nematic liquid crystal order – a Monte Carlo calculation, *Phys. Rev. A* **6**, 426–429.
15. Ermak, D. L. (1975) A computer simulation of charged particles in solution. I. Technique and equilibrium properties, *J. Chem. Phys.* **62**, 4189–4196.
16. Bradač, Z., Kralj, S., and Žumer, S. (2002) Molecular dynamics study of isotropic-nematic quench, *Phys. Rev. E* **65**, 021705–10.
17. Peroli, G. G. and Virga, E. G. (1996) Annihilation of point defects in nematic liquid crystal, *Phys. Rev. E* **54**, 5235–5241.
18. Hohenberg, P. C. and Halperin, B. I. (1977) Theory of dynamic critical phenomena, *Rev. Mod. Phys.* **49**, 435–479.
19. Kleman, M. and Lavrenovich, O. D. (2002) *Soft Matter Physics an Introduction* (Springer, Berlin).
20. Penzenstadler, E. and Trebin, H. R. (1989) Fine structure of point defects and soliton decay in nematic liquid crystals, *J. Phys. (France)* **50**, 1027–1040.

WAVES AT THE NEMATIC–ISOTROPIC INTERFACE: NEMATIC–NON-NEMATIC AND POLYMER–NEMATIC MIXTURES

V. POPA-NITA*
Faculty of Physics, University of Bucharest, P.O. Box MG-11 Bucharest 077125, Romania

T. J. SLUCKIN
School of Mathematics, University of Southampton, Southampton SO17 1BJ, United Kingdom

Abstract. We extend the theory for surface modes at the nematic-isotropic interface[1,2,3,4] to study comparatively two systems: nematic–non-nematic and polymer–nematic mixtures. We employ the dynamical generalization of the Landau–de Gennes model for the orientational (non-conserved) order parameter, coupled with the Cahn Hilliard equation for concentration (conserved parameter), and include hydrodynamic degrees of freedom. The theory uses generalized forms of the Landau–de Gennes and Doi free energy densities to include the coupling between the concentration of the non-nematogen and polymer fluids and the orientational order parameter. Two representative phase diagrams are shown. The method of matched asymptotic expansions is used to obtain a generalized dispersion relation. Further analysis is made in particular cases. Orientational order parameter relaxation dominates in the short wavelength limit, while in the long wavelength limit, the viscous damping processes become important. There is an intermediate region (depending on the temperature) in which the interaction between conserved parameter dynamics and hydrodynamics is important.

Keywords: Liquid crystal, polymer, mixture, orientational order parameter, conserved parameter, dynamics, dispersion relation.

1. Introduction

Most liquid crystals used in technological applications, such as electro-optical devices,[5] are mixtures. Often we have to consider inhomogeneous mixtures between thermotropic nematogen and a non-nematogen or a polymer. The dynamics of such materials involves treating not only the orientational order but also the concentration of non-nematogen or polymer, respectively. However, in

---

* To whom correspondence should be addressed. V. Popa-Nita, e-mail: v.popanita@gmail.com

S. J. Rzoska and V. A. Mazur (eds.), Soft Matter under Exogenic Impacts, 253–268.

a dynamical theory these two coupled variables have a different status. Orientational order is a non-conserved order parameter and normally relaxes diffusively. By contrast, concentration is a conserved order parameter, and relaxes through an induced current. The coupling of a conserved order parameter to a non-conserved order parameter, even when the latter is in some sense the driving force in the system, thus has profound effects on the dynamics. A subset of these effects is the subject of this paper.

In these systems, a biphasic region between the isotropic and the nematic phases appears below the nematic–isotropic transition temperature of the pure nematogen. When the system is thermally quenched from the stable isotropic phase into the biphasic region, fluctuations of concentration and of orientational order occur, and isotropic or nematic droplets can appear. By whatever early stage process, e.g., nucleation or spinodal decomposition, interfaces soon form.

In this paper, we analyze the problem of damping of capillary waves at these interfaces. For mixtures, the physics involves coupling time-dependent Ginzburg–Landau and Cahn–Hilliard equations[6] with hydrodynamic degrees of freedom. To study the nematic–non–nematic mixture, we use a generalized form of Landau–de Gennes free energy density[3] to include the coupling between the concentration of the non–nematic fluid and the orientational order parameter. In the case of polymer–nematic mixture, a generalized form of Doi free energy density[7] is used, with a different coupling term between concentration and orientational order parameter.[4]

The unperturbed or base state of the system is a planar isotropic nematic–isotropic interface in equilibrium. Due to thermal fluctuations, small amplitude monochromatic waves develop at the interface. We use a stability analysis of the equations to obtain their dispersion relation.

The paper is organized as follows. In the next section we introduce the free energies of a non–uniform nematogen–non–nematogen mixture as well as of a polymer–nematic mixture systems. We present the static phase diagrams and derive the interfacial tension between coexisting phases. In Section 3, we give the governing equations. We then present in Section 4 the dispersion relations corresponding to the two regions defined by the typical lengths in the problem. The general dispersion relation and numerical results are presented in Section 5. In Section 6 we draw some conclusions and present directions for future work.

2. Free energy

A thermotropic nematic–non-nematic and a polymer–nematic mixture are characterized by a conserved parameter and a non-conserved parameter. The conserved parameter is the concentration $c = N_1/N$ of the non-nematic or polymer components, respectively, where N_1 is the number of molecules of non-nematic (polymer) component and N is the total number of molecules. The scalar non-conserved parameter is the orientational nematic order parameter[8] S, which has the value 0 in the isotropic liquid and 1 in a fully oriented nematic phase.

Within the mesoscopic approach, the free energy functional is given by

$$F[c,S]=\int\left[f(c,S)+\frac{1}{2}K_c(\nabla c)^2+K_0(\partial_\alpha c)(\partial_\beta S)+\frac{1}{2}K_S(\nabla S)^2\right]dV, \quad (1)$$

where K_c and K_0 are phenomenological coefficients and K_S is the elastic constant.

2.1. NEMATIC—NON-NEMATIC MIXTURE SYSTEM

For a nematic–non–nematic mixture the bulk free energy density *f(c,S)* is given by

$$f=\frac{Nk_BT}{V}[(1-c)\ln(1-c)+c\ln c+\chi c(1-c)]$$
$$+(1-c)\left[\frac{3}{2}a[T-(1-c)T^*]-\frac{3}{4}BS^3+\frac{9}{4}CS^4\right], \quad (2)$$

where k_B is the Boltzmann constant, T is the absolute temperature, $\chi=U_0/k_BT$ is the Flory-Huggins interaction parameter related to isotropic interaction between unlike molecular species.[9]

The second parenthesis in, (2) is the Landau–de Gennes free energy density which governs the isotropic-nematic phase transition. The coupling between c and S results from microscopic considerations[3]. For a pure nematogen, ($c=0$), the free energy describes a weakly first order nematic-isotropic phase transition. For $T=T_{NI}=T^*+B^2/24aC$, the two phases, nematic (S_{nem} = B/6C) and isotropic ($S_{\mathrm{iso}}=0$) coexist in equilibrium. T^* is the undercooling limit temperature of the isotropic phase.

2.2. POLYMER–NEMATIC MIXTURE SYSTEM

Analogues, for a polymernematic mixture the bulk free energy density *f(c,S)* consists of the same two parts[7] as in Eq. (2)

$$f_{\text{mix}}(c) = \frac{Nk_B T}{V}\left[\frac{c}{n_p}\ln c + \frac{1-c}{n_l}\ln(1-c) + \chi c(1-c)\right]$$
$$+\nu(1-c)^2\left[\frac{1}{2}\left(1-\frac{U}{3}\right)S^2 - \frac{U}{9}S^3 + \frac{U}{6}S^4\right], \tag{3}$$

where n_p is the number of segments of the polymer, n_l is the axes ratio of the mesogenic molecules, $U = n_l \nu(1-c)$ and $\nu = U_a / k_B T$ parameterizes the orientation dependent (Maier–Saupe) interactions between the mesogen molecules.[7] The second parenthesis is now the Doi[7,10] free energy density which governs the isotropic–nematic phase transition.

For a pure nematogen, the free energy describes a weakly first order nematic–isotropic phase transition. For $U=U_{NI} = 2.7$, the two phases, nematic ($S_{\text{nem}} = 0.49$) and isotropic ($S_{\text{iso}} = 0$) coexist in equilibrium.

2.3. PHASE DIAGRAMS

The calculation of static phase diagrams requires only the bulk free energy *density. We now non-dimensionalize the free energy. For the nematic–non-*nematic mixture, the orientational order parameter is normalized with respect to its value at the transition $\bar{S} = S/S_{\text{nem}}$, using the reduced temperature $\tau = (T - T^*)/(T_{NI} - T^*)$. The dimensionless free energy density is now $\bar{f} = f/f_0$, where $f_0 = B^4/24^3 C^3$. Omitting the bar notation, the non-dimensional free energy density becomes:

$$f(c,S) = \Gamma[(1-c)\ln(1-c) + c\ln c + \chi c(1-c)] + (1-c)[(\tau+\lambda c)S^2 - 2S^3 + S^4], \tag{4}$$

where $\Gamma = Nk_B T/Vf_0$ and $\lambda = 24aCT^*/B^2$. Now, for a pure nematogen the phase transition temperature is $\tau_{NI} = 1$ and the corresponding value of the nematic orientational order parameter is $S_{\text{nem}} = 1$.

Similar, in the case of polymer–nematic mixture, the dimensionless free energy density has the following form:

$$f(c,S) = \frac{1-c}{n_l}\ln(1-c) + \frac{c}{n_p}\ln c + \chi c(1-c) + \nu(1-c)^2\left[\frac{1}{2}\left(1-\frac{U}{3}\right)S^2 - \frac{U}{9}S^3 + \frac{U}{6}S^4\right]. \tag{5}$$

Now, for a pure nematogen, the phase transition temperature is $\tau_{NI} = 1$.

The equilibrium conditions can be written as

$$\Delta g(c,S)=0;\quad \frac{\partial \Delta g}{\partial c}(c,S)=0;\quad \frac{\partial \Delta g}{\partial S}(c,S)=0, \tag{6}$$

where $\Delta g(c,S)=f(c,S)-f(c_{\rm iso},0)-\mu(c-c_{\rm iso})$ is the difference in Gibbs free energy density between the two phases and $\mu=\partial f(c_{\rm iso},0)/\partial c$ is the chemical potential.

The corresponding phase diagrams are plotted in Figs. 1 and 2. The solid curves refer to the binodal and the dotted lines shows the hidden first order nematic–isotropic transition, where the isotropic and nematic branches of the Gibbs free energy density are equal. The dashed–dotted lines are the spinodals. Figure 1 shows the phase diagram for nematic–non-nematic mixture for $\Gamma=\lambda=\chi=1$, while the corresponding phase diagram for polymer–nematic mixture is plotted in Fig. 2 for $n_l=2$, $n_p=10$, and $\chi/\nu=5$. For temperatures below T_{NI} ($\tau_{NI}=1$), there exists a two-phase coexistence region between an isotropic and a nematic phase (*I*+*N*).

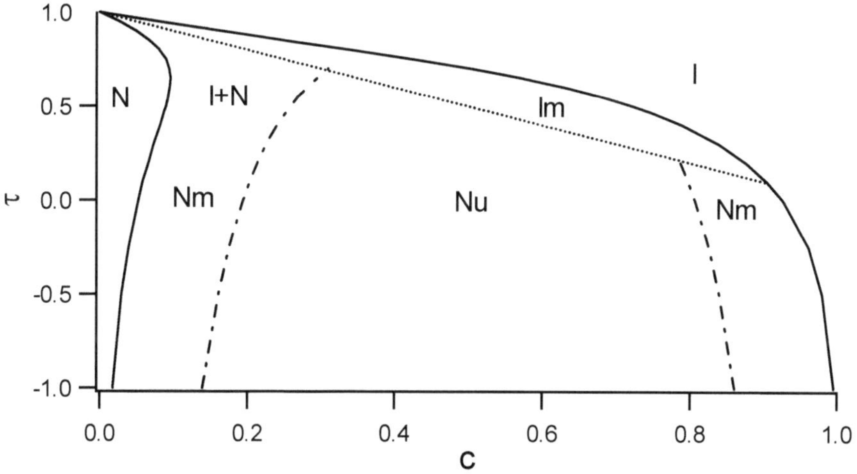

Figure 1. Phase diagram for nematic–non-nematic binary mixture. The solid curve refers to the binodal, the dotted line shows the hidden nematic–isotropic phase transition, and the dashed–dotted line is the spinodal. $\Gamma=\lambda=\chi=1$.

On decreasing the temperature (note that the undercooling limit temperature for isotropic phase T^* in the pure nematogen corresponds to $\tau^*=0$) the biphasic region broadens. Within the biphasic region, there are two different metastable regions: an isotropic metastable (Im) and a nematic metastable (Nm) region, as well as an unstable region of the nematic (Nu). In both phase diagrams we can distinguish two different regions. Near $\tau=1$, the concentration jump at the transition is very small, so that the important effect results from the non-conserved order parameter variation. On decreasing temperature, the concentration jump at the transition is larger and the variation of the conserved parameter becomes important. This effect is more important for nematic–non-nematic system then for polymer–nematic mixture, with the consequences on interfacial tension as well on the dynamics.

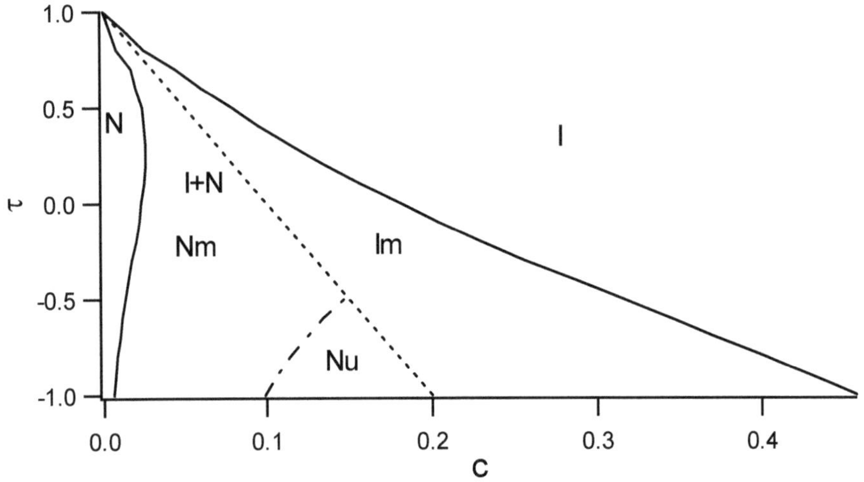

Figure 2. Phase diagram for polymer–nematic binary mixture. The solid curve refers to the binodal, the dotted line shows the hidden nematic–isotropic phase transition, and the dashed–dotted line is the spinodal. $n_l = 2$, $n_p = 10$, $\chi / \nu = 5$.

2.4. INTERFACIAL TENSION

We consider a planar nematic–isotropic interface in equilibrium and take the z-axis perpendicular to the interface. The interfacial tension, defined as the difference, per unit area of the interface, between the free energy of the system and that of the two phases if each were uniform and isolated, can be written

$$\gamma = \int_{-\infty}^{\infty} dz \left[f(c,S) + \frac{1}{2} K_c (d_z c)^2 + K_0 (d_z S)(d_z c) + \frac{1}{2} (d_z S)^2 \right]. \tag{7}$$

The corresponding interfacial tensions γ between the isotropic and nematic phases, for both systems, are plotted in Fig. 3. γ_S is the interfacial tension of an isotropic–nematic interface of the pure nematogen at $\tau = \tau_{NI} = 1$. Again the two regions can be distinguished. Near $\tau = 1$, the interfacial tension is very close to that of a pure nematic. On decreasing temperature, the concentration jump at the transition is larger and its influence on the interfacial tension becomes important. Due to the fact that the coupling term induces a larger difference in concentration in nematic–non-nematic system that in polymer–nematic mixture, the interfacial tension in the first system increases more rapidly with decreasing temperature.

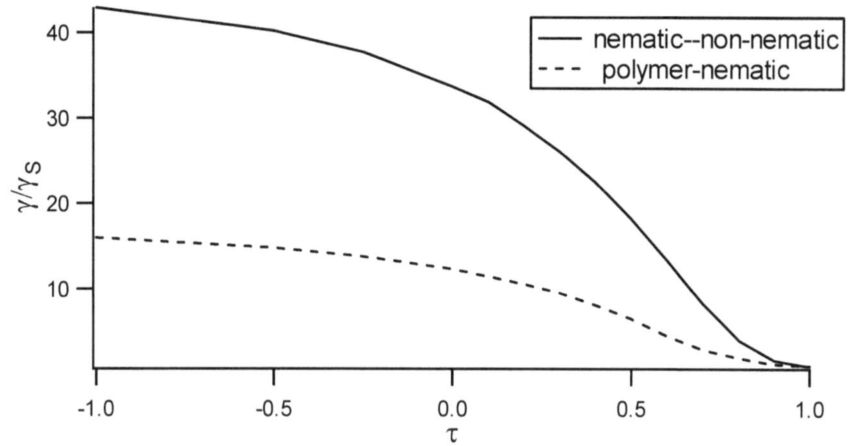

Figure 3. Interfacial tension between the isotropic and nematic phases plotted against the reduced temperature. The solid curve refers to the nematic–non-nematic mixture and the dashed line corresponds to the polymer–nematic mixture.

3. Equations of motion

The equations of motion for an isothermal, incompressible fluid can be written as follows:[11–13]

$$\nabla \bar{v} = 0,$$

$$\rho \frac{d\bar{v}}{dt} = -\nabla p + \eta \nabla^2 \bar{v} - \nabla\left(\frac{\delta F}{\delta \Delta S}\right)\nabla S + \frac{\delta F}{\delta c}\nabla c,$$

$$\frac{dc}{dt} = \Gamma_c \nabla^2 \left(\frac{\delta F}{\delta c}\right),$$

$$\alpha \frac{dS}{dt} = -\frac{\delta F}{\delta S}, \tag{8}$$

where ρ is the density, p is the pressure, Γ_c is transport coefficient assume to be constant, and α is the rotational viscosity of nematogen.

The three typical lengths of the problem are as follows:

(i) The first length is related to the order parameter itself. It strongly varies across the interface over a typical distance known as the microscopic correlation length $l_S = (K_S S^2/f_0)^{1/2}$ that for 5CB has the value 5×10^{-6} cm. This length gives the typical width of the order parameter profile within the interface.

(ii) The second microscopic length is related to the concentration variation inside the interface $l_c = (K_c/f_0)^{1/2}$ that for 5CB has the value 3×10^{-5}cm.

(iii) The third (macroscopic) length is associated with the flow induced by the motion of the interface and can be defined as: $l_\eta = \eta^2/\rho\gamma$. For 5CB, it has the value 1 cm. This length separates the inertial from the viscous regimes; the corresponding value of the Reynolds number is unity.[2] In the following, we shall use $t_\eta = \eta^3/\rho\gamma^2$ as unit time, which is the typical relaxation time of a perturbation of size l_η and for 5CB has the value 10 s.

In the following, the ratio $\varepsilon = l_S/l_\eta = l_c/l_\eta \cong 10^{-5}$ will constitute the small parameter of the theory.

Measuring length in unit of l_η and time in unit of t_η, and considering a two dimensional flow with horinzontal and vertical velocity components u and w in the x and z directions, respectively, we obtain the non-dimensional form of working equations:

$$\partial_x u + \partial_z w = 0,$$

$$\rho \frac{du}{dt} = -\partial_x p + \eta\nabla^2 u - \varepsilon^2(\nabla^2 S + \nabla^2 c)\partial_x S + \left(\frac{\partial f}{\partial c} - \varepsilon^2\nabla^2 c - \varepsilon^2\nabla^2 S\right)\partial_x c,$$

$$\rho \frac{dw}{dt} = -\partial_z p + \eta\nabla^2 w - \varepsilon^2(\nabla^2 S + \nabla^2 c)\partial_z S + \left(\frac{\partial f}{\partial c} - \varepsilon^2\nabla^2 c - \varepsilon^2\nabla^2 S\right)\partial_z c,$$

$$\alpha \frac{dS}{dt} = -\frac{\partial f}{\partial S} + \varepsilon^2\nabla^2 S + \varepsilon^2\nabla^2 c,$$

$$\frac{dc}{dt} = \Gamma_c\nabla^2\left(\frac{\partial f}{\partial c} - \varepsilon^2\nabla^2 c - \varepsilon^2\nabla^2 S\right). \quad (9)$$

In what follows we shall suppose that the stationary planar nematic–isotropic interface (the base state of the system) is situated at $z = 0$, such that the nematic lies in the region $z < 0$ and the isotropic phase in the region $z > 0$. The x axis is taken in the direction of the wave vector $\vec{k}$ of the perturbation along the interface. This is possible without loss of generality, as the system is isotropic in x and y directions. In this way, the wave number k represents the modulus of the two-dimensional wave vector in the plane of the interface.

4. Asymptotic analysis for $\varepsilon \ll 1$

To obtain the dispersion relation we use the method of matched asymptotic expansions.[3,14] The method consists in matching the solution obtained in the outer region, where z is of the order unity, to that in an inner region, in which z is small. In our case, there are two outer regions B and C, of dimension l_η, in which the dominant physics is hydrodynamic: dissipation due to shear flow. These two regions are the deep nematic region $(z \to -\infty)$ and a $(z \to \infty)$

deep isotropic region. In the inner region A, of dimension $l_S = l_c$ both conserved and non-conserved order parameters vary rapidly.

4.1. OUTER REGIONS: B AND C

Here c and S are constant in each phases. The leading order problem in the outer regions is equivalent with the Navier–Stokes equation for an interface between two fluids with the same density but with different viscosities.[15,16] Imposing the boundary conditions at the sharp interface, we obtain the dispersion relation in the leading order in the outer region:[2,3]

$$\Omega_\eta^2 = \left[1 + \frac{k(l_N^2 + l_I^2) - 2k^3}{(l_N + l_I)(k^2 - l_N l_I)}\right]\Omega_0^2, \tag{10}$$

where $\Omega_0^2 = -\gamma k^3/2\rho$ is the capillary wave dispersion relation for ideal fluids and $l_{N,I} = k(1 - \rho\Omega/\eta_{N,I}k^2)^{1/2}$. k is the wave number and Ω is the angular frequency.

4.2. INNER REGION A

The leading order dispersion relation in the inner region in the absence of hydrodynamic degrees of freedom has the following form:[3]

$$\Omega_{Sc} = \frac{\gamma k^2}{\dfrac{\alpha\gamma_S}{K_S} + \dfrac{(c_{\text{iso}} - c_{\text{nem}})^2}{2\Gamma_c k}}. \tag{11}$$

The pure non-conserved order parameter dispersion relation[17] is obtained for $c_{\text{iso}} = c_{\text{nem}}$, and the pure conserved parameter dispersion relation[18] is obtained for $\alpha = 0$.

In Fig. 4 we have plotted, for $\tau = 0.5$, the inner region dispersion relation as well as the particular limits for nematic–non-nematic mixture. The same quantities are plotted in Fig. 5 for polymer–nematic mixture. There are two distinct regimes. In region A_1, relaxation of concentration is the important process, while in the region A_2, relaxation of the orientational order parameter dominates.

There is a transition region between these two regimes that takes place for $\Omega_S = \Omega_c$, which gives,

$$k_c = \frac{(c_{\text{iso}} - c_{\text{nem}})^2 K_S}{\Gamma_c \alpha\gamma_S}, \tag{12}$$

which for nematic–non-nematic mixture gives $k_c = 2.8 \times 10^{-4}\,\mathrm{cm}^{-1}$ with the corresponding critical wavelength $\lambda_c \cong 2.2\ \mu\mathrm{m}$, while for polymer–nematic system we obtain $k_c = 212\ \mathrm{cm}^{-1}$ with the corresponding critical wavelength $\lambda_c \cong 300\ \mu\mathrm{m}$.

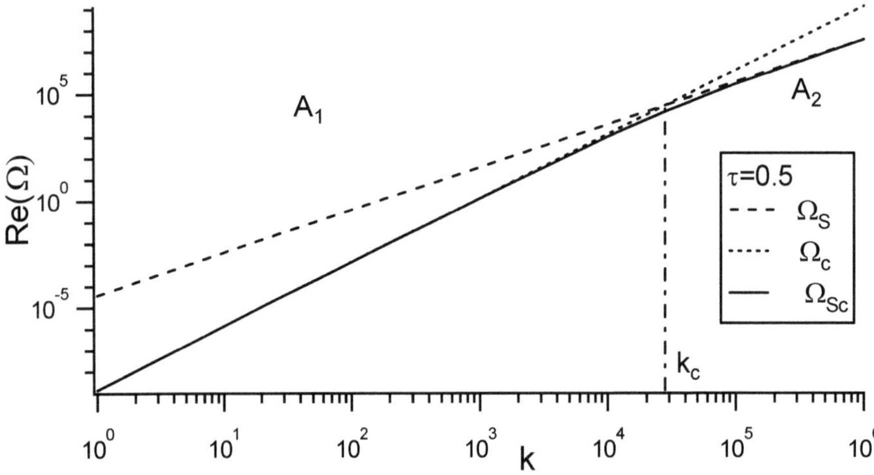

Figure 4. The inner region (leading-order) damping rate (continuous curve) for nematic–non-nematic mixture, defining two distinct regimes. The non-conserved order parameter dispersion relation (dashed curve) and conserved parameter dispersion relation (dotted curve). In the region A_1 relaxation of the concentration is the important process, while in the region A_2 relaxation of the orientational order parameter dominates.

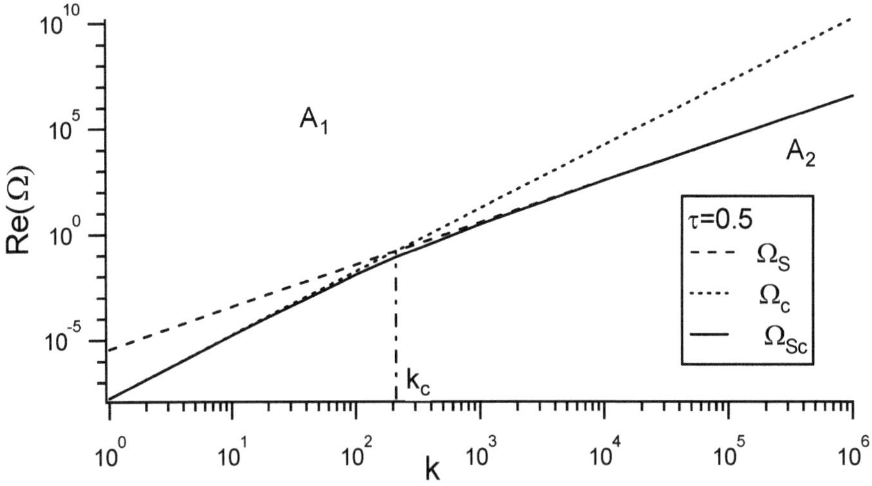

Figure 5. The inner region (leading-order) damping rate (continuous curve) for polymer–nematic mixture, defining two distinct regimes. The non-conserved order parameter dispersion relation (dashed curve) and conserved parameter dispersion relation (dotted curve). In the region A_1 relaxation of the concentration is the important process, while in the region A_2 relaxation of the orientational order parameter dominates.

In Fig. 6 we have plotted the phase diagram in the space (τ, k_c) for both systems. Nematic–non-nematic mixture system is represented by continuous line, while the dashed line represents the polymer–nematic mixture. In the region $k < k_c$, the conserved parameter relaxation is the important process, while in the region $k > k_c$, the non-conserved order parameter relaxation dominates. Near $\tau = 1$ the relaxation of the non-conserved order parameter is the important process for all values of k, except a very small region near $k = 0$. With decreasing temperature, the concentration variation becomes more and more important, and the relaxation of the conserved order parameter is dominant in a larger region. Due to the larger variation in concentration for nematic–non-nematic system, the region in which the conserved parameter relaxation dominates is much larger for this system.

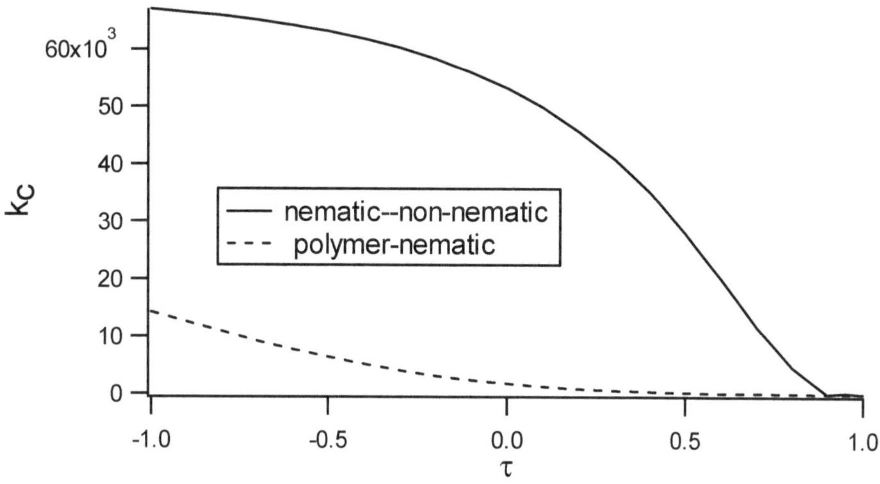

Figure 6. The phase diagram (τ, k_c).

To sum up, we have obtained the leading order dispersion relation in the outer regions B and C of dimension l_η. In this region, the interface is sharp and can be considered as a surface of discontinuity. The physics is governed by the dissipation due to shear flow and the dispersion relation is the classical capillary wave dispersion relation at a sharp interface between two fluids with the same densities and different viscosities. In the inner region A of dimension l_S, without considering the interaction between velocity and the order parameter relaxation, we have obtained the leading order dispersion relation. In this region, the interface is diffusive and the relaxation of the order parameters is the dominant process.

The important point here is that both the leading order solutions for the eigenvalues in the inner and outer regions are approximations of the same eigenvalue problem. Therefore, in the transition region between the inner and outer layers, the two expansions must give the same result. The next step is to combine them into a single expression by matching these two asymptotic expansions.

5. General dispersion relation

Now, we consider the interaction between hydrodynamics and relaxation of the order parameters.[3] Using the method of matching the asymptotic expansions, we obtain the generalized dispersion relation, in unscaled units:

$$\Omega(\Omega-\Omega_{Sc})=\frac{\dfrac{\alpha\gamma_S}{K_S}+\dfrac{(c_{\rm iso}-c_{\rm nem})l_S\gamma_c}{2k\Gamma_c K_c}}{\dfrac{\alpha\gamma_S}{K_S}+\dfrac{(c_{\rm iso}-c_{\rm nem})^2}{2k\Gamma_c}}, \tag{13}$$

where Ω_{Sc} is the leading order dispersion relation in the inner region and Ω_η is the leading order dispersion relation in the outer region.

The real part of the solution and its asymptotic limits given by the Eqs. (10) and (11) are drawn in Fig. 7 (for nematic–non-nematic mixture) and Fig. 8 (for polymer–nematic mixture). Three regions can be clearly distinguished. In the short wavelength limit (region A), the interface is diffuse and the relaxation of the order parameters is the dominant process (dashed curves in Fig. 7 and 8).

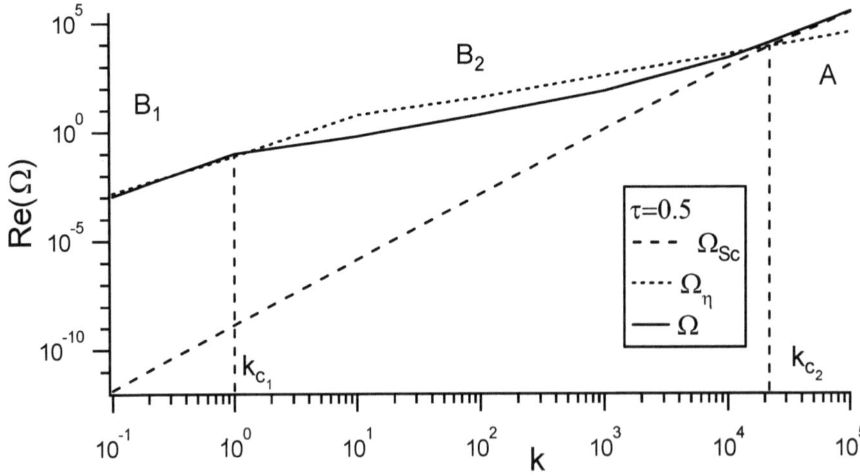

Figure 7. The damping rate for nematic–non-nematic mixture for $\tau = 0.5$. The general dispersion relation (continuous curve), the inner region dispersion relation (dashed line), and the outer region dispersion relation (dotted curve).

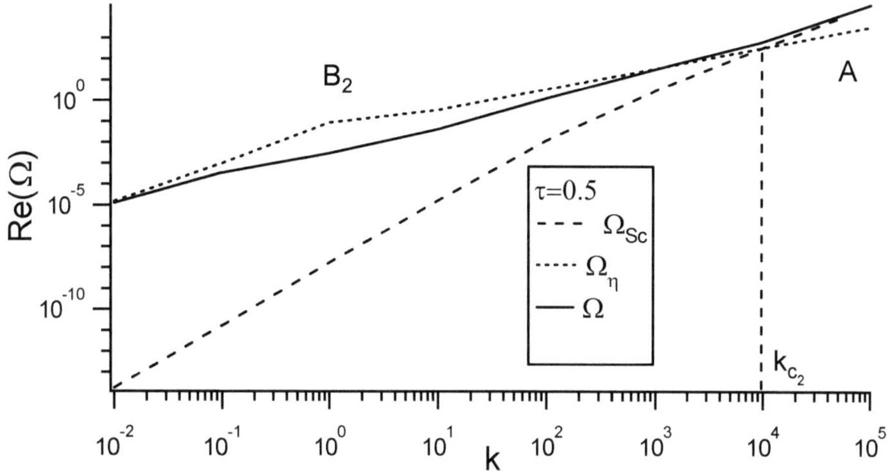

Figure 8. The damping rate for polymer-nematic mixture for $\tau = 0.5$. The general dispersion relation (continuous curve), the inner region dispersion relation (dashed line), and the outer region dispersion relation (dotted curve).

In the long wavelength limit, the viscous damping process in the outer region dominates (dotted curves in Figs. 7 and 8). The transition between these two regions takes place when $\mathrm{Re}(\Omega_\eta) = \mathrm{Re}(\Omega_{Sc})$, which gives for nematic–non-nematic mixture, the critical wave-number $k_{c_2} = 2.2 \cdot 10^4$ cm^{-1} and the corresponding critical wavelength $\lambda_{c_2} = 2.9$ μm. For polymer-nematic mixture, the critical wave-number has the value $k_{c_2} = 9.7 \cdot 10^3$ cm^{-1}, corresponding to a wavelength of $\lambda_{c_2} = 6.5$ μm

Two regions can be further distinguished in the hydrodynamics limit. For nematic–non-nematic mixture, the region B_1 corresponds to $k < k_{c_1} \cong 1$ cm^{-1}, the damping process dominates and the interface mixture behaves as an interface between two isotropic fluids. Finally, the intermediate region B_2 corresponds to $k_{c_1} < k < k_{c_2}$. In this range of the wave numbers, the relaxation of the conserved order parameter plays an important role and cannot be ruled out. For polymer–nematic mixture, the region in which the damping process dictates the physics appears only at very small wave number ($k_{c_1} \cong 0.01$ cm^{-1}), so that the region B_1 does not appear in Fig. 8.

6. Conclusions

In this paper we have examined surface modes at the nematic–isotropic interface in two systems: thermotropic nematic–non-nematic mixture and polymer–nematic mixture. For both systems, we have used a free energy density model as a sum of two terms. The first term is the free energy density of the isotropic mixing for the two components of Flory type. The second term is different: (i) for nematic–non-nematic mixture, we have used a generalized form of the Landau–de Gennes free energy density in which we have included the coupling between c and S from microscopic considerations, while (ii) for polymer–nematic mixture, we have used a generalized form of the Doi model, in which the coupling between c and S is introduced. The two representative phase diagrams generated by these forms of the free energy were presented in Figs. 1 and 2. We have considered a planar nematic–isotropic interface in equilibrium, which constitutes the base state in the inner region of the dynamical system. Minimizing the free energy functional, we have obtained the interfacial tension, that is plotted in Fig. 3 for both systems. Due to the fact that the coupling term induces a larger difference in concentration in nematic–non-nematic system that in polymer–nematic mixture, the interfacial tension in the first system increases more rapidly with decreasing temperature.

To explore the dynamics of this system, we have supplemented the Hess–Olmsted–Goldbart–Qian–Sheng model for the orientational, non-conserved, order parameter with the Cahn–Hilliard equation for the conserved parameter, the concentration. We have assumed an isothermal system, characterized by a scalar order parameter S. In this model, both phases have the same density, but different viscosities. We have considered the equilibrium planar nematic–isotropic interface as the base state of the system. The front was then perturbed with a small amplitude monochromatic plane wave and the linear stability of the front was examined.

In the outer region, in the leading order, we have obtained the classical dispersion relation for the damping of capillary waves, Eq. (10). In the inner region, ignoring the interaction between the hydrodynamics and the order parameter dynamics, in the leading order, the problem is the coupling between the non-conserved dynamics governed by time-dependent Ginzburg–Landau equation and a conserved dynamics described by the Cahn–Hilliard equation. In this approximation, we have obtained the dispersion relation in the inner region, Eq. (11) which we have plotted in Figs. 4 and 5. Two distinct regions can be distinguished. In the small wavelength limit, the relaxation of the non-conserved order parameter is the important process. In the long wavelength limit, the relaxation of the conserved parameter becomes dominant. So, ignoring hydrodynamics degrees of freedom, the addition of a conserved parameter has a profound effect on the dynamics at long wavelengths. The critical wave number which defines the transition between these two regions was analytically obtained and plotted in Fig. 6 as a function of temperature.

Using the method of matching asymptotic expansions, we have obtained the generalized dispersion relation, Eq. (13). The corresponding relaxation rates are plotted in Figs. 7 and 8. Three distinct regions can be distinguished. (i) at very low values of the wave number, in region B_1, the nematic behaves as an isotropic fluid and the dissipation due to shear flow dominates, (ii) at intermediate values of k, in region B2, the influence of the conserved parameter is important, and finally (iii) at large values of k, in region A, the relaxation of the non-conserved order parameter governs the physics.

We mention that the calculation is simplified as it neglected the coupling between the nematic director and the hydrodynamics flow, as well the anchoring effect of director at the interface. We did not use the complete set of Leslie viscosities. Nevertheless, it seems likely that much of the physics will be retained in the general case. However, particularly when the boundary condition at the interface is planar, or when there are anchoring transitions as one goes along the phase coexistence line, there may be interesting new extra physics in the dynamics.

Acknowledgments

V.P.N. thank the School of Mathematics, University of Southampton for scientific hospitality where part of this work was done, P. Oswald for fruitful discussions, and acknowledges support from a Royal Society grant, CNCSIS and CEEX (SIDISANIZ), Romania.

References

1. Popa-Nita, V. and Sluckin, T. J. (2002) *Phys. Rev. E* **66**, 041703.
2. Popa-Nita, V. and Oswald, P. (2003) *Phys. Rev. E* **68**, 061707.
3. Popa-Nita, V., Sluckin, T. J., and Kralj, S. (2005) *Phys. Rev. E* **71**, 061716.
4. Popa-Nita, V., Oswald, P., and Sluckin, T. J. 2005 *Mol. Cryst. Liq. Cryst.* **435**, 215/[875].
5. *Liquid Crystals in Complex Geometries*, edited by G. P. Crawford, S. Zumer (Taylor Francis, London, 1996).
6. Cahn, J. W. and Hilliard, J. E. (1959); (1958) *J. Chem. Phys.* **28**, 258; **31**, 688.
7. Matsuyama, A., Evans, R. M. L., and Cates, M. E. (2000) *Phys. Rev. E* **61**, 2977; *Eur. Phys. J. E* **9**, 79 (2002), **89** (2002).
8. de Gennes, P. G. and Prost, J. (1993) *The Physics of Liquid Crystals* (second edition, Oxford University Press).
9. Flory, P. J. (1953) *Principles of Polymer Chemistry* (Cornell University Press, Ithaca).
10. Doi, M. and Edwards, S. F. (1989) *The Theory of Polymer Dynamics* (Clarendon, Oxford).

11. Hess, S. (1976) *Z. Naturforsch.* **31a**, 1507.
12. Olmsted, P. D. and Goldbart, P. (1992) *Phys. Rev. A* **41**, 4578 (1990); **A46**, 4966.
13. Qian, T. and Sheng, P. (1998) *Phys. Rev. E* **58**, 7475.
14. Holmes, M. H. (1995) *Introduction to Perturbation Methods* (Springler-Verlag).
15. Landau, L. D. and Lifshitz, E. M. (1959) *Fluid Mechanics* (Pergamon Press, Oxford).
16. Levich, V. G. (1962) *Physicochemical Hydrodynamics* (Prentice-Hall, Englewood Cliffs, N.J.).
17. Langer, J. S. (1967) *Ann. of Phys.* **41**, 108; Zittartz, J. (1967) *Phys. Rev.* **154**, 529; Bray, A. J. (1998) *Phys. Rev. E***58**, 1508; W. van Saarloos, (1998) *Phys. Rep.* **301**, 9.
18. Shinozaki, A. and Oono, Y. (1993) *Phys. Rev. E* **47**, 804; Bray, A. J. (1998) *Phys. Rev. E* **58**, 1508.

PART IV: CRITICAL LIQUIDS

GLOBAL PHASE BEHAVIOR OF SUPERCRITICAL WATER – ENVIRONMENTALLY SIGNIFICANT ORGANIC CHEMICALS MIXTURES

SERGEY V. ARTEMENKO and VICTOR A. MAZUR*
Odessa State Academy of Refrigeration, 1/3 Dvoryanskaya Str., 65082 Odessa, Ukraine

Abstract. Recent developments of the global phase equilibria studies of binary mixtures provide some basic ideas of how the required methods can be developed based on global phase diagrams for visualization of the phase behavior of mixtures. The mapping of the global equilibrium surface in the parameter space of the equation of state (EoS) model provides the most comprehensive system of criteria for predicting binary mixture phase behavior. The main types of phase behavior for environmentally significant organic chemicals in aqueous environments are considered using structure–property correlations for the critical parameters of substances. Analytic expressions for azeotropy prediction for cubic EoS are derived. A local mapping concept is introduced to describe thermodynamically consistently the saturation curve of water.

The classes of environmentally significant chemicals (polycyclic aromatic hydrocarbons – PAH, polychlorinated biphenyls – PCB, polychlorinated dibenzo-*p*-dioxins and furans, and selected pesticides) are considered and main sources of the property data are examined. Vapor pressure, heat of vaporization, and critical parameter estimations for pure components were chosen for seeking a correlation between the octanol–water partition coefficients K_{OW} and the EoS binary interaction parameters – k_{12}. The assessment of thermodynamic and phase behavior of representatives for different pollutants is given.

Keywords: Supercritical water, organic chemicals, global phase behavior, octanol–water partition coefficient.

---

*To whom correspondence should be addressed. Victor A. Mazur, e-mail: mazur@paco.net

S. J. Rzoska and V. A. Mazur (eds.), Soft Matter under Exogenic Impacts, 269–286.

1. Introduction

Knowledge about the thermodynamic and phase behavior of chemicals is essential for various applications. To simulate the behavior of different hazardous chemicals in the technological media, models based on equations of state (EoS) for presentation of thermodynamic properties are more preferable. Considering a huge array of existing environmentally significant organic chemicals, which wait in queue to be destroyed or recycled, it is obvious that there is need to develop theoretically sound methods for reliable assessment of their thermodynamic and phase behavior, especially, at supercritical water conditions. Theoretical analysis of the topology of fluid phase equilibria is a very useful tool for understanding the sophisticated phase diagrams observed in multi-component systems.

The first pioneering study of van Konynenburg and Scott[1] has shown that the simple one-fluid van der Waals model can reproduce qualitatively the main phase diagrams of binary fluids. A proposed classification was successful and now is currently used as a basis for description and recognition of different phase behavior types in binary mixtures. Mapping of the global equilibrium surface into the parameter space of the equation of state model provides the most comprehensive system of criteria for predicting binary mixture phase behavior. The influence of critical parameters of compounds on phase topology is visualized via global phase diagrams. Such diagrams are presented not in the pressure–temperature variables but in the space of the equation of state parameters, e.g., the van der Waals geometric – b and energetic – a parameters.

The main goal of the present work is to demonstrate some fundamental links that exist between the studies of the global phase diagrams and the prognosis of the thermodynamic and phase behavior of supercritical water – organic pollutant mixtures.

This paper is organized as follows. In Section 1, theoretical models of global phase behavior are considered and the Soave–Redlich–Kwong (SRK) equation of state is selected. Local mapping of the fundamental equation of state for water on the three parameter SRK model is introduced. Analytical expressions to predict *azeotropy* phenomena in terms of critical parameters of mixture components and K_{12} are derived using the global phase diagram approach. Section 2 presents global phase behavior of water + hydrocarbon blends. Restricted experimental data on high pressure phase equilibria of water + hydrocarbon mixtures are used to identify the binary interaction parameters K_{12} via the octanol–water partition coefficient K_{OW}. In Section 3, we perform the phase equilibria and critical line calculations for binary mixtures supercritical water–organic pollutant. The paper ends with some conclusions and an outlook to further research.

2. Global phase behavior of binary mixtures

The first idea to map a phase equilibria surface on the space of the equation of state parameters belongs to van der Waals. The term global phase diagram (GPD) has been introduced by Furman and Griffiths.[2] The boundaries of the GPD (tricritical points – TCP, double critical endpoints – DCEP, azeotropic lines, etc.) divide the space of model parameters into the regions corresponding to the various phase behavior types. Mapping of the global equilibrium surface into the parameter space of the equation of state model provides the most comprehensive system of criteria for predicting binary mixture phase behavior.

Phase diagrams are a visual representation of the state of a substance as functions of temperature, pressure, and concentrations of components and are, therefore, usually hailed as basic blueprints for chemical reactor design and physical understanding. The p, T-variables and the variable x are of different nature. Pressure and temperature are "field" variables which have the property to be the identical for all coexisting phases. The mole fraction is a "density" variable, which has in general different values for different phases. To understand and classify a wide assortment of phase diagrams the determination of critical points of mixtures is important for a variety of both practical and theoretical reasons. A mixture of given composition may have one, more than one, or no critical points.

Phase behavior types of interest for supercritical water oxidation (SCWO) applications can be characterized as follows[1] (Fig. 1):

- I. a single permanent critical line between both critical points C_1 and C_2;
- II. one critical line connecting C_1 and C_2, another line going from C_m to a critical endpoint;
- III. one critical line going from C_1 to an upper critical endpoint, another line going from C_2 to C_m;
- III-H. a subclass of III having an heteroazeotropic three-phase curve;
- III-A. a subclass of III with a genuine positive azeotropic line;
- IV. one critical line going from C_1 to an upper critical endpoint, a second critical line going from C_2 to a lower critical endpoint, a third line going from C_m to an upper critical endpoint.

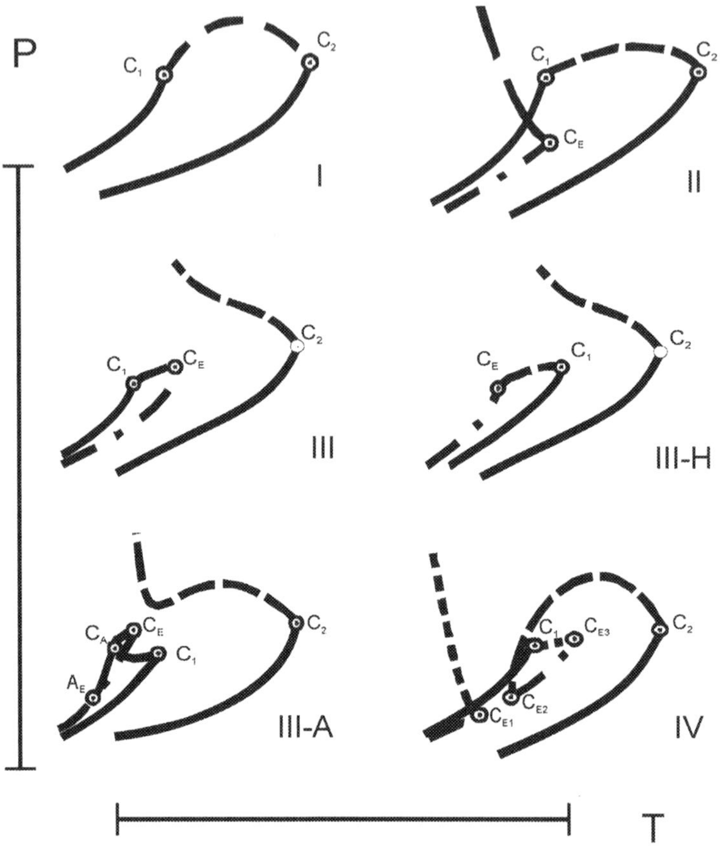

Figure 1. Main types of phase behavior in binary mixtures.

——— – vapor pressure line; – – – – · – vapor–liquid critical line;
— — — · – liquid–liquid critical line; – · – · – · · – three-phase line;
C_i – critical point of pure components; C_A – azeotropic critical point;
C_E – critical end point; and A_E – azeotropic end point.

Global phase diagrams of mixtures are analyzed for simple two-parametric equations of state and the one-fluid approximation mainly for two-component fluids. Global phase diagrams of binary fluids represent the boundaries between different types of phase behavior in a dimensionless parameter space. The boundaries between types in the global phase diagram can be calculated directly using the thermodynamic description of the boundary states obtained from the boundary states principle.[3] The dimensionless coordinates that are used for representation of the boundary states depend on the equation of state model, but normally they are designed similar to those proposed by Scott and van Konynenburg for the van der Waals model.[1] In that case, global phase diagrams for all realistic models have very similar structure, especially for the case of equal sized molecules. For example, the global phase diagrams of such different models as the one-fluid EoS of binary Lennard-Jones fluids[4] and the Redlich–Kwong model[5] are almost identical in their main features including such a sensitive phenomenon as the closed-loops of liquid–liquid immiscibility. Therefore, most of the considerations and conclusions obtained for the one of realistic models will remain valid for another.

2.1. GLOBAL PHASE DIAGRAMS FOR BINARY MIXTURES

We consider here the SRK model[6] as an example in its classical non-modified form:

$$p = \frac{RT}{v-b} - \frac{a\alpha(T)}{v(v+b)}, \tag{1}$$

where R is the universal gas constant, and the EoS parameters a and b of a mixture depend on the mole fractions x_i and x_j of the components i and j and on the corresponding parameters a_{ij} and b_{ij} for different pairs of interacting molecules:

$$\begin{aligned} a &= \sum_{i=1}^{2}\sum_{j=1}^{2} x_i x_j a_{ij}\left(1-k_{ij}\right), \\ b &= \sum_{i=1}^{2}\sum_{j=1}^{2} x_i x_j b_{ij}, \end{aligned} \tag{2}$$

where

$$a_{ii} = 0.4274\left(\frac{R^2 T_c^2}{P_c}\right),$$

$$\alpha(T)=\left\{1+(0.480+1.57\omega+0.176\omega^2)\left[1-\left(\frac{T}{T_c}\right)^{0.5}\right]\right\}^2,$$

$$b_{ii}=0.08664\frac{RT_c}{P_c}. \tag{3}$$

For a normal critical point when two fluid phases are becoming identical critical conditions are expressed in terms derivatives of the molar Gibbs energy in the following way:

$$\left(\frac{\partial^2 G}{\partial x^2}\right)_{p,T}=\left(\frac{\partial^3 G}{\partial x^3}\right)_{p,T}=0. \tag{4}$$

One of most important boundaries in global phase diagrams is given by the tricritical points. This boundary divides the classes I and V, II and IV, or III and IV. The tricritical state is a state where the regions of the liquid–liquid–gas immiscibility shrink to one point, which is named the TCP. Three phases become identical at a TCP. Corresponding critical conditions are

$$\left(\frac{\partial^2 G}{\partial x^2}\right)_{p,T}=\left(\frac{\partial^3 G}{\partial x^3}\right)_{p,T}=\left(\frac{\partial^4 G}{\partial x^4}\right)_{p,T}=\left(\frac{\partial^5 G}{\partial x^5}\right)_{p,T}=0. \tag{5}$$

Another important boundary in global phase diagrams is the locus of double critical end-points (DCEP) that divides type III and IV, or II and III. A DCEP is produced in a type III when the critical curve cuts tangentially the three-phase line in the pressure–temperature diagram. Type IV is characterized by two liquid–liquid–gas curves. One is at high temperatures and is restricted by two critical end points (a lower critical end point – LCEP and an upper critical end point – UCEP). At the upper critical point, the solution becomes immiscible as the temperature is lowered. At a lower critical point, the solution separates into two phases as the temperature is increased. Thermodynamic expressions and mathematical tools for evaluation of the boundaries in the global phase diagram are given in the literature.[5,10]

2.2. LOCAL MAPPING

To define exact equation of state parameters for pure substances, especially for water, the concept of local mapping has been developed. The conventional approach to determination of parameters a, b, and α is derived from critical conditions. But for phase equilibria prediction this parameter set is not sufficient to satisfy the limiting conditions at $x \rightarrow 1$ and $x \rightarrow 0$.

As was shown in Ref. 7 it is necessary to use a three parameter basis for a thermodynamically consistent description of system properties, i.e., simultaneous description of isothermal compressibility, energy, and pressure. Thermodynamic consistency is impossible to attain operating with only two scaling factors.[8] Transition from two-parameters to the multi-parameters models makes phase equilibria calculation more complex and does not necessarily lead to the improvement of results.

To provide a reliable thermodynamic consistency between exact and model equations of state the equalities of pressures, isothermal compressibilities and internal energies should be applied. To satisfy these conditions at a defined point with V–T coordinates the set of Eqs. (6) should be solved:

$$\begin{aligned} &P_{\text{exact}}(T,V) - P_{\text{mod}}(\vec{X},T,V) = 0, \\ &\frac{\partial P_{\text{exact}}(T,V)}{\partial T} - \frac{\partial P_{\text{mod}}(\vec{X},T,V)}{\partial T} = 0, \\ &\frac{\partial P_{\text{exact}}(T,V)}{\partial V} - \frac{\partial P_{\text{mod}}(\vec{X},T,V)}{\partial V} = 0, \end{aligned} \qquad (6)$$

where P_{exact} is calculated from the fundamental EOS for H_2O[9], P_{mod} and P_{SRK} are from the SRK model[6] with adjusted and conventional parameters $\vec{X}$, correspondingly, and $\vec{X}$ is the vector of SRK model parameters. Idea of local mapping is presented in visual form at Fig. 2. Results of calculations, which accurately describe data on saturation curve and supercritical isotherm for water, are presented in Tables 1 and 2.

TABLE 1. Model parameters $\vec{X}$ for the SRK EoS on vapor pressure curve (P_{exp} is an experimental vapor pressure, RD is a relative deviation of calculated pressure from experimental data).

T, K	P_{exp}, MPa	P_{mod}, MPa	RD_{mod}, %	RD_{SRK}, %	a_{H_2O}, 10^5 J·m^3/mol^2	b_{H_2O}, cm^3/mol	α
400	0.2479	0.2479	-3.5e–09	-5.6137	5.4677	20.638	1.469
450	0.9373	0.9373	1.1e–09	-0.7686	5.5861	21.048	1.359
500	2.6475	2.6475	1.3e–09	1.6909	5.6283	21.185	1.259
550	6.1302	6.1302	-1.0e–09	2.5844	5.6694	21.325	1.165
600	12.364	12.364	-1.4e–09	2.2213	5.6697	21.325	1.077

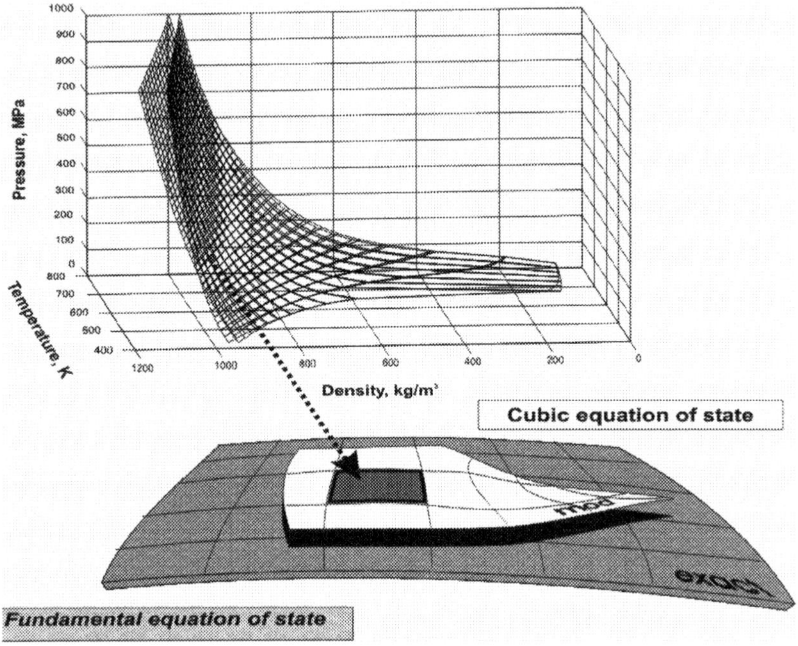

Figure 2. Local mapping concept.

TABLE 2. Model parameters $\vec{X}$ for the SRK EoS on supercritical isotherm $T = 673.15$ K (ρ_{exp} is an experimental density, *RD* is a relative deviation of calculated density from experimental data).

T, K	P, MPa	ρ_{exp}, kg/m^3	ρ_{mod}, kg/m^3	RD_{mod}, %	RD_{SRK}, %	a_{H_2O} J·m^3/mol^2	b_{H_2O}, cm^3/mol	α
673.15	150	751.26	751.26	4.3e–10	29.263	4.0666	14.534	1.00
673.15	200	792.53	792.53	9.2e–10	28.877	4.0234	14.461	1.00
673.15	250	825.11	825.11	-3.6e–09	28.734	3.9606	14.366	0.99
673.15	300	852.34	852.34	-6.2e–10	28.728	3.8858	14.265	0.99
673.15	350	875.90	875.90	-1.2e–10	28.806	3.8031	14.163	0.98
673.15	400	896.80	896.80	4.8e–10	28.939	3.7146	14.061	0.97

2.3. AZEOTROPY CRITERIA

Azeotropy plays a very important role in all technical applications of mixtures and therefore is of special interest. Some criteria for azeotropy in binary fluids can be obtained analytically in the frameworks of global phase diagrams. The corresponding boundary state is called a degenerated critical azeotropic point and represents the limit of the critical azeotropy at $x_i \rightarrow 0$ or at $x_i \rightarrow 1$. As a result of solving the system of thermodynamic equations for a degenerated critical azeotrope:[10, 22]

$$\left(\frac{\partial p}{\partial V}\right)_{T,x} = \left(\frac{\partial^2 p}{\partial V^2}\right)_{T,x} = \left(\frac{\partial p}{\partial x}\right)_{T,V} = 0, \tag{7}$$

one can obtain the following relationship of dimensionless interaction parameters:

$$Z_2 = \mp Z_1 - (1 \pm Z_1)\left(\frac{1 - Z_4}{1 \pm Z_3} - 1\right)\Lambda, \tag{8}$$

where the upper signs + or – correspond to $x_2 = 0$ and the lower signs – to $x_2 = 1$, respectively.

The set of dimensionless parameters for the SRK model has been defined as follows:

$$\begin{aligned} Z_1 &= (a_{22} - a_{11})/(a_{22} + a_{11}), \\ Z_2 &= (a_{22} - 2a_{12} + a_{11})/(a_{22} + a_{11}), \\ Z_3 &= (b_{22} - b_{11})/(b_{22} + b_{11}), \\ Z_4 &= (b_{22} - 2b_{12} + b_{11})/(b_{22} + b_{11}). \end{aligned} \tag{9}$$

The value of the constant Λ depends on cubic equation type. In case of the SRK model the value Λ is given by

$$\Lambda = \frac{\Theta_b}{\Theta_{ab}(\Theta_b - 1)^2} + \frac{1}{\Theta_b + 1}, \tag{10}$$

where $\Theta_{ab} = \frac{\Omega_a}{\Omega_b}$, $\Theta_b = \frac{z_c}{\Omega_b}$.

For the Peng–Robinson model it can be evaluated as

$$\Lambda = \frac{\Xi^{pr}}{\Theta_{ab}(\Theta_b - 1)^2} + \frac{2(\Theta_b - 1)}{\Xi^{pr}}, \tag{11}$$

where $\Xi^{pr} = \Theta_b^{\,2} + 2\Theta_b - 1$.

According to the (8) border between azeotropic and non-azeotropic states on a Z_1–Z_2 plane at fixed values of Z_3 and Z_4 is a straight line. For the Peng–Robinson equation of state $\Omega_a = 0.45724$, $\Omega_b = 0.07780$, and $z_c = 0.307$, the value of Λ is 0.7027, and for the Soave–Redlich–Kwong model, assuming $\Omega_a = 0.42747$, $\Omega_b = 0.08664$, and $z_c = 0.333$, constant Λ is equal to 0.6731.

Knowledge of interaction parameters allows the investigator to estimate a possible azeotrope appearance in aqueous solution of hazardous organic pollutants. Values of interaction parameters k_{ij} and l_{ij} are presented below in Table 3.

3. Global phase behavior of water–hydrocarbon mixtures

If combination rules are known then it is possible to determine the global phase diagram via the ratio of critical parameters of components only. A distribution of reduced critical temperatures and volumes for aqueous solutions of hydrocarbons is given in Fig. 3.

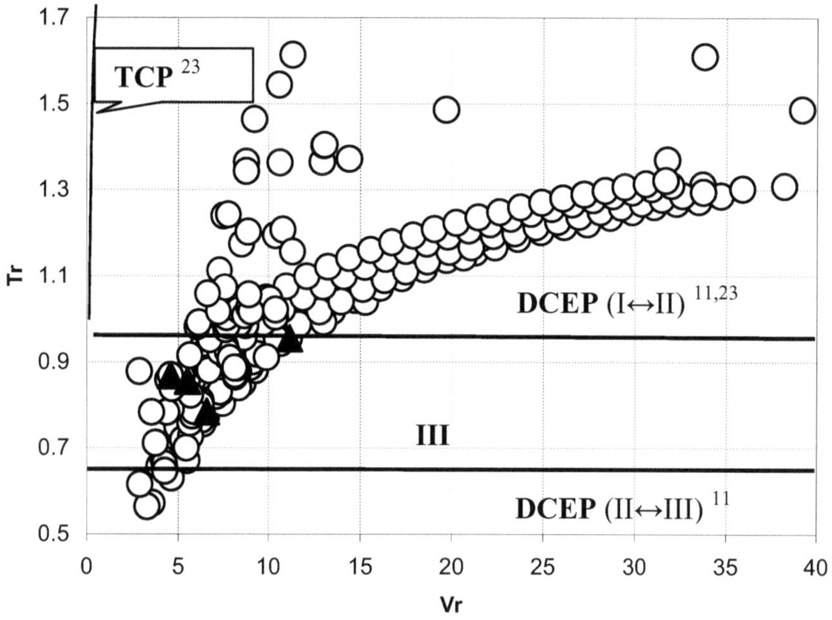

Figure 3. Distribution of hydrocarbon–water mixtures in reduced variables.
▲ – Mixtures performing III type of phase behavior.[13]

Critical parameters of hydrocarbons were taken from correlation.[12] The simple estimation of transition from III type of phase behavior to II type ($T_r = T_{CH_2O}/T_C \approx 0.5$) in terms of pure components for the one-fluid model of the Redlich–Kwong EoS[23] and Carnahan-Starling – van der Waals EoS[11] with the Lorentz–Berthelot combination rules ($k_{12} = 0$) shows serious discrepancies with experimental evidence of III type phase behavior for systems water–alkanes.[13] Hence, more correct estimation of binary interaction parameters is needed to predict phase behavior of organic toxicants in supercritical water. An approach proposed here is based on the search of relationships between well investigated octanol–water partition coefficients for organic pollutants and interaction parameters in the one-fluid models of cubic equation of state.

4. Phase equilibria in supercritical water–organic pollutant mixture

The octanol–water partition coefficient (K_{OW}) is essentially a measure of how good a solvent water is for chemical under study, or alternatively the magnitude of the infinite dilution activity coefficient of the chemical in water. Since there already exist much data on K_{ow} and this parameter has been used to correlate many other properties, it is the parameter that will be used here.

Correlation between log K_{OW} and interaction parameters was presented in the form proposed by Polischuk et al.[15] for homology series of hydrocarbons:

$$l_{12} = \frac{b_{22} - b_{11}}{b_{22} + b_{11}} L_{12}, \tag{12}$$

$$k_{12} = (K_{11} \log K_{ow} - l_{12} \frac{T_{c2}}{T_{c1}})(1 - t) + K_{22} t \log K_{ow}, \tag{13}$$

where $$t = \tan h \left[\left(\frac{T - T_{c1}}{T_{c2} - T_{c1}} \right)^2 \right].$$

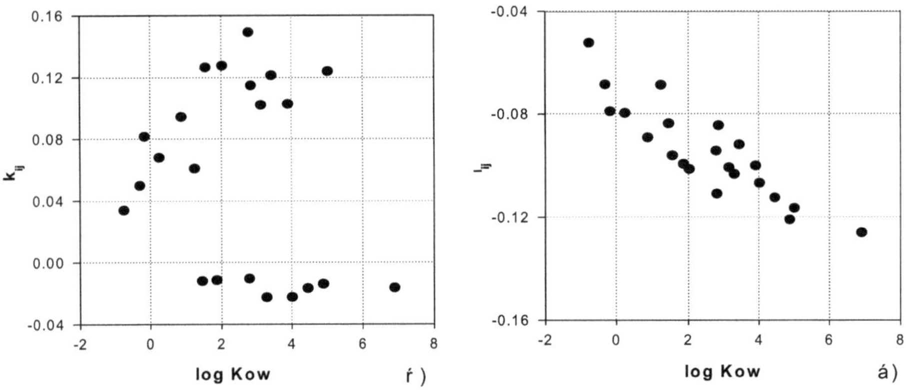

Figure 4. Relationship between interaction parameters and ecological indicator log K_{ow}.

Values of the parameters obtained as a result of data treatment for water solutions of alkanes[13,14] are: $L_{12} = 0.1423$, $K_{11} = 0.017989$, and $K_{22} = -0.008298$.

The region of definition for interaction coefficients is restricted by hydrophilic organic pollutants with log $K_{ow} > 1$. Adjusted parameters k_{12} and l_{12} vs. ecology indicator log K_{ow} are presented in Fig. 4.

Expressions (12) and (13) allow one to establish the interaction parameters k_{12} and l_{12} for insufficiently explored organic pollutants only from value K_{OW}, which is more often presented in the literature.[21] Results of calculations k_{12} and l_{12} for binary mixtures of water–organic pollutants from known log K_{ow} are presented in Table 3.

TABLE 3. Calculated values of k_{12} and l_{12}.

Organic pollutant	Log K_{ow} [21]	k_{12}	l_{12}	Azeotrope
Anthracene	4.4500	–0.0168	–0.1124	
Benzoic acid	1.8700	–0.0115	–0.0994	•
Biphenyl	4.0100	–0.0225	–0.1067	•
Naphthalene	3.3000	–0.0227	–0.1033	•
Phenol	1.4600	–0.0121	–0.0837	•
Pyrene	4.8800	–0.0140	–0.1209	
2,4-D	2.8100	–0.0105	–0.1109	
DDT	6.9100	–0.0164	–0.1259	
Ethylbenzene	3.1500	0.1020	–0.1007	•
Diflouromethane	1.2500	0.0608	–0.0686	•
Toluene	2.7900	0.1491	–0.0942	•

Environmentally significant organic chemicals include: a) PAH – Polycyclic Aromatic Hydrocarbons, b) PCB – polychlorinated biphenyls, c) dioxins and furans, and d) selected pesticides. Data are extracted from existing experimental information on vapor pressure curve, as well as from well known structure–property correlations, presented in.[16] As test systems for phase equilibria calculations in supercritical water–pollutants following classes of organic chemicals were selected: PAH – pyrene; PCB – biphenyl; pesticides – DDT, and dioxins and furans – 2, 4-D.

Prediction of phase equilibria in a supercritical water–benzene system is an example of a macroscopic properties calculation directly from molecular structure, presented in Fig. 5.

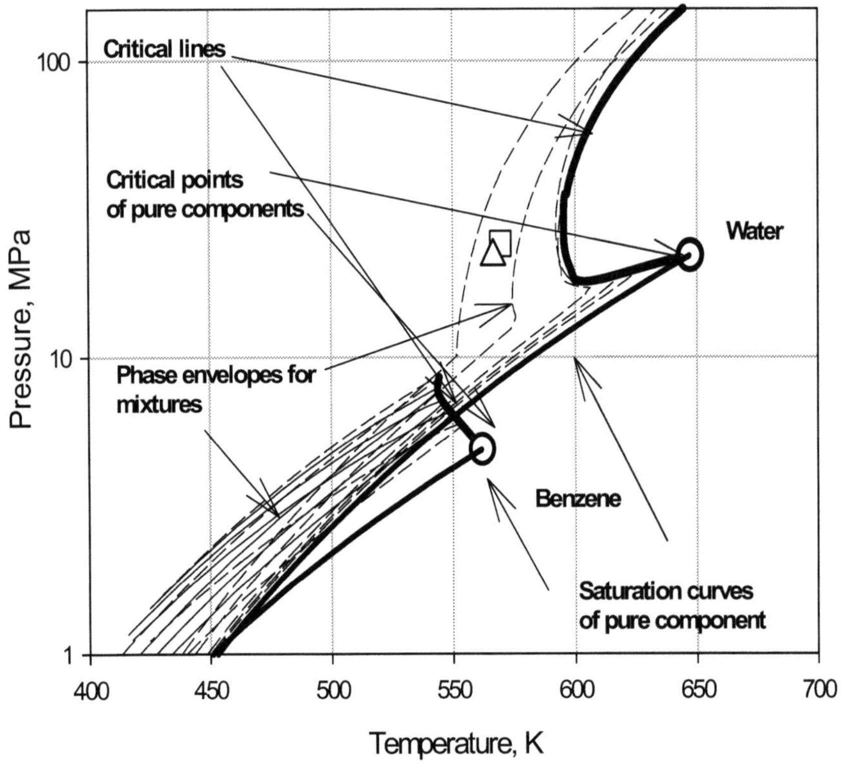

Figure 5. Phase equilibria and critical lines in water–benzene system.
△ – experimental data;[19] □ – experimental data.[20]

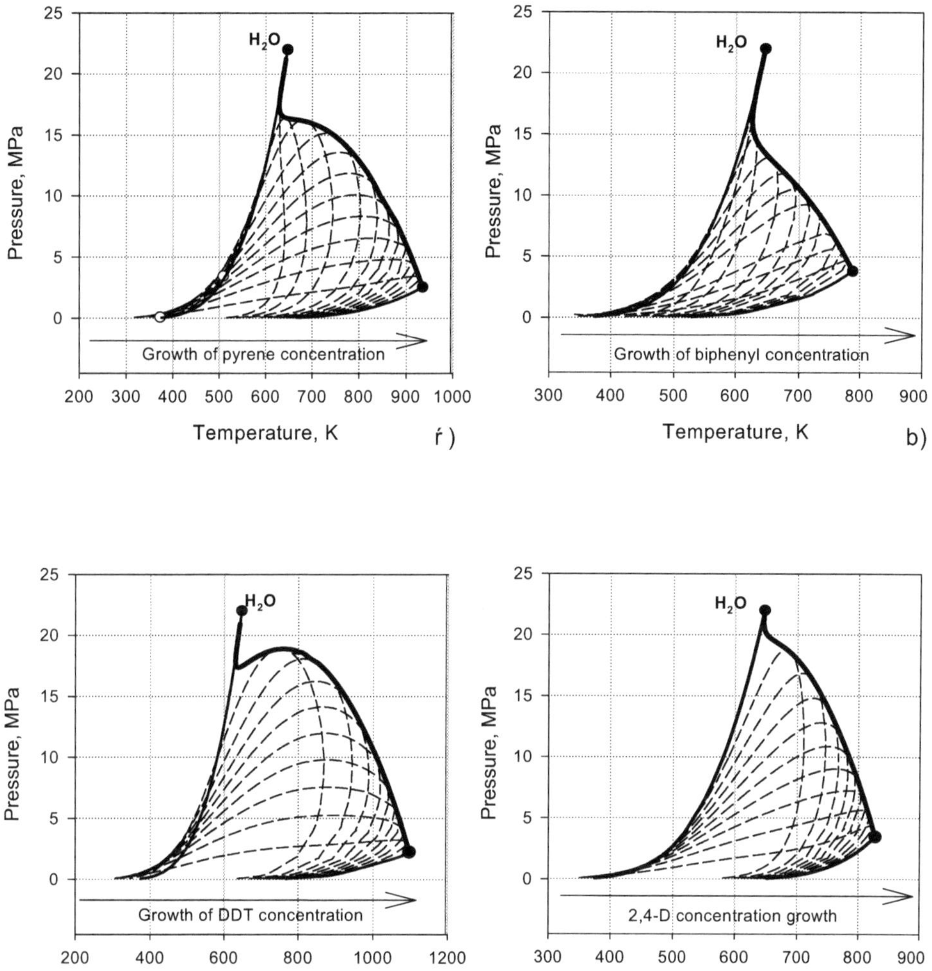

Figure 6. Phase envelopes for mixtures: a) pyrene–supercritical water; b) byphenyl–supercritical water; c) DDT–supercritical water; and d) 2,4-D–supercritical water.

– Phase co-existing curve; △ – critical points of mixture;

– critical lines; ● – critical points of pure components.

Experimental data on phase equilibria at high pressures were taken from,[13] values of log K_{ow} were taken from.[17] Calculation of a standard set of molecular descriptors for benzene was performed using quantum chemical package SPARTAN.[18] The set includes: ionization potential is –9.6527 eV, electron affinity –0.5548 eV, dipole moment – 0.000921 Debye, polarizability – 7.08 a.e., molecular volume – 169.65, molecular surface – 234.13, most negative charge on any non-hydrogen atom – –0.525785 a.u., most positive charge on any hydrogen atom – 0.330911 a.u. All of these descriptors correlated with log $K_{OW} = -2.19$ by means of simple expressions for quantitative structure–properties relationship (QSAR) developed by Verhaar et al..[17]

Results of the calculation of critical lines and isopleths for the given mixture show the presence of two critical lines, which agrees with experiment.[19] Agreement between experimental results and theoretical prediction based on molecular structure data without adjustment of experimental data to the macro properties can be achieved for practical applications.

The phase envelopes for typical environmental pollutants were calculating using interaction parameters taken from Table 3 and presented in Fig. 6.

5. Conclusion

This study is one of the first attempts to establish and demonstrate multiple links existing among the phase equilibria phenomena in supercritical aqueous mixtures of environmentally significant organic chemicals and their models. From the very beginning of these efforts, the global phase diagrams have been a very useful tool for scientists and engineers working in the field of emerging supercritical technologies. There is no doubt that extension of our knowledge about global phase behavior of two- and multicomponent fluids will lead to the creation of reliable engineering recipes for solving the actual problems of hazardous pollutant destruction.

References

1. van Konynenburg, P. H. and Scott, R. L. (1980) Critical lines and phase equilibria in binary van der Waals mixtures, *Phil. Trans. Roy. Soc. London* **298**, 495–540.
2. Furman, D. and Griffiths, R. B. (1978) Global phase diagram for a van der Waals model of a binary mixture, *Phys. Rev A* **17**, 1139–1149.
3. Boshkov, L. Z. (1992) The phase tree of binary fluids: principle, development and thermodynamics, *Proc. Int. Symp. On Solubility Phenomena*, pp.12–13.

4. Mazur, V. A., Boshkov, L. Z., and Murakhovsky, V. G. (1984).Global phase behavior in binary Lennard-Jones mixtures, *Phys. Lett. A* **104**, 415–418.
5. Deiters, U. K. and Pegg, J. L. (1989) Systematic investigation of the phase behavior of binary fluid mixtures. I. Calculations based on the Redlich–Kwong equation of state, *J. Chem. Phys.* **90**, 6632–6641.
6. Soave, G. (1972) Equilibrium constants from a modified Redlich–Kwong equation of state, *Chem. Eng. Sci.* **27**, 1197–1203.
7. Nebelenchuk, V. and Mazur, V. (1991) Transport properties of dense fluids via spherical models of the interaction potential, *Physica A* **178**, 123–148.
8. Mollerup, J. (1998) Unification of the two-parameter equation of state and the principle of corresponding states, *Fluid Phase Equilibria* **148**, 1–19.
9. Wagner, W. and Pruß, A. (2002) The IAPWS formulation 1995 for the thermodynamic properties of ordinary water substance for general and scientific use, *J. Phys. Chem. Ref. Data* **31**, 387–535.
10. Boshkov, L. Z. and Mazur, V. A. (1985) Phase behaviour of the two-component Lennard-Jones fluid, Dep. at the VINITI, No. 6844–B85.
11. Sadus, R. and Wang, Ji. (2003) Phase behaviour of binary mixtures: a global phase diagram solely in terms of pure component properties, *Fluid Phase Equil.* **214**, 67–78.
12. Wakeham, W., Cholakov, G., and Roumiana, S. (2002) Liquid density and critical properties of hydrocarbons estimated from molecular structure, *J. Chem. Eng. Data* **47**, 559–570.
13. Tsonopoulos, C. and Wilson, G. M. (1983) High-temperature mutual solubilities of hydrocarbons and water, *AIChE J.* **29**(6), 990–993.
14. Tsonopulos, C. (1999) Thermodynamic analysis of the mutual solubilities of normal alkanes and water, *Fluid Phase Equil.* **156**, 21–33.
15. Polischuk, I., Wisniak, J., and Segura, H. (2003) Simultaneous prediction of the critical and subcritical phase behaviour in mixtures using equations of state. II Carbon dioxide – heavy *n*-alkanes, *Chem. Eng. Sci.* **58**, 2529–2550.
16. Delle Site, A. (1996) Vapor pressure of environmentally significant organic chemicals: A review of methods and data at ambient temperature, *J. Phys. Ref. Data* **26**, 157–193.
17. Verhaar, H., Ramos, E., and Hermens, J. (1996) Classifying environmental pollutants. 2.: Separation of class 1 (baseline toxity) and class 2 (polar narcosis) type compounds based on chemical descriptors, *Journal of Chemometrics* **10**, 149–162.
18. Hehre, W., Burke, L., and Shusterman, A. (1993) *SPARTAN User's Guide* (Wavefuncion. Inc., Irvine, CA).
19. Alwani, Z. and Schneider, G., (1967) Phase equilibria and critical phenomena in the system benzene – H_2O between 250 and 368 °C up to 3700 bar, *Ber. Bunsenges. Phys. Chem.* **71**, p.633.

20. Connolly, J. (1966) Solubility of hydrocarbons in water near the critical solution temperature, *J. Chem. Eng. Data* **11**, 13.
21. Sangster, J. (1997) *Octanol–Water Partition Coefficients: Fundamentals and Physical Chemistry* (John Willey & Sons Inc., NY).
22. Mazur, V., Boshkov, L., and Artemenko, S. (1998) Global phase behaviour of natural refrigerant mixtures, *Proc. IIR – Gustav Lorentzen Conference: Natural Working Fluids*, 495–504.
23. Kolafa, J. (1999) Azeotropic phenomena in the global phase diagram of the Redlich–Kwong equation of state, *Phys. Chem. Chem. Phys.* **1**, 5665–5670.

PROPERTIES OF WATER NEAR ITS CRITICAL POINT

V. KULINSKII* and N. MALOMUZH
Department for Theoretical Physics, Odessa National University, 2 Dvoryanskaya St, 65026 Odessa, Ukraine

Abstract. The equation of state and dielectric properties of water near its critical point are studied within the picture that the local structure of a system is mainly formed by the dimers, which rotate almost freely. It is shown that the degree of association is about 0.9. The coordinates of the critical point are successfully reproduced. The value and the temperature dependence of the dielectric permittivity, constructed in the canonical framework are in good agreement with the experimental data.

Keywords: Critical point, van der Waals equation of state, chemical equilibrium, dimers.

1. Introduction

Water has many anomalous properties in comparison with other molecular liquids. These anomalies directly or indirectly are connected with the existence of H-bonds. The specificity of interparticle interaction due to formation of H-bonds manifest itself both in equilibrium thermodynamic and kinetic properties. To understand the character of thermal motion in water[1] it was carried out the comparison of the temperature dependencies of the shear viscosity for water and argon. More exactly, in the manner of the principle of corresponding states the reduced values

$$\tilde{\nu}^{(Ar)}\left(\frac{T}{T_c^{(Ar)}}\right)=\frac{\nu^{(Ar)}(T)}{\nu_R^{(Ar)}},\qquad \tilde{\nu}^{(W)}\left(\frac{T}{T_c^{(W)}}\right)=\frac{\nu^{(W)}(T)}{\nu_R^{(W)}},$$

---

*To whom correspondence should be addressed, e-mail: koul@paco.net

S. J. Rzoska and V. A. Mazur (eds.), Soft Matter under Exogenic Impacts, 287–304.

of the shear viscosities, where $\nu_R^{(Ar)}$ and $\nu_R^{(W)}$ are their regular parts near the critical point, were considered as a function of the reduced temperature $t = T/T_c$ ($T_c^{(Ar)} = 150.8$ K and $T_c^{(W)} = 647.15$ K). The corresponding results were obtained by processing the experimental data taken from Ref. 2 and are presented in the Fig. 1.

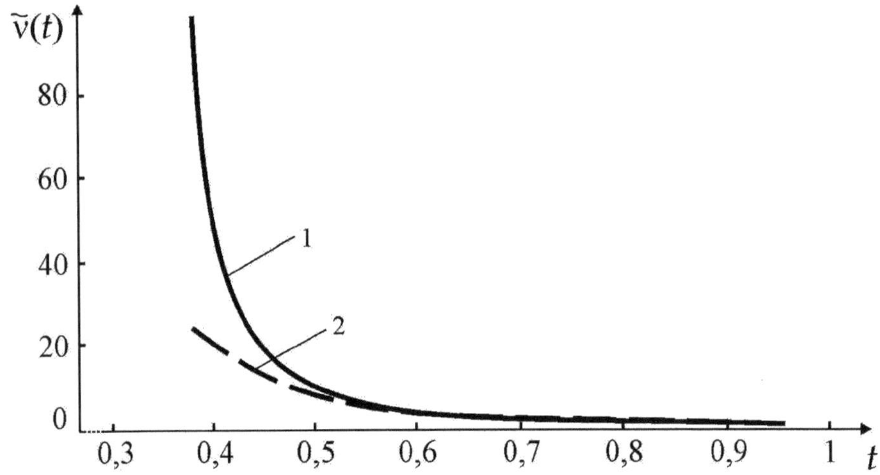

Figure 1. Temperature dependencies of the normalized shear viscosities for: 1 – water and 2 – argon (see the text for explanation).

From here it follows that above the temperature $T_v \approx 320$ K the normalized shear viscosity of water has the argon-like character.

To elucidate the nature of T_v we complete these results by the temperature dependence of the average number $n_H(T)$ of H-bonds per molecule, presented in Ref. 3. There it was found from the analysis of the temperature dependencies of the density and heat capacity of water on the coexistence curve. From Fig. 1 it follows that at $T > T_v$, the average number of H-bonds per molecule is less than two: $n_H(T_v) \approx 2$. Note that at all temperatures above and below the melting point $n_H(T)$ decreases practically linearly (see Fig. 2). This result was obtained in Ref. 4:

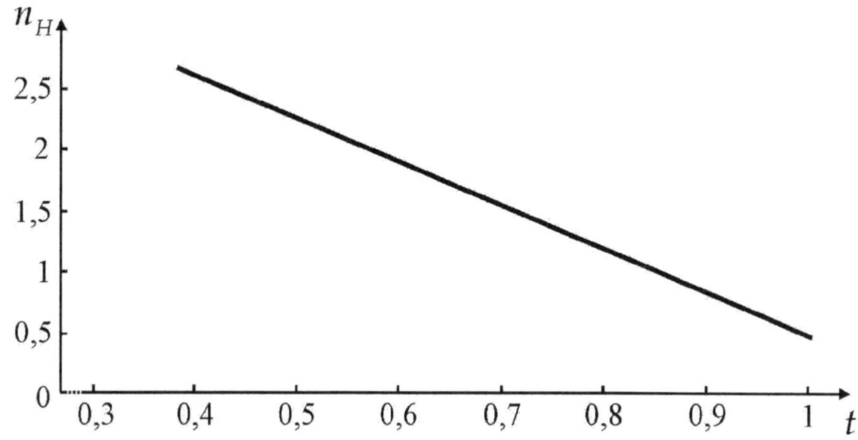

Figure 2. Temperature dependence of the average number of H-bonds per molecule (see the text for explanation).

$$n_H(T) = 4\left(1 - 0.88\frac{T}{T_c^{(W)}} + \cdots\right) \tag{1}$$

At the critical point this formula leads to $n_H\left(T_c^{(W)}\right) \approx 0.5$. This indicates that at $T > T_v$ the integrity of the H-bonds network is broken and it does not play the essential role in the formation of shear viscosity. At the same time, this value of $n_H(T)$ suggests to conclude that practically all water molecules near the critical point are dimerized. Namely in this case $n_H\left(T_c^{(W)}\right) \approx 0.5$.

To examine the conjecture about the existence of the developed H-bond network only at $T < T_v$ we apply to the analysis of the quasi-elastic incoherent neutron scattering data. The formation of the H-bond network with $n_H \approx 3$ essentially influence on the character of thermal motion. According to Refs. 5, 6 in this case: 1) each molecule during time τ_0 oscillates near some local equilibrium position; 2) during the next time interval τ_1 this molecule displaces according to diffusion law to new oscillation position (the corresponding self-diffusion coefficient is assumed to be equal D_1; 3) the oscillation position for a molecule and its nearest surrounding drifts occasionally in the field of thermal hydrodynamic fluctuations, forming in that way the collective part of the self-diffusion coefficient D_0.[6–9] Such a model of thermal motion is known as as the crystal-like one.

In Ref. 9 it has been shown that in the applicability region of the diffusion approach ($D_s \tau_0 \mathbf{k}^2 << 1$, where D_s is the self-diffusion coefficient,

k – wave vector) the differential cross-section of quasi-elastic incoherent neutron scattering takes the view:

$$\left(\frac{d^2\sigma}{d\Omega d\varepsilon}\right)_{\text{inc}} = \frac{b_{\text{inc}}^2 k}{\pi\hbar k_0} e^{-2W}\left(\frac{\gamma_D}{\omega^2+\gamma_D^2} + o\left(\tau_1/\tau_0\right)\right), \tag{2}$$

where

$$\gamma_D \approx \left(D^{(1)} + D_0\right)\mathbf{k}^2 - \tau_0 D^{(1)^2}\mathbf{k}^4 + \tau_0^2 D^{(1)^5}\mathbf{k}^6 + \cdots \tag{3}$$

is the half-width of the neutron peak. The combination

$$D_s = D^{(1)} + D_0 \tag{4}$$

takes meaning of the full self-diffusion coefficient of molecules. The term $D^{(1)} = D_1 \dfrac{\tau_1}{\tau_0+\tau_1}$ should be interpreted as single-particle part of the self-diffusion coefficient. Since the crystal-like representations are correct for $\tau_1 << \tau_0$, then $D_s \approx D_1 \dfrac{\tau_1}{\tau_0}$.

Fitting the experimental data[10,11] for the half-width of the incoherent neutron scattering peak with the help of Eq. (3) one can find the numerical values of all essential parameters: D_s, D_0, τ_0, and τ_0/τ_1. The applicability region of the crystal-like representations for the thermal motion of molecules in water is determined by the inequality $\tau_0/\tau_1 >> 1$. The temperature dependence of τ_0/τ_1, presented in Fig. 2 shows that τ_0 remains greater than τ_1 only at $T \leq T_n$, $T_n \approx 310$ K. Here crosses correspond to experimental data from Ref. 11, squares – data from Ref. 10.

The closeness T_n to T_v is the weighty argument to assert that the H-bond network plays the determinative role in the formation of water properties only up to T_v The inequality D_0/D_s testifies also in the favor of this conclusion.

These and other results allow us to conclude the following: 1) water can be considered as a liquid with the crystal-like thermal motion only for $T < T_v$; 2) the developed H-bond.[12]

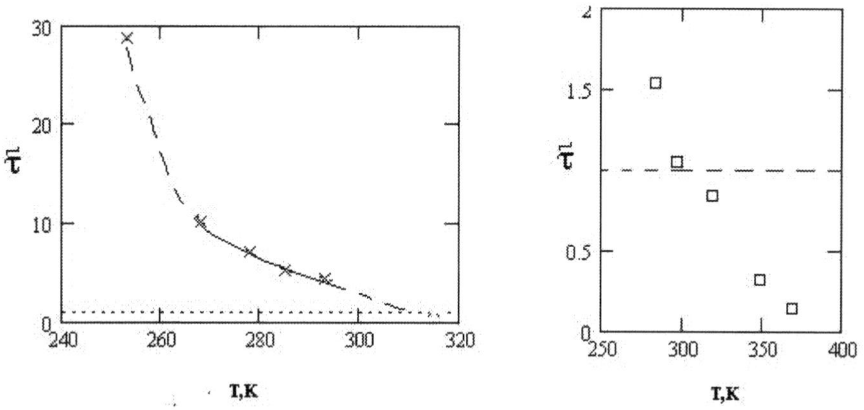

Figure 3. The temperature dependence of the ratio $\tilde{\tau} = \tau_0 / \tau_1$ (see the text for explanation).

Network in water exists only at the temperatures less than T_v. In the opposite case there can exist only linear multimers: dimers, trimers, pentamers and so on; and 3) the kinetic properties of water are less sensitive to the integrity of the H-bond network compared to its structural and thermodynamic properties.

In accordance with said above we conclude, that near the critical point water is practically dimerized fully. The rotating dimers can be considered as an ensemble of subunits determining the main properties of water.

In this paper the main attention will be focused on the mean field version of the equation of state of water in that region of its existence where the dimerization is dominating. It will be shown that the coordinates of the critical point are reproduced with good accuracy within this approach. The behavior of the dielectric permittivity is analyzed. The asymmetry of gas and liquid branches of the coexistence curve is studied.

2. The van der Waals parameters for water

In accordance with said in Introduction we assume that water near critical point is fully dimerized. Thus water dimer $2(H_2O)$ plays the role of a "particle" Therefore we can expect that in near critical region water can be reasonably described by the vdW EOS for the system of dimers. The value of the number density at the critical point within vdW EOS is $n_c = 1/3b$. It is simply related to the diameter of the particle σ:

$$b = \frac{2\pi}{3}\sigma^3. \tag{5}$$

The coordinates of the critical point for pure water are

$$T_c \approx 647\,\text{K}, \qquad P_c \approx 22\ \text{MPa}, \quad \rho_c \approx 322\ \text{kg}/\text{m}^3 \tag{6}$$

For dimerized water $\rho = 2mn$ where m is the mass of the water molecule and n is the number density of dimers. From Eqs. (5) and (6) we find that the diameter of the dimer

$$\sigma_d = 3.1\,\text{Å}\,, \tag{7}$$

which is in agreement with the value of $O-O$ distance for the most energetically favorable configuration of water.[13] Moreover the compressibility factor:

$$Z_c = \frac{P_c}{\rho_c T_c} \tag{8}$$

is $Z_c^{(\text{dim})}$, which is close to the vdW value $Z_{\text{vdW}} = 0.375$.

Note that for water consisting mostly of monomers the corresponding value of the diameter, determined from critical density value is $\sigma \approx$ Å, which is much less then the diameter of water molecule σ obtained in quantum chemical calculation.[14] Also one can find that for such a "monomeric" model $Z_c \approx 0.23$ which is rather low in comparison with the vdW value.

Other critical parameters T_c and P_c depend not only on the geometrical size of the molecules but also on the energy of the interaction, which is characterized by the parameter: [15]

$$a = \pi \int_{\sigma}^{\infty} U(r) r^2 dr\,. \tag{9}$$

The basis for the use the van der Waals EOS in near critical region is due to low density and thus intensive rotation. Indeed, the average spacing between dimers at the critical density is about 3.5 Å and is more than the size of dimer. However, at removing of the critical point along liquid branch the character of the dimer rotation changes: the arbitrary rotation of a dimer passes to the rotation around the H-bond connecting dimer molecules.

Therefore, the interactions between water monomers and dimers are of vdW nature:[16]

$$\begin{aligned} U(r_{12}) &= \langle U(1,2) \rangle = -U_d \left(\frac{\sigma_d}{r_{12}} \right)^6, \\ U_d &= \frac{2}{3} \beta \frac{\left\langle \left\langle d_{\text{dim}}^2 \right\rangle \right\rangle^2}{\sigma_d^6}, \end{aligned} \tag{10}$$

where brackets denote the average with internal partition function of the dimer.

In calculation the vdW parameter a only long distance part $\propto 1/r^6$ have been taken into account. Another vdW part of the potential is due to induced dipole moment is neglected since it is small in comparison with the dipole moment of dimer.

The dipole moment $d_{\dim}$ of the dimer, which can be written as

$$d_{\dim} = \sqrt{\left\langle\left\langle \mathbf{d}^2_{\dim} \right\rangle\right\rangle} = q_{\text{eff}}\sigma_d \tag{11}$$

where q_{eff} is the effective charge which determines the value of the dipole moment and can be written in the form $q_{\text{eff}} = \lambda e$ (e is the electronic charge).

The parameter λ depends on the quantum structure of the dimer. In classical picture, it describes the correlation between the orientation of the dipole moments of water molecules constituting the dimer according to

$$\left\langle\left\langle \mathbf{d}^2_{\dim} \right\rangle\right\rangle = 2\left\langle\left\langle \mathbf{d}^2_{H_2O} \right\rangle\right\rangle + 2\left\langle\left\langle \mathbf{d}^{(1)}_{H_2O} \cdot \mathbf{d}^{(2)}_{H_2O} \right\rangle\right\rangle \tag{12}$$

Therefore dimerized water can be described by vdW EOS:

$$P = \frac{n_d k_B T}{1 - n_d b_d} - a_d n_d^2,$$

where according to Eq. (10):

$$a_d(T) = \frac{2\pi}{9T}\lambda^4 \sigma_d \,. \tag{13}$$

The critical density take the value

$$n_c^{(d)} = \frac{1}{3b_d} = \frac{1}{2\pi\sigma_d^3} \,. \tag{14}$$

The critical temperature is determined by the equation

$$T_c^{(d)} = \frac{8a_d\left(T_c^{(d)}\right)}{27b_d} \tag{15}$$

from which, taking into account Eq. (13), we get

$$T_c^{(d)} = \frac{2\lambda^2\sqrt{2}}{9}\frac{e^2}{\sigma_d} \,. \tag{16}$$

The critical pressure is

$$P_c^{(d)} = \frac{3}{8} n_c^{(d)} T_c^{(d)} = \frac{\lambda^2}{12\pi\sqrt{2}}\frac{e^2}{\sigma_d^4} \,. \tag{17}$$

We choose the value of λ so that to get the best fit for the coordinates of the vdW critical point. Such a fitting gives 0.19 and the coordinates of the critical point are

$$T_c \approx 647\ \text{K}, \qquad P_c \approx 18\ \text{MPa}, \qquad \rho_c \approx 322\ \text{kg/m}^3 \tag{18}$$

The dipole moment of a dimer is $d_{\dim} \approx 2.9D$, which is in good agreement with the value for dimer dipole moment $d_{\dim} \approx 2.6D$.[17] Sure total dimerization is an approximation but above estimates show that it is consistent. Thus the difference in the compressibility factor Z_c is due to the deviation from vdW EOS since the influence of free monomers was neglected. Below we consider the question about degree of dimerization of water near its critical point within the chemical equilibrium picture.

3. The dimerization of water in the near critical region

The dimerization is considered as the H-bond bounding of two H_2O molecules with the energy of H-bond. The molecules both dimerized and free are assumed to form the van der Waals liquid. Therefore, the free energy of the system is the sum:

$$F\left(T,V,N_f,N_d\right)=F_d\left(T,V,N_d\right)+F_f\left(T,V,N_f\right)+F_{fd}\left(T,V,N_f,N_d\right) \quad (19)$$

with

$$F_i/N=\underbrace{f(T)+T\ln n_i}_{F_i^{(id)}}\overbrace{-T\ln(1-n_ib_i)-a_in_i}^{F_i^{(ex)}} \quad (20)$$

where $i=f,\ d$, and $F_i^{(\mathrm{id})}$ is the ideal part and $F_i^{(\mathrm{ex})}$ is the excessive part due to interaction between the particles of the same type.[15] The third term in Eq. (19):

$$F_{fd}=-a_{fd}n_fn_d$$

describes the interaction between monomers and dimers in pair approximation. In dimensionless quantities:

$$\tilde{T}=T/T_c^{(d)}, \qquad \tilde{n}_i=n_i/n_c^{(d)}$$

reduced to the coordinates of the critical point $T_c^{(d)}$, $n_c^{(d)}$ of fully dimerized model given by Eqs. (14), (16), and (17):

$$\frac{F_d}{N_0T_c^{(d)}}=(1-A)\left(f_0\left(\tilde{T}\right)+\tilde{T}\ln\tilde{n}_d-\tilde{T}\ln\left(1-\frac{\tilde{n}_d}{3}\right)-\frac{9}{8\tilde{T}}\tilde{n}_d\right), \quad (21)$$

$$\frac{F_d}{N_0T_c^{(d)}}=2A\left(f_0\left(\tilde{T}\right)+\tilde{T}\ln\tilde{n}_f-\tilde{T}\ln\left(1-\frac{\tilde{n}_f\tilde{\sigma}^3}{3}\right)-\frac{9}{8\tilde{T}}\tilde{\lambda}^4\tilde{\sigma}\tilde{n}_f\right), \quad (22)$$

$$\frac{F_{fd}}{N_0T_c^{(d)}}=-\frac{9}{4\tilde{T}}\tilde{\lambda}^2\left(1+\tilde{\sigma}\right)A(1-A)\tilde{n}_0\cdot \quad (23)$$

Here $N_0=N/2$, where N is the total number of water molecules and we introduce the degree of the dissociation A as following:

$$n_f = 2An_0, \quad n_f = (1-A)n_0. \tag{24}$$

The number density of water molecules $n = N/V$ both free and dimerized is related to $n_0 = N_0/V$ as $n = n_f + 2n_d = 2n_0$. The quantities $\tilde{\sigma} = \sigma_f/\sigma_d$, σ_f is the vdW diameter of free water molecule and $\tilde{\lambda} = \lambda_f/\lambda_d$. Total dimerization limit considered before corresponds to $A = 0$.

The equation for the chemical equilibrium between dimers and monomers is

$$\frac{1-A}{A^2} = n_0 K(T) \exp\left(\beta\mu_f^{(\mathrm{ex})} - \beta\mu_d^{(\mathrm{ex})}\right). \tag{25}$$

where $\mu_f^{(\mathrm{ex})}$, $\mu_d^{(\mathrm{ex})}$ are the excess chemical potentials of the monomers and dimers correspondingly. They can be calculated from the EOS, e.g., Eq. (20):

$$\mu_f^{(\mathrm{ex})} = \frac{\partial\left(F_f^{(\mathrm{ex})}/V\right)}{\partial n_f} = -\tilde{T}\ln\left(1 - \frac{\tilde{n}_f}{3}\right) + \frac{\tilde{T}}{3}\frac{\tilde{n}_f}{1-\tilde{n}_f/3} - \frac{9}{8\tilde{T}}\left(2\tilde{n}_f + \tilde{\lambda}\tilde{n}_d\right), \tag{26}$$

$$\mu_d^{(\mathrm{ex})} = \frac{\partial\left(F_d^{(\mathrm{ex})}/V\right)}{\partial n_d} = -\tilde{T}\ln\left(1 - \frac{\tilde{n}_d}{3}\right) + \frac{\tilde{T}}{3}\frac{\tilde{n}_d}{1-\tilde{n}_d/3} - \frac{9}{8\tilde{T}}\left(2\tilde{n}_d + \tilde{\lambda}\tilde{n}_f\right) \tag{27}$$

The quantity $K(T)$ is the constant of chemical equilibrium.[15] We can approximate it by simple expression

$$K(T) = \exp\left(\frac{E_H}{k_B T}\right), \tag{28}$$

taking into account the energy of H-bond in a dimer only which in $k_B T_c$ units is of order $E = 3/4 k_B T_c$.[18,19] We take $E_H \approx 4 k_B T_c$ in our calculations. Such a large value of the H-bond energy and the results of previous section allows us to conclude that near critical water is dimerized to a great extent. Besides, according to the results of Ref. 3 the average number of H-bonds per molecule is $n_H \approx 0.5$ at the critical point (see Fig. 2). This is indeed the case as the numerical solution shows (see Fig. 4). At the critical point $A_c \le 0.1$. Besides, the inclusion of monomers corrects the value of pressure and slightly raises the value of n_c. Critical temperature changes insignificantly. Thus we may conclude that the value of critical temperature is formed due to the dipole interaction between dimers. The value of the σ_d may be chosen so that to get the value of critical density. This gives $\sigma_d \approx 3.3\,\mathring{A}$. The dipole moment of a dimer changes insignificantly $d_{\mathrm{dim}} \approx 3D$.

Using the expansion in degree of dissociation A, it is easy to get the dependence of the critical point coordinates on this parameter:

$$\tilde{T}_c = 1 + t_1 A + t_2 A^2 + o\left(A^2\right), \tag{29}$$

$$\tilde{n}_c = 1 + n_1 A + n_2 A^2 + o\left(A^2\right), \tag{30}$$

$$\tilde{p}_c = 1 + p_1 A + p_2 A^2 + o\left(A^2\right), \tag{31}$$

where

$$n_1 = \frac{43}{27}, \quad n_2 = \frac{403}{243} + 4\tilde{\lambda}^2 + 4\tilde{\lambda}^2\tilde{\sigma},$$

$$t_1 = -\frac{4}{9}, \quad t_2 = -\frac{4}{243} - \frac{16}{27}\tilde{\sigma}^3 + 3\tilde{\lambda}^2 + 3\tilde{\lambda}^2\tilde{\sigma} + 2\tilde{\lambda}^4\tilde{\sigma},$$

$$p_1 = \frac{20}{9}, \quad p_2 = \frac{548}{243} + 21\tilde{\lambda}^2 + 21\tilde{\lambda}^2\tilde{\sigma} - \frac{16}{27}\tilde{\sigma}^3 + 2\tilde{\sigma}\tilde{\lambda}^4 - 12(1+\tilde{\sigma})\tilde{\lambda}^2,$$

The coexistence curve is calculating according to equality the Gibbs potentials of the coexisting phases:

$$G_g(P,T)/N_0 = G_f(P,T)/N_0 \tag{32}$$

where

$$G_i/N_0 = F(T,n_i)/N_0 + P(T,n_i)/n_i, \quad P = -\left.\frac{\partial F}{\partial V}\right|_T,$$

where F is the equilibrium free energy of the corresponding phase:

$$F(T,n_0) = F(T,2An_0,(1-A)n_0 \,\big|_{A=A(T,n_0)}$$

The above equation is for n_l and n_g densities of equilibrium phases.

Using the results of previous section we are able to calculate the asymmetry of the coexistence curve. The binodal is asymmetrical with the rectilinear diameter if the density is taken as the order parameter:

$$\frac{n_+ + n_-}{2} = c_n + b_n\tau + o(\tau), \quad \tau = \frac{T - T_c}{T_c} \tag{33}$$

with

$$c_n \approx 1.56, \quad b_n \approx -0.3.$$

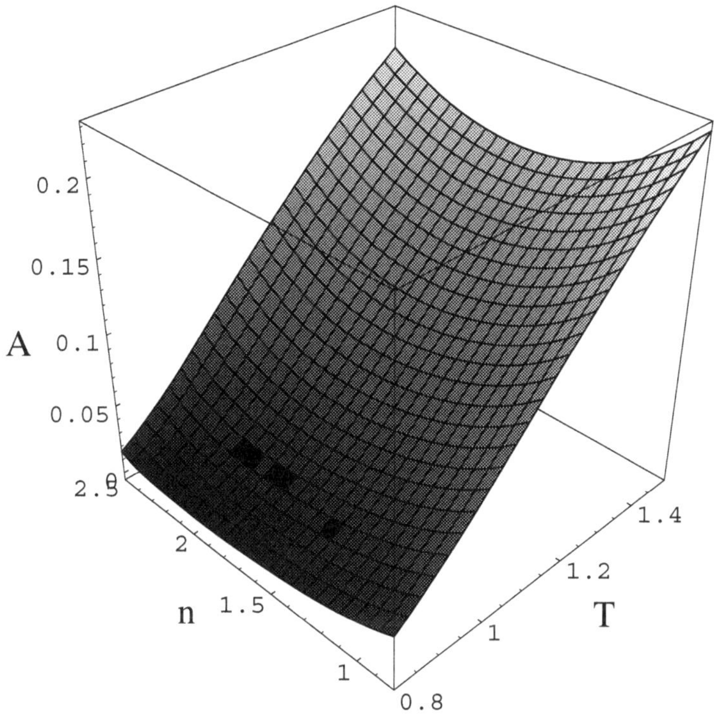

Figure 4. The degree of dissociation A as a function of (dimensionless) density and temperature at $E_H = 3.87\,T_c^{(d)}$, $\tilde{\lambda} = 0.6$, and $\tilde{\sigma} = 0.88$.

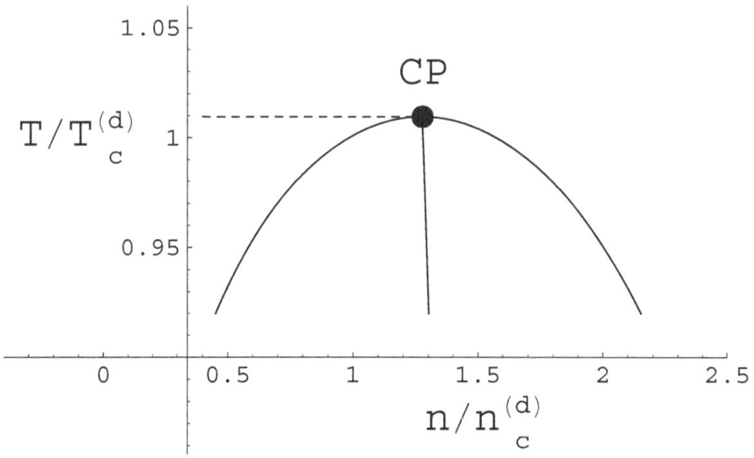

Figure 5. Coexistence curve (solid) in coordinates $\tilde{n} - \tilde{T}$ and the rectilinear diameter.

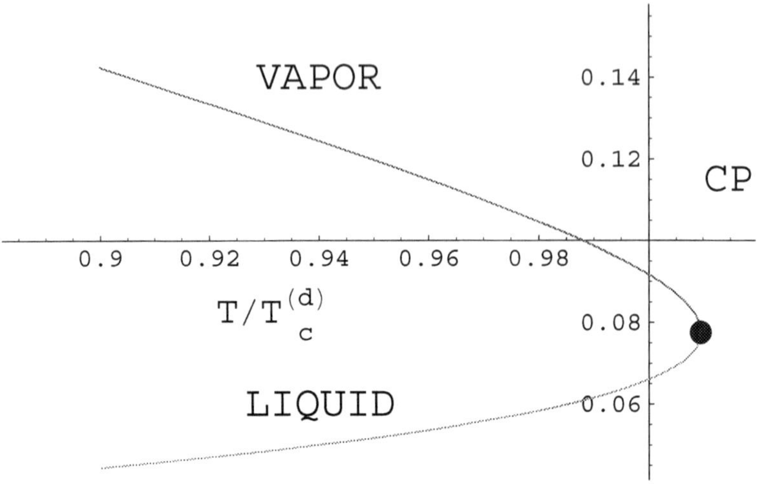

Figure 6. The degree of dissociation along the binodal.

4. Dielectric permittivity

It is interesting to apply the model for the description of the dielectric permittivity of water, which is the dipolar liquid. We also compare the asymmetry of the binodal in terms of static dielectric permittivity. There are extensive data for such a characteristic for both liquid and vapor phases of water.[20] They are represented on Fig. 7 with the dashed line as the guide for an eye.

According to the model proposed and taking into account the results of Ref. 16 for the dielectric permittivity water considered as the mixture of dimers and monomers we can write:

$$\frac{\varepsilon-1}{\varepsilon+2}=\frac{4\pi}{3}\left[\left(\alpha_f+\frac{d_f^2}{3T}\right)n_f+\left(\alpha_d+\frac{d_d^2}{3T}\right)n_d\right]. \tag{34}$$

Neglecting the electronic part of the polarizabilty $(\alpha_f n_c^{(d)} \approx 0.03)$ in dimensionless units Eq. (34) is

$$\frac{\varepsilon-1}{\varepsilon+2}=\frac{1}{\sqrt{2}}\frac{2A\tilde{\sigma}^2\tilde{\lambda}^2+(1-A)}{1+A}\frac{\tilde{n}}{\tilde{T}}. \tag{35}$$

The comparison of the theoretical model (gray line) with the data of Ref. 20 is on Fig. 8.

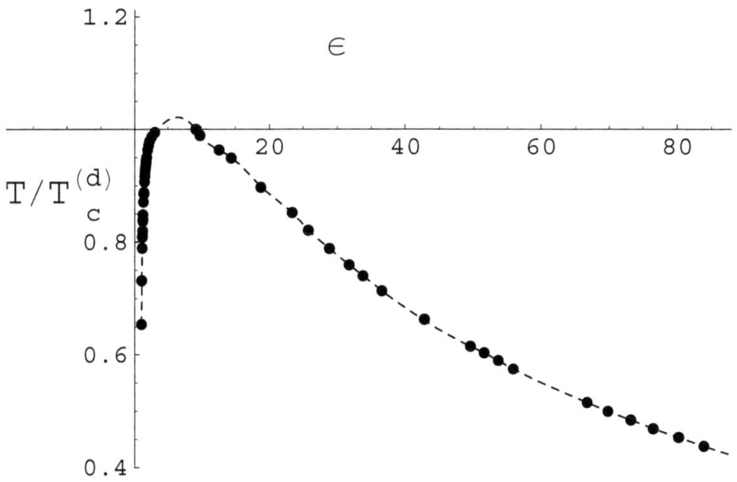

Figure 7. Dielectric permittivity data.

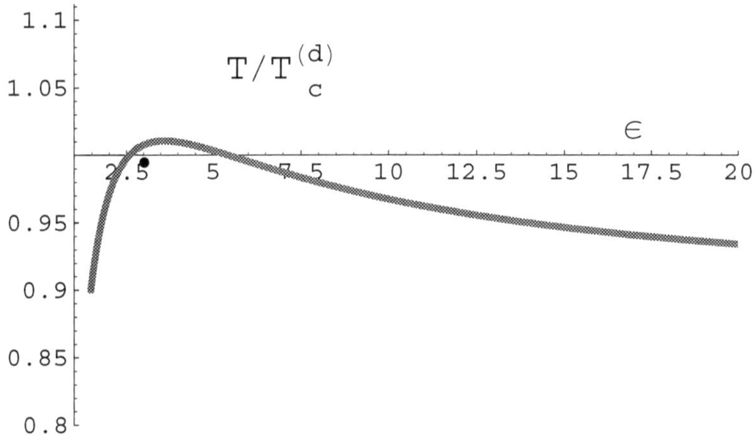

Figure 8. Theoretical curve according to Eq. (34) and experimental points for $\varepsilon(T)$ along the binodal.

We also give the results on the dielectric permittivity according to Onzager model:

$$\varepsilon = \frac{1}{4}\left(1 + x + \sqrt{9 + 2x + x^2}\right), \tag{36}$$

where

$$x = \frac{4\pi}{T} n\left\langle \mathbf{d}^2 \right\rangle = \frac{9}{\sqrt{2}} \frac{\tilde{n}}{\tilde{T}} \frac{1 - A + 2A\tilde{\sigma}^2\tilde{\lambda}^2}{1 + A}.$$

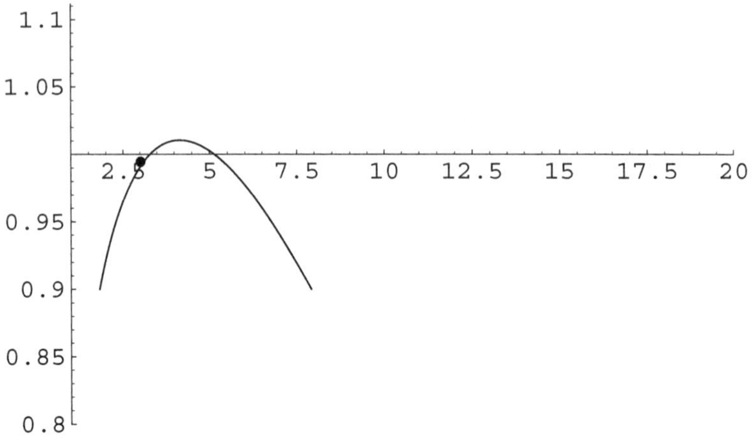

Figure 9. The dielectric permittivity according to Eq. (36) and the data.

Sure the theory does not fit the data. The coexistence curve near critical point is rather flat because of the fluctuational effects, which were not taken into account within the mean-field description used. Nevertheless, we would like to stress the fact that the theory proposed does not contain any *formal* fitting parameters yet it is adequate at least qualitatively. In order to achieve the matching with the experiment one should take into account that the dipole moments d_f, d_d are not those for the single molecule but they depend on the particle correlations. Such a picture is due to Kirkwood and is based on the introduction of the g-factor which accounts for the orientational correlations due to nondipolar interactions (in uncorrelated limit $g = 1$).

In Ref. 21 the experimental data were fitted by the construction of appropriate dependence of g-factor on the parameters of state $g(n, T)$. The Harris–Alder model was used for ε:[22]

$$\frac{\varepsilon - 1}{\varepsilon + 2} = \frac{4\pi}{3}\left(\alpha + g_\varepsilon \frac{d^2}{3T}\right) n, \tag{37}$$

where g_ε is the correlation factor. Such an approach can be interpreted within the concept of "softness" of the dimers, i.e., dependence of the dipole moment on the state of the liquid.

We use the corresponding result for g in order to improve the correspondence between the theory and the data. Sure g factor obtained in Ref. 21 for monomeric water which is not the same as for dimerized one. Nevertheless, we it is natural to assume that the correlations between the dimers are of the same character as for monomers and differ only by the appropriate scale factor. Therefore we can write:

$$\frac{\varepsilon-1}{\varepsilon+2}=\frac{\tilde{g}_\varepsilon}{\sqrt{2}}\frac{2A\tilde{\sigma}^2\tilde{\lambda}^2+(1-A)}{1+A}\frac{\tilde{n}}{\tilde{T}}\,. \tag{38}$$

The results are on Fig. 10.

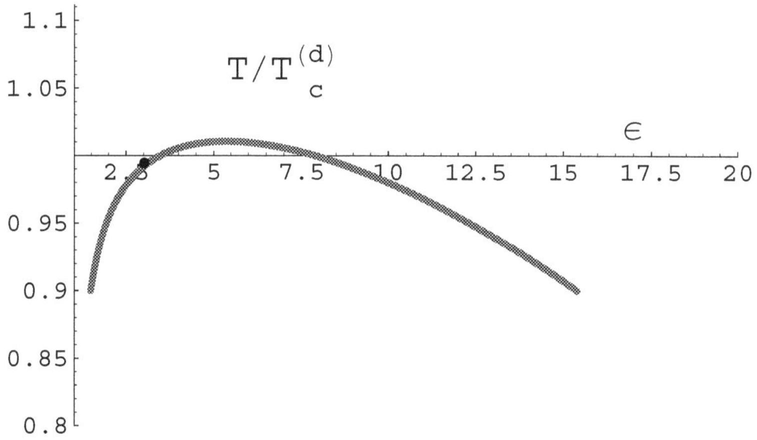

Figure 10. The dielectric permittivity according to Eq. (38) and the data.

The choice of the optimal order parameter in terms for which the binodal restores symmetrical shape is discussed long time. The general approach to this problem was proposed in Ref. 23 and developed in Ref. 24. Within this approach the canonical order parameter which restores the symmetry of coexistence curve is the one which determined by the reduction of the initial Hamiltonian to its canonical form via analytical transformation of initial order parameter. In the case of the critical point the canonical form of the Hamiltonian naturally coincides with the Landau–Ginburg form. For the canonical order parameter field the correlation functions of the coexistent phases are the same. This fact leads, in particular, to the symmetry of the binodal. From such point of view highly asymmetrical shape of the binodal in $\varepsilon - T$ coordinates is explained as the consequence of the differences in correlation functions for the field of dielectric permittivity in liquid and vapor phases. It is obviously caused by different character of orientational correlations in these phases.

5. Conclusion

The main goal of our paper is to describe the properties of water in the vicinity of its critical point using the dimer picture for its structure as a basis. This approach is proved to be quite successful since it allows to find the position of the critical point with satisfactory accuracy. The only fitting parameter of the proposed theory, the dipole moment d_f of a dimer, is close to its handbook value d_h.[2] The difference between d_f and d_h does not exceed 10% and can be easily removed in the approximation of soft dimers. It supposes that the dimer dipole moment is density dependent:

$$d(n) = d_0 (1 + \gamma_1 n + \gamma_2 n^2 + \cdots) \tag{39}$$

where d_0 is the dipole moment of an isolated dimer. The values of the coefficients γ_i in Eq. (9) can be determined from the minimum condition for the internal energy of a dimer interacting with all other surrounding dimmers. This question will be considered separately.

The proposed approach clearly explains the large difference between the positions of the critical points in water and its nearest homologue H_2S. In particular, the value of $T_c^{(H_2O)} - T_c^{(H_2S)} = 273.7\mathrm{K}$ was not interpreted in any previous theories.

The strong support for the dimer picture of water structure near its critical point is given by the calculation of the dielectric permittivity ε. For this purpose we use the approach used in Ref. 16, which is free from any assumptions about properties of the acting electric field. The numerical values of ε are in good agreement with experimental data.[21] It is very important that the dielectric properties and the equation of state for water near its critical point are reproduced quite consistently. The approach developed refines and completes the picture formulated earlier in Refs. 1 and 3 on the basis of density and shear viscosity data.

Within the dimer picture the average number n_H of H-bonds per molecule is close to 0.5. The last is close to that following from the consideration of density[3] and the heat capacity at constant pressure.[4] At the same time in computer simulations[25,26] the value of n_H is essentially larger that is likely connected with definitions of H-bonds, which have no relation to their experimental manifestations.

Since at T_c properties of water and water vapor are identical we should conclude that the last is dimerized too. However, at removing from the critical point the degree of dimerization of the water vapor reduced noticeably.

Acknowledgments

We cordially thank Professors M.-C. Bellissent-Funel, A.V. Chalyi, G.P. Johari, A. Onuki and especially Professors G.G. Malenkov and H.E. Stanley for the fruitful discussion of the results obtained.

References

1. Bulavin, L. A., Malomuzh, N. P., and Shakun, K. S. (2005) How substantial is the role of the H-bond network in water? *Ukr. Phys.* **50**, 653–658.
2. Vargaftik, N. B. (1972) *Handbook of Thermo-Physical Properties of Liquids and Gases*, Moscow, Nauka.
3. Lokotosh, T. V., Malomuzh, N. P., and Zakharchenko, V. L. (2003) Anomalous density and dielectric permittivity effects on the structure of water, *J. Struct. Chem. (Rus.)* **44**, 1101–1010.
4. Lisichkin, Y. V., Novikov, A. G., and Fomichev, N. K. (1989) Vibrational and configurational contributions to the heat capacity of water according to quasielastic scattering of cold neutrons, *J. Phys. Chem. (Rus.)* **63**, 833–835.
5. Singwi, K. S. and Sjolander, A. (1960) Difusive motions in water and cold neutron scattering, *Phys. Rev.* **119**, 863–871.
6. Oskotsky, V. S. (1963) To the theory of quasi-elastic scattering of cold neutrons in liquid, *Fiz. Solid State (SSSR)* **5**, 1082–1085.
7. Fisher, I. Z. (1971) Hydrodynamic asymptote of the velocity autocorrelation function of a molecule in classic liquid, *ZhETP* **61**, 1647–1659.
8. Lokotosh, T. V. and Malomuzh, N. P. (2000) Lagrange theory of thermal hydrodynamic fluctuations and collective diffusion in liquids, *Physica A* **286**, 474–488.
9. Bulavin, L. A., Malomuzh, N. P., and Pankratov, K. N. (2006) Character of thermal motion in water according to data of quasi-elastic incoherent neutron scattering, *J. Struct. Chem. (Rus.)* **47**, 52–59.
10. Blanckenhagen, P. (1972) Intermolecular vibrations and difussion in water investigated by scattering of cold neutrons, *Ber. Bunsenges. (Phys. Chem.)* **76**, 891–903.
11. Teixeira, J., Bellissent-Funel, M. C., Chen, S. H., and Dianoux, J. (1985) Experimental determination of the nature of difussive motions of water molecules at low temperatures, *Phys. Rev. A* **31**, 1913–1917.
12. Frenkel, J. (1955) *Kinetic Theory of Liquids*, NY, Dover Publ.
13. Odutola, J. A. and Dyke, T. R. (1980) Partially deuterated water dimers: Microwave spectra and structure, *J. Chem. Phys.* **72**, 5062–5070.

14. Guillot, B. (2002) A reappraisal of what we have learnt during three decades of computer simulations on water, *J. Mol. Liq.* **101**, 219–260.
15. Landau, L. and Lifshitz, E. (1976) *Statistical Physics*, Vol. **5**, Moscow, Nauka.
16. Kulinskii, V. and Malomuzh, N. (2003) Dipole fluid as a basic model for the equation of state of ionic liquids in the vicinity of their critical point, *Phys. Rev.* E **67**, 011501.
17. Yu, H. and van Gunsteren, W. F. (2004) Charge-on-spring polarizable water models revisited: from water clusters to liquid water to ice, *J. Chem. Phys.* **121**, 9549–9564.
18. Zacepina, G. A. (1974) *Structure and Properties of Water*, Moscow, Moscow University Publ.
19. Harrison, G. B. L., Shakes, T. R., Robinson, C. M., Lawrence, S. B., Heath, D. D., Dempster, R. P., Lightowlers, M. W., Rickard, M. D., Malenkov, G. G., Tytik, D. L., and Zheligovskaya, E. A. (1999) Structure and dynamic heterogeneity of computer simulated water: ordinary, supercooled, stretched and compressed, *J. Mol. Struc.* **82**, 27–38.
20. Fernandez, D. P., Mulev, Y., Goodwin, A. R. H., and Sengers, J. M. H. L. (1995) A database for the static dielectric constant of water and steam, *J. Chem. Phys. Data* **24**, 33–69.
21. Fernandez, D. P., Goodwin, A. R. H., Lemmon, E. W., Sengers, J. M. H. L., and Williams, R. C. (1997) A formulation for the static permittivity of water and steam at temperatures from 238 K to 873 K at pressures up to 1200 MPa, including derivatives and Debye-Huckel coefficients, *J. Chem. Phys. Data* **26**, 1125–1166.
22. Harris, F. E. and Alder, B. J. (1954) Statistical mechanical derivation of Onsager's equation for dielectric polarization, *J. Chem. Phys.* **22**, 1806–1808.
23. Koulinskii, V. and Malomuzh, N. (1997) Canonical formalism for description of critical phenomena in systems isomorphic to simple liquids, *Cond. Matt. Phys. (Ukraine)* **9**, 29–46.
24. Kulinskii, V. (2003) Nonperturbative construction of the Landau–Ginzburg Hamiltonian for the Ising-like systems, *J. Mol. Liq.* **105**, 273–278.
25. Oleinikova, A., Brovchenko, I., Geiger, A., and Guillot, B. (2002) Percolation of water in aqueous solutions and liquid–liquid immiscibility, *J. Chem. Phys.* **117**, 3296–3304.
26. Partay, L. and Jedlovszky, P. (2005) Line of percolation in supercitical water, *J. Chem. Phys.* **122**, (submitted).

FLUCTUATIONAL EQUATION OF STATE AND SLOPES OF CRITICAL CURVES NEAR THE CRITICAL POINT OF SOLVENT

VITALY ROGANKOV* and OLEKSIY BYUTNER
Physics Department, Odessa State Academy of Refrigeration, 1/3 Dvoryanskaya Str., 65082 Odessa, Ukraine

Abstract. A simple, yet reliable, equation of state was developed for fluids ranging from inert gases Ar, Kr, Xe to polar substances NH_3 and H_2O. The equation consists of the van der Waals generalized term as a reference system and a closed-form term approximating a coupling of the molar volume and entropy fluctuations near a phase boundary. Only the coexistence-curve data $P_S(T)$, $\rho_l(T)$, $\rho_g(T)$ and the second virial coefficient B(T) are necessary to determine three temperature dependent coefficients a(T), b(T), c(T) in the wide ranges of P and T. Proposed equation of state is not global and does not incorporate the Maxwell construction to locate a phase boundary. Critical point constraints were used only to obtain the asymptotic values (a_c^0 , b_c^0 , c_c^0) from the side of supercritical region. No critical point constraints were used in the sub-critical region, where the asymptotic values (a_c, b_c, c_c) are determined by the true critical parameters (P_c, ρ_c, T_c) and by the real reduced critical slope $A_c=(T_c/P_c)(dP_S/dT)_c$. This parameter is interrelated for the binary systems with the known Krichevskii parameter and with the initial slopes of the critical curve for a mixture in (P, x)- and (T, x)-planes. A method of analysis for the different types of the critical behavior in binary dilute mixtures of the solvent–cosolvent type with the near critical solvent has been developed. The accurate description of the critical point and coexistence curve data for the pure components provides the correct estimation of binary cross-interaction coefficient by utilizing of the single experimental point of a critical line for the given mixture. The proposed strategy is to determine the correlation between the Krichevskii parameter and the binary cross-interaction coefficient k_{12} in equation $a_{12}=(1-k_{12})(a_1(T)a_2(T))^{1/2}$. The initial slopes of the critical curves near the critical point of solvent have been used to predict reliable values of k_{12}.

Keywords: supercritical solvent, cosolvent, local phase diagram, solubility, critical curve, phase behavior.

---

* To whom correspondence should be addressed. Vitaly Rogankov, e-mail: rogankov @paco.net

S. J. Rzoska and V. A. Mazur (eds.), Soft Matter under Exogenic Impacts, 305–323.

1. Introduction

To study the solubility and phase behavior of a highly asymmetric polymer/ supercritical fluid (SCF)–solvent mixture one needs a development of an effective model. Usually the main SCF–solvent is composed by the small amount of cosolvent or modifier to enhance the solubility. Small solubility of polymers and copolymers in the conventional SCF (CO_2, N_2O, C_2H_4) can be enhanced, for example, by admixture of polar cosolvents, such as acetonitrile (CH_3CN) or methanol (methyl alcohol, CH_3OH) as well as nonpolar n-alkanes, such as pentane (C_5H_{12}), hexane (C_6H_{14}), or heptane (C_7H_{16}). In fact, only a limited number of polymers can be dissolved to any extent in a pure carbon dioxide, nitrous oxide or ethylene.

On the other hand, the above-named organic cosolvents are themselves the effective liquid solvents for polymers at the relatively low pressures. Addition of the small amount of the SCF-cosolvent (CO_2, N_2O) far away from its critical point becomes an important tuning parameter in fractionalization and separation of mixed polymer systems. It can be also used as antisolvent to force liquid–liquid phase transition in the mixture polymer/organic solvent. It is important for the solvent mixture to be in the homogeneous one-phase state above of the critical line (CRL) at the operating conditions P, T. This provides recovering and recycling of the solvent mixture without vapor–liquid separation, in contrast to the pure organic solvent.

Investigation of the binary solvent/polymer molecular binding and the role of cosolvents (the effect of dipole moment, P, T, x – conditions, effective size of the modifiers, solvent–cosolvent and polymer–cosolvent interaction strength) are crucial for understanding the whole dissolution process. The accurate CRL-localization and the reliable prediction of solubility in the system solvent/ cosolvent, especially in the near-critical region of the SCF-component are the necessary prerequisites for any model. In this work we develop the novel method of the local phase diagram (LPD) to represent the sub($T \leq T_{C1}$)- and super($T > T_{C1}$)-critical regions of a mixture in the vicinity of the more volatile SCF-component's critical point. It is based on the recently proposed fluctuational equation of state (FEOS) which is able[1–3] to describe reasonably well the properties of pure components and contains their coexistence curve (CXC)-data as input ones.

The strict equations for three T-dependent coefficients of FEOS in the different one-phase (subcritical liquid, subcritical gas, supercritical fluid) regions will be done in Section 2 for the description of pure components. The LPD-method[4] will be discussed in Section 3 and its results for selected binary mixtures will be presented in Section 4. The proposed semi-empirical model is a rather flexible to account the effective molecular characteristics of low-molecular-mass components and to be extended for the description of polymer components in terms of their quasicritical data.

2. Fluctuational equation of state

The form of a fluctuational EOS (FEOS) has been proposed[1–3] as the thermodynamically-consistent, van der Waals-type superposition of the entropy and internal-energy terms:

$$P = \frac{RT[1-c(T)]}{v-b(T)} - \frac{a(T)}{v^2} = T\left(\frac{\partial P}{\partial T}\right)_v + \left(\frac{\partial e}{\partial \rho}\right)_T \cdot \frac{1}{v^2} \tag{1}$$

in each one of the equilibrium points including the two-phase points where:

$$P_s = T\frac{dP_s}{dT} - \frac{a_s(T)}{v_l v_g} = T\frac{s_g - s_l}{v_g - v_l} + \frac{e_l - e_g}{\rho_l - \rho_g} \cdot \frac{1}{v_g v_l}. \tag{2}$$

It is evident that the Clapeyron equation and equality of the chemical potentials $\mu_l = \mu_g$ are incorporated in the structure of Eq. (2). The well-known undesirable feature of the Maxwell construction is that it employs the analytic isothermal continuation of the pressure P(v) into the two-phase region – is absent in the proposed FEOS. In contrast to majority of the semiempirical vdW-type EOSs, the proposed form is not a global and does not incorporate the Maxwell construction to locate a phase boundary.

The experimental or tabular CXC-data $P_s(T)$, $\rho_l(T)$, $\rho_g(T)$ is considered as the input for calculations of the different sets of T-dependent FEOS-coefficients in the subcritical liquid: a_l, b_l, c_l and subcritical gas: a_g, b_g, c_g regions. Only the second-virial-coefficient B(T)-data must be added to the high-density experimental points for determination of T-dependent coefficients: a, b, c in the SCF-region. Presence of the reduced slope $A_s(T)$ of the vapor-pressure $P_s(T)$-function is the important feature of the derived equations[1]:

$$a_l = \frac{P_s[A_s(T)-1]}{\rho_l \rho_g} = a_s(T), \tag{3}$$

$$a_g = RT^2 \frac{dB(T)}{dT} \quad (T \le T_c), \qquad a = RT^2 \frac{dB(T)}{dT} \quad (T \ge T_c), \tag{4}$$

$$b_l = b_g = \frac{A_s(T)-2}{(\rho_l + \rho_g)[A_s(T)-1]} = b_s(T), \tag{5}$$

$$1 - c_l = Z_l \left[1 + \rho_l (A_s - 1) / \rho_g \right] (1 - b_s \rho_l), \tag{6}$$

$$1 - c_g = \left(Z_g + \rho_g T \, dB / dT \right) (1 - b_s \rho_g), \tag{7}$$

where: $Z_{l,g} = P_s / (\rho_{l,g} RT)$, $A_s = (T / P_s)(dP_s / dT)$, and two generalized vdW-coefficients completely describe the true CXC-data of a substance:

$$P_s = a_1(T) \rho_l \rho_g \left[1 - b_1(T)(\rho_l + \rho_g) \right], \tag{8}$$

The proposed in Eqs. (3–7) transformation of the measurable CXC- and B(T)-data into the FEOS-coefficients does not contain any fitting parameters.

This feature is important for obtaining of the well-defined molecular interpretation of the FEOS-coefficients. There is an alternative, perturbation-theory's meaning of the compressibility factor Z from Eq. (1):

$$Z = P / \rho RT = Z_0 - c(T) / \left[1 - b(T) \rho \right], \tag{9}$$

where the reference system is a generalized vdW-form[2,3]:

$$Z_0 = P_0 / \rho RT = 1 / \left[1 - b(T) \rho \right] - a(T) \rho / RT. \tag{10}$$

Unlike the conventional identification of the repulsive term with the known T-independent Carnahan-Starling EOS for the reference system of hard spheres we adopt the T-dependent coefficient b(T). It can be connected with the effective molecular diameter σ(T):

$$b(T) = (2/3) \pi N_A \sigma^3(T). \tag{11}$$

This parameter thermodynamically represents a coupling of the isothermal changes of molar entropy and volume at any equilibrium state in accordance with Eqs. (1, 2) and (10):

$$(\delta s)_T = \gamma_v (\delta v)_T, \qquad \left(s_g - s_l \right) = \gamma_s (v_g - v_l), \tag{12}$$

where the slopes are: $\gamma_v = (\partial P / \partial T)_v$ and $\gamma_s = dP_s / dT$, respectively.

In this context the small non-dimensional coefficient c(T) in Eq. (9) may be considered[2,3] as the fluctuational, state-dependent correction to the generalized vdW-description of such a coupling in Eq. (10). The same value of the effective molecular diameter σ(T) from Eq. (11) in both coexisting phases is the plausible physical result in Eq. (5) as well as the difference of $c_l(T)$ and $c_g(T)$ in Eqs. (6,7) for these phases.

Taking into account Eqs. (1, 4) the gas and fluid coefficients $a_g(T)$ and $a(T)$, respectively, may be directly connected with the effective pair potential $u(r;T)$:

$$a_g = -2\pi N_A^2 \int_0^\infty u(r;T)\exp\left[-u(r;T)/kT\right]r^2 dr. \quad (13)$$

The molecular interpretation of $a_l(T)$ and $c_l(T)$ is more subtle. In accordance with Eqs. (1–3) $a_l(T)$ thermodynamically represents a coupling of the isothermal changes of internal energy and density. One may consider that T-dependent trends in intermolecular potential energy $u(r)$ are similar for molecular models of normal and polar fluids which are structurally different. In other words, centrally embedded point dipole, for example, will show in the main potential parameters: $\varepsilon_l(T)$ and $\sigma_l(T)$ a T-dependent orientation effect that is very much like a non-spherical but non-polar molecule. If a molecule is both polar and non-spherical, the orientation effects are almost indistinguishable in the $\varepsilon_l(T)$ and $\sigma_l(T)$ but can be, in principle, separable in the temperature trend of the third parameter $\delta_l(T)$ connected with $c_l(T)$. Thus, a single semiempirical orientation parameter $\delta_l(T)$ in the effective $u(r;T)$-function for a liquid may represent either the effects of polarity, non-sphericality or a combination of them.

At the critical point where $A_S(T)\rightarrow A_C$ (the known Riedel's factor of the corresponding states principle (CSP)) the asymptotic values of the subcritical coefficients from Eqs. (3–7) were expressed[1,2] in terms of the true critical parameters:

$$a_c = P_c\left(A_c - 1\right)/\rho_c^2, \quad (14)$$

$$b_c = (A_c - 2)/[2\rho_c(A_c - 1)], \quad (15)$$

$$1 - c_c = Z_c A_c^2 / 2(A_c - 1). \quad (16)$$

No critical point constraints were used in the subcritical region where the FEOS is not analytic and two sets of coefficients are needed to describe the coexisting phases.

On the other hand, the asymptotic values of the supercritical coefficients were defined by applying the standard critical conditions:

$$a_c^0 = 3P_c / \rho_c^2, \quad (17)$$

$$b_c^0 = 1/3\rho_c, \quad (18)$$

$$1-c_c^0 = 8Z_c/3 = Z_c/Z_c^0 \,. \tag{19}$$

The exciting feature is the simultaneous presence in eqs.(17–19) of the universal vdW-parameters $A_c^0 = 4$, $Z_c^0 = 3/8$ and the true critical parameters of a substance. One can exactly determine the jumps of coefficients at the critical temperature T_c by applying the condition: $A_c = A_c^0 = 4$ to Eqs. (14–16).

Critical properties, CSP-parameters and FEOS-coefficients are represented in Table 1 for the typical SCF–solvent (CO_2), polar solvent and/or cosolvent (CH_3OH), non-polar solvents and/or cosolvents (C_5H_{12} and C_6H_{14}). The reduced coefficients: a_l/a_c, b_l/b_c, of a subcritical liquid as the functions of T_c/T for seven selected fluids of generalization: N_2, C_2H_4, C_2H_6, C_3H_8, CO_2, NH_3, H_2O do not intersect and have the same simple shapes, in spite of the complex combinations of increasing or decreasing T-functions: $P_S(T)$, $A_S(T)$, $\rho_l(T)$, $\rho_g(T)$ in Eqs. (3) and (5), respectively.

In accordance with CSP the location of these coefficients varies monotonically with the know Pitzer's factor $\omega = -[1 + \lg P_S/P_C$ at $T/T_C = 0{,}7]$. The general functions have been fitted for the set of basic fluids:

$$a_l = a_c[1 + d_1(\omega)(T_c/T - 1)^{0,4}]\,, \tag{20}$$

$$b_l = b_c[d_2(\omega)\ \ (T_c/T - 1)/(d_3(\omega)\, T_c/T - 1) + 1]\,, \tag{21}$$

where the linear relations are applicable:

$$d_1 = 0{,}28 + 1{,}19\omega,\quad d_2 = -0{,}39 - 0{,}29\,\omega,\quad d_3 = 1{,}61 - \omega\,. \tag{22}$$

Expressions (20–22) yield values a_l, b_l which deviate from those obtained from Eqs. (3, 5) by less than 1. 8 and 1. 6, respectively, for all of the selected fluids including polar ones at all temperatures tested from the triple T_t and up to the critical T_c point. Only the data of Table 1 have been used to predict the reference-system coefficients a_l, b_l for the organic solvents C_5H_{12}, C_6H_{14}, CH_3OH. These components will be used, exclusively, in the subcritical liquid phase.

The possible discrepancies from three-parameter CSP for water, ammonia and such polar component as the methanol, for example, must be related to the polarity and hydrogen-bonds presence. These factors can provide the short-range ordering in a system which leads to the local fluctuations of the thermodynamic properties. In accordance with Eqs. (6, 16) c_c parameter in the critical point is a non-linear combination of two CSP-factors (Z_C, A_C). The reduced c_l/c_c-plots are overlapping for the set of normal liquids: N_2, C_2H_4, C_2H_6, C_3H_8, CO_2 in opposite to plots for polar substances: NH_3, H_2O. It was preferably to describe the third FEOS-coefficient c_l by two variants of the generalized function:

$$c_l = c_c\left(2{,}07\cdot10^{-9}e^{22{,}08T/T_c} - 1{,}68T/T_c - 0{,}71\right)\left(1-T/T_c\right)^{0{,}5} + 1, \quad (23)$$

$$c_l = c_c\left(2{,}07\cdot10^{-9}e^{22{,}08T/T_c} - 11{,}9T/T_c + 4{,}64\right)\left(1-T/T_c\right)^{0{,}5} + 1 \quad (24)$$

for the normal liquids and the polar liquids respectively. The former variant has been used to predict the coefficient c_l for C_5H_{12}, C_6H_{14} and the latter one – for methanol CH_3OH. The available CXC-data for these substances are less extensive than that for the set of selected fluids and the reliable prediction of the FEOS-coefficients is needed, especially, in the critical region. The developed procedure besides conforming the FEOS-coefficients to the CXC-data has the advantage of the realistic $A_S(T)$-slope-representation. Scaling theory postulates that the derivative of A_S with respect to T must be the divergent one to provide the weak singularity of $C_V(T)$ along the critical isochore. Actually, the appearance of the $a_l(T)$- and $c_l(T)$-shapes near T_c confirms the possible divergence of their derivatives. Appearance of the reduced a/a_c^0, b/b_c^0, c/c_c^0 -coefficients for the SCF-CO_2 can be found in[1–3].

TABLE 1. Critical properties and FEOS-coefficients of pure components.

Substance	Carbon Dioxide	Pentane	Hexane	Methanol
T_C, K	304.20	469.69	507.85	512.60
P_C, kPa	7383	3364	3058	8120
ρ_c,mol/dm^3	10.636	3.1959	2.7382	8.625
ω	0.217	0.252	0.303	0.565
Z_C	0.2745	0.2695	0.2645	0.2209
A_C	6.838	7.03	7.27	8.438
a_c, Jdm^3/mol^2	381.12	1986.04	2557.3	811.88
a_c^0, Jdm^3/mol^2	195.86	988.08	1223.6	327.46
b_c, cm^3/mol	38.960	130.51	153.48	50.177
b_c^0, cm^3/mol	31.340	104.30	121.73	38.647
c_c	–0.0993	–0.1044	–0.1148	–0.0573
c_c^0	0.2681	0.2813	0.2947	0.4109

3. Local phase diagram of binary mixtures in near-critical region of solvent

The various types of phase diagrams (PD) in binary mixtures are usually classified according to the nature of their (P^c, T^c)-projections of critical lines (CRL). The importance of the semi-empirical EOSs is that even with the oversimplified van der Waals (vdW) form one can obtain by the so-called "global" PD (GPD) the qualitative agreement with the actual CRL-data (P^c, v^c, T^c, x) over the whole range of T, P, and x [5]. Unfortunately, any attempt to obtain the quantitative prediction for mixtures with an essential difference in components is restricted by the known shortcomings of the cubic EOS. Some of them are connected to the main idea of the global EOS that it is possible to describe the properties of the coexisting phases by the same coefficients a, b, etc. which may be also the T-dependent. If the developed EOS is global, it must predict the location of PD by the known Maxwell construction in the (ρ_l, T)- , (ρ_g, T)-, and (P_s, T)-coordinates. As a rule, this approach does not provide the reliable estimation of the P_s(T)-function and its derivative $\gamma_s = dP_s/dT$ along the coexistence curve (CXC). This derivative is necessary to predict the correlation of the thermal ($v_g - v_l$) and caloric ($s_g - s_l$) properties' jumps across the phase boundary by the Clapeyron equation:

$$\gamma_s \equiv \frac{dP_s}{dT} = \frac{s_g - s_l}{v_g - v_l} = \frac{r_s(T)\rho_l\rho_g}{T(\rho_l - \rho_g)} . \tag{25}$$

There is also an important interrelation[6–8] between the initial slopes of the CRL: (dP^c/dx), (dT^c/dx) and the limiting values of two parameters δ^c, γ_S^c at the critical point of solvent:

$$\delta^c \equiv (\partial P/\partial x)^c_{v,T} = (dP^c/dx)_{x\to 0} - \gamma^c_S (dT^c/dx)_{x\to 0} , \tag{26}$$

where x is the mole fraction of solute. The reduced forms of the Krichevskii (K_c) and Riedel (A_c) parameters, respectively, are

$$K_c = \delta^c / P_{c1} = (1/P_{c1})(\partial P/\partial x)^c_{v,T,x\to 0} , \tag{27}$$

$$A_c = (T_{c1}/P_{c1})\gamma^c_S = (d\ln P_s / d\ln T)^c \equiv A_s(T)_{T\to T_C} , \tag{28}$$

where T_{c1}, P_{c1} are the solvent critical parameters. The thermodynamic Krichevskii parameter K_c determines the rate of change of a mixture's pressure with mole fraction $x_2 = x$ if a small number of solvent molecules is replaced by solute molecules at constant molar volume v and temperature T. It can be obtained by the computer simulation in the (N_1, $N_2\to 0$, V, T)-ensemble, for example, or evaluated from the semi-empirical EOS.[8,9] The addition of a small number of solute molecules can be considered as the fluctuation, which causes the asymptotically linear shifts[6,7] in T^c and P^c of a mixture's CRL with mole fraction x:

$$\overline{P}^c = A_c \overline{T}^c + K_c x (x \to 0), \tag{29}$$

where $\overline{P}^c = P^c(x)/P_{c1} - 1$ and $\overline{T}^c = T^c(x)/T_{c1} - 1$.

The quantitative prediction of the emerging $P^c(x)$ from the solvent's critical point is very important for the study of SCF[7]. P^c, T^c-projection divides the supercritical region ($T \geqslant T_{c1}$) or the subcritical region ($T \leqslant T_{c1}$) of solvent onto the homogeneous and heterogeneous parts of a phase behavior. It is well-known from the experiment as well as from the GPD[5] that ends of a three-phase line (l_1l_2g) with an upper or lower critical end point (UCEP or LCEP) can be localized quite closely to the critical point of a more volatile solvent. In such a case the accurate location of initial CRL-slope in (P,T)-plane by Eq. (29) is especially important to predict the type of a local PD (LPD) near the SCF-solvent critical point. One needs, first of all, in the correct estimates of A_c and K_c because the prediction of (T^c,x)-projection by the semi-empirical EOS is the most accurate procedure in comparison with the (P^c,x)-prediction and, especially, with the (v^c,x)-prediction. One may consider the reliable prediction of the LPD as the first step to predict quantitatively the actual PD of a mixture. Strategy of our work is to derive and use the interrelation between the Krichevskii parameter K_c and the binary cross-interaction coefficient k_{12} in equation:

$$a_{12}(T) = (1 - k_{12})\sqrt{a_1(T)a_2(T)}, \tag{30}$$

where $a_1(T)$, $a_2(T)$ are the T-dependent coefficients of the FEOS (1,2).

At the beginning we will study, the relationship between solute–solvent differences in the FEOS-coefficients, on one hand, and the initial CRL-slopes, on another, for dilute mixtures of nonvolatile liquid solutes (2) in sub- or supercritical solvents (1) with $T_{c1}<T_{c2}$. Many authors have investigated the prediction of phase behavior by using the various modified versions of the vdW-EOS with properly fitted k_{12}-parameter, which may describe the simple forms of CRL itself but poorly predict the PD. To solve this problem some researchers have considered k_{12} as T- , P- , ρ- or, even, x-dependent effectively increasing the number of empirical adjustable coefficients. Other authors have used also the local composition concept in the EOS to account the ordering (depletion or enrichment) of SCF-molecules around a solute-molecule. Although these models give a better description of the PD and solubility, it is, mainly, an empirical modification where the evaluation of a large number of adjustable quantities is necessary.

We will try to improve the correlation of the PD- and CRL-predictions in the frameworks of FEOS by using the conventional mixing rules:

$$a(x,T) = a_1(T)(1 - x)^2 + 2a_{12}(T)x(1 - x) + a_2(T)x^2, \tag{31}$$

$$b(x,T) = b_1(T)(1-x) + b_2(T)x, \tag{32}$$

$$c(x,T) = c_1(T)(1-x) + c_2(T)x, \tag{33}$$

where all coefficients, including k_{12}-parameter from Eq. (30), are phase-state-dependent, in accordance with the FEOS-concept (Section 2). The phase-state influence of solvent on the value k_{12} is especially large because the interrelations derived in this work are different for it's sub- and supercritical ranges, respectively:

$$\left(1-k_{12}\right)=\sqrt{\frac{a_{c1}}{a_2(T_{c1})}}\left[\rho_{c1}b_2(T_{c1})+\frac{1-c_2(T_{c1})}{A_{c1}Z_{c1}}-\frac{K_{c1}}{2(A_{c1}-1)}\right], \tag{34}$$

$$\left(1-k_{12}^0\right)=\sqrt{\frac{a_{c1}^0}{a_2(T_{c1})}}\left[\rho_{c1}b_2(T_{c1})+\frac{1-c_2(T_{c1})}{4Z_{c1}}-\frac{K_{c1}}{6}\right]. \tag{35}$$

In the derivation of Eqs. (34, 35) from FEOS (1) with mixing rules (31–33) only the infinity condition:

$$\frac{RT_{c1}(1-c_1)}{2a_1\rho_{c1}(1-b_1\rho_{c1})^2}=1 \tag{36}$$

for the critical isothermal compressibility: $\chi_{T_c}(T_{c1,x=0})$ of solvent has been used.

This thermodynamic derivative at the solvent critical point diverges and is connected[7-9] with the Krichevskii parameter $\delta^c(x\to 0)$ from Eq. (26) as well as with the solute–solvent direct correlation function's integral C_{12c}^{∞} at the limits: $x\to 0$; $T\to T_{c1}$:

$$\delta^c = \rho_{c1}\chi_{T_c}^{-1}\tilde{v}_{2c}^{\infty} = \rho_{c1}RT_{c1}(1-C_{12c}^{\infty}) \ , \tag{37}$$

where $\tilde{v}_{2c}^{\infty}$ is the solute's partial molar volume at infinite dilution and ρ_{c1} is the solvent's critical density. Presence of C_{12c}^{∞} in Eq. (37) provides, in principle, the possibility to predict the thermodynamic Krichevskii parameter δ^c by the MD-simulations of the molecular short–range solute–solvent correlations. It is known[10] that the direct correlation function integral C_{11} for a pure solvent:

$$1-C_{11} \equiv 1-\rho\int c_{11}(r,\rho,T)d^3r = (\rho RT\chi_T)^{-1} \tag{38}$$

is short-range, finite and well-behaved at the solvent's critical point where the limiting value of C_{12c}^{∞} from Eq. (37) must be also finite one. It is interesting to compare the long-range correlation function's integral for a pure solvent:

$$\Gamma_{11} \equiv \rho \int [g_{11}(r;\rho,T) - 1] d^3 r = \rho RT \chi_T C_{11} \tag{39}$$

with the relevant cluster property[8,9] at the infinite dilution:

$$\Gamma_{12}^{\infty} = \rho_1 \int (g_{12}^{\infty} - 1) d^3 r = \rho_1 RT \chi_{T1} C_{12}^{\infty}. \tag{40}$$

The quantity Γ_{12}^{∞} is the statistical excess number of solvent molecules surrounding a solute molecule with respect to a uniform density ρ_1. The fluctuational integral from Eq. (39) for a pure solvent has a similar meaning. One may consider Eq. (40) as the molecular interpretation of the local composition concept.

The FEOS provides the explicit dependency of C_{12}^{∞} in the near-critical region on the temperature and solvent's density:

$$C_{12}^{\infty} = 1 - \frac{\rho_1 (b_2 - b_1)(1 - c_1) + 1}{(1 - b_1 \rho_1)^2} + \frac{c_2 - c_1}{1 - b_1 \rho_1} + \frac{2\rho_1 (1 - k_{12}) \sqrt{a_1 a_2}}{RT}. \tag{41}$$

The T-dependence is implicit in FEOS-coefficients of components. If the difference of their critical temperatures is large one may take into account the asymptotic generalized vdW-form of FEOS ($c(T) \to 0$, $b(T) \to b_0$, $a(T)$) for a liquid low-temperature solute. The vdW-limit of Eq. (41) at the additional assumption $k_{12} = 0$ is the generalized form of the equation obtained by Petsche and Debenedetti[8]:

$$C_{12}^{\infty,0} = 1 - \frac{1 + \rho_1 (b_2^0 - b_1^0)}{(1 - b_1^0 \rho_1)^2} + \frac{2\rho_1 a_{12}^0}{RT}. \tag{42}$$

The adopted here sequence of limits in the definition of $\delta(x \to 0)$ is the different one. These authors have obtained the similar derivative $\delta^0(x \to 0)$ at fixed $v_1 = V/N_1$ of a solvent:

$$\delta^0 = \lim_{N_2 / N_1 \to 0} \left[(N_1 + N_2) \left(\frac{\partial P}{\partial N_2} \right)_{T,V,N_1} \right] = \delta + \chi_{T1}^{-1} \tag{43}$$

in some contradiction with the thermodynamic definition of δ in Eq. (26) at fixed value: $v = V/(N_1 + N_2)$ in terms of specific volume per molecule.

Debenedetti et al.[8,9] have proposed the important classification scheme to separate the repulsive (r), weakly attractive (w.a.) and attractive (a.) regimes for dilute mixtures with the SCF–solvent:

$$C_{12}^{\infty}(\text{r.}) < 0; \quad \delta(\text{r.}) > \rho RT; \quad \tilde{v}_2^{\infty} \to +\infty; \quad \Gamma_{12}^{\infty} \to -\infty, \tag{44}$$

$$0 \leqslant C_{12}^{\infty}(\text{w.a.}) \leqslant 1; \quad \rho RT \geqslant \delta(\text{w.a}) \geqslant 0; \quad \tilde{v}_2^{\infty} \to +\infty; \quad \Gamma_{12}^{\infty} \to +\infty, \tag{45}$$

$$C_{12}^{\infty}(\text{a.}) > 1; \quad \delta(\text{a.}) < 0; \quad \tilde{v}_2^{\infty} \to -\infty; \quad \Gamma_{12}^{\infty} \to +\infty. \tag{46}$$

It is preferable, from our viewpoint, to use the following sequence of steps at the consistent prediction of the CRL and PD for a binary dilute mixture. Firstly, the experimental initial CRL-slopes (i.e., the single CRL-point P^c, $T^c(x)$) in Eq. (29) are used to predict the reduced Krichevskii parameter: $K_{c1} = \delta^c/P_{c1}$. Then, the pure critical solvent coefficients and the liquid solute coefficients at $T = T_{c1}$ are necessary to determine $k_{12}(\overline{T}^c < 0)$ or k_{12}^0 $(\overline{T}^c > 0)$ by Eqs. (34) or (35), respectively. Next step is the localization of the repulsive, weakly attractive and attractive regimes in the (P,T)-plane by Eq. (41) and conditions (44–46) near the solvent critical point. Then the consistent CRL- and PD-predictions in this region must be obtained by the FEOS to specify the type of LPD for a given mixture.

Different models of EOS will yield different criteria for the classification of mixture's regions as repulsive, weakly attractive and attractive. The similar problem is the borders between different types of CRL predicted by the EOS-based GPD. To provide the quantitative prediction of the FEOS-based LPD, we would like to show how one may systematically determine the binary cross-interaction coefficient k_{12} from the single experimental CRL-point obtained near the critical point of a pure solvent. In this region where a classical global EOS of the vdW-type fails to describe correctly the singular behavior of a system, one needs for a scaling EOS which describes the thermodynamic near-critical properties more accurately. Leung and Griffiths have developed[11] the scaling EOS for description of fluid mixtures near a gas–liquid CRL. One needs the whole experimental CRL of a mixture as input data to develop the scaling EOS. Levelt Sengers has proposed for δ^c from Eq. (26) the term "Krichevskii parameter" and considered[7] a large number of thermodynamic properties of dilute mixtures (CRL, PD, solubility, partial molar volumes and enthalpies) as the results of its value and sign. Especially interesting is the possibility to calculate this K_C parameter from infinite-dilution values C_{12}^{∞} and C_{11} (38):

$$K_c Z_{c1} = \left(C_{11} - C_{12}^{\infty}\right)_{T \to T_c}. \tag{47}$$

At the critical point of a pure solvent $C_{11} \to 1$ and $C_{12}^{\infty} \to C_{12c}^{\infty}$ i.e. Eq. (47) and Eq. (37) are equivalent. We attempt to provide the consistent procedure of such calculation based on the predicted value k_{12} from Eq. (34) at $(dT^c/dx)<0$ or on the value k_{12}^0 from Eq. (35) at $(dT^c/dx)>0$. The result is the LPD, which can be useful to study the relationship between the short-ranged molecular structure of a mixture and its long-ranged scaling behavior near a solvent's critical point.

4. FEOS-based LPD in binary mixtures of SCF/cosolvent: CO_2/C_5H_{12}, CO_2/C_6H_{14}, CO_2/CH_3OH

These systems have been used to predict LPD for the typical binary solvents of the polymer-containing mixtures. Binary fluid mixtures of carbon dioxide with organic solvents are useful[12] a) to achieve lower critical temperatures than that for organic solvent; b) to introduce selectivity and/or reactivity that pure CO_2 alone cannot provide; c) to modulate the fluid density and other properties in some controlled ways and d) to identify alternative processing fluids with reduced amount of undesirable solvents. All mixtures, depending upon the mole composition $x_2 \equiv x$ display critical temperatures $T^C(x)$ that are in between the critical temperatures T_{C1} and T_{C2} of two components, while their critical pressures $P^C(x)$ pass through a maximum.[13,14]

The study of this section is limited by the vicinity of the vapor–liquid type of CRL at temperatures between T_{C1}=304.2 K and T_{C2} for the solvents. The wide range of densities ρ_1 from ideal-gas states ρ_1=0 up to the liquid-like states ρ_1=20 mol/dm^3 has been used in Eq. (41) for the pure SCF–solvent CO_2. Discussion is focused below on the prediction of the most appropriate operating conditions and SCF/cosolvent combinations for the standard SCF-based technologies.[12] A main prerequisite for the choice of the initial pressures and temperatures is that the extraction process is started of conditions (P_0,T_0) at which complete miscibility exists between the primary liquid organic solvent and the SCF-antisolvent or between the primary SCF–solvent and the organic solvent-modifier. We consider the attractive regime from Eq. (46) as an equivalent of the complete homogeneous miscibility and the weakly-attractive regime from Eq. (45) as the range of parameters, respectively, where inhomogeneous miscibility inside a binary solvent exists. The LPD-method provides the prediction of the best operating conditions at which the inhomogeneties of the mole fraction x (composition of a cosolvent) have not to be essential ones in the performance of the polymer/binary solvent extraction process.

As it is shown in Fig. 1 the high-density region of SCF-CO_2 is described reasonably well with the FEOS-coefficients from Table 2, represented for some temperatures. To obtain the values b(T), c (T) at the given value a(T) from Eq. (4) the supercritical experimental data of Ely et al.[15] in the range 305–330 K have been used as well as the tabular data of Altunin[16] in the range 350–500 K. The liquid phase of cosolvents CH_3OH, C_5H_{12}, C_6H_{14} is reliably described by the FEOS-coefficients from Eqs. (20–24).

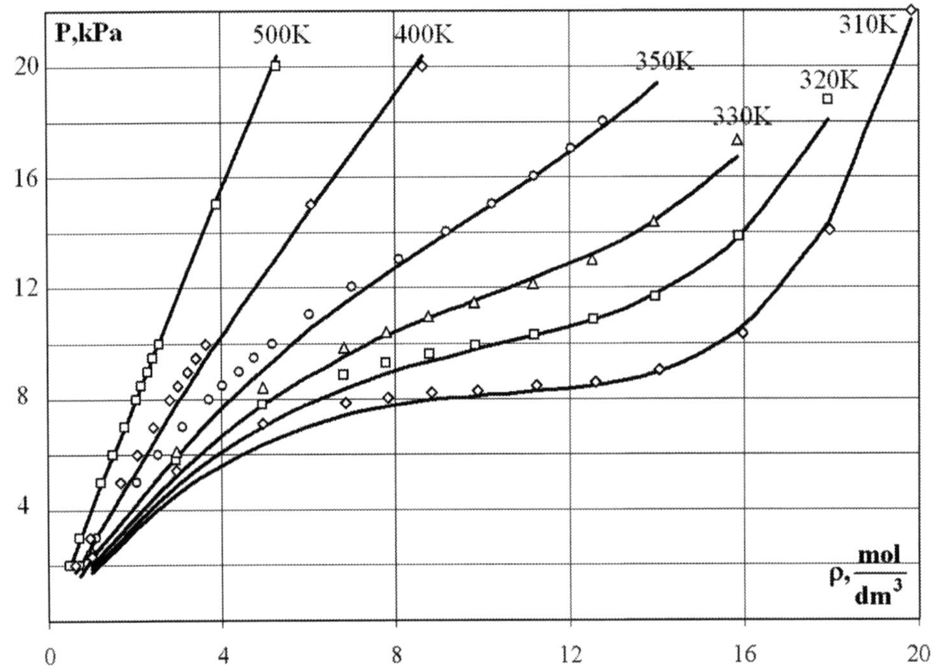

Figure 1. Description of the SCF-CO_2 by FEOS. Experimental data (light circles, squares, diamonds, etc).

TABLE 2. T-dependent supercritical coefficients of carbon dioxide.

T,K	a, Jdm^3/mol^2	b, cm^3/mol	c
304.2	195.86	31.34	0.268
310	191.94	31.05	0.262
320	185.88	30.40	0.257
330	180.47	29.20	0.233
350	171.19	28.00	0.210
400	154.05	27.00	0.150
500	133.17	26.50	0.021

The single-point data on limiting experimental CRL-slopes as well as the predicted by Eq. (29) values of the Krichevskii parameter K_{C1} and the binary cross-interaction coefficient $k_{12}^{0}(T \geq T_{c1})$ from Eq. (35) are represented in Table 3. As is shown in Figs. 2 and 3 the direct correlation functions' integral at infinite dilution C_{12}^{∞} from Eq. (41) provides the useful information on solubility to select the preferable cosolvent for SCF-CO_2 in practical usage of SCF-based technologies. For solutions in carbon dioxide of nonvolatile solutes (cosolvents such as pentane, hexane, or methanol) there is a point of $C_{12}^{\infty,m}$- maximum at each temperature except the near-critical: T=500 K for C_6H_{14} and CH_3OH. However, the values of these maxima are rapidly decreasing if one increases the operating temperature. It is evident that C_5H_{12} and C_6H_{14} have some advantage in comparison to CH_3OH as possible candidates for a cosolvent in the binary solvents with SCF-CO_2 because in the mixture CO_2–CH_3OH the region of attractive behavior and homogeneous miscibility of components is practically absent at all temperatures. An important implication of the developed approach is that attractive regimes in near critical CO_2 become repulsive ones upon interchanging solvent and solute or upon increasing of temperature.

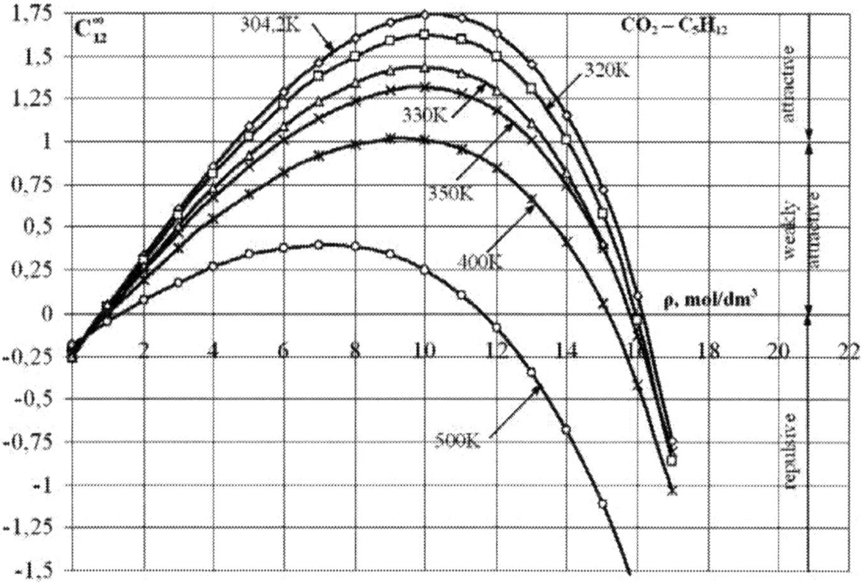

Figure 2. Direct correlation function integral for mixture CO_2–C_5H_{12} at infinite dilution.

Thus, if infinitely dilute n-alkane (C_5H_{12} or C_6H_{14}) in SCF-CO_2 is an attractive system with large solvent contraction upon solute addition, the FEOS-based method predicts that a subcritical organic solvent with traces of CO_2 will exhibit repulsive behavior at near-critical temperatures of n-alkane. The similar result has been reported by Debenedetti et al.[9] for the mixture of inert gases Ne/Xe investigated in the frame-works of the original vdW-model.

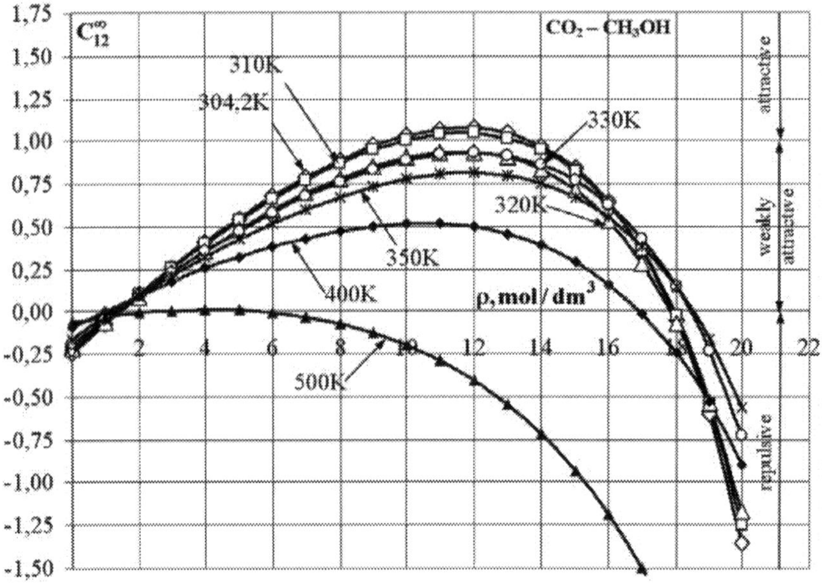

Figure 3. Direct correlation function integral for mixture CO_2-CH_3OH at infinite dilution.

TABLE 3. Limiting experimental CRL-slopes, Krichevskii parameters and recommended binary coefficients k_{12}^0 ; $\Delta P^C/\Delta x$ [kPa], $\Delta T^C/\Delta x$ [K].

SCF	Cosolvent	Ref.	$\Delta T^C/\Delta x$	$\Delta P^C/\Delta x$	K_{C1}	k_{12}^0
$CO_2(1)$	$C_5H_{12}(2)$	[13]	314,1	16794	–4,787	0,274
$CO_2(1)$	$C_6H_{14}(2)$	[14]	264,3	20714	–3,143	0,414
$CO_2(1)$	$CH_3OH(2)$	[14]	254,3	24857	–2,360	0,435

The LPDs are represented themselves in Figs. 4 and 5. The boundary values of density ρ_1 for C_{12}^{∞} between a.- and w.a.-regimes as well as the $C_{12}^{\infty,m}(\rho_1^m, T)$ – data have been used to calculate the relevant pressures $P(\rho,T)$ by FEOS.

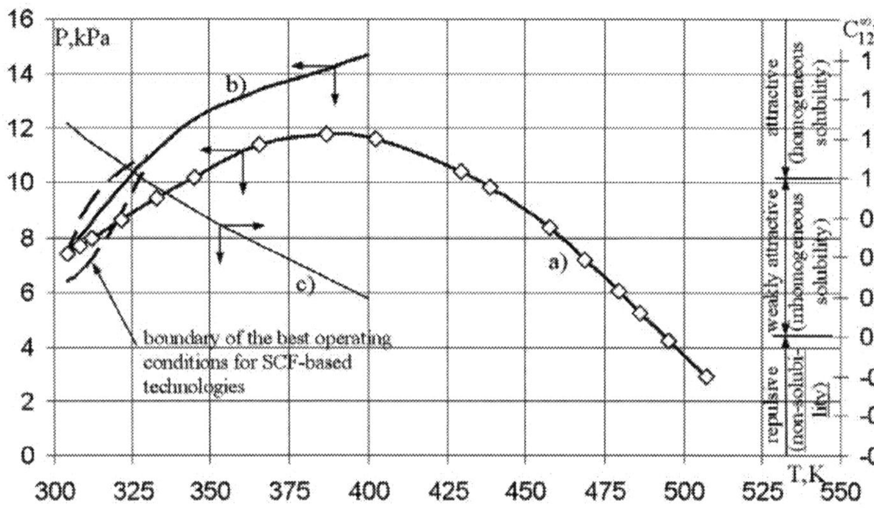

Figure 4. LPD of binary mixture SCF(CO_2)–cosolvent (C_6H_{14}): a) critical curve $P^c(T^c)$;[14] b) maxima-of-pressure curve $P^m(T)$, corresponding to the best miscibility of components; c) decreasing of hexane solubility in carbon dioxide with temperature increasing.

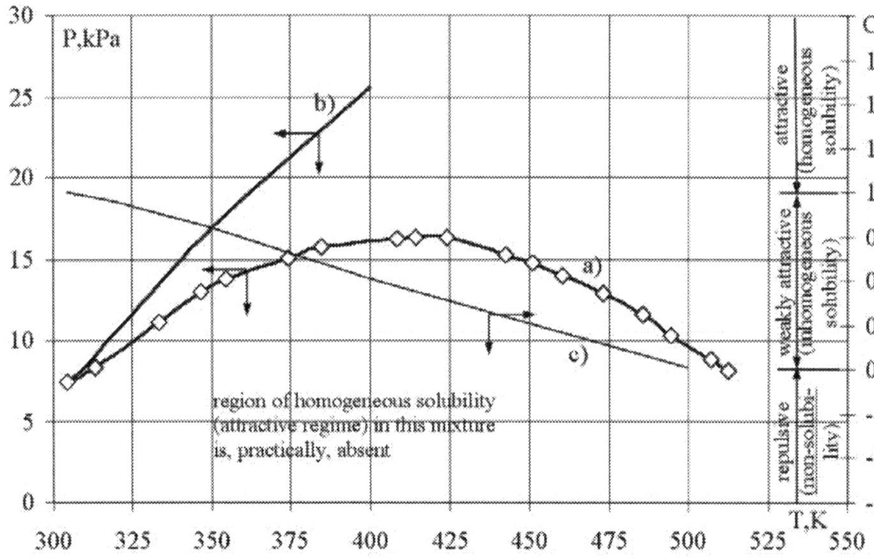

Figure 5. LPD of binary mixture SCF(CO_2)–cosolvent (CH_3OH): a) critical curve $P^c(T^c)$;[14] b) maxima-of-pressure curve $P^m(T)$, corresponding to the best miscibility of components; c) decreasing of methanol solubility in carbon dioxide with temperature increasing.

5. Conclusions

General conclusions is that the best operating conditions for SCF-based extraction processes lies in the rather restricted ranges of P and T, localized above CRL but close to it. It is known that solute solubility is proportional to the density and the solvent strength of SCF can be fine-tuned by changing pressure. However, increasing the density, as it is shown in Figs. 4, and 5, does not guarantee that macromolecule will be better dissolved in a binary system SCF/cosolvent. At each isotherm $T > T_{C1}$ there is maximum of $P^m(T)$ where the solubility is the best one. Simultaneously, the increasing of temperature has a negative effect on the miscibility even if pressure is increased.

The density- and/or pressure-dependent strength of a binary solvent can be understood using the FEOS-results for LPD and the following simplified (see also Ref. 7–9) molecular interpretation. Attractive regime is characterized by relatively long-range solvent (SCF) enrichment around the molecule of solute (cosolvent). In the other words, a local ordering (clusterization) associated with the introducing of a single solute molecule into a solvent is an equivalent of the approaching to the critical point of a pure substance. When a number of solute molecules is increasing only the small clusters can be formed in the given mass of SCF–solvent. This is a weakly-attractive regime where the long-range enrichment is insufficient to overcome the effect of randomization of the small clusters. The miscibility of components is decreased in this case.

References

1. (a) Rogankov, V. B. (1998) Fluctuational equation of state for pure components and mixtures, *Refr. Eng. & Technol.* **59**, 25–29.
   (b) Rogankov, V. B. and Terzey, E. D. (2000) Description of critical curves in mixtures of carbon dioxide with ethane and ethylene on the base of fluctuational equation of state, *J. Eng. Phys. & Thermophys.* **73**, 407–413.
   (c) Rogankov, V. B., Zykalo, A. L., and Bedrova, T. A. (2004) Equation of state for dilute solutions with supercritical fluid-solvent, *Rep. Ukr. Acad. Sci.* 3, 142–148.
2. Rogankov, V. B. and Boshkov, L. Z. (2002) Gibbs solution of the van der Waals-Maxwell problem and universality of the liquid–gas coexistence curve, *Phys. Chem. Chem. Phys. (PCCP)* **4**, 873–878.
3. Mazur, V. A. and Rogankov, V. B. (2003) A novel concept of symmetry in the model of fluctuational thermodynamics, *J. Molec. Liq.* **105**, 165–177.
4. Rogankov, V. B., Byutner, O. G., Bedrova, T. A., and Vasiltsova, T. V. (2005) Local phase diagram of binary mixtures in near-critical region of solvent, accepted to *J. Molec. Liq.* **127**, 53–59.
5. Scott, R. L. and van Konynenburg, P. H. (1970) van der Waals and related models for hydrocarbon mixtures, *Discuss. Faraday Soc.* **49**, 87–97.

6. Krichevskii, I. R. (1975) *Thermodynamics of Critical Infinitely Diluted Solutions*, Khimia, Moscow, p. 120 (in Russian).
7. Levelt Sengers, J. M. H. (1991) Solubility near the solvent's critical point, *J. Supercrit. Fluids* **4**, 215-222.
8. Petsche, I. B. and Debenedetti, P. G. (1991) Influence of solute–solvent asymmetry upon the behavior of dilute supercritical mixtures, *J. Phys. Chem.* **95**, 386-399.
9. Debenedetti, P. G., Petsche, I. B. and Mohamed, R. S. (1989) Clustering in supercritical mixtures: theory, applications and simulations, *Fluid Phase Equil.* **52**, 347–356.
10. Huang, Y.-H. and O'Connell, J. P. (1987) Corresponding states correlation for the volumetric properties of compressed liquids and liquid mixtures, *Fluid Phase Equil.* **37**, 75–84.
11. Leung, S. S. and Griffiths, R. B. (1973) Thermodynamic properties near the liquid-vapor critical line in mixtures of He^3 and He^4, *Phys. Rev. A* **8**, 2670–2679.
12. Kiran, E. and Zhuang W. (1994) Miscibility and phase separation of polymers in near- and supercritical fluids in Supercritical Fluids, Edited by E. Kiran and J. M. H. Levelt Sengers, pp. 2–36.
13. Cheng, H., Fernandez, M. E. P., Zolwegg, J. A., and Street, W. B. (1989) Critical properties of binary mixtures: pentane + carbon dioxide, *J. Chem. Eng. Data* **34**, 319–328.
14. Liu, J., Qin, Z., Wang, G., Hou, X., and Wang, J. (2003) Critical properties of binary and ternary mixtures of hexane + carbon dioxide + methanol, *J. Chem. Eng. Data* **48**, 1610–1613.
15. Ely, J. F., Haynes, W. M., and Bain, B. C. (1989) Isochoric (P,Vm,T) measurements on CO_2 and on ($0.982CO_2 + 0.018N_2$) from 250 to 330 K at pressures to 35 MPa, *J. Chem. Thermodyn.* **21**, 879–894.
16. Altunin, V. V. (1980) *Thermodynamic Properties of Liquid and Gaseous Carbon Dioxide* (GSSD, Moscow).

COMBINED MODELS OF THERMOPHYSICAL PROPERTIES ALONG THE COEXISTENCE CURVE

E.E. USTJUZHANIN*, B.F. REUTOV, V.F. UTENKOV, V.A. RYKOV
Moscow Power Engineering Institute, Krasnokazarmennaya 14, 111250 Moscow, Russia

Abstract. Relationships for thermophysical properties (the densities $-\rho_l,\rho_g$, the refractive indexes $-n_l,n_g$, and saturated pressure $-P_s$) along the coexistence curve (CC) have been considered. These equations are chosen as property models $F(\tau, D, C)$ that include scaling and regular parts. The scaling part, F_{scale}, includes critical characteristics, $D = (\rho_c, T_c, \alpha, \beta, \Delta,...)$, and its structure follows to the scaling theory (ST). The F_{scale} represents measured values $(F_{exp\ k}, \tau_k)$ those are placed in the near critical temperature range $(T_c \cdots T_{scale})$. The regular part, F_{reg}, has to compensate systematic deviations those can appear between F_{scale} and experimental values at temperatures beyond the critical interval. Statistical routines and some criteria have been elaborated to determine adjustable coefficients $C = (C_i)$ with the help of the input data set $(F_{exp\ k}, \tau_k)$. The scaling functions $\psi_{l,g}$ are analyzed and compared with the form that follows from ST. A combined model $P_s(\tau, D, C)$ is proposed to represent experimental data in the temperature interval from the triple point to the critical one. The model is used together with calculated ρ_l, ρ_g, T – data to determine the specific heat of vaporization. The models are built for such substances as R236ea, R134a, R143a, and H_2O. They include effective values of parameters T_c, ρ_c, P_c, α, β, B_{s0}, B_{d0}, those are determined during the input data set treatment. Values β_{opt} and α_{opt} do not coincide with the theoretical values ($\beta = 0.325$, $\alpha = 0.1085$).

Keywords: Coexistence line, scaling theory, density, refractive index, saturation pressure, critical exponents.

---

* To whom correspondence should be addressed. Eugene Ustjuzhanin, e-mail: ustmei@itf.mpei.ac.ru

S. J. Rzoska and V. A. Mazur (eds.), Soft Matter under Exogenic Impacts, 325–338.

1. Introduction

This study analyzes an equation $F(\tau,D,C)$ which reproduces the set of properties $F = (\rho_l,\rho_g,n_l,n_g,P_s,...)$ along CC. We assume that F model is to have a structure

$$F = F_{scale}(\tau,D) + F_{reg}(\tau,C), \tag{1}$$

where $F_{scale}(\tau,D)$ – a scaling part, $F_{reg}(\tau,C)$, – a regular part, D – critical characteristics, C – amplitudes, $\tau = (T - T_c)/T_c$ – a relative distance in the temperature scale.

The scaling term of (1) satisfies the scaling theory (ST) and uses the critical characteristics $D = (\rho_c, T_c, P_c, \alpha, \beta, \Delta,...)$ among them ρ_c – the critical density, P_c – the critical pressure, α, β, Δ – critical indexes. The $F_{scale}(\tau,D)$ has to represent measured values $(F_{exp\ k}, \tau_k)$ those are placed in the critical interval of temperatures $(T_c \cdots T_{scale})$ or in a relative interval $\Delta\tau_{scale}$. RMS deviation S_1 of measured values from $F_{scale}(\tau,D)$ is to be close to an experimental error. The regular part $F_{reg}(\tau,C)$ has to compensate systematic deviations those can appear between $F_{scale}(\tau,D)$ and experimental values at temperatures beyond the critical interval $T < T_{scale}$. The model (1) is discussed in Ref 1–3 and has been used in our works those describe properties $F(\tau,D,C)$ of several refrigerants.

To represent such properties as the order parameter f_s and CC diameter f_d we have built the following model

$$f_s = (\rho_l - \rho_g)(2\rho_c)^{-1} = B_{s0}\tau^{\beta} + B_{s1}\tau^{\beta+\Delta} + B_{s2}\tau^{\beta+2\Delta} + B_{s3}\tau^{2} + B_{s4}\tau^{3}, \tag{2}$$

$$f_d = (\rho_l + \rho_g)(2\rho_c)^{-1} - 1 = B_{d0}\tau^{1-\alpha} + B_{d1}\tau^{1-\alpha+\Delta} + B_{d2}\tau^{1-\alpha+2\Delta} + B_{d3}\tau^{2} + B_{d4}\tau^{3}. \tag{3}$$

A model of the densities, $F(\tau,D,C) = (\rho_l,\rho_g)$, are produced with the help of models (2, 3) in the form

$$\rho_l = (f_d + f_s + 1)\,\rho_c, \tag{4}$$

$$\rho_g = (f_d - f_s + 1)\,\rho_c. \tag{5}$$

Combined Eqs. (2···5) are named as model 2.

As to the history of the problem, the first theoretical step to represent the densities ρ_{lg} on CC was made by Landau.[4,7] He accepted that the reduced free energy density can be written in the form

$$\tilde{F}(T,\rho,\eta) = F_0(T,\rho) + \frac{1}{2}A\eta^2 + \frac{1}{3}B\eta^3 + \frac{1}{4}C\eta^4 + \ldots, \tag{6}$$

where $\tilde{F} = (F/V)/P_c$ – a reduced free energy density, $\eta = (\rho - \rho_c)/\rho_c$ – a relative distance from the critical point, $F(T, \rho) = \rho\mu_0(T) + f_0(T)$ – a regular function, A,B,C – temperature functions, $p = (P - P_C)/P_C$.

An expression for the relative pressure, $(P-P_c)/P_c$, is got from (6) in the form

$$p = a_0\tau + A\eta + B\eta^2 + C\eta^3, \tag{7}$$

It was admitted that following correlations were valid for the functions A and B

$$A = a\tau, \quad B = b\tau. \tag{8}$$

Equations (8) corresponds to such conditions in the critical point as

$$(\partial P/\partial\rho)_T \approx A = 0, \quad (\partial P^2/\partial\rho^2)_T \approx B = 0. \tag{9}$$

The coefficient C was chosen as a positive value because the third derivative was admitted as a positive function in the critical point

$$(\partial P^3/\partial\rho^3)_T > 0, \quad C > 0. \tag{10}$$

Under boundary conditions (8···10) the field function h is considered as

$$h = (\partial\tilde{F}/\partial\eta)_{T,\rho} = (\partial(\tilde{F} - F_0)/\partial\eta)_T = a\tau\eta + b\tau\eta^2 + C\eta^3. \tag{11}$$

If $h = 0$ then the relative density η can be expressed as

$$\eta = 0 \quad \text{at } \tau > 0,$$
$$\eta_{lg} = (\rho_{lg} - \rho_c)/\rho_c = \pm(-a|\tau|/C)^{0.5} - (b\tau)/(2C) = \pm f_s + f_d \quad \text{at } \tau < 0, \tag{12}$$

where the sign "minus" corresponds to η_g, the sign "plus" corresponds to η_l.

It is seen from (12) that the Landau's model includes the properties

$$f_s = (-a|\tau|/C)^{0.5}, \; f_d = -b\tau/(2C). \tag{13}$$

Models (12, 13) have following features: 1) the symmetrical part or f_s has a critical index as "a classical" value of $\beta = 0.5$, 2) the asymmetrical part, (12), of f_d is a linear function of τ, 3) there is no singularity in f_d at $\tau \to 0$.

As it is seen from (12), the Landau' model has a positive amplitude B_{d0}

$$B_{d0} = b/(2C) > 0. \tag{14}$$

It was found[6] from accurate experiments that CC diameter follows to a scaling correlation in the asymptotic region as

$$f_d = B_{d0}\tau^{1-\alpha}, \quad \alpha = 0.11. \tag{15}$$

We consider also extended variant of boundary condition taking into account the following correlation[5] in the critical point

$$(\partial P^3/\partial \rho^3)_T = 0. \tag{16}$$

Under condition (16) there is a transition $C \to 0$ near the critical point. We have accepted that

$$C = c\,|\tau|^{\nu}, \quad 0 < \nu < 1, \quad c > 0. \tag{17}$$

When correlation (17) has been included in (12) then the relative densities η_{lg} are expressed in the form

$$\eta_{lg} = \pm(-a\,(|\,\tau\,|^{1-\nu})/c)^{0.5} + b(|\tau|^{1-\nu})/(2c). \tag{18}$$

Model (18) shows that: 1) f_s includes a critical index as a value of $\beta = (1-\nu)/2 < 0.5$, 2) f_d has a singularity at $\tau \to 0$, 3) there is no linear addend of τ in η_{lg} in the asymptotic region.

Model (18) includes the amplitude B_{d0} as a positive value

$$B_{d0} = b/(2c). \tag{19}$$

A theoretical model of properties, $F = (f_s, f_d)$, was suggested later by Wegner [6] in the form

$$f_s = B_{s0}\tau^{\beta} + B_{s1}\tau^{\beta+\Delta},$$
$$f_d = B_{d0}\tau^{1-\alpha} + B_{d1}\tau^{1-\alpha+\Delta} + B_{d2}\tau, \tag{20}$$

where Δ – a degree of the first non-asymptotic correction term.

Model (20) has to be accurate in a wider interval: non-asymptotic region in comparison with model (13) due to an additional term as the first non-asymptotic correction addend.

A model of (f_s, f_d) was chosen due to Anisimov et al.[7] for H_2O

$$f_s = B_{s0}\tau^{\beta} + B_{s1}\tau^{\beta+\Delta},$$
$$f_d = B_{d0}\tau^{1-\alpha} + B_{d1}\tau. \tag{21}$$

Rabinovich et al.[8] and Shimanskaia et al.[9] have elaborated a model with additional non-asymptotic terms

$$f_s = B_{s0}\tau^{\beta} + B_{s1}\tau^{\beta+\Delta 1} + B_{s2}\tau^{\beta+\Delta 2},$$
$$f_d = B_{d0}\tau^{1-\alpha} + B_{d1}\tau^{1-\alpha+\Delta 1} + B_{d2}\tau^{1-\alpha+\Delta 2}, \qquad (22)$$

where Δ_1, Δ_2 – the first and second non-asymptotic correction terms.

There is a model that is built by Zhelezny et al.[10] and also includes critical characteristics D in the form

$$ln(\rho_l/\rho_c) = B_{l0}\omega^{\beta l(1+Y_1)}, \quad ln(\rho_c/\rho_g) = B_{g0}\omega^{\beta g(1+Y_2)}, \qquad (23)$$

where $\omega = ln(T_c/T)$, $(1 + Y_1)$, $(1 + Y_2)$ – crossover functions.

2. Results

Models (20···22) of $F = (f_s, f_d)$ include critical characteristics $D = (\rho_c, T_c, \alpha, \beta, (\Delta_i))$ and coefficients $C = ((B_{si}),(B_{di}))$. Routine 0 is known as an algorithm of determination of C values. It considers: i) parameters D are values fixed (taken from literature sources), ii) $F(\tau,D,C)$ is a linear function of C that is why amplitudes C are calculated using a weighted least-squares analysis (WLSA) as a statistical treatment of experimental data, $(F_{exp\ k}, \tau_k)$. Routine 0 utilize a single criterion – a minimum of the functional $\Phi(C,D)$ in the form[11]

$$\Phi(C,D) = \sum_{k=1}^{N} w_k (F_{exp\ k} - F(D,C,\tau_k))^2/N = min. \qquad (24)$$

A method of D value selection is described in the literature. Due to it D components are taken as theoretical values. Anisimov,[7] Table 1, has determined average values of $D = (\beta,\alpha)$ those are published in theoretical works in the years 1976…1984. Alekhin[8] got a new theoretical variant (Table 1) 20 years later. We underline that new theoretical value, $\beta = 0.338$, is sufficiently higher than $\beta = 0.325$ suggested by Wegner.[6]

The method is included in Routine 0 that is used by the authors (Anisimov,[7] Aoyama,[12] Defibaugh,[13] Fukushima,[14] Table 1). In the case, the exponents are taken as theoretical data $\beta = 0.325$, $\alpha = 0.1085$, and $\Delta = 0.5$ for (f_s, f_d) models and values of C are calculated with the help of WLSA. Routine 0 uses a single criterion – a minimum of $\Phi(C,D)$ and produces a realization $F = f(D,C,\tau)$.

TABLE 1. Critical characteristics $D = (\beta,\alpha, B_{s0}, B_{d0})$ of models.

Author, Substance	β	α	B_{s0}	B_{d0}	Model
Anisimov (any, theory)	0.3245 ± 0.0005	0.1120± 0.0012			
Alekhin (any, theory)	0.338 ±0.002	0.09 ±0.01			
Fukushima, R143a	0.325	0.1085	1.82649	−4.97503	Wegner
Defibbaugh, R236ea	0.325	0.000	1.80637	0.978	Wegner
Aoyama, R236ea	0.325	0.1085	1.84750	−19.8552	Wegner
Anisimov, H_2O	0.325	0.11	1.975	−1.48	Wegner
Rabinovich, H_20	0.3461	0.132	2.263	1.208	Sheludiak
Shimansky C_6H_6	0.352	0.25	2.036	0.51	Shimansky
Model (2, 3) R236ea	0.34925	0.15074	2.0855	0.2225	Model (2, 3)
Zhelesny, R134a	0.328 (gas) 0.345 (liquid)		1.728 (gas) 2.325 (liquid)		Zhelesny

Our analyses has shown (Table 1) that Routine 0 gives leading amplitudes $C = (B_{s0},B_{d0})$ in rather wide intervals ($C_{middle} \pm \Delta C$). The examples and our tests with Routine 0 show that one step method can produce a set of realizations, ($F(D_j,C_j,\tau)$, $j = 1\cdots L$) if we change border conditions and repeat Routine 0 several times. The boundary conditions include in the case: 1) D – values, 2) sources of experimental data, 3) a number of points in the input data set, 4) temperature intervals and so far.

One more treaty was elaborated by Shimansky et al.[9] The variant considers $D = (\beta, \alpha)$ as adjustable parameters together with C amplitudes and uses a non-linear least-squares analysis (NLSA).

The influence of values of $C = (B_{s0}, B_{d0})$ on $F(D_j,C_j,\tau)$ can be illustrated by one more property that is considered in ST and involved in the work: the scaling function, $\psi_{l,g}$, in the form

$$\psi_{l,g} = /\ (\rho_{l,g} - \rho_c)(\rho_c\tau^{\beta})^{-1}| = |\pm B_{s0} + B_{d0}\tau^{1-\alpha-\beta} \pm B_{s1}\tau^{\Delta 1-\beta} + B_{d1}\tau^{1-\alpha+\Delta 1-\beta} + \cdots|, \tag{25}$$

The sign "minus" corresponds to ψ_g, the sign "plus" corresponds to ψ_l.

Numerical variants of $\psi_{l,g}$ were found and analysed for several liquids (H_2O,Ne,N_2,HFC 134a, HFC 143a, etc.). The value $\psi_{l,g}$ has two terms (Model 0) with leading amplitudes, $C = (B_{s0}, B_{d0})$, in the asymptotic interval ($\Delta\tau_{as} \sim 0.01$)

$$\psi_{l,g} \approx |\pm B_{s0} + B_{d0}\tau^{1-\alpha-\beta}| = |\pm B_{s0} + B_{d0}X| = B_{s0} \pm B_{d0}X, \quad X = \tau^{1-\alpha-\beta}. \tag{26}$$

The following features of a theoretical variant are estimated due to (26):

i) $\psi_{l,g}$ is to be symmetric to the amplitude B_{s0} for the liquid and vapor branches, ii) ψ_l, increases from B_{s0} if X increases, ψ_g, decreases from B_{s0} if X increases, iii) the plot of $\psi_{l,g}$ vs. $X = \tau^{1-\alpha-\beta}$ is to be linear and converge symmetrically to B_{s0}.

An illustration of $\psi_{l,g}$ for R236ea is given in Fig. 1. It shows that it is possible to find out proper parameters of $F(D,C,\tau)$ when conditions, I$\cdots$iii, are fulfilled. It can be seen that Model 0 has a good correlation with experimental values of $\psi_{l,g}$ in $\Delta\tau_{as}$ or $X_{as} = 0.1$. We underline that $B_{d0} = 0.2225$ is positive in the case. Parameters of $F(D,C,\tau) = (\rho_l, \rho_g)$ are calculated for R236ea with the help of Routine 1 (see lower).

Another illustration of $\psi_{l,g}$ for R236ea is given in Fig. 2. Parameters of $F(D,C,\tau)$ and $\psi_{l,g}$ are taken from Aoyama et al.[12] using Routine 0. It can be seen that the Wegner' Model[12] does not correlate with the theoretical conditions i$\cdots$ iii. Experimental values of $\psi_{l,g}$ deviate greatly from the modeling data in placed in $\Delta\tau_{as}$. ψ_l goes down in the vicinity of T_c. We underline that $B_{d0} = -19.8552$ is negative in the case. Routine 0 has only one criterion and the treaty has no chance to get one more realization that will be closer to theoretical conditions I$\cdots$iii. We have involved additional criteria and conditions to build a model, $F_{opt}(\tau,D,C)$, with optimal characteristics. The first condition is connected with the scaling part of (2,3), $F_{scale} = (f_{s\,scale}, f_{d\,scale})$, named Model 1.

Model 1 has to represent non-asymptotic components (the first and the second correction terms with exponents, $\Delta, 2\Delta$). F_{scale} in the form

$$f_{s\,scale} = B_{s0}\tau^{\beta} + B_{s1}\tau^{\beta+\Delta} + B_{s2}\tau^{\beta+2\Delta},$$

$$f_{d\,scale} = B_{d0}\tau^{1-\alpha} + B_{d1}\tau^{1-\alpha+\Delta} + B_{d2}\tau^{1-\alpha+2\Delta}. \tag{27}$$

Model 1 is to represent measured values ($F_{exp\,k}, \tau_k$) those are placed in the critical interval of temperatures $\Delta\tau_{scale}$. The condition is differing from the region of model (22) that is adopted to ($T_c \cdots T_{triple\ point}$) in Ref. 9. Model 1 does not include a linear addend in comparison with the Wegner' model (18).

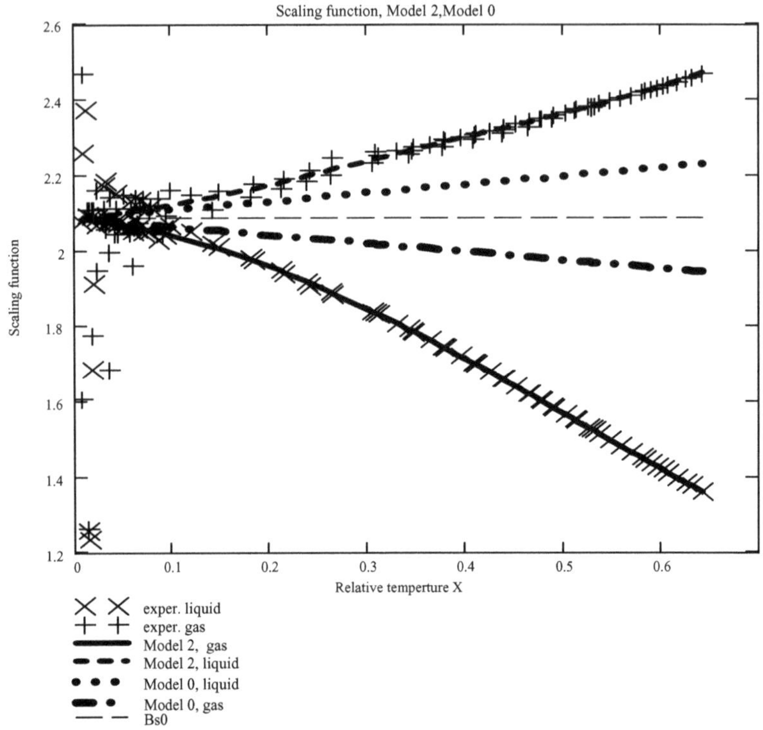

Figure 1. The scaling function (R236ea). $X = \tau^{1-\alpha-\beta}$, Model 2, gas – ψ_g calculated with Model 2 (2, 3); Model 2, liquid – ψ_l calculated with Model 2 (2, 3); Model 0, gas – ψ_g calculated with Model 0 (26), liquid – ψ_l calculated with Model 0 (26), B_{s0}. – a leading amplitude.

Model 1 for density description $F_{scale}(\tau,D,C) = (\rho_l,\rho_g)_{scale}$, is produced with the help of (27) and written as

$$\rho_{scale\,l} = (f_{d\,scale} + f_{s\,scale} + 1)\,\rho_c,$$

$$\rho_{scale\,g} = (f_{d\,scale} - f_{s\,scale} + 1)\,\rho_c. \qquad (28)$$

An approximation quality of Model 1 is characterized by individual deviations and root-mean-square (RMS) deviations of experimental $(\rho_l,\rho_g)_{exp}$ – values from (28)

$$\delta\rho_{scale\,g\,k} = 100\,(\rho_{exp\,g\,k} - \rho_{scale\,g\,k})/\rho_{gk}\,,\quad \delta\rho_{scale\,l\,k} = 100\,(\rho_{exp\,g\,k} - \rho_{scale\,l\,k})/\rho_{lk},$$

$$S_{scale\,g} = (\Sigma\delta\rho_{scale\,g\,k}^{2}/N_1)^{0.5},\quad S_{scale\,l} = (\Sigma\delta n_{scale\,l\,k}^{2}/N_1)^{0.5},\ k = 1\cdots N_1. \qquad (29)$$

Here N_1 – number of experimental points in the interval τ_{scale}.

We have chosen an additional criterion of approximation as RMS deviation S_1 of measured $(\rho_l,\rho_g)_{exp}$ – values from Model 1 in the form:

$$S_1 = ((S_{scale\,g}^2 + S_{scale\,l}^2)/2)^{0.5}. \tag{30}$$

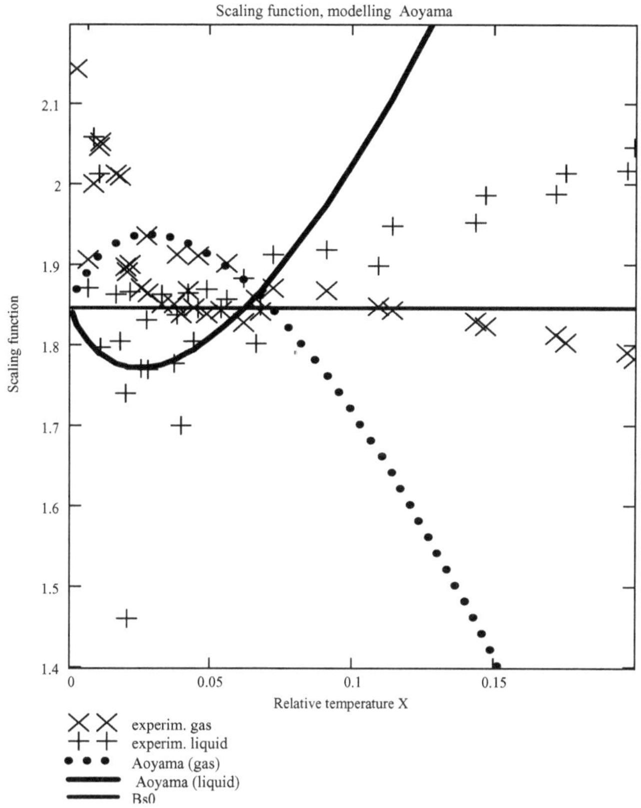

Figure 2. The scaling function for R236ea. $X = \tau^{1-\alpha-\beta}$ Aoyama, gas – ψ_g calculated with Model[12]; Aoyama, liquid – ψ_l calculated with Model[12]; B_{s0}. – a leading amplitude.

Deviations of measured values $(\rho_l,\rho_g)_{exp}$ from Model 2 (4, 5) are also considered as

$$\delta\rho_{g\,k} = 100\,(\rho_{exp\,g\,k} - \rho_{g\,k})/\rho_{gk}, \quad \delta\rho_{l\,k} = 100\,(\rho_{exp\,g\,k} - \rho_{l\,k})/\rho_{lk},$$
$$S_g = (\Sigma\delta\rho_{g\,k}^{\;2}/N_2)^{0.5}, \quad S_l = (\Sigma\delta\rho_{l\,k}^{\;2}/N_2)^{0.5}, \quad k = 1\cdots N_2, \tag{31}$$

where N_2 – a number of experimental points in the input data sets ($\rho_{exp\ lk}$, $\rho_{exp\,gk}$, T) in the interval ($T_c \cdots T_{triple\ point}$).

A combined criterion of approximation is applied to Model 2 in the form of RMS deviation

$$S_2 = ((S_g^2 + S_l^2)/2)^{0.5}. \quad (32)$$

Our tests with Routine 0 elaborate a set of realizations, $(F(D_j, C_j, \tau), j = 1 \cdots L)$ when we shift D – values in a region $\Delta D = (\Delta\rho_c, \Delta T_c, \Delta\beta, \ldots)$ and repeat Routine 0 several times. The average values of D components, D_{middle}, and a region of possible deviations, $\Delta D = (\pm\Delta\rho_c, \pm\Delta T_c, \pm\Delta\beta, \ldots)$, were established. Low borders S_1^{min} and S_2^{min} of the sets $(S_1(B_{soj}), S_2(B_{soj}))$ can be determined. Numerical dependences $(S_1(\rho_{cj}), S_2(\rho_{cj}), S_1(\beta_j), S_2(\beta_j), S_1(T_{cj}), S_2(T_{cj}), \ldots, j = 1 \cdots L)$ were built.

An illustration of paths $S_1(\rho_{c\,j})$ and $S_2(\rho_{c\,j})$ for Models 1, 2 is given in Fig. 3. The numerical dependences are got for R236ea. $D = \rho_c$ was shifted ($\rho_c = var$) during the test. Low borders S_1^{min} and S_2^{min} are shown too as A and B.

Our analyses has estimated some general features of the realizations, $(F(D_j, C_j, \tau), j = 1 \cdots L)$: 1) values of S_1 and S_2 do not coincide for the majority of realizations (F_j), 2) it was impossible to find the variant $F(D_j, C_j, \tau)$ that delivered $S_1 = S_1^{min}$ and $S_2 = S_2^{min}$ at once. The last circumstance has been one of the reasons to look for an optimal realization $F_{opt} = f(D_{opt}, C_{opt}, \tau)$ with compromise values, $S_1(D_{opt}), S_2(D_{opt})$.

It was admitted that the realization $F_{opt} = f(D_{opt}, C_{opt}, \tau)$ is optimal if it delivers RMS deviations those follow to a compromise condition

$$S_c = ((S_1^2 + S_2^2)/2)^{0.5}, \quad S_c(D_{opt}, C_{opt}) - S_c^{min} < \varepsilon, \quad (33)$$

where ε – ambiguity error; S_c^{min} – a lower border of the set (S_{ci}).

We have decided to include the leading amplitudes, B_{s0}, B_{d0}, to D characteristics fixed. We have adopted WLSA to calculate fitting coefficients C using conditions: i) $D = (\rho_c, T_c, \alpha, \beta, \Delta, B_{s0}, B_{d0})$ are fixed parameters, ii) criterion (24) is accepted to determine C values.

Routine 1 was elaborated as multistage of input data treatment to find an optimal realization. It includes several steps: i) to select $D_0 = (\rho_{c\ middle}, T_{c\ middle}, \beta = 0.325, \alpha = 0.1085, B_{s0\ middle}, B_{d0\ middle})$ as initial fixed parameters, ii) to form $\Phi(C,D)$ as (25) and to calculate $C = C_0$ employing LSQA (25), iii) to determine the criterion, $S_c(D_0, C_0)$, and to analyse conditions (33); if the analysis shows that $S_c(D_0, C_0)$ satisfies condition (33) then it is admitted that $F = f(D_0, C_0, \tau) = F_{opt} = f(D_{opt}, C_{opt}, \tau)$ and the data treatment is finished.

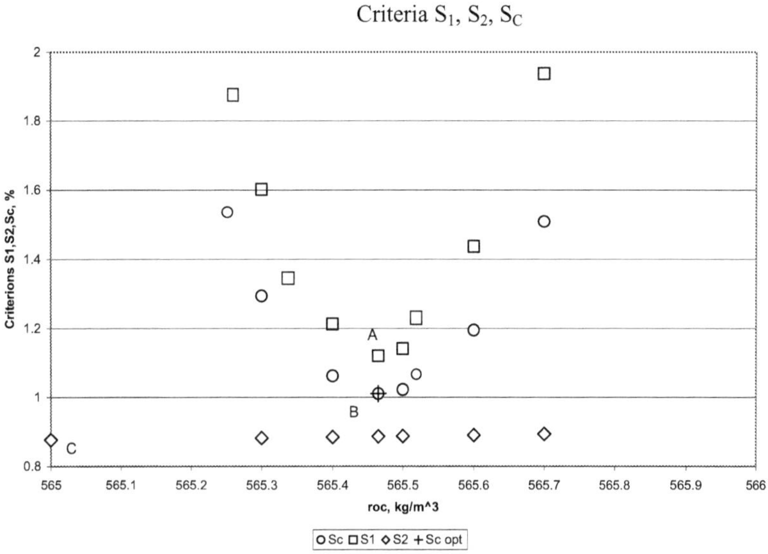

Figure 3. The influence of $\rho_c = var$ on the criterions, S_1, S_2, and S_c, of Models 1, 2 for the densities of R236ea. S_1^{min}, S_2^{min}, and S_c^{min} are shown as *A*, *B*, and *C*.

If conditions is not satisfied, the data treatment is continued: i) a new fixed parameter (for example, it is ρ_c) is chosen (ρ_c is shifted as $\rho_{cs} = \rho_{c\ middle} + \Delta\rho_{cs}$), $D_{1s} = (\rho_{cs}, T_{c\ middle}, \beta_0, \alpha_0, B_{s0\ middle}, B_{d0\ middle})$. Steps ii) and iii) are repeated and a realization C_{1s} is determined together with $S_c(D_{1s},C_{1s})$ and $\Delta S_c = S_c(D_{1s},C_{1s}) - S_c(D_0,C_0)$. A shift, $\Delta\rho_1$, is to be fined to reduce $S_c(D_0,C_0)$ as

$$\Delta\rho_1 = -\lambda(S_c(D_{1s}C_{1s})/(\Delta S_c/\Delta\rho_{cs})), \tag{34}$$

where λ – a fitting parameter.

An illustration of the path $S_c(\rho_{c\,j})$ is given in Fig. 3.

Routine 1 allows to obtain Model 2 as (4, 5) that has optimal characteristics, D_{opt}, C_{opt}. An example of Model 2 for the densities on CC of R236ea is given in Table 2.

TABLE 2. Parameters of Model 2 for the densities of R236ea.

ρ_c, kg/m^3	T_c, K	α	β	Δ
565.465	412.45	0.15074	0.34925	0.5
B_{s0}	B_{s1}	B_{s2}	B_{s3}	B_{s4}
2.0855	-0.006733	−0.421108	−0.348829	0.926804
B_{d0}	B_{d1}	B_{d2}	B_{d3}	B_{d4}
0.2225	1.947753	−1.467006	−0.411682	0.986974

A similar Model 2 is elaborated to represent the experimental results on refractive indexes, n_l,n_g,T – data, and $F = (f_{ns},f_{nd})$ in the combined form

$$f_{ns}= B_{ns0}\tau^{\beta} + B_{ns1}\tau^{\beta+\Delta 1} + B_{ns2}\tau^{\beta+\Delta 2} + B_{ns3}\tau^{2}+ B_{ns4}\tau^{3},$$
$$f_{nd}= B_{nd0}\tau^{1-\alpha} + B_{nd1}\tau^{1-\alpha+\Delta 1} + B_{nd2}\tau^{1-\alpha+\Delta 2} + B_{nd3}\tau^{2}+ B_{nd4}\tau^{3}. \quad (35)$$

Equations $F = (n_g,n_l)$ can be determined with the help of known (35) as

$$n_l = (f_d+f_s+1)\, n_c, \quad n_g = (f_d - f_s+1)\, n_c. \quad (36)$$

The scaling part of (35) also consists of three terms and represents experimental values in the interval τ_{scale}. Routine 1 was applied for multistage n_l,n_g,T – data treatment as it was done in the case of $F = (\rho_l,\rho_g)$.

An example of Model 2 for the refractive indexes on CC of R236ea is given in Table 3.

TABLE 3. Parameters of Model 2 for for the refractive indexes the densities of R236ea.

n_c	T_c /K	α	β	Δ
1.094	412.45	0.15074	0.34925	0.5
B_{ns0}	B_{ns1}	B_{ns2}	B_{ns3}	B_{ns4}
0.18360	-0.006182	−0.027412	−0.011005	0.037445
B_{nd0}	B_{nd1}	B_{nd2}	B_{nd3}	B_{nd4}
0.02284	0.223722	−0.227489	−0.054554	0.213751

An illustration of paths $S_1(\beta_j)$ and $S_2(\beta_j)$, for Models 1, 2 is given in Fig. 4. Values of $S_c(\beta_j)$ are shown also. The numerical dependences are got during the tests when Model 2 was built for the refractive indexes of R236ea and β was shifted ($\beta = var$).

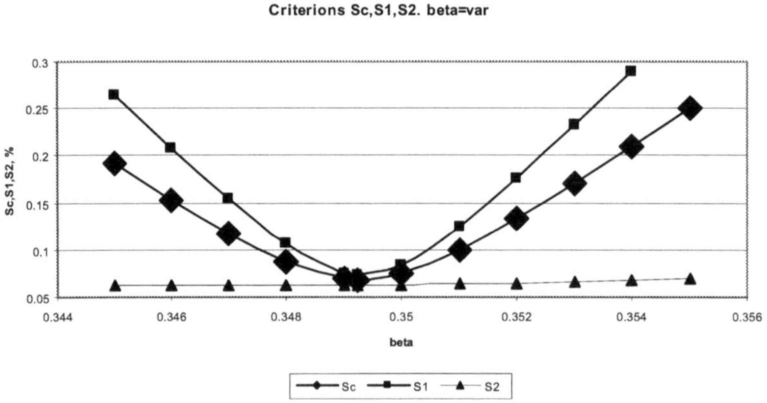

Figure 4. The influence of $\beta = var$ on the criteria, S_1, S_2 and S_c, of Models 1, 2 for the refractive indexes of R236ea.

An expression of the saturation pressure, P_s, was built in the combined form:

$$ln(P_s/P_c) = B_{p0}(1 - 1/t) + B_{p1}\tau + B_{p2}\tau^{2-\alpha} + B_{p3}\tau^{2-\alpha+\Delta} + B_{p4}\tau^3 + B_{p5}\tau^5, \quad (37)$$

This equation, $P_s = F(\tau,D,C)$, consists of scaling and regular parts and includes critical characteristics and regular amplitudes, B_{pi}–fitting coefficients, $t = T/T_c$.

An example of Model 2 for the saturation pressure of R143a is given in Table 4.

TABLE 4. Parameters of saturation pressure equation for HFC 143a.

P_c/MPa	T_c/K	α	Δ	B_{p0}
3.761	345.815	0.22	0.5	8.50
B_{p1}	B_{p2}	B_{p3}	B_{p4}	B_{p5}
1.17781	5.67519	−7.64488	8.55837	−9.18576

3. Discussion

The analysis of realizations, optimal variants of the combined equations and results of a comparison with literature data allow us to make a conclusion that Model 2 can be useful to approximate experimental $\rho_l,\rho_g,T-$ data and $n_l,n_g,T-$ data including a broad neighborhood of the critical point. Numerical variants of Model 2 for R134a, R143a, and R236ea[1,2,3,15,16] improve known data in the region $0.005 < \tau < 0.1$ where known results report only a few data with significant errors. The models represent experimental data in the regular area with high accuracy. The equations are based on effective values of critical characteristics, $D_{opt} = (\rho_c,T_c,\alpha,\beta,\Delta,\ldots)$, those have been determined with the help of the routines and experimental data sets. Values of D_{opt} give new information about the critical characteristics P_s Eq. (37) represents experimental data with low error. RMS deviation of the input data set for HFC 134a is determined as 0.04% in the temperature range from triple till critical point.[1,15,16]

Acknowledgment

The work is supported by The Russian Fund of Base Researches.

References

1. Ustjuzhanin E. E., Reutov B. F., and Kusubov K. B. (2000) The Coexisting Curve of the Refrigerant HFC 134a: Some Scaling Models in a Wide Critical Region. In *Proc. of Fourteenth Symposium on Thermophysical Properties* (Boulder, Colorado, USA, June 25–30).

2. Ustjuzhanin, E., Yata, J., Reutov, B., Grigoriev, B., and Jakovenko. K. (2003) A scaling model for refracting index of HFC 134a and HFC 143a on the coexistence curve. In *Proc. of Fifteenth Symposium on Thermophysical Properties* (Boulder, Colorado, USA, June 25–30).
3. Ustjuzhanin, E., Yata, J., Reutov, B., Grigoriev, B., and Magee, J. (2002) Thermodynamic properties of HFC 143a in a wide region of the critical temperature. In *Proc. of 16-th European Conference on Thermophysical Properties* (Imperial College, London, 1–4 Sept.).
4. Landau L. and Lifshiz M. (1972) *Theoretical Physics* (in russian) (Nauka, Moscow) p. 327.
5. Novikov, I. I. (1995) Substance Properties in the critical region (theory and experiment). *Thermophys. of High Temperture* (rus). 2001. V. 39, N 1, pp. 47–552.
6. Wegner, C. (1985) *Int. J. Thermophys.* **11**, 421–429.
7. Anisimov, M. A. et al. (1990) *Thermodynamics of Individual Substances in the Critical State* (rus) (Moscow, Energoizdat).
8. Rabinovich, V. A. and Sheludiak, Yu. E. (1995) *Thermophys. of High Temperture* in russian) V. **33**, N4, 546–552.
9. Shimanskaya, E. T., Shimansky, Yu. I., and Oleinikova, A. V. (1996) *Int. J. Thermophys.* **17**, 641–649.
10. Zhelezny, V. P. (1994) Scaling principles in the prediction of thermodynamic properties on CC. In *Proc. of Twelfth Symposium on Thermophysical Properties* (Boulder, USA).
11. Alekhin, O. D. et al. (2003) Values of critical indexes for inhomogeneous equilibrium of liquids. In *Proc. of 15-th Symposium on Thermophysical Properties* (June 22–27, Boulder, USA).
12. Aoyama, H., Kischizawa, G., and Sato, H. (1996) Vapor–Liquid Coexistence Curves in the critical region and the critical temperature and densities 1,1,1-trifluoroethane (R-143a), and 1,1,1,2,3,3-hexafluoropropane (R-236ea). *J. Chem. Eng. Data* **41**, 1046–1051.
13. Defibaugh, D. R., Gillis, K. A., Moldover, M. R., Schmidt, J. W., and Weber, L. A. (1996) Thermodynamic properties of CF_3–CHF–CHF_2,1,1,1, 2,3,3-hexafluoropropane. *Fluid Phase Equilibr.* **122**, 131–155.
14. Fukushima, M. (1993) *Trans. JAR* **10**(1), 87–93.
15. Ustjuzhanin, E., Magee, J., Yata, J., Reutov, B., Grigoriev, B., Jakovenko, K. (2003) Scaling Models for Thermodynamic Properties of HFC 134a and HFC 143a on the Coexistence Curve. In *Proc. of Fifteenth Symposium on Thermophysical Properties* (Boulder, Colorado, USA, June 25–30), 57–71.
16. Ustjuzhanin, E., Stankus, S., Reutov, B., Lipatov, A., and Vavilov, S. (2004) Scaling models for thermophysical properties of HFC 236ea on the coexisting curve. *Proc. of 25-th Conference on Thermophysical Properties* (Nagano, Japan, 19–24 oct.), 37–40.

INTERMOLECULAR POTENTIAL FOR SIMPLE LIQUIDS AND GASES IN THE HIGH PRESSURE REGION

V.YU. BARDIC*, L.A. BULAVIN, V.M. SYSOEV
Dept. of Molecular Physics, Kyiv University, 6 acad. Glushkov prospect, Kiev 03022, Ukraine
N.P. MALOMUZH*, K.S. SHAKUN
Dept. of Theoretical Physics, Odessa Univesity, 2 Dvoryanskaya str., Odessa, 65026, Ukraine

Abstract. The modified form of the repulsive potential has been derived by using of the equation of state treatment in the mean field approximation, such parameters of the potential as the molecular size and the steepness parameter being considered as the effective ones. The functional form of a repulsive potential is discussed on the base of results obtained by an analysis of statistical valid equations of state, high frequency asymptotic of the depolarized light scattering and the processing of temperature dependences for one-particle contributions to the self-diffusion coefficient.

Keywords: Repulsive potential, steepness parameter, equation of state, high frequency asymptotic of the depolarized light scattering, one-particle contributions to the self-diffusion coefficient.

1. Introduction

The simple potential models with parameters obtained from experiments have a wide use in solving of many problems. As a rule a functional form of potentials is based on theoretical concepts about the main contributions to considered interactions. Despite recent trends are toward increased use of more complex potential forms with a great number of parameters, it is of interest to improve the functional form of simple potential models, so it efficiently simplifies solving of different tasks. In this paper, we consider the repulsive potential in the framework of soft-sphere model, being guided by the results obtained from three different approaches. First of them is based on the treatment of the

*To whom correspondence should be addressed. E. Bardic, Dept. of Molecular Physics, Kyiv University, 6 acad. Glushkov prospect, Ukraine, Kyiv, 03022, e-mail:vital@univ.kiev.ua

S. J. Rzoska and V. A. Mazur (eds.), Soft Matter under Exogenic Impacts, 339–350.

statistical grounded equation of state which has the fitted constants connected with intermolecular potential parameters and describes adequate isothermal compression of dense systems in the high pressure region or in the range of thermodynamics variables where the isothermal compressibility β_{to} is low. We suggest that the role of an intermolecular potential corresponding to repulsive forces is determinant in this region. The central idea of the study involves an assumption that a repulsive part of the intermolecular potential is a homogeneous function with the homogeneity exponent m, the Euler's theorem about the homogeneous functions being applied to the potential function

$$U_r\left(s\bar{r}_1,\ldots,s\bar{r}_n\right)=s^{-m}U_r\left(\bar{r}_1,\ldots,\bar{r}_n\right). \tag{1}$$

For example, the function $U\left(\bar{r}_1,\ldots,\bar{r}_n\right)=\sum_{1\le i\ell\, j\le N}\varphi\left(\bar{r}_i,\bar{r}_j\right)$ that describes the sum of pair potentials in the frame of soft-sphere model

$$\varphi(\bar{r})=\varepsilon\left(\frac{\sigma}{r}\right)^m \tag{2}$$

must obey this theorem. In this case the partition function of a system on the grounds of similarity

$$\bar{r}\to q\bar{r},\quad q^3=\frac{V}{V_0}=\frac{\rho_0}{\rho},\qquad \rho_0=\frac{N}{V_0} \tag{3}$$

can be presented as a function of a variable $((V^*)^{m/3}, kT)$, the generalized form of the equation of state for this system being described by the expression

$$\frac{PV}{NkT}=1+\varphi\left(\frac{\left(V^*\right)^{m/3}}{kT}\right). \tag{4}$$

In processing of experimental PVT-data analysis for argon we calculated the value of steepness parameter m as a fitted constant of the equation of state. The obtained results gave a good agreement with both high frequency asymptotic of the depolarised molecular light scattering and the processing of temperature dependences for one-particle contributions to the self-diffusion coefficient for simple liquids. In particular, we show that the asymptotic law for $I^{VH}(\omega)$ at $\omega\to\infty$ is connected with the steepness parameter m of the repulsive potential and consider the character of the high frequency asymptotic for spherical molecules. The value of the parameter m was evaluated in processing experimental data analysis for the intensity of scattering in argon for very wide frequency interval (0.400 cm^{-1}). We have studied the correlation between the one-particle component of the self-diffusion coefficient and the repulsive potential, by fitting the theoretical value of the steepness parameter to the corresponding experimental data, the value of this parameter has been determined.

2. The equation of state approach

The consecutive derivation of the equation of state has been carried out by using of the thermodynamic perturbation theory[1,2] based on scaling transformation of the partition function (3). Using the thermodynamic reference state rather than the reference system allows establishing connection between the parameters of the intermolecular potential and the thermodynamic properties of a system under this consideration. As mentioned above, we have suggested that the potential function corresponding to repulsive forces is a homogeneous function with the homogeneity exponent *m* (Euler's theorem, expression (1)). One can say the homogeneity exponent *m* contains information about the connection of potential parameters with thermodynamics properties of substances. According to the papers[1,2] an expression for the first approximation of the free energy can be written as the average perturbed energy with the canonical distribution function of a reference state

$$F(V) - F(V_0) = 3NkT \cdot \ln q - \langle \Delta U \rangle, \tag{5}$$

where, $F(V_0)$ is the free energy of a reference state and $q = \sqrt[3]{V_0/V}$. To differentiate of the expression for the free energy with respect to volume, it is necessary to express the potential function $U_r(\bar{r}_1, ..., \bar{r}_n)$ as a function of volume. Thus, the perturbation of the repulsive potential function should be represented as

$$\langle \Delta U_r \rangle = \left\langle \frac{\partial U_r}{\partial \sigma} \cdot \Delta\sigma \right\rangle, \tag{6}$$

where $\Delta\sigma = \dfrac{\sigma' - \sigma}{\sigma}$, reduce $\sigma' = \sqrt[3]{\dfrac{V_0}{V}}\sigma$ to the form $\Delta\sigma \approx \dfrac{1}{3}\Delta V$, and write expression (6)

$$\langle \Delta U_r \rangle = -\frac{1}{3}\Delta V \Sigma \langle \nabla_i U_r(\vec{r}_i) \cdot \vec{r}_i \rangle. \tag{7}$$

Being differentiated the free energy expression (5) with respect to *V* and taking into account homogeneity of the potential function (4) write an equation for pressure

$$P = \frac{NkT}{V} - \frac{1}{3V_0} \cdot \left[\Sigma \langle \nabla_i U_r(\vec{r}_i) \cdot \vec{r}_i \rangle \right]. \tag{8}$$

Using the virial theorem we transform equation of state (Eq. (8)) to thc two-constant equation of state in the form

$$P - \frac{NkT}{V} = P_r \left(\frac{V_o}{V} \right)^{m/3}, \tag{9}$$

where P_r – the pressure caused by the repulsive forces between molecules in a reference state, m – the exponent of homogeneity or the steepness parameter that defines the functional form of the repulsive potential in the framework of soft-sphere model (2).

In order to analyze the experimental PVT-data on isothermal compression and in processing of analysis to obtain the value of the parameter m we used the experimental results.[3] It turned out that Eq. (9) reveals a good fit to the experimental data for a number of inert gases and gases nitrogen in the range of the thermodynamic variables where the isothermal compressibility is low, that is corresponding to the pressure interval 100–1000 MPa and temperatures 308.14–673.14 K. As far as it was interesting to prove namely the extrapolation capabilities of the equation of state the values of parameters m and P_r were considered as fitting constants. The analysis of the results obtained by using Eq. (9) was carried out in two stages. At first, the parameters m and P_r were calculated at three reference points P_0, P_1, and P_2 (P_0 is the pressure in a reference state, usually, we choose any of those points for P_0 where the isothermal compressibility became as small as desired or parameter $\Delta V << 1$). The values of the parameters obtained in this way were then substituted into Eq. (9) in order to calculate the volumes on isotherms. Just for illustrative purposes we showed only the value of average deviation $(\delta = (V_e - V_t)/V_t \cdot 100\%)$ for argon, where V_e is the experimental value[3] of volume, V_t – the value calculated from Eq. (9), so the value $\langle \delta \rangle$ does not exceed 0.78% on isotherm T=673.14 K, on isotherm T=308.14K, $\langle \delta \rangle$ takes the value 0.02%, in the stated above pressure range. Thus, the analysis of PVT-data for gaseous argon shows that Eq. (9) correctly describes the experimental data on isothermal compression and yields fairly accurate results upon extrapolation. The processing of experimental results led to the temperature dependence of the steepness parameter m, so in the temperature interval 308.14–673.14K, m takes the values from 14.89 to 10.10. It should be noticed that a similar processing was carried out for a number of inert gases,[4] for liquid argon[5] and fullerenes.[6] As it was shown in the paper,[4] the processing of PVT-data based on the equation of state led to the different values of the steepness parameters m being varied from 10.5 to 17.5 for a number of inert gases (from neon to krypton), that does not match with the value m=12 generally accepted for many potential models. In connection with this one should refer to the paper,[5] in which the treatment of a theoretically-based equation of state for liquids gave for argon in the liquid phase the value of the steepness parameter m being equaled 28. This fact seems to be to some extent surprising but an excellent agreement of the experimental data with the equation of state results encourages. Besides that, the majority of modern papers devoted to the equations of state investigations yield the empirical equation of state containing a great number of fitting parameters whose physical sense remained obscure and which could be apt as merely interpolation formulae for treating of the experimental data.

3. The high frequency asymptotic of the molecular light scattering

As it was shown,[7–9] the high-frequency asymptotic of molecular light scattering was being formed by the collision-induced mechanism. The analysis of the binary collision contributions into the spectrum $I^{VH}(\omega)$ of the depolarized light scattering leads to the following expression:

$$I^{VH}\underset{\omega\to\infty}{(w)} \sim \omega^{p} \exp\underset{\omega}{\left((\omega\tau_{*})^{q} \right)}, \tag{10}$$

where

$$p = \frac{1}{3} - 2\left(\frac{7}{9m+6} + \frac{2}{m+2} \right), \tag{11}$$

$$q = \frac{2m}{3m+2} \quad \text{and} \quad \tau^{*} = 16.32 \frac{\sigma}{m} \left(\frac{\mu}{2kT} \right)^{\frac{1}{2}} \left(\frac{\varepsilon}{kT} \right)^{\frac{1}{m}}. \tag{12}$$

Here σ, m, and ε are the parameters of the potential (2). From physical point of view the high frequency asymptotic of the molecular light scattering spectra at the high shifts of frequencies is generated by the quickness of the process passing in a system. One can consider molecular collisions as a combination of soft and hard ones. Some of them are caused by the attractive and the repulsive forces, acting on distances which are comparable with the average interparticle spacing. As a result of soft collisions is many-particle interactions. With satisfactory accuracy hard collisions can be considered as a sequence of hard binary collisions, in this case simultaneous interaction between three and more molecules is improbably. The binary hard collisions are responsible for the high frequency behaviour of spectra under the assumption that they are characterised by the smallest time of interaction in a system. Thus, if one takes into account adduced arguments two types of the characteristic times can be introduced

$$\tau_{S} = \frac{\sigma}{V_{T}} \quad \text{and} \quad \tau_{H} = \frac{\tau_{S}}{m}, \tag{13, 14}$$

which describe soft and hard collisions. The first of them τ_S is the time interval corresponding to the smooth variation of the relative velocity in the binary collision, τ_H – characterizes more fast variation of the relative velocity near the stop point and therefore it is called hard collision time. Correspondingly to this the formula for τ^* can be rewritten in the form

$$\tau_* = a^{1/q}\tau_H, \tag{15}$$

where
$$\alpha^{\frac{1}{q}} = 16.32\left(\frac{\varepsilon}{kT}\right)^{1/m}. \tag{16}$$

Then the expression (10) could be transformed

$$I^{VH}_{\omega\to\infty}(\omega) \sim \omega^p \exp\left(-a\left(\omega\tau_H\right)^q\right). \tag{17}$$

The expression (17) gives the prediction for the leading term of high-frequency asymptotic. We expect also, that this expression can be applied to the molecular light scattering spectra in liquids. To motivate this assertion it would be well to assume: 1) binary hard collisions unlike soft collisions are correctly defined both in gases and liquids and 2) it is improbably that three or more particles are involved in hard collision simultaneously. To verify a qualitative agreement between theory and experiment and also to evaluate the optimal values of the parameter m, entering the repulsive part of the potential, we used experimental data on the depolarized light scattering for argon.[10] Calculations were carried out in the frequency interval $4.6 \cdot 10^{12} - 1.7 \cdot 10^{13}\ s^{-1}$ had a minimal experimental error. The best coincidence is observed with such values of the parameters: $a = 8.67$, $q = 0.648$, $\tau_H = 6 \cdot 10^{-14}$ s, at that the value of the steepness parameter m was defined from Eq. (5). Its value equals 24 for liquid argon, that is very closed with the values of this parameter obtained from the equation of state in paper.[5]

4. Self-diffusion of liquid molecules

According to the general theory,[11,12] the self-diffusion coefficient D is presented as a sum

$$D = D_1 + D_c \tag{18}$$

of the collective, D_c, and one-particle, D_l, components. The algorithm of calculations of the collective component was described in detail in work.[13] At the same time, the one-particle component of the self-diffusion coefficient is not studied sufficiently; in fact, we know a single work dealing with this issue.[14]

An important stage in studying the diffusive processes is a correct account of intermolecular interactions, based on the proper choice of an adequate model form of the intermolecular potential. On the other hand, the potential function can be reproduced making use of the self-diffusion experimental data and comparing them with theoretical calculations, as it is done, for example, when calculating the effective pair potential from the known radial distribution function or when constructing an authentic potential curve of intermolecular interactions on the basis of the method of inverse scattering transform. The character of the decrease of the one-particle contribution to the molecule-velocity autocorrelation function (MVACF) at small times is governed by the behavior of repulsive forces at distances shorter than the molecule diameter. Really, at small times, the MVACF is determined by the expansion series

$$\left\langle \vec{v}(t)\vec{v}(0)\right\rangle = \left\langle \vec{v}^{\,2}(0)\right\rangle + \left\langle \vec{v}''(0)\vec{v}(0)\right\rangle \frac{t^2}{2} + \cdots \qquad (19)$$

where, $v(t)$ is the velocity of a selected liquid molecule. The expression for the autocorrelation function[15] at small t can be written down as

$$\varphi(t) = \varphi_1(t) = \frac{3kT}{M}\left(1 - \frac{1}{2}\frac{t^2}{\tau_0^{\,2}} + \cdots\right), \qquad (20)$$

where k is the Boltzmann constant, T is the absolute temperature, M is the molecule mass, and

$$\tau_0^{\,-2} = \frac{\left\langle \Delta U(\vec{r})\right\rangle}{3M} \qquad (21)$$

or approximately

$$\varphi_1(t) \approx \frac{3kT}{M}\exp\left(-\frac{1}{2}\frac{t^2}{\tau_0^{\,2}}\right). \qquad (22)$$

For the numerical evaluation of τ_0, we use a spherically symmetric intermolecular potential in the form

$$U(r) = 4\varepsilon\left[\left(\frac{\sigma}{r}\right)^m - \left(\frac{\sigma}{r}\right)^6\right], \qquad (23)$$

whence we obtain the following expression for $\Delta U(r)$

$$\Delta U(r) = \frac{4\varepsilon}{\sigma^2}\left[m(m+1)\left(\frac{\sigma}{r}\right)^{m+2} - 42\left(\frac{\sigma}{r}\right)^8\right]. \qquad (24)$$

Using expressions (21) and (24) and the mean field approximation, where

$$\left\langle \left(\frac{\sigma}{r}\right)^m \right\rangle \approx \left(\frac{\sigma}{\langle r\rangle}\right)^m \approx \left(n\sigma^3\right)^{m/3} \qquad (25)$$

we obtain

$$\tau_0^{\,-2} = \frac{4\varepsilon}{3M\sigma^2}\left[m(m+1)\left(n\sigma^3\right)^{(m+2)/3} - 42\left(n\sigma^3\right)^{8/3}\right] \qquad (26)$$

A comparison of the results of computer simulation[14] with the normalized autocorrelation functions calculated according to Eq. (20) and Eq. (22) is presented in Fig. 1. The calculations were carried out for the values of the steepness parameter $m = 18$ and 24 on the isotherm $T = 100$ K. As it is seen from Fig. 1, the dependence of the normalized autocorrelation function of the one-particle contribution $\varphi_1(t)$, which decays by the polynomial law, intersects the data of molecular dynamics[14] at the definite point $x_* = \dfrac{t}{2\tau_M} = 0.78$, if the steepness parameter $m = 24$, where τ_M is the Maxwell relaxation time of viscous stresses in the liquid. Thus, summing up both plots, the MVACF can be written down as

$$\varphi(t) = [\varphi_1(t) - \varphi_c(t)] \cdot \theta(t^* - t) + \varphi_c(t), \tag{27}$$

where
$$\begin{cases} \theta(t^* - t) = 0, & t^* - t < 0 \\ \theta(t^* - t) = 1, & t^* - t > 0. \end{cases}$$

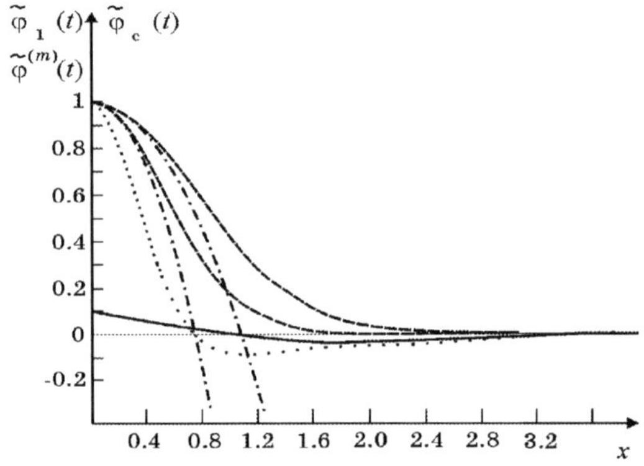

Figure 1. Time dependences of the autocorrelation functions. The solid curve corresponds to the normalized autocorrelation function of a Lagrange particle,[11] the dash-dotted curves to the dependences $\varphi_l(t)$ that decay by a polynomial law, the dashed curves to exponential dependences, and the dotted curve to a dependence obtained on the basis of the molecular dynamics data.[14]

The function $\varphi_c(t)$ is the collective component of the MVACF, and $t_* = 2\tau_M x^*$. In accordance with formula (27), the one-particle contribution to the self-diffusion coefficient of a molecule is determined by the expression

$$D_1 = \frac{1}{3} \int_0^{t^*} [\varphi_1(t) - \varphi_c(t)] \mathrm{d}t. \tag{28}$$

The values of the self-diffusion coefficient D of liquid argon, taking into account formula (28), are quoted in Table 1. In this table, the notation D_P is used for the self-diffusion coefficient calculated taking into account the dependence $\varphi_l(t)$ in the form of a quadratic polynomial, D_G in the case of the Gaussian exponent dependence, and $D_{\exp}$ for the experimental value of the self-diffusion coefficient.[16]

TABLE 1. Self-diffusion coefficient of liquid argon at T=100 K for various values of the slope parameter m.

m	$t^* \cdot 10^{13}$, s	$D_G \cdot 10^5$, cm²/s	$D_P \cdot 10^5$, cm²/s	$D_{\exp} \cdot 10^5$, cm²/s
28	3.44	2.54	2.395	2.47
24	3.47	2.61	2.416	
18	3.81	2.72	2.61	
12	5.03	3.17	3.298	

The best agreement between theoretical values and experimental data takes place for the value m=24. A comparison of the theoretically calculated temperature dependence of the self-diffusion coefficient with experimental data[16] is presented in Fig. 2. We note that the computer simulation of the MVACF, carried out making use of the Lennard-Jones potential (12:6), resulted in a more abrupt decay of this function in comparison with the polynomial dependence (22). Moreover, formula (22) reproduces the results of computer simulation,[16] obtained for the steepness parameter values of 24 and 28, with quite a good accuracy, which is in a qualitative agreement with the processing of the equation of state results[5] and the light scattering data mentioned above.

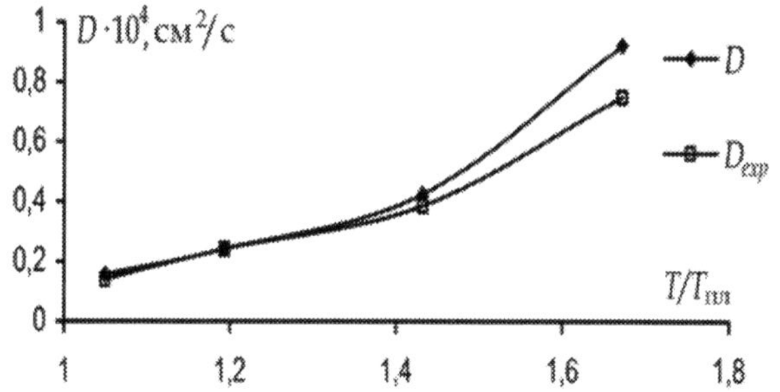

Figure 2. Comparison of theoretical and experimental[16] temperature dependences of the self-diffusion coefficient of liquid argon. T_m is the melting temperature.

Thus, taking into account the data reported above and the results of works,[4,5] we may say that the slope of the potential function is an individual characteristic of the specific substance considered in the framework of the inverse-power potential model, and the value of this parameter is sensitive to temperature and depends on the aggregate state of the substance. Summarizing all those circumstances indicating for an effective character of the functional form, at the next stage of study we tried to improve the power model for the repulsive potential.

5. Repulsive potential with effective steepness parameter

In the general case, considering variation of the steepness parameter with temperature, intermolecular potential in the high pressure region can be written in the form of power series

$$U(r) = -4\varepsilon\left(\frac{\sigma}{r}\right)^6 + \sum_{m>6} a_m\left(\frac{\sigma}{r}\right)^m. \tag{29}$$

In the subsequent discussion we focused our attention namely on the repulsive part of the potential

$$U(r) = \sum_{m=6} a_m\left(\frac{\sigma}{r}\right)^m. \tag{30}$$

It should be noticed that alertation coefficient a_m with temperature and pressure is expounded as variation of the effective molecular size of a molecule

$$U(r) = \sum_{m>m_0} a_m\left(\frac{\sigma}{r}\right)^m = 4\varepsilon \sum_{m>m_0}\left(\frac{\sigma_m(P,T)}{r}\right)^m. \tag{31}$$

For a dense system, when the number particle density is inessential, considering the mean field approximation $\langle U'(r)\rangle = U'(\langle r\rangle)$ and $\langle r\rangle = n^{-1/3}$ the expression (30) can be rewritten

$$U(r) = 4\varepsilon \sum_{m=6} \alpha_m \left(\sigma n^{1/3}\right)^m. \tag{32}$$

Actually, in this case the molecule distribution in the given vicinity is identical to that for the dense packing, then in accordance with the mean field approximation we transform the general expression for pressure which presents the virial theorem

$$P = nkT - \frac{1}{6}n^2 \int_V \mathrm{d}VrU'(r)g(r) \tag{33}$$

to the form

$$P = nkT - \frac{z}{6}n\,\langle r\rangle\, U'(\langle r\rangle), \tag{34}$$

where $\langle r\rangle$ – average intermolecular distance, z – the number of molecules in the first coordination sphere. Considering that $\langle r\rangle = n^{-1/3}$ and the formula (30), express Eq. (34) by a formula

$$P \approx nkT + \frac{4zn\varepsilon}{6} \sum_{m>m_0} m\tilde{a}_m n^{m/3}. \tag{35}$$

where

$$\tilde{a}_m = \alpha_m \cdot \sigma^m, \quad \alpha_m = \frac{a_m}{4\varepsilon} \tag{36}$$

or

$$\frac{P}{nkT} = 1 + \frac{2z\varepsilon}{3kT} \sum_{m>m_0} m\tilde{a}_m n^{m/3}. \tag{37}$$

Thus, the members of the series (37) with values $m \approx m_0$ make the main contribution to P/nkT in the low pressure region, for example, generally accepted value for the exponent of the repulsive potential $m_0 = 12$. In contrast to this, in the high pressure region, the members with values $m > m_0$ have a dominant role. In this sense Eq. (37) completely conforms with assymptotic Eq. (4). A knowledge of the equation of state in the wide pressures and temperatures interval allows to calculate values of coefficients $\tilde{a}_m$ in Eq. (37). Coefficients $\tilde{a}_m$ have been fitted by the least square method using available experimental PVT-data.[17] According to this Eq. (37) can be performed to the form

$$\frac{P}{nkT}=1+\frac{z}{6kT}\left(8\tilde{a}_8 n^{8/3}+12\tilde{a}_{12}n^{12/3}+16\tilde{a}_{16}n^{16/3}\right). \quad (38)$$

The value of the effective molecular size σ is defined by a formula

$$\sigma(T)=\left(\frac{\tilde{a}_m}{4\varepsilon}\right)^{1/m}. \quad (39)$$

The variation of the steepness parameter m and the effective molecular size σ of liquid argon are quoted in Table 2.

TABLE 2. Values of the steepness parameter m and the effective molecular size σ at different temperatures.

T, K	308.14	200	140
m	15.98	21.2	23.46
σ, A	3.002	3.204	3.21
$\tilde{a}_m$	$\tilde{a}_8 = 2.8\cdot10^{-6}$	$\tilde{a}_8 = 2.8\cdot10^{-4}$	$\tilde{a}_9 = 1.13\cdot10^{-5}$
	$\tilde{a}_{12} = 8.19\cdot10^{-5}$	$\tilde{a}_{14} = 1.73\cdot10^{-3}$	$\tilde{a}_{16} = 2.027\cdot10^{-5}$
	$\tilde{a}_{16} = 0.995$	$\tilde{a}_{21} = 0.981$	$\tilde{a}_{24} = 0.945$

As a result of approximation in terms of the results (38) and (39), we have obtained the expression for the repulsive potential

$$U(r)=4\varepsilon\left[\tilde{a}_{m1}\left(\sigma n^{1/3}\right)^{16}+\tilde{a}_{m2}\left(\sigma n^{1/3}\right)^{21}+\tilde{a}_{m3}\left(\sigma n^{1/3}\right)^{24}\right], \quad (40)$$

where values of parameters m_i are being fitted by minimization of average deviation the data calculated by Eq. (38) from the experimental PVT-data of liquid argon.

6. Conclusion

The statistical mechanics approaches have given a possibility to establish a relation between the parameters of the intermolecular potentials and thermodynamic properties of substances. Setting such a relation allowed to calculate the homogeneity parameter m (the steepness parameter for soft-sphere model) as a fitting constant. The processing of the obtained results testifies that the generally-excepted power model of the repulsive potential is inapplicable for the high-pressures region, based on results indicated above it would be correctly supposed that different parts of the potential function are being realized in the high pressures region (P>100 MPa) on isothermal compression and at large shifts of frequencies of depolarized molecular light scattering.

In this paper, we derived the form of the repulsive potential taking into account the effectiveness of intermolecular parameters with using the mean field approximation and treatment of the equation of state. It should be noticed that the proposed form of the repulsive potential is being tested by molecular dynamics simulations.

References

1. Zwanzig, R. W. (1954) High temperature equation of state by a perturbation method, 1. Nonpolar gases, *J. Chem. Phys.* **22**(8), 1420–1426.
2. Zwanzig, R. W., Kirkwood, J. G., Stripp, K. F., and Oppenheim, I. (1953) Radial distribution functions and the equation of state of monoatomic fluids, *J. Chem. Phys.* **21**(7), 1268–1275.
3. Robertson, S. L., Babb, S. E., and Scott, G. J. (1969) Isotherms of argon to 10000 bars and 4000C, *J. Chem. Phys.* **50**(5). 2160–2166.
4. Atamas, A. A., Bardik, V. Yu., and Sysoev, V.M. (2000) Correction of the repulsive potential parameters *Ukr. J. Phys.* **45**, 1184–1187.
5. Bardik, V. Yu., and Sysoev, V. M. (1998) Equation of state for liquid argon in the high-density limit, *Low Temp. Phys.* **24**, 602-603; *Fiz. Niz.Temp.* **24**, 797–799.
6. Bulavin, L. A., Bardic, V. Yu., Prylutskyy, Yu. I., and Sysoev V. M. (2002) On the connection of elastic properties for fullerene systems with parameters of model intermolecular potential, *Ukr. J. Phys.* **47**, 486–48.
7. McTague, J. P., and Birnbaum, G. (1968) Collision-induced light scattering in gaseous argon and krypton, *Phys.Rev. Lett.* **21**(4), 661–664.
8. Volterra, V., Bucaro, J. A., and Litovitz, T.A. (1971) Two mechanisms for depolarized light scattering from gaseous argon, *Phys. Rev. Lett.* **26**, 55–57.
9. Barocchi, F., and McTague, J. P. (1975) Collisions-induced light scattering in gaseous CH_4, *Phys.. Lett. A*. **53**, 488–490.
10. Levine, H. B., and Birnbaum, G. (1968) Collisions-induced light scattering in gaseous *Phys. Rev. Lett.* **20**(9), 439–441.
11. Lokotosh, T. V., and Malomuzh, N. P. (2000) Lagrange theory of thermal hydrodynamic fluctuations and collective diffusion in liquids, *Physica A* **286**(3–4), 474–488.
12. Lokotosh, T. V., and Malomuzh, N.P. (2001) Manifestation of collective effects in the rotational motion of molecules in liquids, *J. Mol. Liq.* **93**(1–3), 95–108.
13. Bulavin, L. A., Lokotosh, T. V., Malomuzh, N. P., and Shakun, K. S. (2004) Collective contributions to the selfdiffusion process in liquids, *Ukr. Fiz. J.* **49** (6), 556–561.
14. Rahman, A. (1964) Correlation in the motion of atoms in liquid argon, *Phys. Rev. A* **136**, 405–411.
15. March, N. H., and Tosi, M. P. (1976) Atomic dynamics of liquids, Macmillan Press, p.296.
16. *CRS Handbook of Chemistry and Physics: Ready-Reference Book of Chemical and Physical Data*; 67-th ed/ Ed.-in-chief R.C. West (Boca Raton: CRS Press).
17. Tegeler, Ch., Span, N., and Wagner, W. (1999) A new equation of state for Argon covering the fluid region for temperatures from the melting line to 700 K at pressures up to 1000 MPa *J. Phys. Chem. Ref. Data* **28**(3).

HOMOGENEOUS NUCLEATION AND GROWTH FROM HIGHLY SUPERSATURATED VAPOR BY MOLECULAR DYNAMICS SIMULATION

N. LÜMMEN, B. FISCHER, T. KRASKA*
Department of Physical Chemistry, University Cologne
Luxemburger Str. 116, D-50939 Köln, Germany

Abstract. The formation of particles and aerosols is an important step in several industrial as well as environmental processes. Its formation can be an undesired phenomenon, for example, in thermal separation processes, in turbines or the discharge of environmentally problematic aerosols from industrial processes. However, it can also be the central process in case of the production of nanoparticles or disperse systems such as pharmaceutical nanodispersions. The atmosphere and hence the climate is influenced by the presence of several different aerosols. While primary aerosols are particles that are emitted into the atmosphere, secondary aerosols are formed in the atmosphere by nucleation and growth. One possible way of particle and aerosol formation is homogeneous nucleation caused by density fluctuations in supersaturated systems. This is the first step followed by surface growth and initial coalescence processes. The initial steps happen on the nanometer scale and can be investigated by molecular simulation, which enable the modeling and analysis based on molecular interactions. Here, different molecular dynamics methods for the investigation of nucleation and particle growth are discussed. The focus is on various substances such as argon, alkanols, and metals. The simulations start at highly supersaturated states in the metastable region of the phase diagram.

Keywords: Condensation, nucleation, coalescence, agglomeration, supersaturated vapor, metastable states, metal nanoparticles, liquid clusters, iron, platinum, methanol, argon.

1. Introduction

1.1. HOMOGENEOUS NUCLEATION

The transition from one phase state to another can proceed via different mechanisms depending on the state of the original phase. In case of a metastable phase of a pure substance homogeneous nucleation can take place.[1–3] Due to density fluctuations in the metastable phase a cluster can be

* To whom correspondence should be addressed. E-mail: T.Kraska@uni-koeln.de

S. J. Rzoska and V. A. Mazur (eds.), Soft Matter under Exogenic Impacts, 351–377.

formed which may grow to a stable particle. The question whether a cluster becomes stable or evaporates again is determined by the nucleation barrier. The height of this barrier depends on the temperature, the supersaturation, and in this context on the molecular interactions in the substance. The size of a cluster at the top of the nucleation barrier is called critical cluster size. In Fig. 1 the work of cluster formation is plotted schematically for different states of the original phase state. This state is characterized by the supersaturation defined as the ratio of the present partial pressure $p_i(T)$ of the nucleating substance before nucleation and the saturated vapor pressure $p_{\text{vap}}(T)$ at the same temperature:

$$S = \frac{p_i(T)}{p_{\text{vap}}(T)}. \qquad (1)$$

For $S < 1$ a phase is stable, for $S = 1$ it is saturated, and $S > 1$ it is supersaturated. A supersaturated phase can be either metastable or unstable. In Fig. 2 the phase diagram of a pure substance is shown together with the stability limits.[3–6] The boundary between these states is the spinodal. Homogeneous nucleation takes place only in the metastable region while a system approaching the unstable region splits instantaneously by spinodal decomposition.

For homogeneous nucleation a relatively high supersaturation is necessary until the system crosses the nucleation barrier. In many natural systems like the atmosphere small solid or liquid particles are present. These particles can act as heterogeneous.

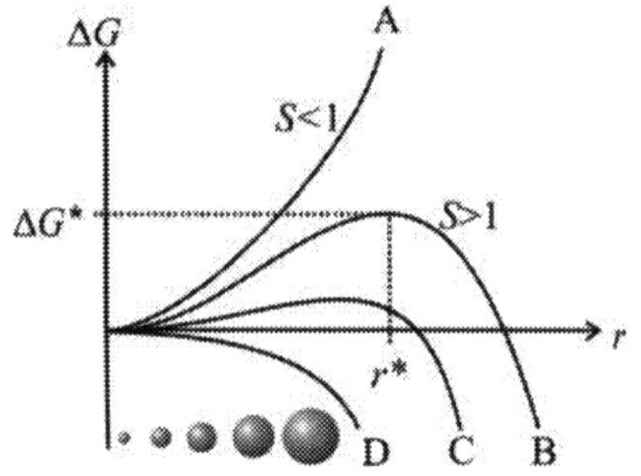

Figure 1. Work of cluster formation for different supersaturations. *A*: S<1; *B*,*C*,*D*: S>1; *B*,*C*: metastable state; *D*: unstable state.

The modeling of the homogeneous nucleation is subject to research since many decades. The widely used so-called classical nucleation theory (CNT) has been proposed by Volmer and Weber[7] and Becker and Döring[8] about 80 years ago. The CNT is very appealing for applications because it describes the nucleation rates based on macroscopic thermodynamics properties such as the liquid density and the surface tension. Since these properties are usually not known for small clusters like the critical nuclei, the bulk phase properties are used as input for the CNT. Further assumptions within the so-called capillary approximation[9] are a sharp interface between the liquid and the vapor phase and the use of the perfect gas state for the vapor phase. The CNT basically combines two contributions to the free energy of cluster formation, one being the bulk contribution the other the surface contribution.

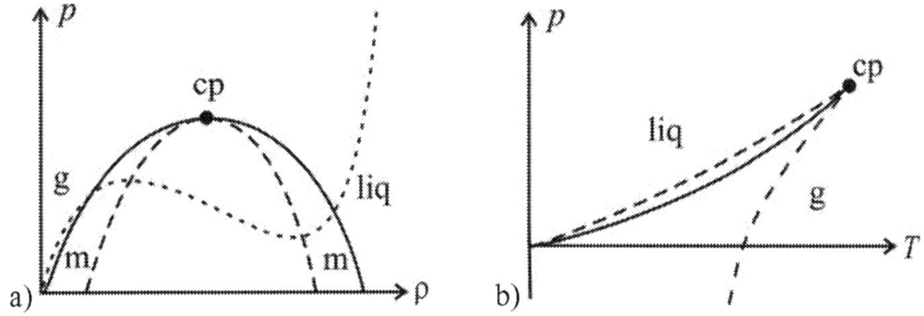

Figure 2. Phase diagram of a pure substance including the stability limits. a) Pressure–density projection; b) pressure–temperature projection. solid curve: equilibrium curve (coexistence curve and vapor pressure curve); long dashed: spinodal; short dashed: isotherm; *g*: gas phase; *liq*: liquid phase; *cp*: critical point; *m:* metastable. nuclei, which lower the height of the nucleation barrier. Hence, such heterogeneous nucleation requires a lower supersaturation compared to homogeneous nucleation in order to yield particle formation.

As the bulk phase properties depend on the third power of the cluster radius and the surface properties on the second power of the radius, CNT is mathematically a competition between a quadratic and a cubic function. For stable systems $S < 1$ the surface energy required to form a cluster dominates and particle formation is not favored. Hence, the free energy of cluster formation is a monotonically increasing function with the radius of a possible cluster and resembles a positively curved quadratic function. In supersaturated states the cubic term leads to a maximum, the critical work of cluster formation at the critical cluster size. Increasing supersaturation decreases the critical work and the critical cluster size. However, CNT is not able to detect the stability limit, the spinodal. It may therefore predict nucleation at high supersaturation where actually spinodal decomposition takes place.

There are several further developments of the classical nucleation theory developed in recent decades.[10–13] However, as long as these models are based on the capillary approximation, the same limitations apply. In addition numerical methods such as the thermodynamic density functional theory and molecular simulations have been applied for the investigation of nucleation.[14,15] While density functional theory is an integral equation method, molecular simulations are methods to directly calculate the partition function of a system based on the molecular interactions. Two molecular simulation methods have been applied to investigate nucleation: Monte Carlo simulations[16,17] and molecular dynamics simulation.[18,19] In order to investigate the dynamics of a nucleating system directly one has to use molecular dynamics simulation.

1.2. COALESCENCE

Once small stable clusters have been formed by nucleation they can grow by surface growth until the supersaturation is abolished. In addition to surface growth, the condensation of single atoms on the surface of an existing cluster, it is possible that particles collide and form a larger particle. These are rather a stochastic events than thermodynamically controlled like surface growth. Depending on the state conditions of the colliding particles and their size the development of the particle after a collision can be different. For large solid particles or at low temperature agglomeration is more likely. The two colliding particles stick together at their surface and form agglomerates with a morphology very different from a sphere. The more particles agglomerate the more dendrimer-like are the resulting particles. Such agglomerates can be characterized on two length scales, the size of the primary particles resulting from nucleation and surface growth and the size of the complete agglomerate. For small solid particles or particles at high temperature it is more likely that the particles fuse and form a new cluster with a spherical morphology. This process is called coalescence. Coalescence does not only affect the morphology but also the atomic order in a particle since the atoms are reorganized during fusion.[20] The restructuring of the atomic order is determined by the diffusion of the atoms in the particle but also at the surface of the clusters. It is therefore expected that a coalescence process, which entirely takes place in the solid state, is slower than coalescence during that the particles are heated up above their melting poin.[21]

2. Methods

2.1. MOLECULAR DYNAMICS SIMULATION AND POTENTIAL MODELS

Within molecular dynamics simulation Newton's equation of motion of each atom or molecule in the force field of all other atoms is solved numerically. The value of the increment for the numerical integration depends on the potential model describing the interactions between the atoms and molecules. In order to conserve the energy during numerical integration a time step in the order of a few femtoseconds or even lower is required. For fast modes such as the intermolecular vibration typically one femtosecond is used as time increment. The force acting on each atom or molecule is calculated by integrating the potential model. The reliability of a molecular simulation depends crucially on the potential model as only input property. Potential models vary from simple models such as the hard-sphere fluid to accurate models obtained from quantum mechanics calculations.[22–24] The majority of today's simulations are performed with semiempirical potential models based on interaction sites representing one atom or a small group of atoms (united atom approach). Each site is typically modeled by a Lennard-Jones potential with a r^{-6} attraction and a r^{-12} repulsion term. The sites within a molecule are connected either by rigid bonds or by flexible bonds described by a quadratic spring potential. Other terms regularly used include angle-bending and dihedral terms, as well as cross-terms between them. In case of polar molecules different approaches are possible. One can assign partial charges modeled by the Coulomb potential to the sites, which represent the charge distribution in a molecule. Alternatively, one can assign a point dipole, quadrupole, or higher multipole moments to the molecule. Such point multipole potentials consist of an angle dependent term and a r^{-n} potential where n depends on the order of the multipole moments. Both approaches modeling the charge distribution have been applied successfully in numerous molecular dynamics investigations. Furthermore, depending on the polarizability of the molecules induction can take place in dense liquid phases, which effectively increases the dipole moment. Polarization effects are often accounted for by an increased dipole moment of the molecule in the liquid phase in molecular simulations. Recently developed potential models account for these multibody contributions to the potential energy.[25,26] In larger molecules such as small oligomers or especially in polymers additional contributions affecting the conformation are included in the potential models. These are angle potentials and torsion potentials, which involve three and four sites in chains. The semiempirical potential models are usually developed by using experimental data as well as quantum mechanical calculations as input. The structure of a molecule, its equilibrium angles, and bond lengths can be obtained from experimental data[27] or alternatively from quantum mechanics on a relatively low level with sufficient accuracy.[28] The calculation of the charge distribution is more difficult since different values can be obtained depending on the calculation method.[29] The potential models can be further optimized by comparing the results of molecular simulations to experimental data of, for example, vapor pressures or liquid densities. Another important input is the local order in a liquid, which can be obtained experimentally by diffraction and compared to molecular simulations. The experimentally obtained structure factor is related to the pair correlation function obtained from simulation by Fourier transformation. Less often dynamical properties are included in the development of potential models. Especially non-equilibrium simulations have been performed rarely for the optimization of potential models. Since the potential models are mainly adjusted to equilibrium data the question rises whether they are accurate for describing non-equilibrium processes such as nucleation and particle growth which are topics of interest here.

The potential models sketched so far are developed for molecular fluids such as organic molecules or large biomolecules. For inorganic substances as semiconductors or metals other approaches are necessary. In case of metals the delocalized conduction electrons are responsible for a multibody interaction.

This multibody interaction is related to properties of the metals which cannot be described by pairwise additive potentials like the Lennard-Jones potential. These are, for example, elastic properties like the Cauchy-Relation being the ratio of the elastic constants C_{12} to C_{44}. In Lennard-Jones-systems this ratio is equal to 1 but those of metallic systems deviates up to 30% to smaller values.[30] Other deviations are the ratio of vacancy formation energy to cohesive energy[31] and the experimentally observed contraction of the surface layers along the surface normal[32] for which pair potentials show an expansion. Although these deviations can be worked around by analytical corrections to pair potential functions,[33] they apply only to one specific lattice type and are neither feasible for studies of structural phase transitions nor nanoparticle growth. For metals a semiempirical potential model has been developed which effectively describes the multibody interaction by the delocalized electrons. This model has been introduced about two decades ago inspired by density functional theory. The so-called embedded atom method (EAM)[34–36] has been originally developed for calculations of hydrogen embrittlement in metals and is today widely used in the modeling of properties of metals. It is suitable for simulations on the time scale of the nucleation and particle growth process for given supersaturation and captures the main properties of the metal–metal interaction. EAM is related to the quasi-atom[37] or effective medium[38,39] theories. It is composed of two contributions, a pairwise interaction potential of the atomic cores and a functional of the local electron density. The latter term captures the multibody interactions, which are caused by the valence electrons while the first term describes the pairwise repulsion of the positively charged cores. The local density of the delocalized electrons is calculated from the contributions of the valence electrons of all surrounding atoms in between which a certain atom is embedded. This local electron density at the site of an atom i is simply the superposition of radial symmetric atomic electron density functions. It should be noted that this approach includes the effect of the delocalized electrons on the binding energy effectively. Since the evaporation of atoms from a cluster happens under given conditions only for clusters typically smaller than 10 atoms, as shown later, the cluster size dependent transition from an insulator to a metallic cluster has no effect on the behavior of the investigated systems here. In recent years several further developments have been accomplished[40–43] based on the original EAM. The EAM-potential used here is a recent parameterization developed for modeling bcc bulk iron [40]. The EAM-potential for the study of Pt-cluster formation from the gas phase was taken from the work of Zhou et al.[44]

2.2. TEMPERATURE CONTROL

During the condensation of a vapor the condensation heat is set free and increases the temperature of the system. The origin of this heat is the decrease of the potential energy in the system when vapor atoms, that are in large distance from each other, approach to form a liquid. In the liquid state the atoms or molecules are at distances somewhere around the minimum of the interaction potential and the difference in potential energy to the vapor phase is transformed into kinetic energy. This happens in an adiabatic system, which practically means that the process is so fast that the heat cannot be removed in given time. In molecular dynamics simulations the situation is similar. In the basic molecular dynamics ensemble, the NVE ensemble with constant number of particles N, constant volume V, and constant the total energy E, the temperature of the newly formed clusters increases. As a result the vapor pressure of the new cluster increases and the cluster likely will evaporate again. In order to obtain a stable cluster growing to a larger particle it is necessary to remove energy from the system. In experiments this heat removal happens mainly by collisions of the cluster with carrier gas atoms, which are regarded as inert in the context of the nucleation process. Such inert gas should neither condense at given conditions nor adsorb at the clusters or influence the cluster formation beyond its function as heat bath. In experiments on the nucleation of water, for example, argon is used as carrier gas.[45] Another carrier gas used is helium for cryogenic nucleation experiments of argon or nitrogen.[46,47] In molecular dynamics simulations several different approaches have been proposed to control the temperature of a system. The general idea is to couple the system with a heat bath being a thermostat. The exchange of energy between the nucleating system and the heat bath can be realized directly. The simplest approach is to rescale the atomic or molecular velocities in each simulation step in order to reach a given system temperature. This way of thermalizing a system represents an isokinetic ensemble, which differs from a NVT ensemble by the fact that the temperature is exactly constant and does not fluctuate as required in a NVT ensemble. Examples for direct thermostats, which lead to a NVT ensemble, are the Nosé–Hoover method[48–50] or the Andersen method.[51] In the Nosé–Hoover method the forces are affected instead of the velocities. An additional degree of freedom is introduced into the system and its development is also calculated by numerical solution of the corresponding equation of motion similar to those of the atoms and molecules. This additional degree of freedom is the thermostat having a fictitious mass. As the Nosé–Hoover method affects the forces a fluctuation of the temperature is possible. Within the Andersen method[51] stochastic collisions of the atoms are performed according to the velocity distribution for given temperature. Since these collisions are sampled from a physical velocity distribution temperature fluctuations are included. All these thermostats work more or less well for molecular dynamics simulations of equilibrium states. In case of non-equilibrium states with inhomogeneous particle distributions in space such methods can become disadvantageous. The reason is that the methods treat all atoms equally and affect them in a more or less direct way. In nucleating systems, however, heat is set free in the nucleating clusters but not in the vapor phase. So there are several small hot spots in a large significant colder system. This has minor effects in cases where weakly interacting systems such as Lennard-Jones argon is investigated because the temperature increase due to cluster formation is small. However, in systems with very high interaction energy such as metals a thermostat affecting all atoms homogeneously can lead to a significant cooling down of the vapor phase caused by the very high temperature of the clusters. As a result the vapor phase may even freeze.[52] In order to avoid such effects one can mimic the experimental system by introducing an inert carrier gas.[53–56] This inert gas itself has to be connected to a direct molecular dynamics thermostat discussed above, but since it does not condense under given conditions the problems mentioned above do not appear. The nucleating substance is then thermalized by collisions with the inert gas atoms only. This method allows the clusters to heat up without forcing the vapor phase to cool down. It should be noted that cluster cooling by other means than collisions with carrier gas atoms like thermal radiation or emission of thermal electrons are neglected. The heat removal by the carrier gas atoms may therefore be regarded as an effective approach at very high cluster temperature.

Another method for removing the heat without a carrier gas and which avoids the problems mentioned above uses the NVE ensemble.[57] Within this approach energy is removed from a homogeneous vapor phase in a single step while the simulation of nucleation and growth is performed in a NVE ensemble. This method works if enough energy is removed from the system that the heat generated by nucleation and growth is not able to compensate this energy. An advantage of this method is that the forming clusters are completely unperturbed, not even a carrier gas can affect the growth. It is also interesting for investigating adiabatic nucleation.

2.3. DETERMINATION OF NUCLEATION RATES

As in experiments there are different methods for the estimation of nucleation rates from molecular simulation data. These are at first the Yasuoka–Matsumoto method[19,53] which is based on an analysis of the cluster statistics and secondly the estimation of nucleation rates from the onset of the nucleation in a given chamber geometry. Experimentally in the onset method the so-called critical supersaturation is measured at which in a given volume nucleation takes place. The nucleation rate can be calculated from the chamber geometry while the nucleation onset is measured for different temperatures.[46,47] In molecular dynamics within the onset method one simply picks the simulation time until a cluster forms which remains stable in the system. Since at this time only one cluster is formed in the simulation box one can calculate the nucleation rate as one divided by the onset time and the box volume. Experimentally a more direct method for measuring the nucleation rate is the two-piston chamber.[58,59] Within this method a system is supersaturated by a fast expansion, which initiates nucleation. Shortly after the expansion the system is pressurized again in order to suppress further nucleation. The nuclei formed during the time in the supersaturated state continue to grow and can be detected by light scattering when they are in the micrometer size. The number of the micrometer size droplets corresponds to the number of nuclei assuming that no coalescence takes place which can be estimated from the particle density in the chamber and the degree of polydispersity of the droplets.

In molecular dynamics simulation the cluster statistics can be evaluated by a quasi-stationary method proposed by Yasuoka and Matsumoto.[19,53] Within this method the number of clusters larger than a certain threshold value is plotted against the simulation time as shown schematically in Fig. 3. The second domain of these curves represents a linear increase in the number of clusters larger than the threshold value. If for different threshold values the linear domains are nearly parallel the system is quasi-stationary. The increasing number of clusters can only origin from nucleation if one assumes that there is neither coalescence nor particles breaking up in the system. The nucleation rate can then be calculated from the slope of the linear domain divided by the box volume. While in simulations the quasi-stationary method is applicable only to systems with a significant cluster statistics such as large systems or highly supersaturated systems, the onset method is required in cases with a low supersaturation leading only to a single nucleation event in the given box dimension and simulation time. Consequently the reproducibility of the simulations is better in cases where the quasi-stationary method can be used than in cases where the onset method is required.[57] The onset method needs many simulation runs at the same state conditions to be carried out to obtain reasonable statistics.

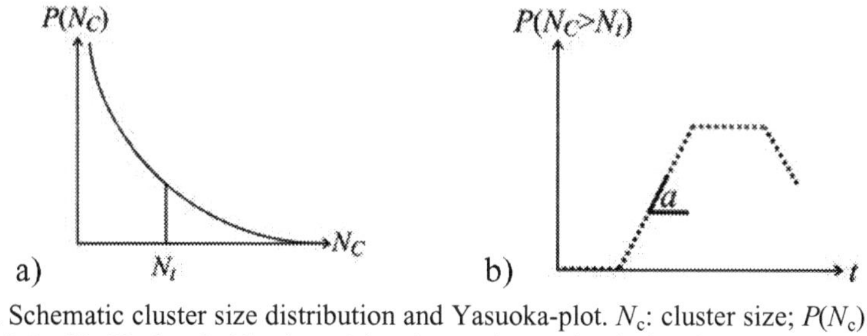

Figure 3. Schematic cluster size distribution and Yasuoka-plot. N_c: cluster size; $P(N_c)$: amount of clusters of site N_c; N_t: threshold value of the cluster size; t: simulation time; a: slope of the second domain.

2.4. NUCLETION THEOREMS

Important relations for the analysis of experimental nucleation rate data are the nucleation theorems. The first nucleation theorem was proposed by

Kashchiev[60,61] and further developed by several authors.[62–66] The nucleation theorem is very useful because it allows the determination of cluster properties from experimental data without using any model. The theorem relates the slope of the nucleation isotherms in the double logarithmic plot to the size of the critical nuclei at given conditions. Within this scheme properties of the critical clusters such as surface tension and density are not required as in the capillary approximation. The derivative of the logarithm of the nucleation rate with respect to the logarithm of the supersaturation at constant temperature yields the excess number N^* of atoms in the critical cluster.[62] In order to obtain the complete number of molecules in the critical cluster one has to add the background gas phase density N_{gas}. This is the number of monomers of the nucleating substance in the volume corresponding to the volume of the critical cluster. However, for the low-density gas phases considered here this contribution can be neglected:

$$\left.\frac{\partial \ln J}{\partial \ln S}\right|_T = N^* + N_{gas} + c. \tag{2}$$

Here J is the nucleation rate representing the number of nuclei formed per time and volume. The left-hand-side of this equation can be calculated from experimental data for the nucleation rates. The addend c on the right hand side is related to the pre-exponential factor in the equation for the nucleation rate (Eq. (3)). Kashchiev[60] obtained the value c=1. This is consistent with a thorough derivation of the nucleation theorem,[64] based on stationary solutions to the cluster size distributions. Within this approach the cluster size changes only by monomer loss and capture. The value c=1 can be traced back to a linear dependence of the condensation rate on the supersaturation. The CNT yields c=2 being inconsistent with the nucleation theorem.[64] In general the parameter c enters the equation for the nucleation rate as exponent of the supersaturation S in the pre-exponential factor:

$$J = A\, S^c \exp\left(-\frac{B}{(\ln S)^2}\right). \tag{3}$$

Here A and B are parameters representing the physical properties of the system and the state variables. The physical meaning of these parameters can be obtained by comparing Eq. (3) with the different nucleation theories. In this work we treat A and B as adjustable parameters for the correlation of the nucleation rates obtained from the simulations.

In addition to the analysis of the critical cluster size using the so called first nucleation theorem we employ the second nucleation theorem. It relates the temperature dependence of the nucleation rate to the excess energy $E_x(N^*)$ of the critical clusters compared to the corresponding bulk phase. Since this excess energy is related to surface energy of the clusters, one can use the second or energetic nucleation theorem for testing the capillary approximation. Such analysis has been suggested by Ford[64] and Vehkamäki and Ford[66] who analyzed experimental data of octane. The size of the critical cluster in these systems is in the order of 30 to 80 molecules for given conditions. The energetic nucleation theorem reads:[66]

$$E_x\left(N^*\right) = k_B T^2 \left(\left(\frac{\partial \ln J}{\partial T} \right)_S - \left(\frac{\partial \ln J_0}{\partial T} \right)_S \right). \tag{4}$$

Here J_0 is the pre-exponential factor being $A \cdot S^c$ for the correlation of Eq. (3) with c=1. The derivative of J_0 with respect to T at constant S is zero. Therefore, only the temperature dependence of the parameter B enters Eq. (4).

2.5. FINITE SIZE EFFECTS

It is necessary to check finite size effects because despite using periodic boundary conditions the investigated systems consist of a limited number of atoms. In case of argon for low supersaturation a bimodal temperature distribution has been found in microcanonical nucleation simulations using a finite simulation box with periodic boundary conditions.[57] With increasing supersaturation this bimodal temperature distribution vanishes. The bimodal distribution corresponds to a dynamic coexistence of a vapor-dominated state and a cluster-in-vapor state. Such behavior of a bimodal temperature distribution has also been reported for the solid–liquid transition in small argon clusters.[67]

Another effect of the finite size of a system is related to the nature of the two states, which are in dynamic equilibrium. Rao et al.[68] have shown that in finite size systems a vapor state can be stabilized which would be non-stable for infinite size at the same conditions. Similar observations have been reported by Binder and Kalos[69] in studies of a lattice model. Consequently non-stable equilibria are possible in finite size systems. In a recent work using the Ising-model Binder[70] has shown that a phase coexistence of finite systems near the coexistence curve of the corresponding infinite system exists. It is argued that a cluster has to have a minimum size in order to exist, which is in contrast to the lever rule which would allow the existence of arbitrary small liquid droplets in the vapor phase if the average molar volume of the system is arbitrary close to the vapor branch of the coexistence curve.

Finally one finds an effect of the system size on the probability of coalescence processes. While virtually no coalescence appears in small systems, the larger the system the more coalescence events are possible. The reason is that due to the small number of atoms in a small system only one major cluster can be formed. The larger the system the more clusters can be formed which can eventually coalesce.

3. Results

3.1. GROWTH CURVES

A first analysis of the simulation data is available from the growth curves. Here different ways of representing the cluster growth are possible. The plot of the largest cluster in the system against the simulation time allows a differentiation between surface growth and coalescence. A cluster is detected by the Stillinger criterion saying that two atoms belong to a cluster if their distance is smaller than a certain value.[71] This Stillinger distance is typically 1.5 times the diameter of the atoms. In Fig. 4a such diagram is shown for the nucleation of iron from a supersaturated vapor.[55] One can recognize the domain of surface growth up to approximately 2 to 3 ns simulation time. Then the growth continues mainly by coalescence or agglomeration caused by collisions of the clusters. This is visible as jumps in the growth curves. Although there is still some surface growth after approximately 4 ns the growth is dominated by cluster collisions. The dependence of the surface growth on the temperature is thermodynamically controlled and behaves systematically: with increasing carrier gas temperature surface growth becomes slower. Deviations to this systematic dependence are related to different developments of the monomer temperature of the nucleating substance and the cluster temperature that can vary due to the development of the system. The collision of two clusters is on the other hand a stochastic process and therefore not directly influenced by the temperature of the carrier gas. An indirect influence is given by the temperature on the nucleation rate, because for higher nucleation rates more primary particles are generated and more collisions are possible. This, however, affects only the number of collisions but not the temperature dependence of the growth curves. In recent simulations on argon nucleation[57] it has been shown that systems consisting of few thousand atoms are sufficient for the determination of the nucleation rates provided a high enough supersaturation.

Another way of representing the growth of particles is the average cluster size distribution shown in Fig. 4b. This plot gives a more systematic development of the cluster size and does not pronounce the cluster collisions too much. One can also see that the averaged cluster size exhibits a more systematic dependence on the temperature than the plot of the largest cluster does. This is because it includes more information than just the largest cluster.

3.2. CLUSTER SIZE DISTRIBUTION

A more detailed insight into the development of the system is available from the cluster size distribution. Before a particle formation process is initiated either only monomers are present or an equilibrium distribution of mainly monomers and a small fraction of dimers and oligomers with decreasing amount is given. The latter situation is given especially for associating substances such as methanol. After the supersaturation is initiated clusters are formed by nucleation, which leads to a cluster size distribution function that decreases monotonously with the cluster size.

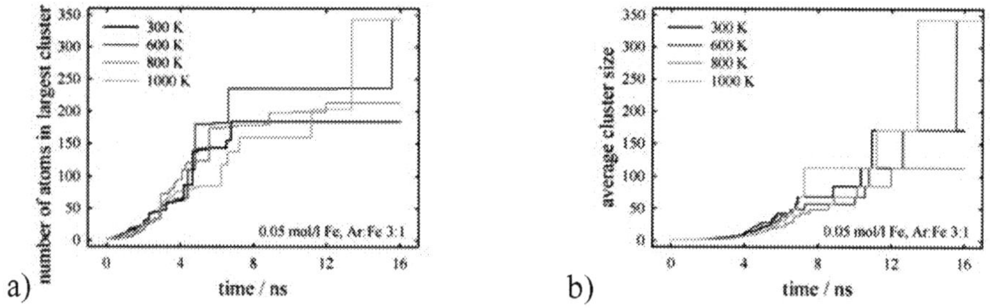

Figure 4. a) Size of the largest cluster in the system. b) Average cluster size.

This is shown in Fig. 5 for methanol[72] and in Fig. 6 for iron and platinum. After some time for both substances one can recognize the formation of a maximum in the cluster size distribution. Since the major species is still the monomer the cluster size distributions also have a minimum, which is somewhere around the dimer, or the trimer in the simulations performed here. After some further time the cluster size distribution resembles a logarithmic-normal distribution.

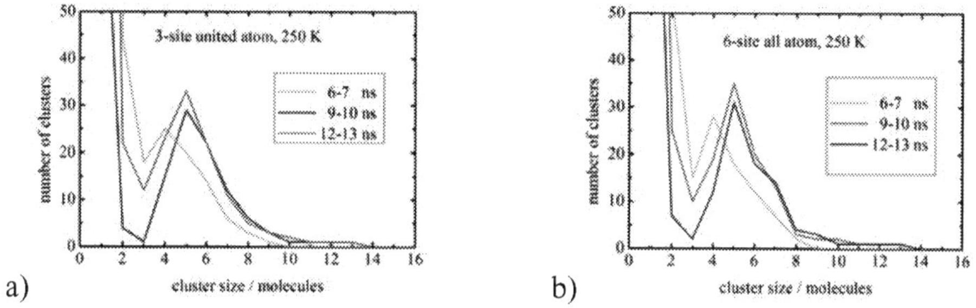

Figure 5. Cluster size distributions for a carrier gas temperatures of 250 K. a) 3-Site van Leeuwen potential model. b) 6-Site all atom AMBER potential model.

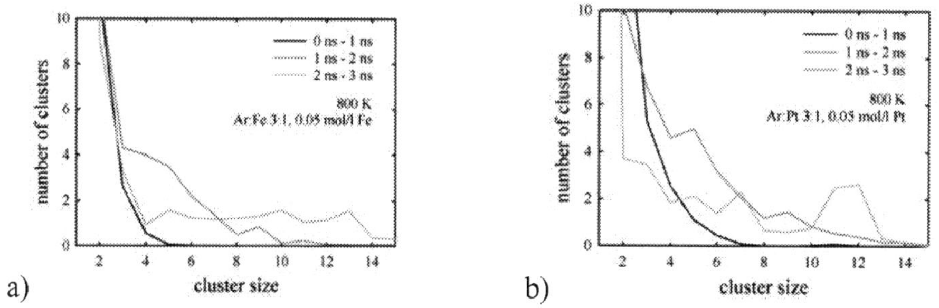

Figure 6. Cluster size distributions for a) iron and b) platinum.

3.3. NUCLEATION RATES

The nucleation rates have been obtained mainly by the quasi-stationary method of Yasuoka and Matsumoto[19] and in case of argon in addition with the onset analysis. The results for argon shown in Fig. 7 also indicate that the two methods are consistent. The nucleation rate at 70 and 80 K are obtained by the onset of the first cluster while the other nucleation rates at lower temperature and therefore higher supersaturation are calculated from the quasi-stationary method. In case of the highly supersaturated iron and platinum systems the quasi-stationary method is used only.

The nucleation rates obtained with the adiabatic method[57] described above for argon are plotted in Fig. 7. Here the simulation data are correlated with a function including a reciprocal temperature dependence. In case of these adiabatic simulations the supersaturation is mainly calculated from the target temperature of the jump into the metastable region and the number of monomers at the beginning of the simulation being the complete number of argon atoms in the simulation box. In some cases of slow nucleation the state conditions at the onset of nucleation are used for the calculation of the supersaturation. The data obtained here with the adiabatic simulation method seems to be in agreement with other theoretical data in the literature.[19,73,74] More difficult is again the comparison with experimental data. For argon some data points at much lower nucleation rate are available from onset experiments.[47] One can check for agreement only by extrapolating the correlation of the simulation data towards the experimental region. This comparison using a reciprocal temperature dependence in the correlation equation[57] exhibits very good agreement. For the size of the critical cluster there are some data points available from the Monte Carlo investigation by Senger et al.[74] plotted in Fig. 7b.

The results for iron are plotted in Fig. 8a together with experimental data from the literature.[75] First one can recognize that the simulation data seem to scatter quite a lot. A detailed analysis shows that the reason for that scattering is the temperature fluctuation of the iron monomers in the system.[56] The temperature of the monomers of the nucleating substance is typically above the carrier gas temperature but it can develop differently for different simulations. Since the monomer temperature is related to the supersaturation of the system it leads to different supersaturations, which appear as scattered data

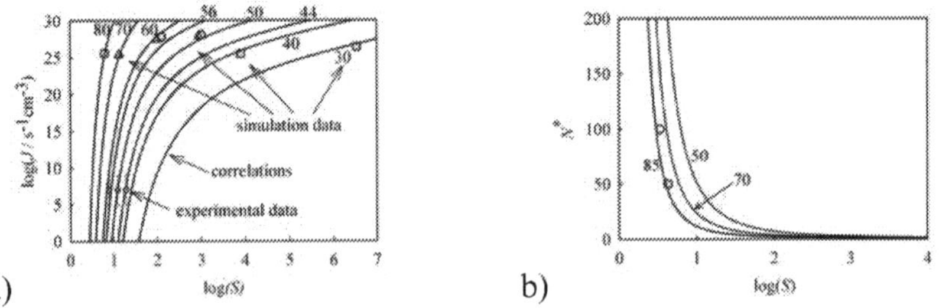

a)

Figure 7. a) Nucleation rates for argon obtained from simulations[57] together with experimental data.[47] b) Critical cluster size obtained from the correlation shown in a). The two data points are theoretical values taken from the literature.[74]

if plotted for the same carrier gas temperature. However, for the analysis of the consistency between the simulation data and the experimental data, here all data are included in the correlation for each carrier gas temperature. The correlation equation used for iron is Eq. (3) in logarithmic form:

$$\ln J = \ln A + c \ln S - \frac{B}{(\ln S)^2}. \tag{5}$$

It can be used for the analysis of the size and surface energy of the critical cluster which can also serve as a consistency check of experimental and simulation data. For this analysis the nucleation theorems are used. If they give reasonable values for the critical cluster size and the surface energy of the critical cluster one can conclude that experimental and simulation data are consistent. In case of iron the critical cluster size of 2 to 3 (Fig. 8b) is in agreement with the analysis of experimental observations.[75] The surface energy obtained from the second nucleation theorem is in the same order of magnitude as one can estimate from the atomic interactions.[56] In case of platinum there are no experimental data for the nucleation rate given. Therefore, we have correlated the nucleation rates not for the carrier gas temperature but for the monomer temperature with a simple correlation function[76]:

$$\log J = a + bT + \log S - \frac{cT + d}{(\log S)^2}. \tag{6}$$

Here a, b, c, and d are fitting parameters. The result is shown in Fig. 9a. One can see that this simple function is able to fit the data well. The derivative in the double logarithmic plot yielding the critical cluster size is plotted in Fig. 9b. Although there are no experimental data to compare to one can conclude by analogy to the data for iron and by comparing for other experimental data for silver[77] that the order of magnitude of the critical cluster size obtained for platinum is reasonable. In case of a platinum monomer temperature of about 1000 K the critical cluster size obtained from the nucleation theorem is approximately 2 which is consistent with the observation of the cluster size fluctuation during the simulation runs (4–5).[76] The same is true for the simulation at a monomer temperature of 800 K which gives critical cluster sizes of 4 from the nucleation theorem compared to 6–7 obtained from the cluster size fluctuation. This also shows that the two independent methods of analyzing the critical cluster sizes yield consistent results.

In Fig. 10a the nucleation rates for two different methanol-models at three different densities are shown and compared to the prediction with the classical nucleation theory and experimental data of Strey et al..[78] The estimated nucleation rates of the simulation are consistent with experimental data being both below the prediction with the classical nucleation theory. The simulation results scatter about one to two orders of magnitude.

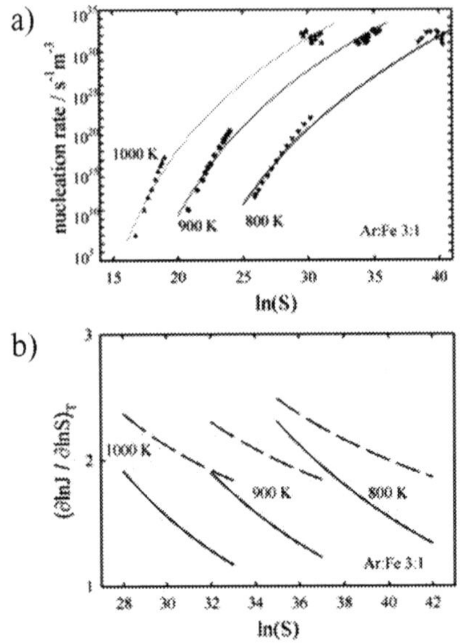

Figure 8. a) Nucleation rates for iron obtained from the simulation together with experimental data.[75] b) Critical cluster size obtained from the correlation shown in a); dashed lines: c = 1, solid lines: c = 0.

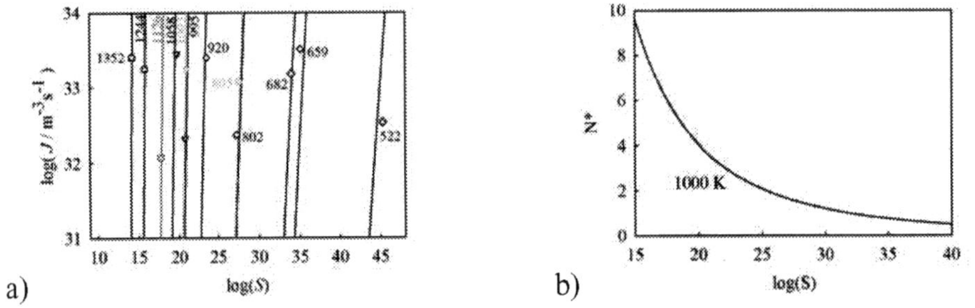

Figure 9. a) Nucleation rates for platinum obtained from the simulation. b) Critical cluster size obtained from the correlation shown in a).

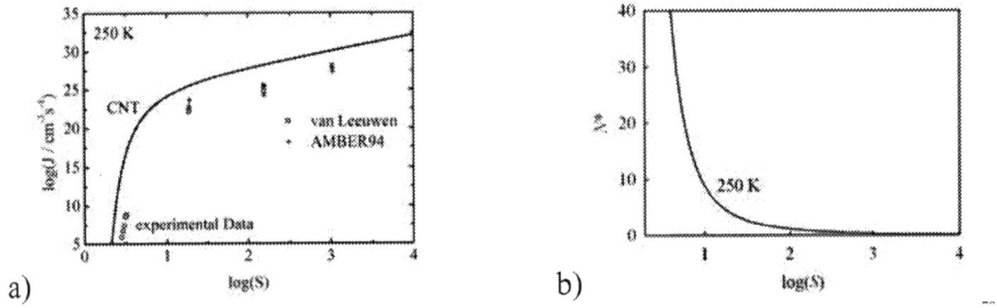

Figure 10. a) Nucleation rates for methanol obtained from the simulation together with experimental data.[78] The solid curve is the prediction by the classical nucleation theory (CNT). b) Critical cluster size obtained from a correlation of the simulation data at 250 K and the experimental data shown in a) using the nucleation theorem.

Within this scattering the results for the two different models overlap. The slope of simulation results with respect to supersaturation is in agreement with the theoretical prediction. The size of the critical clusters is shown in Fig. 10b obtained again from an empirical correlation of the data.[72]

3.4. STRUCTURE DEVELOPMENT

In case of solid-like particles the formation of the atomic order is of special interest. Within molecular simulations different approaches for the investigation of the structure are possible. The usual analysis of the atomic order uses the pair correlation function. It represents the probability to find an atom at distance r from any selected central atom. Figure 11a shows pair correlation functions at the end of the simulation at different temperatures for iron nanoparticles. The narrow peaks represent the structure of a perfectly ordered fcc crystal with the iron fcc lattice constant corrected by 1% in order to get agreement for the first peak. This correction may be related to an inaccuracy of the potential model but also to the fact that a nanoparticle has a different, lattice constant than the corresponding bulk crystal.[79] One can see that the pair correlation function exhibits pronounced maxima which agree fairly well with the fcc structure. This is a hint that the iron particle is in a fcc structure and hence solid-like at the end of the simulation. For further analysis the common neighbor analysis method is employed.[80–82] This is a geometric analysis of the environment of each single atom, which gives one, or a set of three-digit signatures, which can be used to identify the structural environment. Figure 11b shows the development of the amount of atoms in different structures over the course of a particle growth simulation for iron. Here all subtypes for icosahedral, fcc, hcp, and bcc structures are summarized. One can recognize that once a cluster is formed it takes the icosahedral structure. During the cluster growth and the cooling down of the cluster by collision with the carrier gas atoms the amount of atoms in icosahedral structure increases. At about 4 ns in Fig. 11b a few atoms appear in the closed packed structures fcc and hcp which become dominant on cost of the icosahedral structure at about 8 ns simulation time. This is a structural transition from icosahedral to close packed due to cluster growth.

Besides the structure formation due to particle growth, the cooling down by collisions with the carrier gas atoms can lead to solidification. Such process is shown in Fig. 12 for a cluster with 642 atoms taken from a full nucleation and growth simulation. At approximately 12 ns simulation time the almost completely unstructured liquid-like cluster simultaneously forms icosahedral and mainly closed packed fcc and hcp structures on cost of the number of atoms in disordered structure. This is accompanied by a temperature increase of about 200 K. In Fig. 12 also some snapshots of the cluster configurations over the course of the structure formation are shown. One can clearly recognize the unstructured cluster at 12 ns simulation time that develops into an icosahedron at 15.27 ns. The latter snapshot already exhibits the fivefold symmetry with triangular faces.

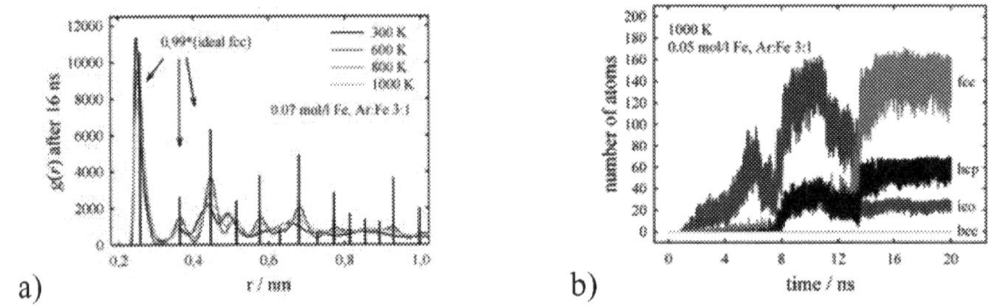

Figure 11. a) Radial distribution functions $g(r)$ for simulation runs at Ar:Fe 3:1 and 0.07 mol/l Fe for different inert gas temperatures. For comparison the $g(r)$ for an ideal iron fcc lattice scaled down by 1% is plotted with the results. b) Time development of the four main structure types as analyzed with the common neighbor analysis; fcc: face centered cubic; hcp: hexagonal closed packed; bcc: body centered cubic; ico: icosahedral.

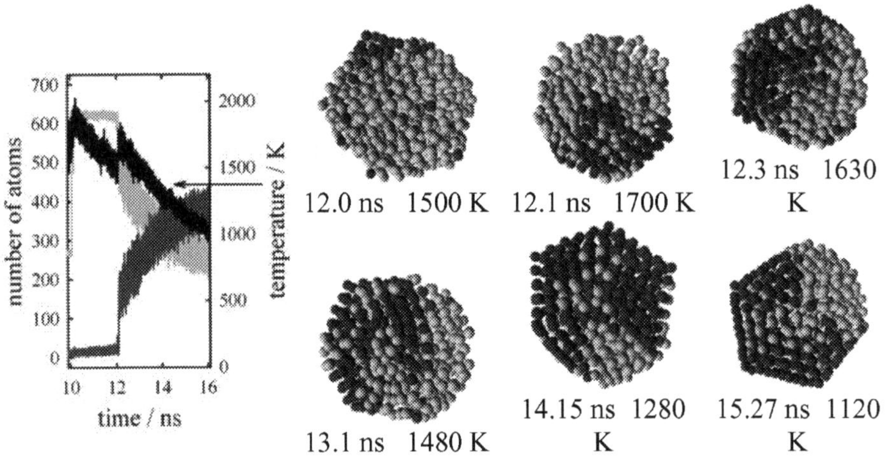

Figure 12. Structure development in a cluster while it cools down. Left: number of atoms in structured (dark gray) and unstructured (light gray) environment and cluster temperature vs. time. The dark gray atoms in the snapshot of the cluster are in structured surrounding.

3.5. COALESCENCE PROCESSES

The fusion of clusters, the coalescence, is a further important growth process. It has been investigated by molecular dynamics simulations so far either in adiabatic simulations[20] or in simulations using a direct thermalization method such as velocity scaling.[83] However, it has been found that using a carrier gas for thermalization, which is much closer to the experimental situation, can give a quite different development of the coalescence process.[84] In order to identify and analyze coalescence processes one can use the surface fraction describing the fraction of atoms or molecules in the cluster surface related to the total amount of atoms or molecules in a cluster. It can be calculated from the atom configurations with the cone algorithm.[85] In Fig. 13 the surface fraction is plotted for the largest cluster as function of its size. The symbols in Fig. 13 represent the lowest possible surface fraction corresponding to the most spherical cluster morphology. The simulation data show deviations from this minimum surface fraction but tend towards it. In this plot one can identify two separate steps, one being agglomeration (jump in cluster size with nearly constant surface fraction) the other one the fusion (decrease in surface fraction at constant cluster).

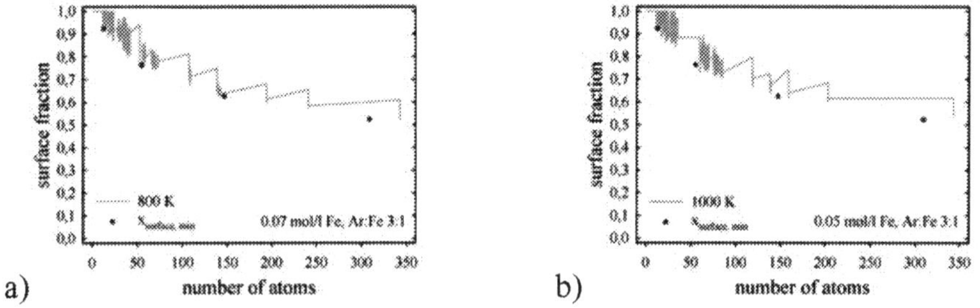

Figure 13. Surface fraction of the largest cluster in the system vs. its size for a) 800 K and 0.07 mol/l Fe and b) 1000 K and 0.05 mol/l Fe.

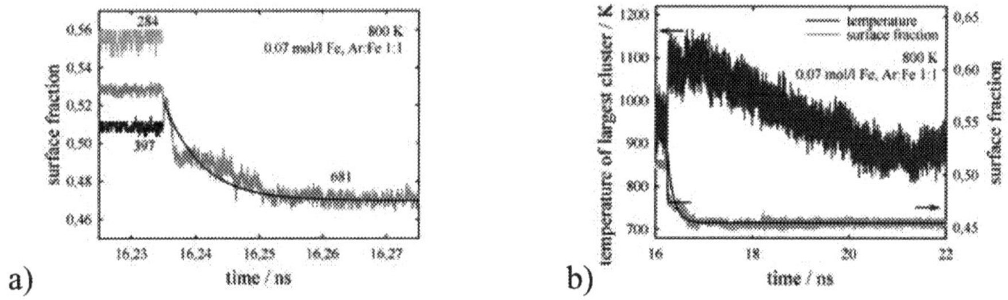

Figure 14. a) Surface fraction change during a coalescence event. Numbers in the diagram indicate the cluster sizes. The solid curve is an exponential decay function fitted to the simulation data. b) Surface fraction and temperature development corresponding to a) but on a different time axis. The solid curves are exponential decay functions fitted to the simulation data.

Such plot can be used to identify coalescene processes. The plot of the surface fraction as function of the simulation time as shown in Fig. 14a in a relative high resolution can be used for the analysis of the coalescence kinetics. This diagram shows the surface fractions of both clusters before the collision and the cluster size weighted average surface fraction of both clusters, which corresponds to the cluster after the collision. Immediately after the collision the surface fraction decreases rapidly roughly along an exponential decay. The fact that the surface fraction does not follow the exponential function perfectly is attributed to the fact that the clusters are already structured before the collision. After some further simulation time the surface fraction reaches a limiting value. The relaxation time of this coalescence process, which can be calculated from the fitting of the exponential function, is in the order of 5 ps, which is very short in this context. As the snapshot of the cluster at 16.4 ns (Fig. 15c) shows, the coalescene process is not yet finished at the point of the limiting surface fraction in Fig. 14a. The cluster is only fused at the surface leading to dumbbell shape. The same plot is shown in Fig. 14b on a longer time scale. Here it appears that the limiting value is only a holding point followed by a further decrease of the surface fraction towards a second limiting value. This second limiting value corresponds to an oval shape as the snapshot in Fig. 15d shows. The relaxation time for the second exponential decay of the surface fraction is already in the order of hundreds of ps simulation time. The limiting surface fraction is still above the minimum surface fraction of a spherical cluster with the same number of atoms. The transition from the oval shape to the spherical shape is expected to last much longer than possible to investigate by molecular dynamics simulation. A reason for the much longer time scale is the low cluster temperature (Fig. 15b) and the fact that the cluster is structured. The edges of structured clusters act as diffusion barriers that slow down any coalescence by surface diffusion.

Figure 16 shows the development of the structure during the coalescence process on two different time scales. Shortly after the collision (Fig. 16a) much of the structure vanishes due to the increasing temperature. On a longer time scale (Fig. 16b) the cluster temperature decreases again which leads to an increasing amount of the different types of structure, especially the closed packed fcc structure. Another interesting coalescence process is reported in

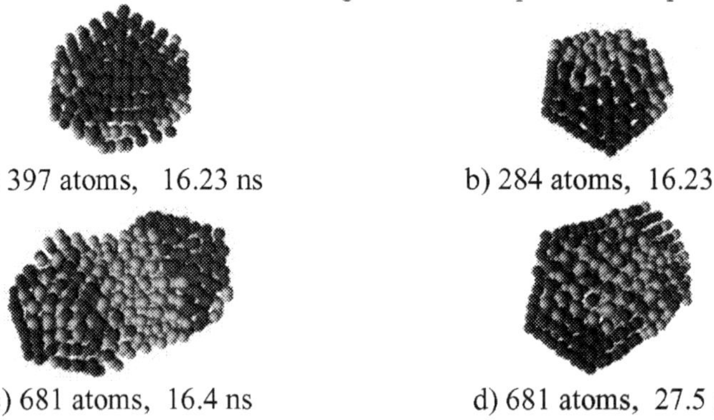

a) 397 atoms, 16.23 ns b) 284 atoms, 16.23 ns

c) 681 atoms, 16.4 ns d) 681 atoms, 27.5 ns

Figure 15. In all subfigures dark gray colored atoms represent atoms in ordered structure and light gray colored atoms those in unordered environment. a) The largest cluster in the system before the collision, which happened at about 16.235 ns. b) The collision partner of the cluster in a) at the same time. c) The coalescing cluster at 16.4 ns. d) The same cluster as in c) but 11.1 ns later.

Fig. 17 in detail.[76] In this case two platinum clusters with 97 and 165 atoms collide, followed by another collision with a tetramer. The first collision leads to a temperature increase of about 300 K from about 400 K to 700 K (Fig. 17a).

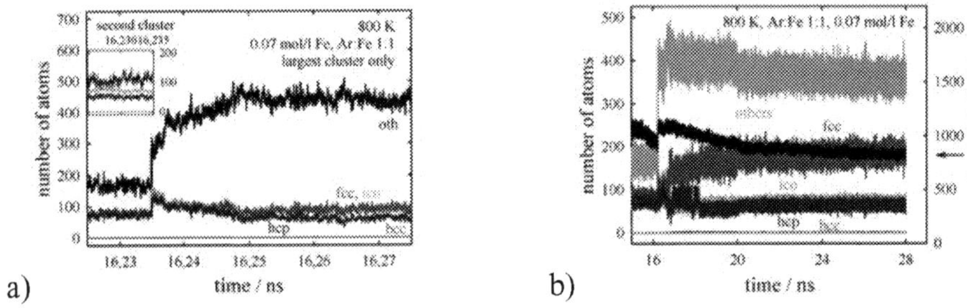

Figure 16. a) Time development of the structure components of the largest cluster and the collision partner before and during the first 400 ps after the collision. b) Time development of the structural components of the largest cluster before and after the collision process on a larger time scale. Also plotted is the time development of the cluster's temperature.

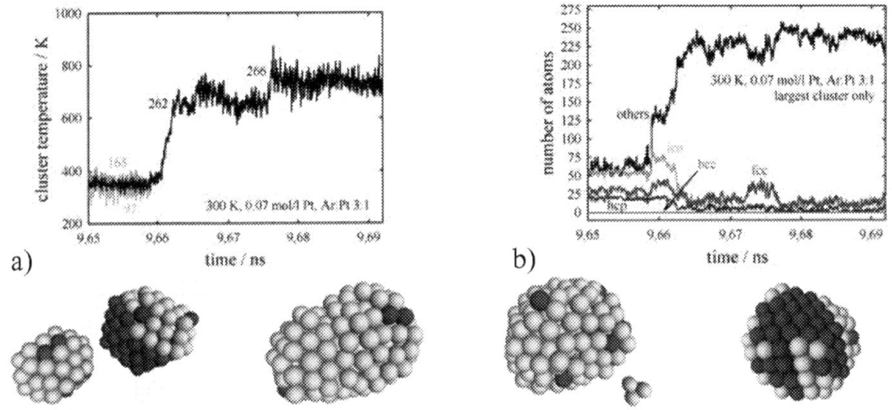

c) 9.6596 ns d) 9.6648 ns e) 9.6755 ns f) 14.9 ns

Figure 17. a) High temporal resolution of the heating up of the largest cluster after the collision of the two clusters of 165 and 97 atoms. The newly formed cluster of 262 atoms is hit by a tetramer about 150 ps later. b) Development of the structural composition for the same time range as in a). c)–f) Snapshots of the clusters taking part in the coalescence process c) before the collision, d) shortly after, e) snapshot prior to the collision with the tetramer, f) restructured cluster several ns later.

The tetramer appears to be quite hot because the collision with the 262-atom cluster leads to a temperature increase of additional 100 K. The second collision with the tetramer does affect the surface fraction only minor. It is however interesting that the tetramer does significantly affect the structure formation although it is very small and no structure formation due to cluster size effects may be expected. At about 200 ps after the initial collision the cluster again forms structure that is mainly fcc. The heat set free during the collision of the clusters is spent on rearrangement of the atomic structure. The mainly icosahedral structure (Fig. 17c) is lost during the coalescence process

(Figs. 17d and e) and turned into fcc structure (Fig. 17f). Hence, one can conclude that coalescence processes have a significant effect on the particle structure and can lead to different structures and geometries than surface growth.

4. Conclusions

In several investigations we have shown the feasibility of the direct simulation of the nucleation and particle growth from a supersaturated vapor. A wide range of substances from metals to noble gases and associating fluids has been treated. Different simulation methods have been further and newly developed. For the majority of the simulations a carrier gas has been employed for removing the condensation heat. In case of argon an adiabatic ensemble has been used. Direct molecular dynamics simulation of nucleation requires a relatively high supersaturation. This leads to nucleation rates being typically above the values of the available experimental data but the consistency of experimental and simulation data over a wide range of supersaturation has been shown. All methods and investigations yield results, which are consistent with other theoretical investigations and with experimental data where available. In addition several properties besides the nucleation rates are available which are difficult to obtain from experiments. In this context molecular dynamics simulation provides a complementary tool for the investigation of nucleation and particle growth. It also supplements the other theoretical methods such as thermodynamic density functional theory or molecular Monte Carlo simulations. Molecular dynamics simulation can contribute especially to the analysis of the detailed kinetics of structure formation mimicking the corresponding experimental situation. It also allows to analyze the properties of the critical clusters. After having proven the feasibility of the method and the evaluation of its predictive capabilities the next step is the application of the method to new systems not investigated experimentally yet. This is especially of interest for systems that are experimentally very difficult or rather impossible to investigate due to experimental limitations or substance dependent difficulties such as toxicity or chemical stability.

Acknowledgments

This work has been supported by the Ministerium für Schule, Forschung und Wissenschaft of NRW state (Germany) within the joint project VerMoS, the German Research Foundation (DFG) within the Special Priority Program SPP-1155 by grants Kr 1598/19-1 and 19-2 as well as by the regular grant Kr 1598/24-1. The support of the NATO for attending the advanced research workshop is acknowledged by one of us (TK).

References

1. Kashchiev, D. (2000) *Nucleation, Basic Theory with Application* (Butterworth-Heinemann, Oxford).
2. Seinfeld, J. H., and Pandis, S. N. (1998) *Atmospheric Chemistry and Physics* (John Wiley & Sons, New York).
3. Debenedetti, P. G. (1996) *Metastable Liquids* (Princeton Univ. Press, Princeton NJ).
4. Beegle, B. L., Modell, M., and Reid, R. C. (1974) Thermodynamic stability criterion for pure substances and mixtures, *AIChE J.* **20**, 1200–1206.
5. Wismer, K. L. (1922) The pressure–volume relation of superheated liquids, *J. Phys. Chem.* **26**, 301–315.
6. Kraska, T. (2004) Stability limits of pure substances: An investigation based on equations of state, *Ind. Eng. Chem. Res.* **43**, 6213–6221.
7. Volmer, M. and Weber, A. (1926) Keimbildung in übersättigten Gebilden, *Z. Phys. Chem. (Leipzig)* **119**, 277–301.
8. Becker, R. and Döring, W. (1935) Kinetische Behandlung der Keimbildung in übersättigten Dämpfen. *Ann. Phys.* **24**, 719–752.
9. Gibbs, J. W. (1948) On the equilibrium of heterogeneous substances, *Trans. Connect. Acad.* **3**, 108–248, 343–524 (1876–1878), reprinted in *The Collected Works of J. Willard Gibbs. Vol. 1: Thermodynamics* (Yale University Press, New Haven), first publ. 1906, pp. 55–353.
10. Reiss, H., Kegel, W. K., and Katz, J. L. (1997) Resolution of the problems of replacement free energy, 1/S, and internal consistency in nucleation theory by consideration of the length scale for mixing entropy, *Phys. Rev. Lett.* **78**, 4506–4509.
11. Reiss, H., Kegel, W. K., and Katz, J. L. (1998) Role of the model dependent translational volume scale in the classical theory of nucleation, *J. Phys. Chem. A* **102**, 8548–8555.
12. Girshick, S. L. and Chiu, C. P. (1990) A new expression for the rate of homogeneous nucleation from an ideal supersaturated vapour, *J. Chem. Phys.* **93**, 1273–1277.
13. Reguera, D. and Reiss, H. (2004) Fusion of the extended modified liquid drop model for nucleation and dynamical nucleation theory, *Phys. Rev. Lett.* **93**, 165701.
14. Zeng, X. C., and Oxtoby, D.W (1991) Gas–liquid nucleation in Lennard-Jones fluids, *J. Chem. Phys.* **94**, 4472–4478.
15. Talanquer, V. and Oxtoby, D. W. (1994) Dynamical density functional theory of gas–liquid nucleation, *J. Chem. Phys.* **100**, 5190–5200.
16. Wolde, P. R. and Frenkel, D. (1998) Computer simulation study of gas-liquid nucleation in a Lennard-Jones system, *J. Chem. Phys.* **109**, 9901–9918.
17. Bin Chen, J., Siepmann, J. I., Oh, K. J., and Klein, M. L. (2002) Simulating vapor-liquid nucleation of *n*-alkanes, *J. Chem. Phys.* **116**, 4317–4329.
18. Rao, M., Berne, N. J., and Kalos, M. H. (1978) Computer simulation of the nucleation and thermodynamics of microclusters, *J. Chem. Phys.* **68**, 1325–1336.

19. Yasuoka, K. and Matsumoto, M. (1998) Molecular dynamics of homogeneous nucleation in the vapour phase. I. Lennard-Jones-fluid, *J. Chem. Phys.* **109**, 8451–8462.
20. Zachariah, M. R. and Carrier, M. J. (1999) Molecular dynamics computation of gas phase nanoparticle sintering: A comparison with phenomenological models, *J. Aerosol Sci.* **30**, 1139.
21. Lewis, L. J., Jensen, P., and Barrat, J.-L. (1997) Melting, freezing, and coalescence of gold nanoclusters, *Phys. Rev. B* **56**, 2248–2257.
22. Brooks, B. R., Bruccoleri, R. E., Olafson, B. D., States, D. J., Swaminathan, S., and Karplus, M. (1983) CHARMM: A program for macromolecular energy, minimization, and dynamics calculations, *J. Comp. Chem.* **4**, 187–217.
23. Cornell, W. D., Cieplak, P., Bayly, C. I., Gould, I. R., Merz, K. M., Ferguson, D. M., Spellmeyer, D. C., Fox, T., Caldwell, J. W., and Kollman, P. A. (1995) A second generation force field for the simulation of proteins, nucleic acids, and organic molecules, *J. Am. Chem. Soc.* **117**, 5179–5197.
24. Sun, H. (1998) COMPASS: An ab initio force-field optimized for condensed-phase applications — overview with details on alkane and benzene compounds, *J. Chem. Phys. B* **102**, 7338–7364.
25. Gao, J., Habibollazadeh, D., and Shao, L. (1995) A polarizable intermolecular potential function for simulation of liquid alcohols, *J. Phys. Chem.* **99**, 16460–16467.
26. Kaminski, G. A., Stern, H. A., Berne, B. J., and Friesner, R. A. (2004) development of an accurate and robust polarizable molecular mechanics force field from ab initio quantum chemistry, *J. Phys. Chem. A* **108**, 621–627.
27. Weiner, S. J., Kollman, P. A., Case, D. A., Singh, U. C., Ghio, C., Alagona, G., Profeta Jr., S., and Weiner, P. (1984) An new force field for molecular mechanical simulation of nucleic acids and proteins, *J. Am. Chem. Soc.* **106**, 765–784.
28. Leach, A. R. (2001) *Molecular Modelling* (Prentice Hall, Harlow, England).
29. Bayly, C. I., Cieplak, P., Cornell, W. D., and Kollman, P. A. (1993) A well-behaved electrostatic potential based method using charge restraints for deriving atomic charges: The RESP model, *J. Phys. Chem.* **97**, 10269–10280.
30. Ercolessi, F., Parrinello, M., and Tosatti, E. (1988) Simulation of gold in the glue model, *Philos. Mag.* **58**, 213–226.
31. Carlsson, A. E. (1990) Beyond pair potentials in elemental transition metals and semiconductors, *Sol. Stat. Phys.* **43**, 1–91.
32. Gupta, R. P. (1981) Lattice relaxations at a metal surface, *Phys. Rev. B* **23**, 6265–6270.
33. Johnson, R. A. (1972) Relationship between two-body interatomic potentials in a lattice model and elastic constants, *Phys. Rev. B* **6**, 2094–2100.
34. Daw, M. S. and Baskes, M. I. (1983) Semiempirical, quantum mechanical calculation of hydrogen embrittlement in metals, *Phys. Rev. Lett.* **50**, 1285–1288.
35. Daw, M. S. and Baskes, M. I. (1984) Embedded atom method: Derivation and application to impurities, surfaces, and other defects in metals, *Phys. Rev. B* **29**, 6443–6453.

36. Foiles, S. M., Daw, M. S., and Baskes, M. I. (1986) Embedded atom-method functions for the fcc metals Cu, Ag, Au, Ni, Pd, Pt, and their alloys, *Phys. Rev. B* **33**, 7983–7991.
37. Stott, M. J. and Zaremba, E. (1980) Quasiatoms: An approach to atoms in nonuniform electronic systems, *Phys. Rev. B* **22**, 1564–1583.
38. Nørskov, J. K. (1982) Covalent effects in the effective-medium theory of chemical binding: Hydrogen heats of solutions in the *3d* metals, *Phys. Rev. B* **26**, 2875–2885.
39. Nørskov, J. K. and Lang, N. D. (1980) Effective-medium theory of chemical binding: Application to chemisoprtion, *Phys. Rev. B* **21**, 2131–2136.
40. Meyer R., and Entel, P. (1998) Martensite–austenite transition and phonon dispersion curves of $Fe_{1-x}Ni_x$ studied by molecular-dynamics simulations, *Phys. Rev. B* **57**, 5140–5247.
41. Wen, M., Xu, X.-J., Fukuyama, S., and Yokogawa, K. (2001) Embedded-atom-Method functions for the body-centered-cubic iron and hydrogen, *J. Mater. Res.* **16**, 3496–3502.
42. Johnson R. A. and Oh, D. J. (1989) Analytic embedded atom method model for bcc metals, *J. Mater. Res.* **4**, 1195–1201.
43. Lee, B.-J., Baskes, M. I., Kim, H., and Cho, Y. K. (2001) Second nearest-neighbor modified embedded atom method potentials for bcc transition metals, *Phys. Rev. B* **64**, 184102.
44. Zhou, X. W., Wadley, H. N. G., Johnson, R. A., Larson, D. J., Tabat, N., Cerezo, A., Petford-Long, A. K., Smith, G. D. W., Clifton, P. H., Martens, R. L., and Kelly, T. F. (2001) Atomic scale structure of sputtered metal multilayers, *Acta Mater.* **49**, 4005–4015.
45. Wölk, J. and Strey, R. (2002) Homogeneous nucleation of H_2O and D_2O in comparison: The isotope effect, *J. Phys. Chem. B* **105**, 11683–11701.
46. Fladerer, A. (2002) *Keimbildung und Tröpfchenwachstum in übersättigtem Argon-Dampf* (PhD-thesis, University Cologne).
47. Iland, C. (2004) *Experimente zur homogenen Keimbildung von Argon und Stickstoff* (PhD-thesis, University Cologne).
48. Nosé, S. (1984) A molecular dynamics method for the simulation in the canonical ensemble, *Mol. Phys.* **52**, 255–268.
49. Hoover, W. G. (1985) Canonical dynamics: Equilibrium phase space distributions, *Phys. Rev. A* **31**, 1695–1697.
50. Hoover, W. G. (1986) Constant pressure equations of motion, *Phys. Rev. A* **34**, 2499–2500.
51 Andersen, H. C. (1980) Molecular dynamics at constant pressure and/or temperature, *J. Chem. Phys.* **72**, 2384–2393.
52. Erhart, P. and Albe, K. (2004) The role of thermostats in modeling vapor phase condensation of silicon nanoparticles, *Appl. Surf. Sci.* **226**, 12.
53. Yasuoka, K. and Matsumoto, M. (1998) Molecular dynamics of homogeneous nucleation in the vapour phase. II. Water, *J. Chem. Phys.* **109**, 8463–8470.
54. Westergren, J., Grönbeck, H., Kim, S.-G., and Tománek, D. (1997) Noble gas temperature control of metal clusters: A molecular dynamics study, *J. Chem. Phys.* **107**, 3071–3079.

55. Lümmen N. and Kraska, T. (2004) Investigation of the formation of iron nano particles from the gas phase by molecular dynamics simulation, *Nanotechnology* **15**, 525–533.
56. Lümmen, N. and Kraska, T. (2005) Homogeneous nucleation of iron from supersaturated vapor investigated by molecular dynamics simulation, *J. Aerosol Sci.* **36**, 1409–1425.
57. Kraska, T. (2006) Molecular dynamics simulation of nucleation of argon from supersaturated vapour phase in the NVE ensemble, *J. Chem. Phys.* **124**, 054507–054520.
58. Wagner, P. E. and Strey, R. (1984) Measurements of homogeneous nucleation rates for *n*-nonane vapor using a two-piston expansion chamber, *J. Chem. Phys.* **80**, 5266–5277.
59. Strey, R., Wagner, P. E., and Viisanen, Y. (1994) Feature article: "The problem of measuring homogeneous nucleation rates and molecular contents of nuclei: Progress in the form of nucleation pulse measurements", *J. Phys. Chem.* **98**, 7748.
60. Kashchiev, D. (1982) On the relation between nucleation work, nucleus size, and nucleation rate, *J. Chem. Phys.* **76**, 5098–5102.
61. Kashchiev, D. (2003) Thermodynamically consistent description of the work to form a nucleus of any size, *J. Chem. Phys.* **118**, 1837–1851.
62. Viisanen, Y., Strey, R., and Reiss, H. (1993) Homogeneous nucleation rates for water, *J. Chem. Phys.* **99**, 4680–4692.
63. Strey, R. Wagner, P. E., and Viisanen, Y. (1994) The problem of measuring homogeneous nucleation rates and the molecular contents of nuclei: Progress in the form of nucleation pulse measurements, *J. Phys. Chem.* **98**, 7748–7758.
64. Ford, I. J. (1997) Nucleation theorems, the statistical mechanics of molecular clusters, and a revision of classical nucleation theory, *Phys. Rev. E* **56**, 5615–5629.
65. Oxtoby, D. W. and Kashchiev, D. (1994) A general relation between the nucleation work and the size of the nucleus in multicomponent nucleation, *J. Chem. Phys.* **100**, 7665–7671.
66. Vehkamäki, H. and Ford, I. J. (2001) Excess energies of *i*- and *n*-octane molecular clusters, *J. Chem. Phys.* **114**, 5509–5513.
67. Jellinek, J., Beck, T. L., and Berry, R. S. (1986) Solid–liquid phase changes in simulated isoenergetic Ar_{13}, *J. Chem. Phys.* **84**, 2783–2794.
68. Rao, M., Berne, N. J., and Kalos, M. H. (1978) Computer simulation of the nucleation and thermodynamics of microclusters, *J. Chem. Phys.* **68**, 1325–1336.
69. Binder, K. and Kalos, M. H. (1980) "Critical clusters" in a supersaturated vapor: theory and Monte Carlo simulation, *J. Stat. Phys.* **22**, 363–396.
70. Binder, K. (2003) Theory of the evaporation/condensation transition of equilibrium droplets in finite volumes, *Physica A* **319**, 99–114.
71. Stillinger, F. H. (1963) Rigorous basis of the Frenkel-band theory of association equilibrium, *J. Chem. Phys.* **38**, 1486–1494.
72. Fischer, B. and Kraska, T. (2001) Investigation of the nucleation of methanol from supersaturated vapour by molecular dynamics simulations, *Mol. Phys.* submitted.
73. Wonczak, S., *Molekulardynamische Simulationen von Argon-Clustern* (PhD-thesis, University Cologne)

74. Senger, B., Schaaf, P., Corti, D. S., Bowles, R., Pointu, D., Voegel, J.-C., and Reiss, H. (1999) A molecular theory of the homogeneous nucleation rate. II. Application to argon vapor, *J. Chem. Phys.* **110**, 6438–6450.
75. Giesen, A., Kowalik, A., and Roth, P. (2004) Iron atom condensation interpreted by a kinetic model and a nucleation model approach, *Phase Transitions* **77**, 115–129.
76. Lümmen, N. and Kraska, T. (2005) Molecular dynamics investigation of homogeneous nucleation and cluster growth of platinum clusters from supersaturated vapour, *Nanotechnology* **16**, 2870–2877.
77. Nuth, J. A., Donnelly, K. A., Donn, B., and Lilleleht, L. U. (1986) Experimental studies of the vapor phase nucleation of refractory compounds. III. The condensation of silver, *J. Chem. Phys.* **85**, 1116–1121.
78. Strey, R., Wagner, P. E., and Schmeling, T. (1986) Homogeneous nucleation rates for *n*-alcohol vapors measured in a two-piston expansion chamber, *J. Chem. Phys.* **84**, 2325–2335.
79. Yokozeki, A. and Stein, G. D. (1978) A metal cluster generator for gas-phase electron diffraction and its application to bismuth, lead, and indium: Variation in microcrystal structure with size, *J. Appl. Phys.* **49**, 2224–2232.
80. Honeycutt, J. D. and Andersen, A. C. (1987) Molecular dynamics study of melting and freezing of small Lennard-Jones clusters, *J. Phys. Chem.* **91**, 4950–4963.
81. Clarke, A. S. and Jónsson, H. (1993) Structural changes accompanying densification of hard-sphere packings, *Phys. Rev. E* **47**, 3975–3984.
82. Cleveland, C. L., Luedtke, W. D., and Landman, U. (1999) Melting of gold clusters, *Phys. Rev. B* **60**, 5065–5077.
83. Hendy, S., Brown, S. A., and Hyslop, M. (2003) Coalescence of nanoscale metal clusters: Molecular-dynamics study, *Phys. Rev. B* **68**, 241403.
84. Lümmen, N. and Kraska, T. (2005) Molecular dynamics investigations of the coalescence of iron clusters embedded in an inert-gas heat bath, *Phys. Rev. B* **71**, 205403.
85. Wang, Y., Teitel, S., and Dellago, C. (2005) Melting of icosahedral gold nanoclusters from molecular dynamics simulations, *J. Chem. Phys.* **122**, 214722.

HOW TO GENERATE AND MEASURE NEGATIVE PRESSURE IN LIQUIDS?

ATTILA R. IMRE
KFKI Atomic Energy Research Institute, 1525 Budapest, POB. 49, Hungary; E-mail: imre@sunserv.kfki.hu

Abstract. It is not very widely known that liquids can have very big tensile strength, therefore they can endure deep negative pressures. In this review paper we are describing several experimental methods to generate negative pressure in liquids in the hope that we can convince a few people to make experimental (or at least theoretical) studies in this under-represented field of the physics.

Keywords: Negative pressures, liquids, critical solutions, blends.

1. Introduction

While for a gas, the smallest possible pressure is $p=0$ (absolute vacuum), liquids and solids can endure negative pressure[1–5]– it is a quite known fact for solids but not-so-known for liquids.[1] The reason of this ability is the tensile strength of these materials; for a lot of liquids – as well as for solids – states with –100 MPa is not impossible ones. One should know that for condensed matters, the scalar pressure often has to be replaced by a stress (or pressure) tensor.[2] Scalar pressure can be used when all off-diagonal elements (shear) are zero and all diagonal elements are equal. By pulling a solid triaxially – due to the anisotropicity of the solids – it is difficult to avoid non-zero shear elements, therefore for stretched solids the "negative pressure" is not a general descriptor. For liquids, the triaxial pull is the first problem; it is not very easy to realize it. Because liquids have more or less isotropic structure, therefore shear can be avoided in several cases. In this paper we would like to describe several well-established methods to generate and estimate negative pressure in liquids.

2. Theoretical background

For a lot of researcher even the idea that the pressure can have negative value is strange, therefore here we have a short chapter to demonstrate that there is no theoretical argument against it. We are using a simple van der Waals fluid for this demonstration.

The van der Waals equation (actually introduced by James Thomson in 1871, a couple of years before van der Waals[6]) is one of the simplest equation of state state to describe coexisting liquid and vapor phases. The form of the equation is

S. J. Rzoska and V. A. Mazur (eds.), Soft Matter under Exogenic Impacts, 379–388.

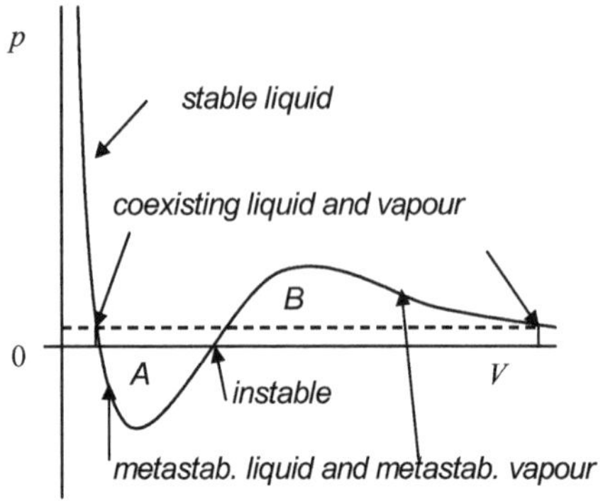

Figure 1. Van der Waals isotherm for a sub-critical temperature. See explanation in the text.

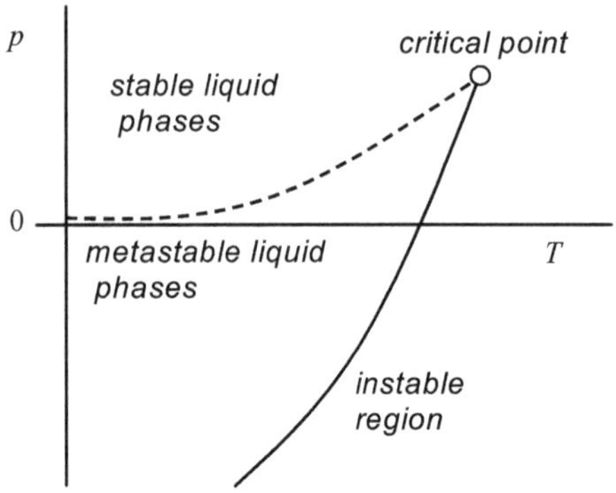

Figure 2. Schematic phase diagram and liquid–vapor spinodal for a van der Waals fluid. Dashed line represents the vapor pressure curve, solid line represents the L–V spinodal.

$$\left(p+\frac{a}{V_{\mathrm{m}}^{2}}\right)\left(V_{\mathrm{m}}-b\right)=RT\cdot$$

Here p is the pressure, T is the temperature, V_{m} is the molar volume, a and b are substance-dependent constants. For each temperature one might plot an isotherm on the pV-space (Fig. 1); some of these isotherms (the low temperature ones) can cross the $p=0$ line. Between the two local extrema (called spinodals), the compressibility is negative, which is a violation of the stability criterion;[4] therefore the part between these two points do not represent any existing state. The volume (density) of the coexisting liquid and vapor phases can be calculated by the Maxwell construction; the isotherm has to be cut in a line parallel with the V-axis in a way that area A and area B has to be equal. The left and right crossing points represent the liquid and vapor phases, respectively. With the Maxwell construction we got rid of the "non-existing" part, but we also lost two other parts of the curve; the ones between the coexistence points and the extrema. These parts represent the so-called metastable states; the left one is the overheated or stretched liquid state, the right one is the supersaturated vapor state. In this paper we are interested in the left one.

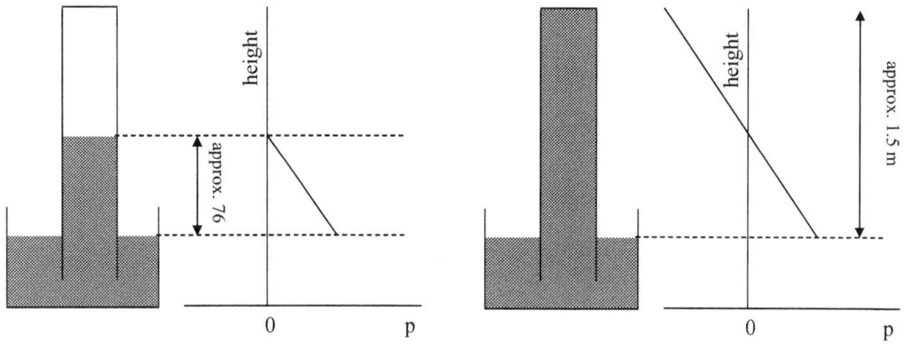

Figure 3. Comparison of the experiments of Toricelli (left) and Huygens (right). The pressure–height dependence is demonstrated on the diagrams.

On Fig. 2 one can see the schematic phase diagram of a van der Waals fluid together with the so-called liquid–vapor (L–V) spinodal in the pT-space. In pT-space the L–V spinodal (the local minimum on Fig. 1) form a line while the two coexisting phases are also represented by a common line (the vapor pressure line, also called saturation line). Above the vapor pressure the liquids are stable; between the vapor pressure curve and the spinodal they are metastable with respect to the L–V phase transition (i.e., they might boil or cavitate at any moment) while below the spinodal no liquid phase can exist.

We can conclude that liquids can endure some tension (it can be quite big for certain liquids) but – just like solids – they will "break" (cavitate, boil) when the tension is too big. Usually this break happens well before the spinodal by homogeneous or heterogeneous nucleation. The latter is caused by contamination; by carefully cleaning and degassing the liquid one can approach the theoretical tensile strength.

In Section 3, a historical overview will be given for the earliest results dealing with stretched liquids.

3. Historical overview

Stretched liquids can be found in the Nature; liquid closed into tiny inclusions in minerals can be extremely stretched (–80 MPa)[7] while the sap in the xylem of the trees can show moderate tension (–4 atm)[8], so one might say that the history of negative pressure states is as long as the history of Nature. The history of negative pressure states in the laboratory is a little bit shorter. The first experiment dealing with stretched liquid was done by Huygens in

1663.[9] This is a hardly-known experiment. Huygens tried to reproduce the famous Toricelli experiment but sometimes instead of getting a 76 cm tall mercury column, he observed that the mercury remained hanging in the whole length of his glass tube (approx. 1.5 m) (see Fig. 3). His experiment was hardly reproducible (even a small disturbance was enough to break the hanging

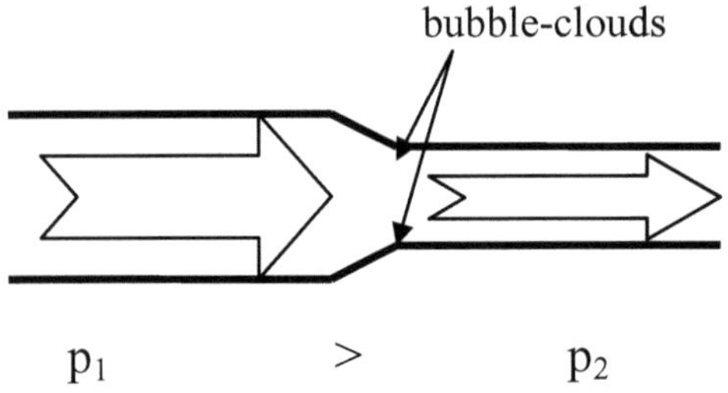

Figure 4. Pressure-drop in a channel due to the Bernoulli law marking the place of cavitation.

column) and – at that time, due to the lack of the knowledge about cohesion, adhesion and metastability – totally unexplainable; it was reproduced and explained only more than one and a half decade later by Donny.[3] With our current knowledge, it is obvious that when he carefully turned the glass tube (totally filled with bubble-free mercury) sometimes the mercury column stuck to the glass wall (by adhesion) and the whole column was stretched by its own weight. From the right diagram on Fig. 3 it is clear that the pressure in the upper half of the column was under $p=0$; at the top, the pressure was –1 atm. Although this method is very good for demonstration (it is straightforward, the pressure value can be easily estimated and – by knocking on the wall – the nature of the metastability can be easily shown), one cannot use it for experiments. To generate –1 MPa in an aqueous solution one need a 100 m long (preferably thermostated) glass tube. To study water close to the room temperature spinodal (for example, around –100 MPa) a 10 km long, vertical tube would be necessary; the laboratory would be taller than the Mount Everest.

The other historical lab example cannot be connected with a date; starting from the XIXth century, the phenomenon has been very often reported.[3] A liquid flowing through a narrowing would exhibit a pressure drop (Fig. 4), as it is stated in the Bernoulli law which is (in vertical tube, laminar flow, without gravitation)

$$p_1 - p_2 = \frac{1}{2}(\rho v_2^2 - \rho v_1^2),$$

where $p_1 - p_2$ is the pressure drop, ρ is the density while v_1 and v_2 are the flow speeds in the two part of the tube. Choosing the flow speed and the channel size properly, the pressure can be easily dropped below the vapor pressure – or even below zero pressure – which would initiate bubbling (cavitation). Due to the remarkability of the bubble clouds, this method is also a good demonstration but to generate high tension the speed should be very high and the flow would be turbulent very soon. There are two additional problems with this method:

– in real liquids the off-diagonal (shear) elements of the pressure tensor could be non-zero, therefore the scalar pressure cannot be used to describe these states properly.

– although the location of the negative pressure zone is fixed, the liquid will pass it very soon; i.e., a chosen part of the liquid will be under negative pressure only for short time.

We do hope that these two examples are clear enough to convince the sceptics about the existence and experimental availability of negative pressure states. In the next chapter we will describe methods to generate deep, long-lasting static tensions.

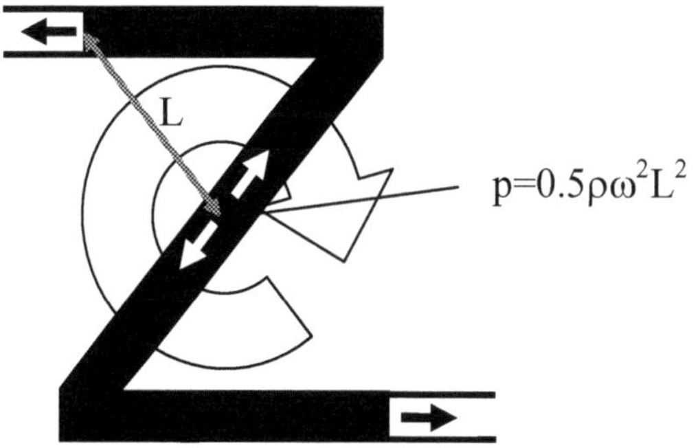

Figure 5. Centrifugal method to generate negative pressure in an open-ended, liquid-filled Z-shaped rotating tube. The direction of the rotation is clockwise in this case.

4. Experimental methods

In this chapter, two main groups of methods will be described:

- centrifugal methods
- Berthelot methods

The centrifugal method is using the centrifugal force to generate negative pressure in a rotating, usually Z-shaped tube (Fig. 5).

The pressure changes along the tube; the deepest tension can be reached at the middle. This method was used first by Reynolds[3] using a J-shaped tube. On Fig. 5 one can see a partially liquid-filled, Z-shaped tube, open at both ends. Rotating the tube in the plane of the figure around the middle, the centrifugal force tries to tear the two half of the liquid column at the middle. The pressure can be calculated as $\Delta p = \frac{1}{2}\rho\omega^2 L^2$, where Δp is the difference between the internal and external pressure.

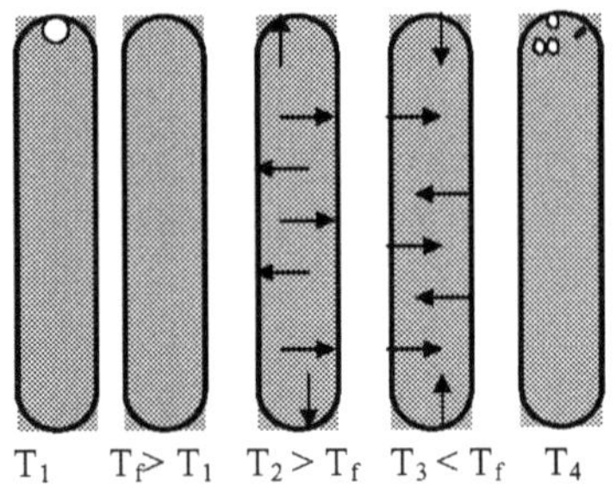

Figure 6. Straight-glass Berthelot method. For detailed explanation, see the text.

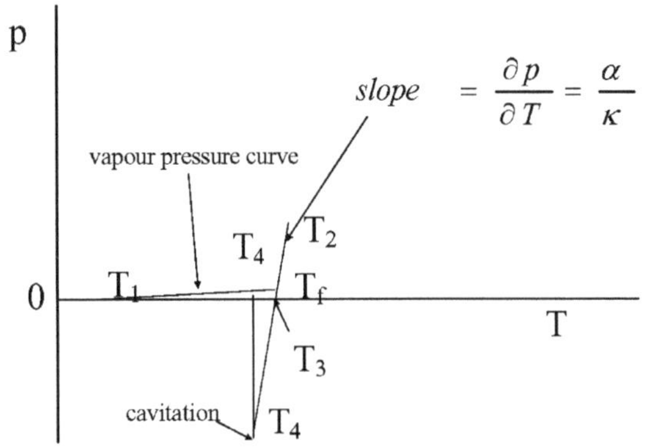

Figure 7. The change of pressure in a Berthelot tube by changing the temperature.

For a water filled tube with L=5 cm, several 1000 rpm would be required to reach a few MPa negative pressure. This method was mainly used by Briggs; he reached a few tens of MPa's in various liquids (for example, –28 MPa for water).[10]

Under the names of "Berthelot methods" four different, but closely related methods will be discussed: (a) straight glass tube method, (b) metal tube method, (c) spiral glass method, and (d) inclusion method.

The first one will be described in details, while for the other ones only the differences will be mentioned. It is true for all methods that relatively big, long-lasting (minutes-to-days) tension can be created by them.The straight glass tube Berthelot method[3] would require a thick-wall, preferably bottle-shaped glass tube with closed bottom (for example, we used Pyrex tubes with 3 mm thick wall, 3 mm inner diameter in the main body and 1 mm inner diameter in the neck, 10 cm long main body, 5 cm long neck).[11,12] The container has to be filled

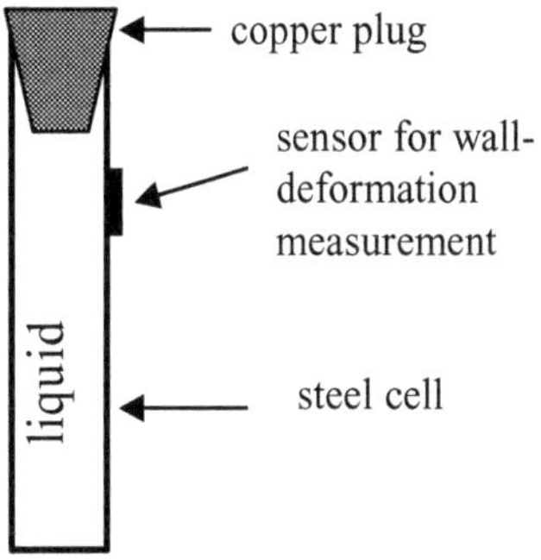

Figure 8. Schematic draw for a steel Berthelot tube closed with a copper plug and with a mounted deformation sensor to measure the difference of the internal and external pressure through wall deformation.

by clean, degassed liquid and flame sealed under vacuum (the liquid should be cooled or frozen during the sealing) very close to the original liquid level. At the end one will have a sealed glass container with a lot of liquid, containing only a small vapor phase (Fig. 6, first tube; Fig. 7, T_1). Heating the tube, the liquid will soon fill the whole place, being the heat expansion coefficient of a normal liquid much bigger than for a normal solid. In the presence of the bubble, the pressure of the liquid will be equal to the vapor pressure. The temperature where the bubble completely disappears is called filling temperature (Fig. 6, second tube; Fig. 7, T_{fill}). By further heating (Fig. 6, third tube; Fig. 7, T_2), the pressure can increase very rapidly (for water around room temperature the rate is around 1 MPa/K) and the container would break soon. Cooling the system down, the pressure would decrease at the same rate. After a while the liquid would try to shrink, but the adhesion will keep it at the walls, therefore the liquid will be isotropically stretched, i.e., the pressure will be negative (Fig. 6, fourth tube; Fig. 7, T_3). Obviously deep cooling could induce very deep tension which can break the liquid; during the break one or more bubble would be formed and the pressure would jump back to the (positive) vapor pressure (Fig. 6, fifth tube; Fig. 7, T_4). Stopping the cooling before cavitation one can obtain and keep a well-defined p–T state with $p<0$ for several minutes or even several hours (try to avoid mechanical disturbances, they can cause sudden boiling); almost any kind of optical measurement can be done during that time. T_{fill} and the location of the T_2–T_4 line depend on T_1 and on the initial liquid and gas volumes. The ratio of this volume can be tuned during the sealing but the gas flame is not very good for fine tuning. This method has at least two big advantages; it can be done quite easily in every laboratory (at least when one need to reach only a few bar) and – due to the transparency of the glass – one can see the sample and immediately notice cavitation. One big disadvantage is that once the tubes are sealed they will always move along the fixed T_2–T_4 line (Fig. 7); when we would like to study something located at a different p–T pair, we need to make new tube. When the studied phenomenon is located close to the cavitation limit we should try to reach it by doing the T_1–T_4 cycle again and again in the hope that we can push the cavitation limit a little bit deeper; this approach works, but only with limited success. The other disadvantage that the pressure cannot be measured; one has to estimate it by using the compressibility and thermal pressure coefficients of the glass and the liquid; for this estimation one should use extrapolated values obtained at positive pressures. Because the compressibility and thermal expansion coefficient change very dramatically close to the liquid–vapor spinodal therefore this assumption can be used only at moderate negative pressures. To avoid the first problem and to enable us to measure the negative pressure directly, one can use a metal tube (Fig. 8), instead of a glass. In this way the wall thickness can be much smaller and the wall deformation can be measured. The wall deformation depends on the difference of the external and internal pressure; the cell can be calibrated with positive pressure in- and outside and the during the negative pressure session from the positive external temperature and from the wall deformation one can estimate the inner pressure. Obviously metal tubes cannot be sealed by flame; a soft (usually copper) plug has to be squeezed into it for sealing. The most detailed study of metal Berthelot tubes was conducted by Ohde.[5,13] With metal tube we can measure pressure directly but we cannot see the sample and cannot make any optical measurement. Inserting a window would create a weak point; the liquid would break easily at the vicinity glass/metal connection. Unlike the glass tubes, the metal Berthelot tubes can be used more than once and even the filling temperature can be changed a little bit by squeezing the plug into the cell a little bit deeper after a few measurements. The spiral-shaped glass Berthelot tube[14] can solve both problems; the pressure can be measured directly and the sample can be studied optically; unfortunately the productions of these cells are very difficult. Curved metal tubes have been used for pressure measurements for quite a long time; these are the so-called Bourdon tubes. While a metal Bourdon tube need to have only one loop, a glass Berthelot–Bourdon tube should have at least seven or nine loops. The tube has to be sealed just like the straight ones; the internal pressure – after initial calibration – can be measured by measuring the tiny movement (opening in case of pressure increase and closing in case of pressure decrease) of the two straight legs. Although theoretically this is the best method, due to the experimental difficulties it is hardly used.

The last method is the inclusion method. The size of a Berthelot tube is not necessarily macroscopic. As it has been mentioned before, tiny inclusions in a mineral can be filled by liquid; by heating and/or cooling the whole mineral the inclusion will act like a miniature Berthelot tube. One can use a natural inclusion to study the trapped fluid[7] but it is also possible to produce synthetic inclusions to study different liquids.[15,16] The biggest advantage of this method is that only a very small amount of liquid has to be used and that amount can be cleaned very well; in that way the heterogeneous nucleation can be more or less

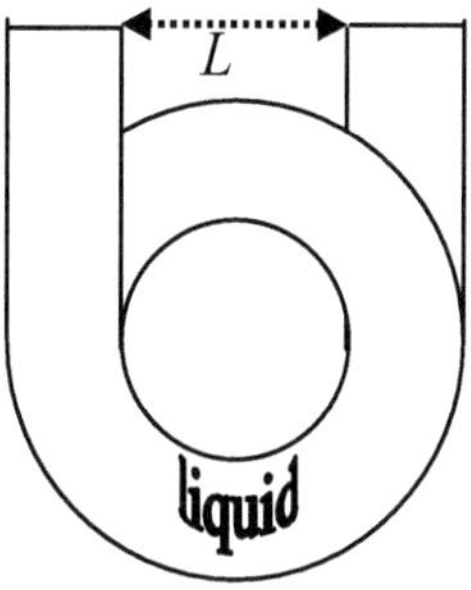

Figure 9. Spiral-shaped Berthelot–Bourdon tube. The pressure can be monitored by measuring the distance (L) between the two legs or by fixing one of them by measuring the movement of the unfixed one.

suppressed. The deepest reported negative pressure in liquid was obtained in an inclusion cell; –180 MPa was reached in water (see the review of C.A. Angell in Ref. 5). Probably all problems needs different approach; one cannot make a "general-purpose" negative pressure cell. But we believe that one of the five methods described here can be used for a lot of study connected to the structure and behavior of liquids. We would like to encourage researchers to extend their pressure studies below $p=0$ – although they might have experimental difficulties – and also encourage theoreticians to extend their calculations into that region; for them, there would not be any difficulties to do that.

5. Conclusions

Liquids can endure negative pressure. The deepest reported negative pressure was –180 MPa in water. Due to the metastability of states below $p=0$ it is very difficult to reach such a deep value but moderate tensions (a few MPa's) can be reached quite easily. In this paper five different experimental methods have been described in detail to encourage experimentalist to extend their pressure studies below $p=0$ and to convince theoreticians that it would be useful to extend their calculations/simulations into this regions.

Acknowledgments

The author thanks Prof. W.A. Van Hook who supervised his initial steps in this field. This work was partially supported by the Hungarian Science Foundation (OTKA) under contract number T 043042. The author was also supported by a Bolyai Research Fellowship of the Hungarian Academy of Science.

References

1. Imre, A.R., Drozd-Rzoska, A., Kraska, T., Martinás, K., Rebelo, L.P.N., Rzoska, S.J., Visak, Z.P., and Yelash, L.V. (2004) Phase equilibrium in complex liquids under negative pressure, in Nonlinear Dielectric Phenomena in Complex Liquids (Eds.: Rzoska, S.J. and Zhelezny, V.P.), NATO Science Series, Kluwer, Dordrecht, 117–125.
2. Imre, A., Martinás, K., and Rebelo, L.P.N. (1998) Thermodynamics of negative pressures in liquids, *J. Non-Equilib. Thermodyn.*, **23**, 351–375.
3. Trevena, D.H. (1987) *Cavitation and Tension in Liquids*, Adam Hilger, Bristol. *The reference recalls historical facts related to negative pressure studies.*
4. Debenedetti, P.G. (1996) *Metastable Liquids: Concepts and Principles*, Princeton University Press, Princeton, N.J.
5. Liquids Under Negative Pressure (Eds.: Imre, A.R., Maris, H.J., and Williams, P.R.), NATO Science Series, Kluwer, Dordrecht, 2002.
6. Thomson, J. (1871) Speculations on the continuity of the fluid state of matter, and on relations between the gaseous, the liquid, and the solid states, *Report of the Meeting of the British Association for the Advancement of Science*, **41**, 30–33.
7. Roedder, E. (1967) Metastable superheated ice in liquid–water inclusions under high negative pressure, *Science*, **155**, 1413–1416.
8. Steudle, E. (1995) Trees under tension, *Nature*, **378**, 663–664.
9. Kell, G.S. (1983) Early observations of negative pressures in liquids, *Am. J. Phys.*, **51**, 1038–1041.
10. Briggs, L.J. (1950) Limiting negative pressure of water, *J. Appl. Phys.*, **21**, 721–722
11. Imre, A. and Van Hook, W.A. (1994) Polymer-solvent demixing under tension. Isotope and pressure effects on liquid–liquid transitions. VII. Propionitrile–polystyrene solutions at negative pressure, *J. Polym. Sci. B*, **32**, 2283–2287.
12. Imre, A. and Van Hook, W.A. (1997) Continuity of solvent quality in polymer solutions. Poor-solvent to Θ-solvent continuity in some polystyrene solutions, *J. Polym. Sci. B.*, **35**, 1251–1259.
13. Ohde, Y., Komori, K., Nakamura, T., Tanzawa, Y., Nishino, Y., and Hiro, K.J. (2001) Effects of gas transports in metals on negative pressures in water in Mo/Cu Berthelot tubes, *J. Phys. D: Appl. Phys.*, **34**, 1717–1726.
14. Henderson, S.J. and Speedy, R.J. (1980) A Berthelot–Bourdon tube method for studying water under tension, *J. Phys. E.*, **13**, 778–782.
15. Zheng, Q., Durben, D.J., Wolf, G.H., and Angell, C.A. (1991) Liquids at large negative pressures: water at the homogeneous nucleation limit, *Science*, **254**, 829–832
16. Alvarenga, A.D., Grimsditch, M., and Bodnar, R.J. (1993) Elastic properties of water under negative pressures, *J. Chem. Phys.*, **98**, 8392–8396.

INDIRECT METHODS TO STUDY LIQUID–LIQUID MISCIBILITY IN BINARY LIQUIDS UNDER NEGATIVE PRESSURE

ATTILA R. IMRE*
KFKI Atomic Energy Research Institute, Materials Department, 1525 Budapest, POB. 49, Hungary

ALEKSANDRA DROZD-RZOSKA, SYLWESTER J. RZOSKA
Institute of Physics, Silesian University, ul. Universytecha 4, 40-007, Katowice, Poland

THOMAS KRASKA
Institut für Physikalische Chemie, Universität zu Köln, Luxemburger Str. 116, D-50939 Köln, Germany

Abstract. Liquids below the vapor pressure are metastable. A part of this metastable region is located at negative pressure. In spite of this metastability, these states can be reached experimentally and their lifetime can be long enough for a lot of experimental investigations. Homogeneous solution of partially miscible liquids can exhibit liquid–liquid phase transition by splitting into two liquid phases caused by the change of state parameters such as the pressure. The extensibility of the liquid–liquid critical curve towards negative pressure has been proposed first in the early fifties and finally realized experimentally in the nineties. The reason of this long delay is the experimental difficulty to work with metastable liquids. One has to avoid contamination, mechanical disturbance, etc., in order to suppress sudden boiling caused by cavitations and in order to keep the sample in liquid state. Due to these experimental difficulties, methods which can track liquid–liquid phase equilibrium lines into the metastable region without actually going below zero pressure are very valuable. This tracking is possible by studying the so-called pre-transitional anomalies. These anomalies usually can be seen only in the immediate vicinity of the phase transitions. Fortunately this "immediate vicinity" can be sometimes quite wide. In this paper, several methods mainly dielectric ones will be discussed and compared. Also, we will discuss the danger to get artificial liquid–liquid locus by these methods.

Keywords: Binary mixture, liquid–liquid miscibility, negative pressure, metastable state.

* To whom correspondence should be addressed. Attila R. Imre; e-mail: imre@sunserv.kfki.hu

S. J. Rzoska and V. A. Mazur (eds.), Soft Matter under Exogenic Impacts, 389–398.

1. Introduction

From everyday experience, by mixing two liquids one might obtain a homogeneous mixture (like water + ethanol) or a heterogeneous one (like water + oil). In reality, the distinction is not so sharp. Some binary mixture can be homogeneous in a certain pressure and temperature range and can become heterogeneous when changing these or other state variables. Such systems are called "liquids with limited miscibility".

Most binary liquids with limited miscibility can be divided into two parts: strongly and weakly interacting ones (see, for example, Ref. 1). In this paper all the studied systems are weakly interacting. Probably, the most characteristic difference is that for the weakly interacting systems like most non aqueous organic solutions (for example, the two systems – nitrobenzene/n-hexane and polystyrene/methyl acetate – studied here), the mixtures are homogeneous at some intermediate temperature and can become heterogeneous upon heating or cooling. For strongly interacting systems like a lot of aqueous solutions (for example, water/methyl pyridines[2]), there are two homogeneous regions, one located at high and one at low temperature, and one heterogeneous region in the middle (mention an example here). The concentration dependence of the liquid–liquid (L–L) phase transition is shown in Fig. 1a. In certain systems only one of these two branches is visible, for example, in case of type II and V mixtures in the classification of binary phase behavior due to Scott–van Konynenburg.[3,4] The lower line has a maximum while the upper one has a minimum. These two extrema are critical points: the lower one being a maximum is called upper critical solution temperature (UCST) and the higher one being a minimum is called lower critical solution temperature (LCST). These critical points form curves in the *p*–*T* space. These critical curves can be a continuous with joined UCST and LCST as in case of type III mixtures or they can be virtually two independent lines as in case of type IV mixtures. For these two classes (III and IV) the low or the high temperature branches can be hidden by freezing or by thermal degradation and the systems can be wrongly classified as types V or II. Starting from the early fifties there are some considerations that L–L critical curve can be extended to negative pressure.[2,5] In that case the two virtually separated L–L branches can be continuously extended into the region of negative pressures and finally might merge, unifying the type III and type IV mixtures topologically. To study these phenomena a method is required to track down the critical curve into the negative pressure region. In this book, there is a review paper dealing with the experimental problems to create and measure negative pressure.[6] Here we focus on some specific problems which can affect the reliability of indirect methods.

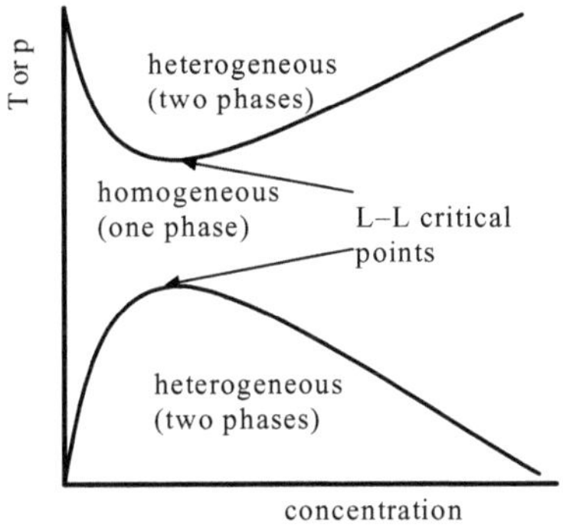

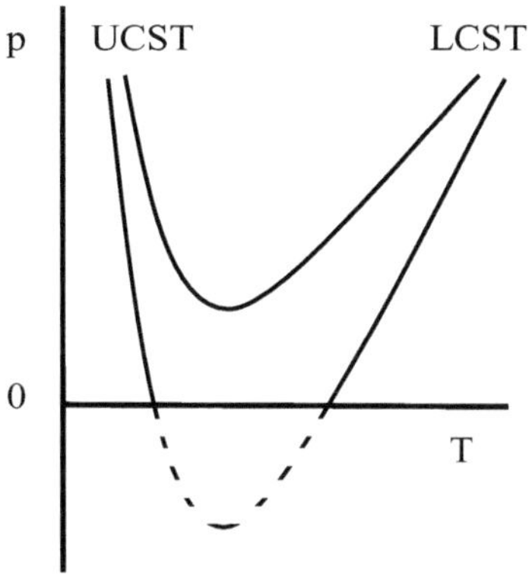

Figure 1. a (left): Schematic representation of liquid–liquid phase diagrams for weakly interacting binary liquid mixtures. b (right): The two different types of pressure dependencies of the L–L critical points.

2. Negative pressure

It is a not very well-known fact, that liquids – just like solids – can have considerable tensile strength. It means that they can endure stretching. Applying isotropic, triaxial stretch one can generate absolute negative pressure. Experimental realization of the triaxial stretch in a liquid can be difficult [6–10] but not impossible. It gives access to the study of several phenomena including L–L phase transition of liquids at negative pressure. Liquids below the vapor pressure are metastable with respect to the liquid–vapor (L–V) phase transition, i.e., they might boil any time. The value of the vapor pressure is always positive, therefore a liquid under negative pressure cannot be stable in principle. A liquid can be metastable down to the so-called L–V spinodal, which is the thermodynamic stability limit for the liquid state. Liquids at negative pressure, are very closely related to the overheated liquids. Actually they are two, non-separated parts of the same metastable region. The only difference is the way how to generate these two states either by increasing the temperature or by decreasing the pressure. Some liquids are quite strong, for example, water can endure more than –100 MPa at room temperature.[10] Theoretically, even lower values can be reached but in the reality the homogeneous or heterogeneous bubble nucleation[7–10] can break the liquid and cause boiling. After forming a macroscopic vapor phase the pressure of the liquid reaches the positive vapor pressure. The probability of the nucleation is bigger with deeper negative pressures, therefore it might be difficult to study L–L phase transitions at stronger tensions. There is only a handful of published experiments[11–16] dealing with the direct observation of the L–L phase transition in binary liquids under

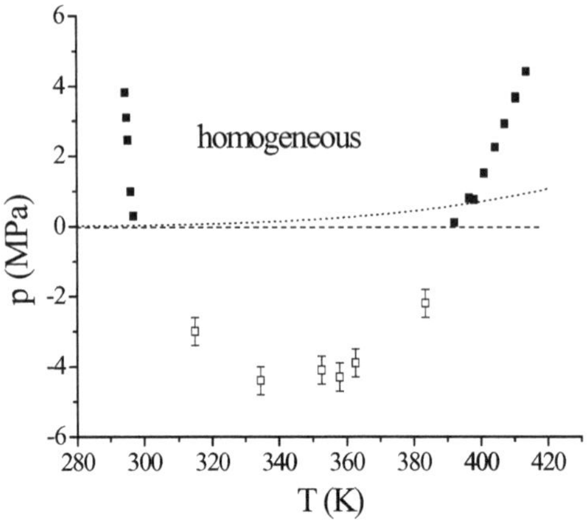

Figure 2. Liquid-liquid locus of polystyrene/methyl acetate mixture (2 w%, M_w=2,000,000). Black marks: direct data by optical measurement, [12] hollow marks: extrapolated data from isotherm conductivity measurements, dotted line: vapor pressure curve of the methyl acetate.

negative pressure. In these experiments the so-called straight glass tube Berthelot method was used. Details about the method are described in several publicatio[11–16] and are omitted here. Although this method is not very difficult, it has two drawbacks: first, it is very time consuming to collect data in this way, secondly only moderate negative pressures can be reached by using this method. The very few direct observations reached –5 MPa for organic solutions [11–15] and –22 MPa for aqueous solutions [16]. Although it might be possible to go below this limit and observe L–L phase transition directly at even stronger tensions, there are other ways to study the location of the L–L loci. In the next Chapter we will discuss some of these methods.

3. Indirect methods

Several phenomena show anomalous behavior, for example, when approaching a phase transition or a critical point. For the sake of simplicity, we studied mainly mixtures with critical composition corresponding to the maximum in Fig. 1a. Therefore all pre-transitional anomalies are critical anomalies. The exponents of the scaling behavior of critical anomalies are well-known from theories and from experiments. Although most of the related studies are isobaric (constant p, changing T) one can observe a similar anomaly approaching a critical point on an isothermal path, sometimes even bigger and clearer. Tracking one of this behavior as close to the transition as possible, one might be able to get the location of the phase transition by extrapolation. The mathematical form of the anomaly can be obtained from theory or from measurements at positive pressure. The general for of these anomalies are

$$Y = A + BX^{\omega},$$

where Y is the studied property exhibiting the anomaly, A and B are constants, X is the distance from the critical point in the changing variable, usually in reduced units, and ω is the critical exponent for the relation of the investigated properties.

In this chapter, two groups of indirect methods are described: dielectric ones and scattering ones, while the emphasis will be on the first group.

3.1. DIELECTRIC METHODS

Under the name of "dielectric methods", three more or less separate methods are treated:

- Electric conductivity method
- Dielectric permittivity method
- Nonlinear dielectric effect (NDE) method

All organic liquids have some electrical conductivity associated with the residual ionic impurities. The anomalous behavior of the conductivity by approaching the L–L phase transition from the homogeneous phase is known for long time, although the theoretical explanation is still not clear.[17] The anomaly is not very big, therefore the method is not very sensitive, although it is quite simple. In our yet unpublished experimental data polystyrene/methyl acetate system was studied and the merging point of the LCST and UCST has been found at around –4 MPa. In direct measurement of L–L phase transition at negative pressure we were able to go down to approximately –2 MPa which was not sufficiently deep to see this merging. Further analysis of the data showed that the conductivity method can be used down to –10 MPa. Below this value the error of the extrapolation is bigger than –2 MPa. We use this value for the error as a border line. The results of the polystyrene/methyl acetate solution (2 w% polystyrene, M_w=2,000,000) are shown in Fig. 2.

The dielectric permittivity has a slightly stronger anomaly, therefore it can be more sensitive. It cannot be used in macromolecular systems, because according to our measurements, there is a continuous drift in the permittivity, probably due to the changing configuration of the polymeric chain in the strong electromagnetic field. Therefore, we used this method to study L–L phase transition in small molecule mixtures, namely mixtures of nitrobenzene in normal hexane, which has only a UCST branch. The details of the experiment and the results are presented in Ref. 18. According to our results, the method can be used down to –20 MPa. The electric conductivity and the static dielectric permittivity can be measured simultaneously in one experiment.[17]

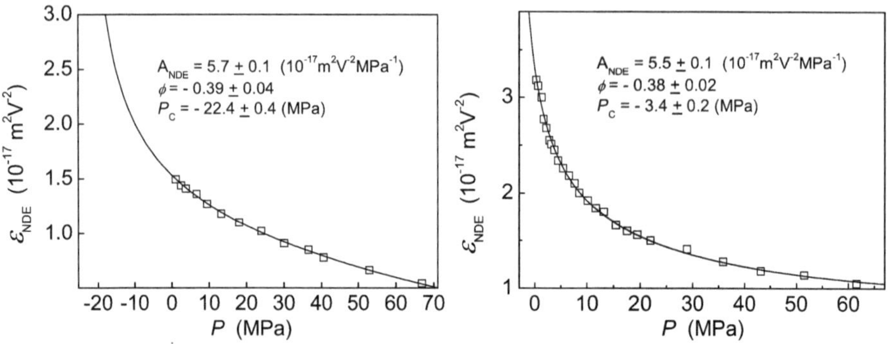

Figure 3. Anomaly of the nonlinear dielectric effect (NDE) measured on two isotherms. The form of the anomaly is $\varepsilon_{NDE} = [A_{NDE}(P-P_c)^{\phi}]_{critical} + [C+BP]_{background}$. The obtained critical pressures can be seen in the inserts. a (left): $T_c = 26.7^{\circ}$C, b (right): $T_c = 19.5^{\circ}$C.

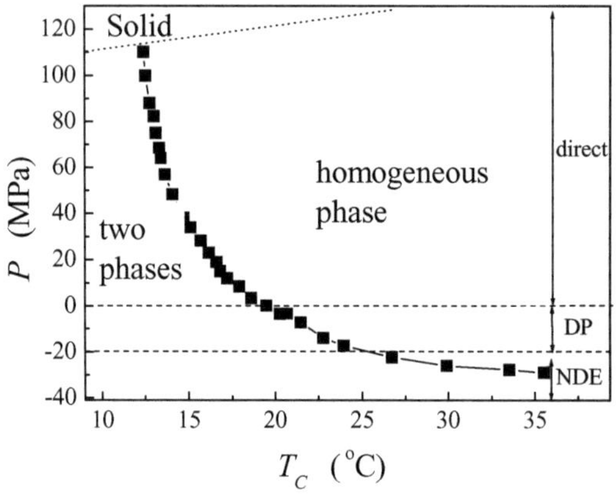

Figure 4. Liquid-liquid locus of nitrobenzene/*n*-hexane mixture (0.43 mole fraction of nitrobenzene). From 0 to 110 MPa the data was obtained directly, from – 20 MPa to 0 by the extrapolation of Dielectric Permittivity (DP), and below –20 MPa by extrapolation of nonlinear dielectric effect.

The third dielectric method is based on the anomaly of the so-called nonlinear dielectric effect (NDE) which is the high-field counterpart of the dielectric permittivity.[19] With macromolecular solutions we had the same drift reported in the dielectric permittivity experiments, therefore we tested this method with the same nitrobenzene – normal alkane mixture. Down to –20 MPa the permittivity and the NDE give the same result within the error but the NDE could be extended into much stronger tensions. In this system, the deepest measured point was at –37 MPa but theoretically this method can be applied down to –50 MPa.[1, 20] The main importance of this method that –50 MPa is very close to the L–V stability line for many organic liquid at room temperature, therefore in this way the whole metastable liquid region can be mapped.

The form of the anomaly with one "near" and one "far" extrapolation measured in NDE is shown in Fig. 3. The results obtained by dielectric permittivity, by NDE measurements, and by extrapolation for the introbenzene/*n*-hexane system are shown (Fig. 4).

3.2. SCATTERING METHODS

Under the name of "scattering methods," we here report very briefly about the usability of dynamic light scattering (DLS) and small-angle neutron scattering (SANS) to determine the L–L locus in certain systems at negative pressure. These methods can be used to measure the hydrodynamic radius and the radius of gyration in macromolecular solutions.[21] These two quantities – just like the three dielectric ones – show pre-transitional anomaly. Therefore, they can be applicable to determine an L–L critical point by extrapolating data measured at positive pressure. Although our results are very preliminary, it seems to be possible to track down the critical curve below –10 MPa, i.e., slightly below than by the conductivity method. Therefore, these two methods can be more useful to study macromolecular systems, although the required equipment is more sophisticated and more expensive than for the one required for conductivity measurement.

3.3. ARTIFACTS

All methods based on extrapolation can produce not only errors but also artifacts. Error means that the obtained L–L critical curve can be quantitatively deviated from the real curve. Artifact means that the method suggests a L–L critical curve in places where it should not exist.

One of the reasons for such artifacts is that the extrapolation cannot distinguish between real critical points and the apparent continuation or traces of thermodynamic states into state region where they actually do not exist. An example is the so-called Widom line[22] which is a trace of phase equilibrium lines into the homogenous region. In the context of the behavior studied here, Widom lines are "ridges" starting from the critical point being the absolute maximum of the L–L coexistence region. To have a Widom line, one needs an end point on the phase equilibrium line. When the measurement ends far away from such extension of a higher order thermodynamic state into the homogenous region such as the Widom line, the extrapolation might artificially recognize it a critical line. Possible forms of end points on the L–L locus – which can be the origin of Widom lines – are given in references.[23–25]

The second reason, that having an extrema at the critical state – like the minimum on Fig. 1b – an isobaric route just passing below the minimum would also yield some anomalous behavior which could be mistakenly fitted by a scaling function and might give artificial L–L critical points. A very classical example is the case of water/3-methyl pyridine mixture when upon heating, one can see some cloudiness (i.e., anomaly in light scattering) but the cloudiness disappears for further heating without any phase transition. In that case the presence of an upside down parabola-shaped L–L locus hidden below 30 MPa is the reason of the opacity, but there is no Widom line involved.

While in the case of the Widom line we passed a ridge, in this second case we would pass the slightly elevated foothill.

TABLE 1. Comparison and applicability of the reviewed methods. The approximate limit is the limit of extrapolation where the error will be bigger than 2 MPa.

Method	Limit (approximate)	Equipment	Applicability
Conductivity	–10 MPa	medium	all nonionic (including macromolecular ones)
Dielectric permittivity	–20 MPa	medium	nonionic, small molecule mixtures
Nonlinear dielectric effect	–50 MPa	difficult/expensive	nonionic, small molecule mixtures
Scattering (DLS and SANS)	probably below –10 MPa	difficult/expensive	macromolecular ones (even ionic ones)

4. Conclusions

There are three main conclusions:

- Liquid–liquid critical curves can be extended to negative pressure. Crossing zero pressure or the vapor pressure curve a L–L critical curve exhibits no discontinuity.
- Direct measurement of critical curves at negative pressure is very difficult, but some dielectric and scattering method can give us an estimate abut the location of the parts hidden at negative pressure. However, the picture is very "fuzzy" and a combination of different methods is suggested in order to obtain reliably results. A comparison of these methods is shown in Table 1.
- To avoid artifacts one need to know at least the topology of the studied L–L locus. This can be supported by equation of state calculation which should give at least qualitatively correct behavior.

Acknowledgment

This work was partially supported by the Hungarian Science Foundation (OTKA) under contract number T 043042, by the German Science Foundation (DFG) under grant Kr1598/10-1/2, by the bilateral Hungarian-Polish Scientific and Technological Grant under contract number PL-10/03, and by a joint grant of the German Science Foundation (DFG) and the Hungarian Academy of Science. A.R.I. was also supported by a Bolyai Research Fellowship.

S.J. Rzoska was supported by the Polish Com. For Sci. Res. (KBN, Poland), grant no. 2PO3B 034 25 for years 2003–2005 (grant resp. S.J. Rzoska).

References

1. Imre, A.R., Drozd-Rzoska, A., Kraska, T., Martinás, K., Rebelo, L.P.N., Rzoska, S.J., Visak, Z.P. and Yelash, L.V. (2004) Phase equilibrium in complex liquids under negative pressure, in *Nonlinear Dielectric Phenomena in Complex Liquids* (Eds.: Rzoska, S.J. and Zhelezny, V.P.), NATO Science Series, Kluwer, Dordrecht, 117–125.
2. Schneider, G.M. (1972) Phase behavior and critical phenomena in fluid mixtures under pressure, *Ber. Bunsen Ges.* **76**, 325–331.
3. Scott, R.L. and Konynenbur, P.H. (1970) Static properties of solutions. Van der Waals and related models for hydrocarbon mixtures, *Disc. Far. Soc.* **49**, 87–97.
4. van Konynenburg, P.H. and Scott, R.L. (1980) Critical lines and phase-equilibria in binary van der Waals mixtures, *Phil. Trans. R. Soc. London* **298A**, 495–540.
5. Timmermans, J. and Lewin, J. (1953) A forgotten theory: The "negative saturation curve," *Discuss. Faraday Soc.* **15**, 195–201.
6. Imre, A.R. (2007) How to generate negative pressure, *NATO Sci. Series II*, Soft Matter under exogenic impacts, Springer, Berlin, pp. 379–387.
7. Trevena, D.H. (1987) *Cavitation and Tension in Liquids*, Adam Hilger, Bristol.
8. Debenedetti, P.G. (1996) *Metastable Liquids: Concepts and Principles*, Princeton University Press, Princeton, N.J.
9. Imre, A., Martinás, K. and Rebelo, L.P.N. (1998) Thermodynamics of negative pressures in liquids, *J. Non-Equilib. Thermodyn.* **23**, 351–375.
10. *Liquids Under Negative Pressure* (Eds.: Imre A.R., Maris H.J. and Williams P.R.), NATO Science Series, Kluwer, Dordrecht, 2002.
11. Imre, A. and Van Hook, W.A. (1994) Polymer-solvent demixing under tension isotope and pressure effects on liquid–liquid transitions. VII. Propionitrile–polystyrene solutions at negative pressure, *J. Polym. Sci. B.* **32**, 2283–2287.
12. Imre A. and Van Hook W. A. (1997) Continuity of solvent quality in polymer solutions. Poor-solvent to Θ-solvent continuity in some polystyrene solutions. *J. Polym. Sci. B.* **35**, 1251 1259.
13. Imre, A. and Van Hook, W.A. (1998) Liquid-liquid equilibria in polymer solutions at negative pressure, *Chem. Soc. Rev.* **27**, 117–123.
14. Visak, Z.P., Rebelo, L.P.N. and Szydlowski, J. (2002) Achieving absolute negative pressures in liquids: Precipitation phenomena in solution, *J. Chem. Edu.* **79**, 869–873.

15. Rebelo, L.P.N., Visak, Z.P. and Szydlowski, J. (2002) Metastable critical lines in (acetone+polystyrene) solutions and the continuity of solvent-quality states, *Phys. Chem. Chem. Phys.* **4**, 1046–1052.
16. Visak, Z.P., Rebelo, L.P.N. and Szydlowski, J. (2003) The "hidden" phase diagram of (water+3-methylpyridine) at large absolute negative pressures, *J. Phys. Chem. B* **107,** 9837–9846.
17. Malik, P., Rzoska, S.J., Drozd-Rzoska, A. and Jadzyn, J. (2003) Critical behavior of dielectric permittivity and electric conductivity in temperature and pressure studies above and below the critical consolute point, *J. Chem. Phys.* 118, 9357–9363.
18. Drozd-Rzoska, A., Rzoska, S.J. and Imre, A.R. (2004) Liquid–liquid phase equilibria in nitrobenzene–hexane critical mixture under negative pressure, *Phys. Chem. Chem. Phys.* **6**, 2291–2294.
19. Drozd-Rzoska, A., Rzoska, S.J. and Czuprynski, K. (2000) Phase Transitions from the isotropic to liquid crystalline mesophases studies by linear and nonlinear dielectric permittivity, *Phys. Rev. E.* **61**, 5355–5360.
20. Drozd-Rzoska, A. Rzoska, S.J., Imre, A.R. and Paluch, M. (2005) Negative pressure related critical behavior in nitrobenzene–hexane mixture, *Fluid Phase Equilibria,* in print.
21. Van Hook, W.A., Wilczura, H., Imre, A., Rebelo, L.P.N. and Melnichenko, Y. (1999) Correlation radii for polystyrene (PS) in poor and theta solvents from dynamic light scattering and small angle neutron scattering. New data for PS/acetone. Remarks on PS/acetone, PS/cyclohexane, and PS/methylcyclohexane, *Macromolecules* **32**, 7312-7318.
22. Xu, L., Kumar, P., Buldyrev, S.V., Chen, S-H., Poole, P.H., Sciortino, F. and Stanley, H.E. (2005) Relation between the Widom line and the strong-fragile transition in systems with a liquid–liquid phase transition, *Proc. Nat́ l Acad. Sci.* **102,** 16558–6562.
23. van Pelt, A., Peters, C.J., de Swaan Arons, J. and Meijer, P.H.E. (1993) Mathematical double points according to the simplified-perturbed-hard-chain theory, *J. Chem. Phys. Phys.* **99**, 9920–9929.
24. Imre, A.R., Drozd-Rzoska, A., Kraska, T., Rzoska, S.J. and Yelash, L.V. (2003) Liquids under absolute negative pressure, in *Complex Systems in Natural and Social Sciences* (Eds.: E. Hideg, K. Martinás, M. Moreau, D. Meyer), ELFT, Budapest, 93–107.
25. Imre, A.R. and Kraska, T. (2005) Stability limits in binary fluids mixtures, *J. Chem. Phys.* **122**, 064507–064514.

PART V: BIO-LIQUIDS AND RELATED PROBLEMS

CRITICAL PROPERTIES OF SOFT MATTER AT RESTRICTED GEOMETRY AS EMERGING PROBLEM: FUNDAMENTALS AND BIOLOGICAL APPLICATIONS

ALEXANDER V. CHALYI*,
Physics Department, National Medical University
13, Shevchenko Blvd., Kiev 01601, Ukraine,
e-mail:avchal@nmu.kiev.ua

LEONID A. BULAVIN, KYRYLO A. CHALYY,
Department of Molecular Physics, Physics Faculty,
Kiev Taras Shevchenko National University,
6, Acad. Glushkov Boulevard, Kiev 03022, Ukraine

LYUDMILA M. CHERNENKO,
Institute of Surface Chemistry, National Academy of Sciences,
17, General Naumov Str., Kiev 03164, Ukraine

ALEXEI N. VASIL'EV, ELENA V. ZAITSEV
Department of Theoretical Physics, Physics Faculty,
Kiev Taras Shevchenko National University,
6, Acad. Glushkov Boulevard, Kiev 03022, Ukraine

Abstract. Fundamental and applied aspects of critical behavior of liquids at restricted geometry are discussed, namely: correlation functions of confined multicomponent systems near the critical state; heat capacity in confined liquid ^{4}He; neutron optics in confined liquids near the critical point; kinetic model of synaptic transmission which is isomorphic to the critical phenomena in finite-size liquid mixtures.

Keywords: Confined multicomponent systems, critical point, correlation functions, heat capacity, neutron optics, synaptic transmission.

1. Introduction

Last decades gave us at least two outstanding achievements in experimental and theoretical physics: (a) revolutionary discoveries in the field of nanotechnology, (b) principle solution the problem of the second order phase transitions and critical

* To whom correspondence should be addressed.

S. J. Rzoska and V. A. Mazur (eds.), Soft Matter under Exogenic Impacts, 399–412.

phenomena. Unification of these directions gave and is still giving a tremendous rising of scientific interest to mesoscale systems (see, for example, Ref. 1–3). Here we present studies of finite-size effects on the critical behavior of physical properties in mesoscale liquid systems continuing our previous studies.[4–9] The topics under consideration are as follows: (a) correlation properties of multicomponent confined liquid mixtures; (b) heat capacity of confined liquid ^{4}He; (c) neutron optics of liquids at restricted geometry; (d) biological application of critical phenomena to the process of cell-to-cell communication (synaptic transmission).

2. Correlation functions of confined multicomponent systems

To find the basic pair correlation functions we consider a confined multicomponent mixture with cylindrical geometry ($l >> R$, l is the length of cylindrical axis, R is the radius of cylinder). It is reasonable to start from the system of integral Ornstein–Zernike (OZ) equations presented in a matrix form as follows:

$$\hat{G}(\vec{r}) = \hat{F}(\vec{r}) + \int \hat{F}(\vec{r}_1)\hat{G}(\vec{r} - \vec{r}_1)\, d\vec{r}_1 \,. \tag{1}$$

Here $\hat{G}(\vec{r})$ and $\hat{F}(\vec{r})$ are the matrices of pair and direct correlation functions, respectively. After presenting correlation functions in form of series on Bessel functions one has the following equation for Fourier harmonics:

$$\hat{G}_{(m)}(k) = \hat{F}_{(m)}(k) + \pi R^2 J_1^2(\mu_m)\hat{F}_{(m)}(k)\hat{G}_{(m)}(k) \,. \tag{2}$$

Using such a transformation

$$\hat{f}_{(m)}(k) = \pi R^2 J_1^2(\mu_m) \cdot \hat{S}_{(m)}^{-1}(k)\hat{F}_{(m)}(k)\hat{S}_{(m)}(k) \,, \tag{3}$$

a single equation for new matrices is received

$$\hat{g}_{(m)}(k) = \hat{f}_{(m)}(k) + \hat{f}_{(m)}(k)\hat{g}_{(m)}(k) \,. \tag{4}$$

For matrices $\hat{g}_{(m)}(k)$ and $\hat{f}_{(m)}(k)$ of diagonal form the matrix $\hat{S}_{(m)}(k)$ can be formed as a direct sum of unit eigenvectors $\vec{e}_{(m),i}(k)$ of matrix $\hat{F}_{(m)}(k)$ and diagonal elements of matrix $\hat{f}_{(m)}(k)$, i.e., coincides with matrices $\hat{S}_{(m)}(k)$, eigenvalues $\xi_{(m),i}(k)$ of matrix $\hat{F}_{(m)}(k)$. We suppose to find $\hat{S}_{(m)}(k)$, $\hat{f}_{(m)}(k)$, and eigenvalues $\xi_{(m),i}(k)$ in form of expansion with respect to parameter $k_m^2 = k^2 + (\mu_m/R)^2$:

$$\hat{S}_{(m)}(k) = \hat{S}_{(m)}^{(0)} - k_m^2 \hat{S}_{(m)}^{(2)}, \quad \hat{f}_{(m)}(k) = \hat{f}_{(m)}^{(0)} - k_m^2 \hat{f}_{(m)}^{(2)},$$

$$\xi_{(m),i}(k_m) = \xi_{(m),i}^{(0)} - k_m^2 \xi_{(m),i}^{(2)} \,.$$

It is easy to show that $\hat{f}_{(m)}^{(0)} = \hat{S}_{(m)}^{(0)T} \hat{C}_{(0)} \hat{S}_{(m)}^{(0)}$, where $\hat{C}_{(0)}$ is the matrix of zero spatial moments of direct correlation functions and matrix $\hat{S}_{(m)}^{(0)}$ is formed as direct sum of eigenvectors $\vec{e}_i^{(0)}$ of matrix $\hat{C}_{(0)}$. Eigenvalues $\xi_{(m),i}^{(0)}$

($i = 1, 2, \ldots, N$, where N is the number of mixture components) that correspond to these eigenvectors are the roots of equation

$$\det |\hat{C}_{(0)} - \xi^{(0)}_{(m),i}\hat{E}| = 0. \tag{5}$$

Both eigenvalues $\xi^{(0)}_{(m),i}$ and matrix $\hat{S}^{(0)}_{(m)}$ in fact do not depend on index m. On the other hand, for matrix $\hat{f}^{(2)}_{(m)}$ we get

$$f^{(2)}_{(m),ij} = (\xi^{(0)}_{(m),j} - \xi^{(0)}_{(m),i})\beta_{(m),ij} + \lambda_{(m),ij}, \tag{6}$$

where $\lambda_{(m),ij}$ are elements of symmetric matrix $\hat{S}^{(0)T}_{(m)}\hat{C}_{(2)}\hat{S}^{(0)}_{(m)}$ and for non-diagonal elements $\beta_{(m),ij} = \dfrac{\lambda_{(m),ij}}{\xi^{(0)}_{(m),i} - \xi^{(0)}_{(m),j}}$, meanwhile all diagonal elements of matrix $\hat{\beta}_{(m)}$ are zero. Diagonal elements $\xi^{(2)}_{(m),i}$ of matrix $\hat{f}^{(2)}_{(m)}$ are determined by relation $\xi^{(2)}_{(m),i} = \lambda_{(m),ii}$, and coincides with diagonal elements of matrix $\hat{S}^{(0)T}_{(m)}\hat{C}_{(2)}\hat{S}^{(0)}_{(m)}$. Taking into account all these relations one can find finally

$$g_{(m),i}(k) = \frac{f_{(m),i}(k)}{1 - f_{(m),i}(k)} = \frac{1}{\xi^{(2)}_{(m),i}} \cdot \frac{1}{k^2 + (\mu_m/R)^2 + \kappa^2_{(m),i}} - 1, \tag{7}$$

where $\kappa^2_{(m),i} = \left(1 - \xi^{(0)}_{(m),i}\right) / \xi^{(2)}_{(m),i}$. After inverse Fourier transformation for harmonics of pair correlation functions expansion we get

$$g_{(m),i}(z) = \frac{\exp(-\sqrt{(\mu_m/R)^2 + \kappa^2_{(m),i}}\,|z|)}{2\xi^{(2)}_i\sqrt{(\mu_m/R)^2 + \kappa^2_{(m),i}}}. \tag{8}$$

Knowing harmonics, one can find basic pair correlation functions:

$$g_{ij}(\rho, z) = \delta_{ij} \sum_{i=1}^{N} \sum_{m=1}^{\infty} \frac{\exp(-\sqrt{(\mu_m/R)^2 + \kappa^2_{(m),i}}\,|z|)}{2\xi^{(2)}_{(m),i}\sqrt{(\mu_m/R)^2 + \kappa^2_{(m),i}}} \cdot J_0\left(\frac{\mu_m \rho}{R}\right). \tag{9}$$

Pair correlation functions of density fluctuation of multicomponent liquid are the linear combination of basic correlation functions. We should mention that this result is in a good agreement with available data for simple and binary liquids.[10,11]

Equation of the OZ form can be used to analyze the correlation functions of multicomponent mixture in porous media. Indeed we can transform this equation to differential OZ matrix equation as follows:

$$\Delta\hat{G}(r) - \hat{C}^{-1}_{(2)}\left(\hat{E} - \hat{C}_{(0)}\right)\hat{G}(r) = -\hat{C}^{-1}_{(2)}\hat{F}(r), \tag{10}$$

where $\hat{E}$ is unit matrix and matrices of spatial moment $\hat{C}_{(n)}$ depend on porosity parameter η (i.e., fraction of filled volume in porous media):

$$\hat{C}_{(n)} = \frac{1-\eta}{(n+1)!}\int \hat{F}(r) r^n d\vec{r}\,, \quad (n{=}0{,}2). \tag{11}$$

Using method of spectral expansion[8] one can find the following expression for pair correlation functions of density fluctuations of multicomponent liquids in porous media:

$$\hat{G}(r) = \sum_{m=1}^{N} \hat{\gamma}_m \frac{\exp(-\kappa_m r)}{r}, \tag{12}$$

where κ_m^2 is eigenvalue of matrix $\hat{\Lambda} = \hat{C}_{(2)}^{-1}(\hat{E} - \hat{C}_{(0)})$ and matrices

$$\hat{\gamma}_m = \hat{C}_{(2)}^{-1} \prod_{\substack{i=1,\\ i\neq m}}^{N} \frac{\hat{E}\kappa_i^2 - \hat{\Lambda}}{\kappa_i^2 - \kappa_m^2}. \tag{13}$$

Presence of porous media affects the critical behavior of liquid systems. At the close vicinity of critical state all the correlation functions of infinite systems can be approximated by such an expression of the OZ type:

$$G_{ij}(r) \sim \frac{\exp(-\kappa r)}{r} \tag{14}$$

with parameter κ is approaching zero according to the scaling law

$$\kappa^2 = \kappa_0^2 \left(\frac{Z - Z_C}{Z_C}\right)^{2\lambda} \tag{15}$$

when thermodynamic parameter Z becomes equal its critical value Z_C, λ is the critical index. If one compares the critical value Z_C for an infinite system and for a mixture in porous media, it occurs that the shift of critical parameter takes place with the following dependence on porosity parameter η: $\Delta Z / Z_c \propto \eta^{1/2\lambda}$.This dependence is similar to dependence of the critical temperature shifts on system's linear size L for confined liquids[5,6] and presents a good agreement with experimental data.[10,11]

3. Heat capacity of confined liquid helium

The properties of confined liquid helium in the vicinity of λ-transition point will be examined in terms of the heat capacity temperature dependence. The system to be considered hereafter has a reduced geometry in form of plane-parallel layer with the typical thickness from a few hundred angstroms and over. Such range of sizes could be associated with upper border of nanoscale and substantial part of microscale and could be referred to as the mesoscale. Films represent an example of so-called "well-defined geometry" that allows to perform direct analytical calculations. Verification of the validity of the proposed here theoretical results is conducted by comparing it with the high-resolution experimental data.[12–15]

Let us consider the geometry of thin liquid systems in the form of a plane-parallel layer $D \times D \times H$. The thickness of the layer H is supposed to be much smaller than distance D in XY directions. We will study the case corresponds to situation when the correlation length ξ becomes comparable or even larger than H but still much smaller than D.

In the case of planar geometry the heat capacity $C_{plan.}(\tau,H)$

$$C_{plan.}(\tau,H) \sim \left[\left(\left(\frac{\pi\xi_0}{H}\right)^{1/\nu}+1\right)\tau+\left(\frac{\pi\xi_0}{H}\right)^{1/\nu}\right]^{-\alpha}. \qquad (16)$$

Here $\tau=|T-T_c|/T_c$ is the temperature variable, $\alpha=-0.0127$ and $\nu=0.6705$ are the critical exponents for ^{4}He.

Let us examine the limiting cases, namely $H\rightarrow\infty$ and $\tau=0$. Eq. (16) appears reasonable and leads to the expected expressions:

$$C_{plan.}(\tau,H\rightarrow\infty)\sim\tau^{-\alpha},$$

which obviously demonstrates a classic bulk behavior, and

$$C_{plan.}(\tau=0,H)\sim(\pi\xi_0)^{-\alpha/\nu}H^{\alpha/\nu},$$

which shows that the heat capacity remains finite at bulk T_λ ($\tau=0$) if $H<\infty$. A new transition temperature $T_c(H)$ indicating the location of the heat capacity maximum in the liquid film system is described by the following formula: $T_c^*(H)=T_c[1+(\pi\xi_o/H)^{1/\nu}]^{-1}$. The liquid helium film's experimental data[12–15] and the current analytical estimations are combined in the Table 1. Points $\Delta\tau_E$ F1–F7 belong to the size range below 1 μm and represent the experimental results by Gasparini Group.[13,15] Points F8 and F9 correspond to the film thickness 19 and 57 μm and represent the results by Lipa Group.[12,14] Here, in the table $\Delta\tau_E=(T_\lambda-T_m)/T_m$ are the experimental temperature shifts of the helium heat capacity maximum T_m from its bulk value T_λ, with taking into account that $T_\lambda>T_m$. The shifts $\Delta\tau_T$ are calculated in accordance with the proposed theoretical expression: $\Delta\tau_T=(\pi\xi_o/H)^{1/\nu}$. This equation demonstrates an agreement with the finite-size scaling theory predictions: $\Delta\tau=aL^{-1/\nu}$, where a is a constant depending on the geometry. In the case considered above, the linear size of a system L is treated as the film thickness H and, as a result, the expression for the scaling coefficient a reads: $a=(\pi\xi_o)^{1/\nu}$. The comparison shows that the theoretical values $\Delta\tau_T$ in most cases (F1, F2, F5, F6, and F9) underestimate the shift of the new transition temperature on 1.8% in average. It was especially exciting to compare the theoretical evaluation $\Delta\tau_T$ with the result F9 of most advanced confined helium experiment (CHEX).[14] CHEX has performed a high-resolution measurement (±1 nK) of the finite-size effect using a sample of liquid helium, confined to a planar geometry with thickness 57 μm. For this size it appears that $\Delta\tau_T$ is about 2.1% lower than $\Delta\tau_E \approx 2{,}5\cdot 10^{-8}$ that seems to be a adequate result.

TABLE 1. Experimental and theoretical data for shifts of ^{4}He heat capacity maximum T_m from its bulk value T_λ.

Film thickness H	Shift $\Delta\tau_E$, experiment	Shift $\Delta\tau_T$, theory	Experimental points/[Ref.]
483 Å	9.356×10^{-4}	9.337×10^{-4}	F1/[13]
1074 Å	2.860×10^{-4}	2.835×10^{-4}	F2/[13]
2113 Å	1.273×10^{-4}	1.033×10^{-4}	F3/[13]
3189 Å	5.3×10^{-5}	5.594×10^{-5}	F4/[15]
5039 Å	2.920×10^{-5}	2.827×10^{-5}	F5/[13]
6918 Å	1.808×10^{-5}	1.762×10^{-5}	F6/[13]
9869 Å	1.278×10^{-5}	1.037×10^{-5}	F7/[13]
19 μm	1.2×10^{-7}	1.260×10^{-7}	F8/[12]
57 μm	2.5×10^{-8}	2.447×10^{-8}	F9/[14]

It is possible to compare the contributions to the shift of transition temperature caused by the gravity effect (δT_g) and by the finite-size effect (δT_{fs}) in terms of direct calculations. Let us introduce the "effect's ratio" $\Omega(H)=(\delta T_g/\delta T_{fs})\times100\%$ that gives an opportunity to check which effect dominates for any given film thickness H:

$$\Omega(H) = \frac{\gamma}{(\pi\xi_0)^{1/\nu}\,T_\lambda} H^{(\nu+1)/\nu} \times 100\%,$$

where coefficient γ=1.273 μK/cm. It is assumed[16] that data are unaffected by gravity if $|T-T_\lambda|\geq 10\cdot\delta T_\lambda$. According to this statement, the confined helium experimental results could be considered as unaltered by the gravity if $\Omega(H)\leq10\%$. It gives an exact value of the maximum film thickness for earth-bound experiments as follows: H_L=50,29 μm. Thus, results obtained due to the current theoretical approach are reasonably matching to confined ^{4}He heat capacity experimental data over the wide range of system's sizes from 50 nm up to about 60 μm for planar type of geometry.

4. Neutron optics of confined liquids near the critical point

4.1. NEUTRON REFRACTION IN CONFINED LIQUIDS

The problem of neutron propagation and scattering in finite-size liquids is reduced in neutron optics to the Schrödinger equation which has the following form of the electrodynamical wave equation for the neutron wave function ψ (see, for example, Ref. 17):

$$\Delta\psi(\vec{r}) + k_0^2 n^2(\vec{r})\psi(\vec{r}) = 4\pi b_{coh}\rho'(\vec{r})\psi(\vec{r}). \tag{17}$$

Here the Fermi pseudo-potential $U(\vec{r}) = 2\pi\hbar^2 b_{coh}\rho(\vec{r})m^{-1}$ is used where $\hbar = h/2\pi$, b_{coh} is the coherent scattering length (amplitude), $\rho(\vec{r})$ is the local numerical density of nuclei, $\rho'(\vec{r})$ is the fluctuation of neutron density, m is the neutron mass, $k_0 = 2\pi/\lambda$ is the wave vector, λ is the de Broglie wavelength of neutrons. The neutron analog of the optical refractive index n in Eq. (17) is given by the formula

$$n = n_c - s\Delta\rho/2, \qquad n_c = 1 - b_{coh}\rho_c\lambda^2 N_A(2\pi M)^{-1},$$

where ρ_c is the critical density of nuclei, N_A is the Avogadro number, M is the molar mass, $\Delta\rho$ is the density deviation from the critical density, $s = b_{coh}\rho_c\lambda^2 N_A(\pi M)^{-1}$ is the constant value.

In order to study the influence of boundary surfaces, the following formula was derived with taking into account results obtained in:[18]

$$n = n_c - (s/2)\Delta\rho_0\left[1 - \exp\left(-\frac{\Delta x - \lambda_e}{\xi}\right)\right], \tag{18}$$

where $\Delta\rho_0$ and ξ are, correspondingly, the order parameter and correlation length in a bulk phase, Δx is the distance from a boundary surface, λ_e is the extrapolation length (for practical calculations it is often taken $\lambda_e=0$). As is seen from Eq. (18), the neutron refractive index has its maximal (minimal) values in the center of restricted volume for negative (positive) coherent correlation length. In analogy with the crystal optics, the neutron refractive index may be characterized by the corresponding ellipsoid: $\frac{x^2}{n_x^2} + \frac{y^2}{n_y^2} + \frac{z^2}{n_z^2} = 1$ which has two equal axes ($n_x = n_y$) and one unequal axis n_z. Parameters of the ellipsoid for a cylindrical pore with radius $R > \xi$ are as follows:

$$n_x = n_y = n_c - (s/2)\Delta\rho_0(\tau,h)\left[1 - \exp\left(-\frac{R - \rho^* - \lambda_e}{\xi(\tau,h)}\right)\right],$$

$$n_z = n_c - (s/2)\Delta\rho_0(\tau,h).$$

Ratio of these parameters is $\frac{n_x}{n_z} = \frac{n_y}{n_z} = 1 - \exp\left(-\frac{R - \rho^* - \lambda_e}{\xi(\tau,h)}\right)$, where $\rho^* = (x^2+y^2)^{1/2}$ is the distance from cylinder axis to an arbitrary point at xy-plane. Results of computer simulations[19] shows that for distances from wall surface

$R-\rho^* \geq (3/4)\xi$, multiplier $1-\exp\left(-\frac{R-\rho^*-\lambda_e}{\xi(\tau,h)}\right)$ approaches 1 and neutron refractive index n equals its bulk value: $n \to n_0 = n_c - (s/2)\Delta\rho_0$. Thus, on the cylinder axis all three refractive indexes become equal *(n_x=n_y=n_z)* and instead of ellipsoid one has a sphere of the neutron refractive indexes.

As a result of above-mentioned analogy between nonhomogeneous liquid near the critical point and crystals, a phenomenon similar to the light double-refraction could be expected in such anisotropic liquids.

In a slitlike pore $0 \leq x \leq L_1$, $0 \leq y \leq L_2$, $0 \leq z \leq L_3$ with a nonhomogeneous fluid near its critical isotherm one has the following formula for the angle of refraction:

$$tg\alpha = -\frac{b_{coh}\lambda^2 L_1}{2\pi n_c} h^{1/\delta}\left[\frac{F_2(0,w)}{\delta h} + bL^{\beta\delta/\nu}\frac{\partial F_2(0,w)}{\partial w}\right]. \tag{19}$$

It follows from Eq. (19) that a neutron beam incident on a sample in the horizontal direction is deflected (a) downward just like a light beam for substances with $b_{coh} < 0$ (for example, in H_2, H_2O, see a beam 1 on Fig. 1a); (b) upwards for substances with $b_{coh} > 0$ (for example, in C_6H_6, CO_2, N_2, see a beam 2 on Fig. 1a). These results confirm calculations of the neutron refraction in bulk liquids.[18]

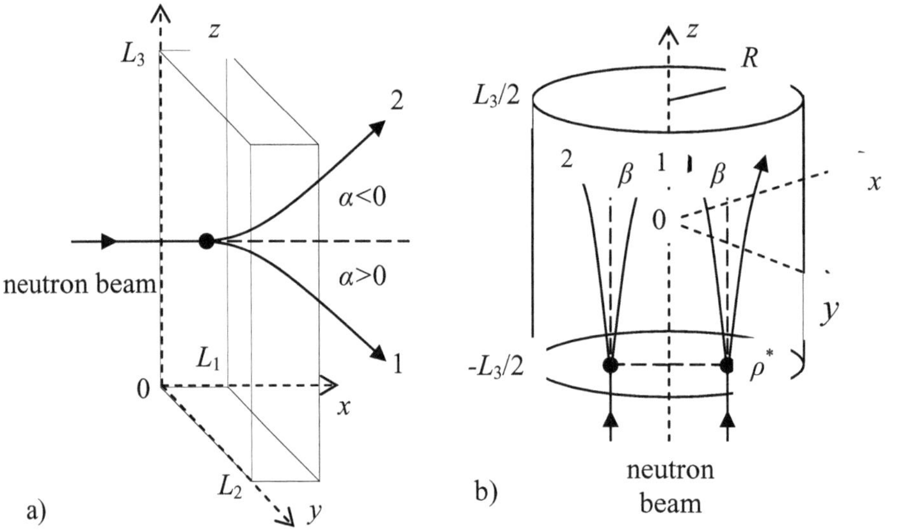

Figure 1. Neutron refraction in a) slitlike pores, b) cylindrical pores.

Maximal values of the angle of refraction $(tg\alpha)_{max}$ take place on two levels

$$h_{\max}^{1,2} = qh_n\left[1+\sqrt{1+\frac{(\delta-1)}{q}}\right].$$

Here $q = \dfrac{[F_2'(0,w)]^2}{F_2(0,w)F_2''(0,w)}$ and a quantity $h_n = -b_h L^{-\beta\delta/\nu}$ with $b_h = \dfrac{F_2(0,w)}{\delta b F_2'(0,w)}$ corresponds to positions of flex points at the height profile of the neutron refractive index.

Another experimental geometry of a cylindrical pore is also examined with effects of boundary surfaces taken into account (Fig. 1b).
In this case such an expression for the angle of refraction is found:

$$tg\beta = -\frac{sL_1}{2}\frac{\Delta\rho_0(\tau,h)}{\xi(\tau,h)}\left[1-\frac{\Delta\rho(\rho^*,\tau,h)}{\Delta\rho_0(\tau,h)}\right],$$

which allows to study peculiarities of the neutron refraction in the vicinity of the critical isotherm in "strong" fields ($\tau << h^{1/\beta\delta}$), the critical isochore in "weak" fields ($h << \tau^{\beta\delta}$), and the coexistence curve. An interesting consequence connecting with influences of boundary surfaces is a possible effect of focusing (for b_{coh} <0) and defocusing (for b_{coh} >0) of neutron beams in the critical region (see Fig. 1b).

4.2. PROPAGATION AND COHERENT SCATTERING OF NEUTRONS

Using an iteration procedure for solving the wave Eq. (17) it is possible to find the neutron wave function ψ_0 describing the propagation of neutrons in liquids (Fig. 2).

The coefficients of transmission T and reflection R in confined liquids, as in bulk liquid systems,[18] demonstrate a periodic behavior with quantity

$$\alpha = k_0\int_{-L_0}^{L_0} n(h)dh = 2k_0 n_c L_0 - \frac{k_0 s}{2}\int_{-L_0}^{L_0}\Delta\rho(\tau,h)dh, \qquad (20)$$

which is a neutron analog of the optical path.

It is necessary to mention that an additional factor of limitation $S_2=L/\lambda$ (with respect to $S_1=L/\xi$) appears in problems of propagation and scattering of electromagnetic waves and particles in finite-size systems. The periodic dependence of the first principal contribution in Eq. (20) on $\alpha_0 = 2k_0 n_c L_0$ contains exactly this factor of limitation because $\alpha_0 = 2\pi S_2$, where $S_2=2L_0/\lambda_c$ with $\lambda_c = \lambda / n_c$ being the de Broglie wavelength of neutrons. Additional contribution $\Delta\alpha$ of the second term in (20) gives a modulation of coefficients T and R with much larger wavelength $\lambda^* = \lambda_c (s\Delta\rho)^{-1} \approx 10^6\lambda_c$.

The differential neutron scattering cross-section in slitlike and cylindrical pores is obtained in the following form using an expression for the wave function ψ_1 of the singly-scattered neutron beam:

$$\left\langle \frac{d\sigma}{d\Omega} \right\rangle^P \sim \left[\left(\kappa_0^2 \tau^{2\nu} + \pi^2/4L_0^2 \right) + k_{xy}^2 \right]^{-1} \left[(\pi^2/4L_0^2) + k_z^2 \right]^{-1}, \tag{21}$$

$$\left\langle \frac{d\sigma}{d\Omega} \right\rangle^C \sim \left\{ 1 + \frac{a^2}{2} \left[(1 - 4/\mu_1^2) k_{xy}^2 + 2a^2 \left(\kappa_0{}^2 \tau^{2\nu} + \mu_1^2/a^2 \right) k_z^2 \right] \right\}^{-1}. \tag{22}$$

Here $k_{xy} = (4\pi/\lambda)(n_c - s\Delta\rho/2)\sin(\theta_{xy}/2)$ and $k_z = (4\pi/\lambda)\ (n_c - s\Delta\rho/2)$ $\sin(\theta_z/2)$ are the corresponding wave vectors. As is seen from (21) and (22), size-dependent contributions to the neutron cross-sectionm realize only at directions of spatial limitation (an axis z for a slitlike pore or axes x,y for a cylindrical pore), while dependence on thermodynamic variables takes place for other directions with linear size $L >> \xi$.

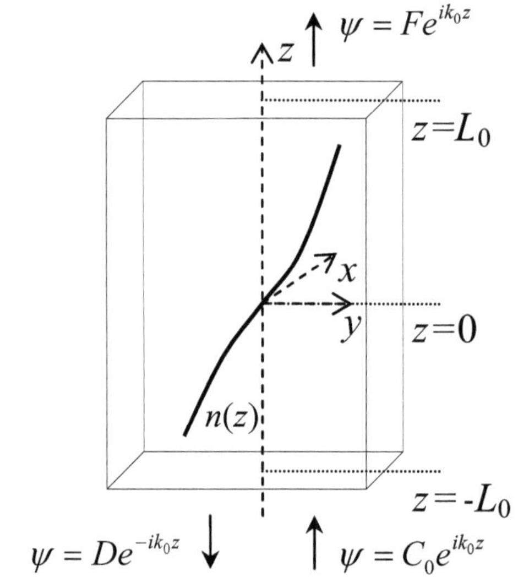

Figure 2. Geometry of the problem of neutron propagating in confined liquids.

Another conclusion is as follows: maxima of the neutron cross-section $\left\langle \frac{d\sigma}{d\Omega} \right\rangle$ should be experimentally observed at the temperature T_M, density ρ_M, and external field h_M in accordance with formulae:

$$T_M = T_c\left[1 + f(S_1^*)^{1/\nu}\right]^{-1}, \rho_M = \rho_c\left[1 + f(S_1^*)^{\beta/\nu}\right]^{-1}, h_M = h_c\left[1 + f(S_1^*)^{\beta\delta/\nu}\right]^{-1}.$$

Depending on the type of geometry of a liquid volume, a function $f(S_1^*)$ has such values: $f^P(S_1^*) = \pi\xi_0/2L_0$ for a slitlike (plane) pore; $f^C(S_1^*) = \mu_1\xi_0/a_{cyl}$ for a cylindrical pore.

Detailed study of size-dependent effects in the coherent neutron scattering near the critical point shows that the differential single-scattering neutron cross-section in confined liquids ($L^{-1} >> \xi^{-1}$) is described by the following formula:

$$\left\langle \frac{d\sigma}{d\Omega} \right\rangle^P \sim \left[\frac{\pi^2}{4L_0^2} + k_{xy}^2 \right]^{-1+\eta/2} \sim L^{2-\eta} \left(1 + \text{const} \frac{L^2}{\lambda^2} \right)^{-1+\eta/2}, \tag{23}$$

It is easy to see that (a) the neutron cross-section (23) does not depend on the proximity of the temperature and field variables to the critical point; (b) for $S_2 << 1$, i.e., small sizes L or/and large wavelength λ, the light-scattering intensity (neutron cross-section) depends on a system's linear size; (c) in the opposite case $S_2 >> 1$, i.e., for $L >> \lambda$, intensity (cross-section) does not depend on system's linear size, being dependent only on the wavelength $\lambda : \left\langle \frac{d\sigma}{d\Omega} \right\rangle \propto \lambda^{2-\eta}$. This result which is valid for confined systems ($S_1 \leq 1$) and simultaneously for relatively small wavelength ($S_2 >> 1$) confirms the well-known scaling relation for the dependence of light-scattering intensity on the wave vector $I^{-1} \propto k^{2-\eta}$ in the close vicinity of the critical point of bulk systems (see, for example, Ref. 20–23).

5. Kinetic model of synaptic transmission

The problem of cell-to-cell communication is of the same fundamental importance in the alive nature as the intermolecular interaction is in the non alive world. As was shown in, Ref. 4,9 the process of synaptic transmission is isomorphic to the critical phenomena in confined liquid binary mixtures near the critical mixing point. This study is aimed at creating the physical model of transmission through a synaptic cleft at the cell-to-cell communication with taking into account such real processes: (a) time-dependent release of mediator (acetylcholine) molecules (ACh) to synaptic cleft, (b) receptor binding of ACh – formation of mediator–receptor complexes (AChR), (c) decay of ACh and AChR by action of a specific ferment – acetylcholinesterase (AChE).

The corresponding experimental studies show that the dependence of ACh concentration release into synaptic cleft upon the time t of medulla stimulation tends to reach saturation which can be approximated by the function $\sin \Omega t$ in the first quarter. Using such a simple approximation for the source function describing the intensity of ACh release, we obtain the following system of nonlinear differential equations where variables x, y, z correspond to the concentrations of ACh, AChR, AChE, respectively:

$$\frac{dx}{dt} = M \cos \Omega t - k_1 x(a-y) - k_2 xz - k_3 x,$$
$$\frac{dy}{dt} = k_1 x(a-y) - k_4 xyz\,,$$
$$\frac{dz}{dt} = k_5 x - k_2 xz - k_4 xyz\,.$$

The first equation describes such processes: first term gives the velocity of ACh release to synaptic cleft (with taking into account that velocity is a derivative of the source function, i.e., proportional to $\cos \Omega t)$, second term – velocity of receptor binding of ACh (a is the total amount of receptors on the postsynaptic membrane), third term – velocity of ACh decay under the action of AChE, and fourth term describes the process of ACh removing from the synaptic cleft due to diffusion. first term in the second equation corresponds to formation of mediator–receptor complexes (AChR), second term describes the velocity of AChR decay due to AChE. The third equation corresponds to such processes: velocity of AChE release after free ACh and AChR splitting (first term), second and third terms have the same meanings as in two previous equations.

To describe the effect of saturation for velocity of ACh release, it is reasonable to use the Taylor expansion for $\cos \Omega t$. As a result, one can obtain the following system of nonlinear differential equations:

$$\frac{dx}{dt} = M(1 - \frac{\Omega^2}{2} t^2) - k_1 x(a-y) - k_2 xz - k_3 x, \tag{24}$$
$$\frac{dy}{dt} = k_1 x(a-y) - k_4 xyz, \tag{25}$$
$$\frac{dz}{dt} = k_5 z - k_2 xz - k_4 xyz. \tag{26}$$

Considering the first term in Eq. (24) as a small perturbation, the stationary solution of system (24)–(26) has the following form: $x_0 = 0$, $y_0 = 0$, $z_0 = g$, where g is a certain concentration of AChE. This solution should satisfy the system (24)–(26) in the case when $k_5 = 0$. Really, for large t AChE has to split all ACh and AChR and its velocity of release becomes zero. Partial solution of the corresponding lineralized system (24)–(26) is as follows:

$$\delta x_{ps} = A_1 + B_1 t + C_1 t^2 + D_1 t^3, \quad \delta y_{ps} = A_2 + B_2 t + C_2 t^2 + D_2 t^3,$$
$$\delta z_{ps} = A_3 + B_3 t + C_3 t^2 + D_3 t^3\,.$$

Here

$$A_1 = \frac{2(k_1 a + k_2 g + k_3)M - M\Omega^2}{2(k_1 a + k_2 g + k_3)^3}, \qquad B_1 = \frac{M\Omega^2}{2(k_1 a + k_2 g + k_3)^2},$$
$$C_1 = -\frac{M\Omega^2}{2(k_1 a + k_2 g + k_3)}, \quad D_1 = 0, \quad A_2 = 0,$$

$$B_2 = k_1 a \frac{2(k_1 a + k_2 g + k_3)M - M\Omega^2}{2(k_1 a + k_2 g + k_3)^3}, \qquad C_2 = k_1 a \frac{M\Omega^2}{2(k_1 a + k_2 g + k_3)^2},$$

$$D_2 = -\frac{M\Omega^2}{2(k_1 a + k_2 g + k_3)}, \quad A_3 = 0, \quad B_3 = 0, \quad C_3 = 0, \quad D_3 = 0. \tag{27}$$

Solution of the characteristic equation for the corresponding homogeneous equation is $\lambda_1 = -\zeta$, $\lambda_2 = k_5$, $\lambda_3 = 0$, where $\zeta = k_1 a + k_2 g + k_3$.

Thus, general solution of the system (24)–(26) has such a form:

$$x = x_0 + A_1 + B_1 t + C_1 t^2 + F_1 \exp(\lambda_1 t) + F_2 \exp(\lambda_2 t) + F_3,$$
$$y = y_0 + B_2 t + C_2 t^2 + D_2 t^3 + G_1 \exp(\lambda_1 t) + G_2 \exp(\lambda_2 t) + G_3,$$
$$z = z_0 + H_1 \exp(\lambda_1 t) + H_2 \exp(\lambda_2 t) + H_3.$$

Here A_i, B_i, C_i, D_i are given in Eq. (27) and F_i, G_i, H_i can be found from initial conditions. The proposed kinetic model of synaptic transmission under normal conditions has a singular point being a strange attractor. If $\lambda_2 = k_5 \to 0$, a phase portrait of this kinetic model is a stable tore. It corresponds to our primary ideas about the process of synaptic transmission, namely: for large time t all the ACh molecules will be removed from the synaptic cleft due to diffusion and hydrolysis over AChE ($x_0 = 0$), all the mediator–receptor complexes will be splitted ($y_0 = 0$), all the receptors will be deactivated and AChE will be free and ready for a new action ($z_0 = g$).

6. Conclusions

It has been shown that the theoretical studies of finite-size effects can provide a deeper understanding of fundamental and applied aspects of critical phenomena in confined liquids. Experimental studies of such systems exhibiting size-dependent critical phenomena and phase transitions should be, without any doubt, very useful to obtain principal results for verification the basic ideas of the theory of finite-size soft-matter systems.

References

1. Fisher, M. E. (1971) Critical Phenomena. *Proceedings of the International School of Physics "Enrico Fermi"*, edited by M.S. Green (Academic, New York).
2. Privman, V. (ed.) (1990) *Finite Size Scaling and Numerical Simulation of Statistical Systems*, (World Scientific, Singapore).
3. Binder, K. (1992) Phase Transitions in Reduced Geometry, *Annu. Rev. Phys. Chem.* **43**, 33–59.
4. Chalyi, A. V. and Chernenko, L. M. (1993) Phase Transitions in Finite-size Systems and Synaptic Transmission, in *Dynamic Phenomena at Interfaces, Surfaces and Membranes*, edited by D. Beysens, N. Boccara, G. Forgacs (Nova Sci. Publ., New York).

5. Chalyi, A. V. (1993) Critical Phenomena in Finite-size Systems, *J. Mol. Liq.* **58**, 179–191.
6. Chalyi, A. V., Chalyi, K. A., Chernenko, L. M., and Vasil'ev, A. N. (2000) Critical Behavior of Confined Systems, *Cond. Matt. Phys.* **3**, 335–358.
7. Chalyi, K. A., Hamano, K., and Chalyi, A. V. (2001) Correlating properties of a Simple Liquid at Criticality in a Reduced Geometry, *J. Mol. Liq.* **92**, 153–164.
8. Vasiliev, A. N. (2003) Pair Correlation in Multi-component Anisotropic Liquid, *Theor. Math. Phys.* **135**, 714–720.
9. Chalyi, A.V., Chalyy, K. A., Chernenko, L. M., and Vasil'ev, A. N. (2004) Critical Phenomena in Confined Binary Liquid Mixtures, *NATO Sci. Series II. Math. Phys. Chem.* **157**, 143–152.
10. Formisano, F. and Teixeira, J. (2000) Critical Fluctuations of a Binary Fluid Mixture Confined in a Porous Medium, *Eur. Phys. J.* **E1**, 14.
11. Dierker, S. B. and Wiltzius, P. (1991) Statics and dynamics of a critical binary fluid in a porous medium, *Phys. Rev. Lett.* **66(9)**, 1185–1188.
12. Nissen, J. A., Chui, T. C. P., and Lipa, J. A. (1993) The Specific Heat of Confined Helium near the Lambda Point, *J. Low Temp. Phys.* **92(5–6)**, 353.
13. Mehta, S., Kimball, M. O., and Gasparini, F. M. (1999) Superfluid Transition of ^{4}He for Two-Dimensional Crossover, Heat Capacity and Finite-Size Scaling, *J. Low Temp. Phys.* **114**, 467.
14. Lipa, J. A., Swanson, D. R., Nissen, J. A., Geng, Z. K., Williamson, P. R., Stricker, D. A., Chui, T. C. P., Israelsson, U. E., and Larson, M. (2000) Specific Heat of Helium Confined to a 57-μm Planar Geometry near the Lambda Point, *Phys. Rev. Lett.* **84**, 4894–4897.
15. Diaz-Avila, M., Gasparini, F. M., and Kimball, M. O. (2004) Behavior of ^{4}He Near T_λ in Films of Infinite and Finite Lateral Extent, *J. Low Temp. Phys.* **134(1/2)**, 613–618.
16. Weichman, P. B., Harter, A. W., and Goodstein, D. L. (2001) Criticality and Superfluidity in Liquid ^{4}He under Non-equilibrium Conditions, *Rev. Mod. Phys.* **73**, 1.
17. Chalyi, A. V. (1992) Non-homogeneous Liquids near the Critical Point and the Boundary of Stability, *Sov. Sci. Rev. A. Phys.* **16**, 1–103.
18. Sugakov, V. I., Chalyi, A. V., and Chernenko, L. M. (1990) Neutron Critical Opalescence in a Non-homogeneous Liquid in the Gravitational Field, *Ukrainian Phys. J.* **35**, 1036–1041.
19. Brovchenko, I. and Oleinikova, A. (2005) Molecular Organization of Gases and Liquids at Solid Surfaces, *Handb. Theor. Comput. Nanotechnol.* **62**, 1–98.
20. Fisher, M. E. (1965) The Nature of Critical Points, in *Lectures in Theoretical Physics* (University of Colorado Press, Colorado).
21. Stanley, H. E. (1971) *Introduction to Phase Transitions and Critical Phenomena* (Clarendon Press, Oxford).
22. Patashinskii, A. Z. and Pokrovskii, V. L. (1979) *Fluctuation Theory of Phase Transitions* (Pergamon, Oxford).
23. Anisimov, M. A. (1987) *Critical Phenomena in Liquids and Liquid Crystals* (Nauka, Moscow).

WATER–BIOMOLECULE SYSTEMS UNDER EXTREME CONDITIONS: FROM CONFINEMENT TO PRESSURE EFFECTS

M.-C. BELLISSENT-FUNEL
Laboratoire Léon Brillouin (CEA-CNRS), CEA Saclay
91191 Gif-sur-Yvette Cedex (France)

Abstract. Water is a unique liquid, many of whose properties are critical for the continued support of life. In living systems, essential water-related phenomena occur in restricted geometries in cells, and at active sites of proteins and membranes or at their surface. The effects of hydration on equilibrium protein structure and dynamics are fundamental to the relationship between structure and biological function. In particular, the configuration of water molecules near the hydrophilic–hydrophobic interfaces is of considerable relevance. The structure and dynamics of water confined in model systems developing hydrophilic interactions are compared with that of bulk water as determined by neutron scattering. It is well known that hydration, internal dynamics, and function in proteins are intimately associated. Studies of dynamics of water molecules at surface of a C-phycocyanin protein are presented.

Keywords: Proteins, water, surface effects, neutron scattering.

1. Introduction

Water is the most abundant fluid on earth and the major constituent of organisms. At molecular level, in its bulk form, liquid water exhibits three features, the tetrahedral symmetry of the H_2O molecule, the large number of hydrogen bonds formed between neighboring molecules, and the very short characteristic lifetime (between 10^{-13} and 10^{-12} s) of these bonds. The explanation of such a short time is that hydrogen bond lifetimes are determined by the proton dynamics.[1] In particular, large amplitude librational movements are responsible for the making and breaking of hydrogen bonds. The solvent abilities of water arise primarily from its tendency to form hydrogen bonds and its dipolar character.[2] The high dielectric constant ($\varepsilon \sim 80$), already alluded to, results from its dipolar character. How are these properties modified when liquid water is put under pressure or in contact with hydrophilic or hydrophobic interfaces or both as it occurs in biological macromolecules. Water constitutes 70% mass of cell[3] and plays a major role in the stability and function of proteins and biological complexes. The structural properties of water are now mostly well understood in some range of temperatures and pressures using

S. J. Rzoska and V. A. Mazur (eds.), Soft Matter under Exogenic Impacts, 413–431.

different techniques such as X-ray diffraction,[4–6] neutron scattering,[7–15] molecular dynamics (MD),[16–22] and Monte Carlo (MC)[23] simulations. The number of models of its structure is very large and the number of experimental results enormous. The polymorphism of water is indeed remarkable and besides all the crystalline phases[24] one must add two amorphous solids, the low-density amorphous ice (LDA) and the high-density amorphous ice (HDA). Recently, a very high-density amorphous ice (VHDA) has been obtained. The microscopic structure of these forms of amorphous ice prepared by different techniques[25,26] has been studied both by X-ray[27,28] and neutron scattering.[29–31] In this paper, we recall the structural and dynamic properties of water in its bulk form as functions of temperature and pressure. We define what we call confined or interfacial water. We compare the structural and dynamic properties of confined water with that of bulk water. We report also on some unusual nucleation of water confined in a porous silica glass and on dynamic properties of water confined in the same silica glass and at surface of a C-phycocyanin protein.

2. Theory

For a molecular liquid, the structure factor may be split into two parts:[32]

$$S_{\mathrm{M}}(Q)=f_1(Q)+D_{\mathrm{M}}(Q) \tag{1}$$

where $f_1(Q)$ is defined as the molecular form factor

$$f_1(Q)=\left[b_{\mathrm{O}}^2+4b_{\mathrm{O}}b_{\mathrm{D}}\,j_0(Qr_{\mathrm{OD}})\times\exp(-\gamma_{\mathrm{OD}}Q^2)+2b_{\mathrm{D}}^2\,j_0(Qr_{\mathrm{DD}})\times\exp(-\gamma_{\mathrm{DD}}Q^2)\right]/(b_{\mathrm{O}}+2b_{\mathrm{D}})^2 \tag{2}$$

The $D_{\mathrm{M}}(Q)$ function contains all the intermolecular contributions. Thus, at large Q ($Q > 8$ Å^{-1}), the main contribution to $S_{\mathrm{M}}(Q)$ comes from $f_1(Q)$ whereas $D_{\mathrm{M}}(Q)$ goes to zero. Moreover, at $Q = 0$, $f_1(Q)$ is equal to unity and the interference function $S_{\mathrm{M}}(Q)$ has the thermodynamic limit $S_{\mathrm{M}}(0)=\rho_{\mathrm{M}}k_B T\chi_T$, where ρ_{M} is the density, k_B the Boltzmann constant, T the temperature, and χ_T the isothermal compressibility. The total pair correlation function $g(r)$ is related to the Fourier transform of $S_{\mathrm{M}}(Q)$ by the relation

$$4\pi r\rho_{\mathrm{M}}(g(r)-1)=\frac{2}{\pi}\int_0^{\infty}Q(S_{\mathrm{M}}(Q)-S_{\mathrm{M}}(\infty))\sin QR\,dQ \tag{3}$$

where $S_{\mathrm{M}}(\infty)=(b_{\mathrm{O}}^2+2b_{\mathrm{D}}^2)/(b_{\mathrm{O}}+2b_{\mathrm{D}})^2=0.3346$ is the asymptotic value of $f_1(Q)$ at large Q and ρ_{M} is the molecular density.

The function $g(r)$ is a combination of the different pair partial correlation functions and includes both the intramolecular and intermolecular distances. It is more convenient to remove the intramolecular terms by subtracting the molecular form factor from $S_{\mathrm{M}}(Q)$ to obtain $D_{\mathrm{M}}(Q)$, which may be Fourier transformed in order to obtain the pair correlation function $g_L(r)$ for intermolecular terms only:

$$d_L(r) = 4\pi r\rho_M(g_L(r)-1) = \frac{2}{\pi}\int_0^\infty QD_M(Q)\sin QRdQ \tag{4}$$

In this way, no truncation effect arises; moreover, the first peaks due to the intramolecular distances disappear in the low *r* range.
The composite $g_L(r)$ function is thus defined as:

$$g_L(r) = 0.489g_{DD}(r) + 0.421g_{OD}(r) + 0.090g_{OO}(r) \tag{5}$$

where the dominant contributions concern the DD and OD spatial correlations.

3. The microscopic structure of water

3.1. SUPERCOOLED WATER AT NORMAL PRESSURE, AS A FUNCTION OF TEMPERATURE[33]

3.1.1. Structure factors $S_M(Q)$

According to the temperature range of the experiment, two kinds of D_2O samples have been used: (i) pure liquid D_2O and in order to increase the degree of supercooling of water, (ii) D_2O in emulsions with sorbitan as a surfactant. The emulsion was constituted either with deuteroheptane or with a mixture of deuteroheptane and CCl_4. The average size of the water droplets in the emulsion was around some μm.

Figure 1 gives the structure factors of supercooled water,[33] at respectively, –10.5°C and –31.5°C as compared with that of low-density amorphous ice. When the temperature is decreased it appears that the main peak position Q_0 of the structure factor of supercooled water tends towards the value 1.69 Å that is characteristic of low-density amorphous ice.[29] The composite pair correlation functions of supercooled water[33] at –10.5°C and –31.5°C are displayed in Fig. 2 and compared with that of low-density amorphous ice.[29]

At –31.5°C, in the small-*r* range, some features characteristic of low-density amorphous ice show up. In particular, the O–D hydrogen bond distance at 1.82 Å and the D–D intermolecular distance at 2.31 Å are present. In the large-*r* range one observes additional small oscillations that are not present at –10.5°C and an out of phase behavior of the broad oscillations for values of *r* higher than 10 Å. These features are similar to those observed in low-density amorphous ice.

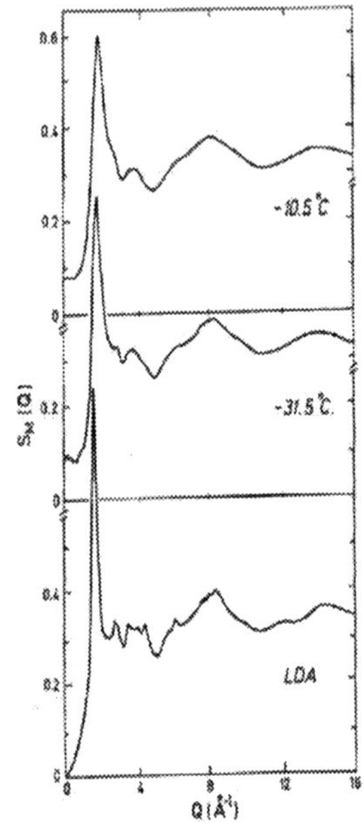

Figure 1. Structure factors $S_M(Q)$ of supercooled D_2O, at normal pressure, as compared with that of low-density amorphous ice (LDA) at 77 K.[29]

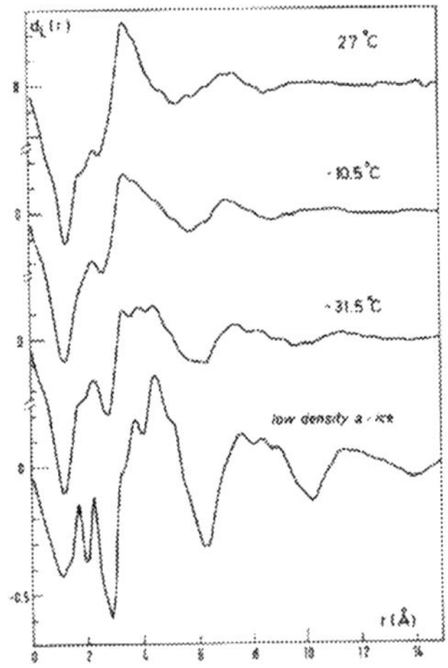

Figure 2. Pair correlation functions $d_L(r)$ of supercooled D_2O, at normal pressure, as compared with that of low-density amorphous ice at 77 K.[29]

3.1.2. *Pair correlation functions $d_L(r)$*

At normal pressure and low temperature, the measurements confirm the increasing spatial correlations in deeply supercooled water as the temperature is decreased and the tendency to evolve towards the structure of low-density amorphous ice. In supercooled water, the structure present between 4.5 and 6.0 Å is connected to the formation of tetrahedrally coordinated patches.[34] It is clear that such patches are not present in the high-pressure liquid water, either because the hydrogen bonds are broken, i.e., the molecular energy is on average too low, or more likely because the hydrogen bond network is distorted.

3.2. LIQUID WATER UNDER PRESSURE[35]

The microscopic structure of liquid water under pressure has been studied at pressures up to 6000 bar and in a range of temperatures from 53 down to –65 °C. The experiments have been performed by neutron diffraction with the same samples as those described in Section 3.1.1. A scheme of the Ti–Zr pressure cell is given in Fig. 3. This cell is made with a "null" Ti–Zr alloy which scatters neutron incoherently as the mean coherent scattering length for the Ti and Zr atoms is zero for the composition of the alloy: 34% titanium and 66% zirconium. The dimensions of the cell (internal diameter = 5.70 mm, wall thickness = 5.22 mm) have been determined in order that the cell supports pressures up to 10 kbar.

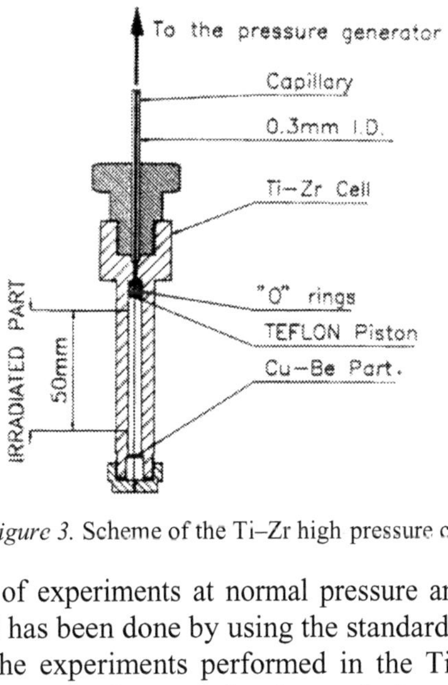

Figure 3. Scheme of the Ti–Zr high pressure cell.

The data analysis of experiments at normal pressure and performed in a thin walled vanadium cell has been done by using the standard procedure.[33] The data analysis relative to the experiments performed in the Ti–Zr pressure cell has been done by using a procedure previously applied to He–Xe mixture[36] and which allows us to correct for the multiple scattering either in the cell alone or both in the cell and the sample.

3.2.1. Structure factor $S_M(Q)$ and the evolution of the position of the main diffraction peak Q_0 as functions of temperature and pressure

Figure 4 gives the structure factors of liquid water under 6000 bar and, at respectively, –4°C and 53°C[35] as compared with that of high-density amorphous ice (HDA).[29] Under a pressure of 6000 bar, when the temperature is increased up to 53°C, it appears that the main peak position Q_0 of the structure factor of liquid water tends towards the value of 2.20 Å which is characteristic of that of high-density amorphous ice.[29] A very slight increase of the main peak position (from 2.16 to 2.20 $Å^{-1}$) is obtained when the temperature is raised from –4°C to 53°C. When the pressure is increased, from normal pressure to 6000 bar, the effect of temperature becomes smaller which confirms that the effect of pressure is to reduce the anomalies of liquid water that evolves to a normal associated liquid.[37] Figure 5 gives the evolution of the main diffraction peak position Q_0 of the structure factor: (a) at constant pressure as a function of temperature and (b) at constant temperature as a function of pressure.

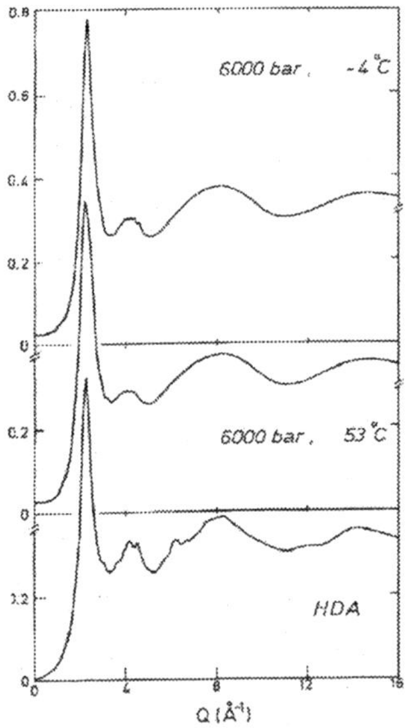

Figure 4. Structure factors $S_M(Q)$ of liquid D_2O, under 6000 bar, as compared with that of high-density amorphous ice (HDA) at 77 K.[29]

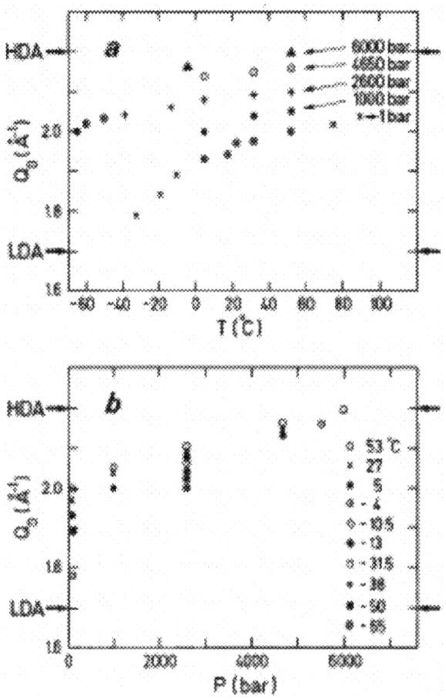

Figure 5. Evolution of the main diffraction peak position Q_0 of the structure factor (a) at constant pressure as a function of temperature and (b) at constant temperature as a function of pressure.

3.2.2. Pair correlation functions $d_L(r)$

Figure 6 gives the composite pair correlation functions of water under 6000 bar and, at respectively, –4°C and 53°C, as compared with that of high-density amorphous ice. Upon increasing the temperature up to 53°C it appears that there is a slight evolution of the $d_L(r)$ functions that present the characteristics of that of high-density amorphous ice.[29] All the peaks are broader, showing that interatomic correlations become weaker as the pressure is increased at high temperature. A very broad peak is present at 3.34 Å and a minimum exists at 5.0 Å, i.e., in the region where there is a maximum for the low-density form. The O–D hydrogen bond distance and the D–D intermolecular distance are shifted respectively to 1.84 and 2.36 Å. Moreover, the peaks observed by X-ray[38] at 2.8 Å and around 4.6 Å appear as small bumps.

On one hand, at normal pressure and low temperature, the measurements confirm the increasing spatial correlations in deeply supercooled water as the temperature is decreased and the tendency to evolve towards the structure of low-density amorphous ice. On the other hand, the structural study of liquid water, at 6000 bar and respectively –4°C and 53°C,[35] confirms the tendency that the high-density amorphous ice appears as the limit of liquid water at high temperature and high pressure. Under a pressure of 6000 bar, the highest pressure to be reached with our pressure experimental set-up, the $d_L(r)$ function of liquid water at 53°C exhibits features which are characteristic of high-density amorphous ice. If for the supercooled water the structure present between 4.5 and 6.0 Å is connected to the formation of tetrahedrally coordinated patches, it is clear that such patches are not present in the high-pressure liquid water, either because the hydrogen bonds are broken, i.e., the molecular energy is on average too low, or more likely because the hydrogen bond network is distorted.

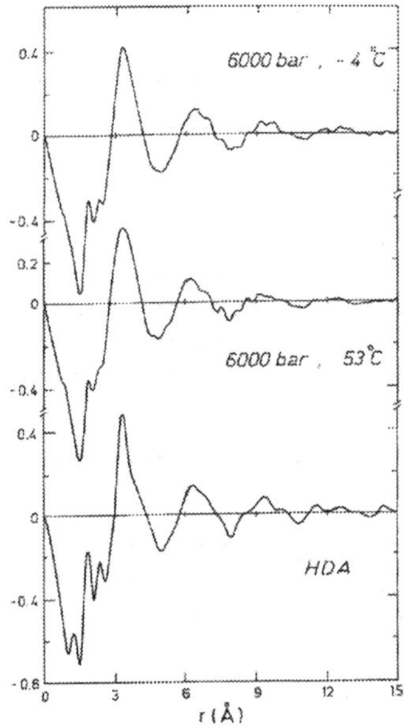

Figure 6. Pair correlation functions $d_L(r)$ of liquid D_2O, under 6000 bar, as compared with that of high- density amorphous ice (HDA) at 77 K.[30]

4. Dynamic properties of bulk water

Quasi-elastic neutron scattering technique is a powerful and unique tool for studying the self-dynamics of water; actually the large incoherent scattering cross section of the protons yields unambiguous results about the individual motions of water molecules. It gives access to the correlation function for the atomic motions which are explored over a space domain of the order of a few Å and for times of the order of 10^{-12} s (Ref. 39). This space and time domain is similar to that of CMD that makes the comparison between neutron scattering and CMD justified. The correlation function can be calculated for various models for the assumed motion of the diffusing particles (e.g., Brownian motion, jump diffusion, diffusion in a confined space, rotational motion, etc.) and the microscopic properties of the environment of the scattering atoms, e.g., the residence time, the jump length, the diffusion coefficients, the H-bond lifetime become available.

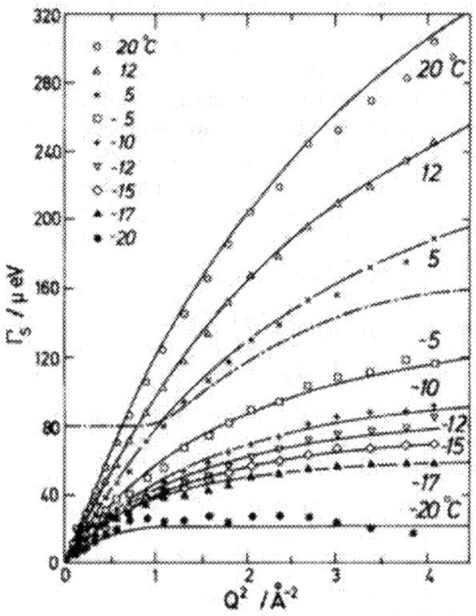

Figure 7. Linewidths of the quasi-elastic incoherent neutron scattering of bulk water at different temperatures.[40] The solid lines are the resulting fitting with the randomjump diffusion model.[39] The plateau at large Q is proportional to the residencetime of the molecules. The dashed line represents the Lorentzian linewidth for water bound to C-phycocyanin at room temperature.[41]

Another quantity that can be obtained is the density of states of mobile protons, in particular the translational and librational modes of water.

The traditional way of analyzing the quasi-elastic region of the spectra assumes separation and decoupling of three types of proton motions: vibrations, center-of-mass translations (i.e., motion of the whole molecule), and proton rotation around the center of mass (oxygen atom for water). The proton vibrational contribution in this region (which are beyond experimental resolution) is approximated by a Debye–Waller factor, where $\langle u^2 \rangle$ is the mean square amplitude of proton vibrations. We have studied by quasi-elastic neutron scattering the individual motions of water molecules in liquid water[40] as a function of the temperature and down to –20°C in the supercooled state. Translational diffusive motions are generally well described by a Lorentzian function. Figure 7 represents the linewidths of the quasi-elastic incoherent neutron scattering of bulk water for a series of nine temperatures ranging from 20°C down to –20°C. Experimental data are well accounted for by the random jump diffusion model.[39] The plateau at large Q is inversely proportional to the residence time τ_0 of the molecules. The dashed line represents the Lorentzian linewidth for water bound to C-phycocyanin at room temperature.[41] Water rotational relaxation has been successfully described using a diffusion model proposed by Sears for hindered rotations over a sphere.[40] The characteristic time for the hindered rotations can be associated with the hydrogen bond lifetime τ_1.

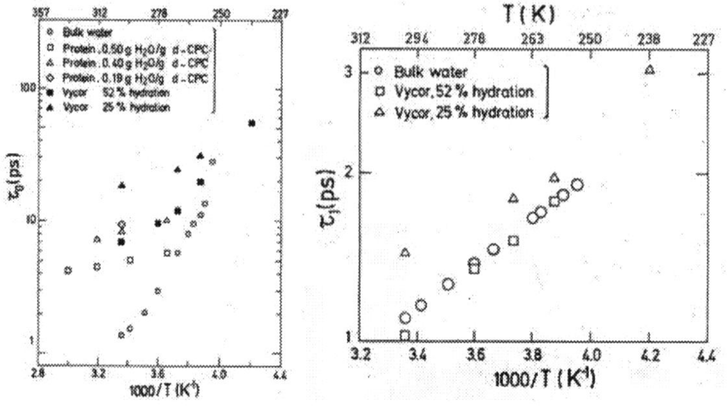

Figure 8. (a) (left side) Arrhenius plote of the residence Time τ_0 for bulk water and for hydrated Vycor and C-phycocyanin protein.

(b) (right side) Arrhenius plot of the H bond lifetime τ_1 for bulk water and for two levels hydration of Vycor.

Figures 8 a and 8 b represent respectively the residence time τ_0 and the H bond life-time τ_1 of water molecules, for bulk and confined water.

5. Microscopic structure of confined water

In many technologically important situations, water is not in its bulk form but is instead attached to some substrates, or filling small cavities. Common examples are water in porous media, such as rock or sand stones, and water in biological material as in the interior of cells or attached to surfaces of biological macromolecules and membranes. This is what we define here as the "confined" or the "interfacial water."

Water in confined space has attracted a considerable interest in the recent years. It is commonly believed that the structure and dynamics of water are modified by the presence of solid surfaces, both by a change of hydrogen bonding and by modification of the molecular motion that depends on the distance of water molecules from the surface. Among the hydrophilic model systems, let us mention clay minerals,[42,43] and various types of porous silica.[44,45] In the last case, the authors have interpreted their results in terms of a thin layer of surface water with more extensive H-bonding, lower density and mobility, and lower nucleation temperature as compared to bulk water.

The structure of water confined in a Vycor glass[46] has been studied by neutron diffraction. Vycor glass is a porous silica glass, characterized by a quite sharp distribution of cylindrical interconnected pores, and hydrophilic surfaces. We present here some of our results[47] which have been obtained as functions of temperature and level of hydration from full hydration (0.25 g water/g dry

Vycor) down to 25%. Based on the information that the dry density of Vycor is 1.45 g/ml, the porosity 28%, and the internal cylindrical pores of cross sectional diameter 50 Å, the 50% hydrated sample has three layers of water molecules on its internal surface and a 25% hydrated sample corresponds roughly to a monolayer coverage.

5.1. FULLY AND PARTIALLY HYDRATED VYCOR

Results for two levels of hydration of Vycor demonstrate that the fully hydrated case is almost identical to the bulk water and the partially hydrated case is of little difference. However, the three site–site radial correlation functions are indeed required for a sensible study of the orientational correlations between neighboring molecules.[48] An interesting feature is that we observe in differential analysis measurements, supercooling of water to a much lower temperature of −27°C in a partially hydrated sample compared to a fully hydrated sample (−18°C).[47]

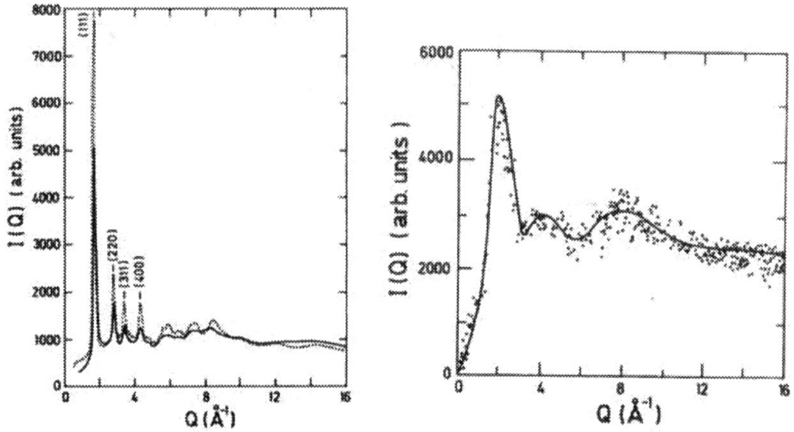

Figure 9. (*a*) (left side) Spectrum of cubic ice (−198 °C) (dotted line) compared with that of confined D_2O at −100°C from fully hydrated Vycor (full line).

(b) (right side) Spectrum of confined D_2O at −18°C from fully hydrated Vycor after subtraction of Bragg peaks. There is 23% liquid water.[47]

It seems that the confinement of the water favors the nucleation of cubic ice that appears superimposed on the spectrum of liquid water and whose proportion can be deduced from the intensity of the (111) Bragg peak. The proportion of cubic ice increases with decreasing temperature. In fact, at −100°C, the spectrum of confined water looks similar to that of cubic ice (Fig. 9a). This is in sharp contrast to bulk water that always nucleates into hexagonal ice. In Fig. 9b, we show a spectrum which gives a clear evidence of water which is present below the Bragg peaks at –18°C obtained by subtraction of the weighted spectrum of the same sample cooled down to –100°C. The main diffraction peak position is close to 2 $Å^{-1}$. It looks that confined water from fully hydrated Vycor behaves like water under pressure.[35]

5.2. MONOLAYER WATER COVERAGE AT VYCOR SURFACE

Results relative to a 25% hydrated Vycor sample indicates that, at room temperature, interfacial water has a structure similar to that of bulk supercooled water[33] at a temperature of about 0°C which corresponds to a shift of about 25 K. Therefore the structure of interfacial water is characterized by an increase of the long-range correlations that corresponds to the building of the H-bond network as it appears in low-density amorphous ice.[29] There is no evidence of ice formation when the sample is cooled from room temperature down to –196°C (liquid nitrogen temperature).[49]

6. Translational motions of interfacial water

6.1. A SIMPLE MODEL FOR DYNAMICS

A complete study of the self-dynamics of water close to some well-defined hydrophilic surface as the Vycor surface has been performed by quasi-elastic neutron scattering. It has been done for levels of hydration ranging from full hydration down to the lowest one (25%), corresponding to one monolayer coverage of water molecules. The effect of temperature has also been studied. We report here on the main results.[50] The short-time diffusion (few picoseconds) of water molecules close to the Vycor surface has been described in terms of simple models for all the studied samples.[50] At short times, the water molecules, close to some hydrophilic surface, perform very local rotational jumps characterized by D_t and τ_1 like in bulk water, but with a longer residence time τ_0 on a given site before diffusing to an adjacent site along the surface with a diffusion coefficient equal to D_{local}. This diffusion is limited to some volume estimated as spherical. For the 25% hydrated sample, the diffusion coefficient measured by NMR[51] appears to be smaller than D_t, which is smaller than D_{local}. This is due to the fact that NMR technique measures the long-time and long-range diffusion coefficient.

The effect of the temperature has been followed down to –35°C. The radius of the spherical volume of confinement varies between 5 and 2 Å; it decreases when the temperature is lowered which means that water molecules are more localized at low temperatures. The observed trend seems reasonable. The values obtained for D_{local} are low which demonstrate the influence of the hydrophilic groups when one reaches a monolayer coverage of water molecules. Moreover, these values are close to the values of the diffusion coefficient of water molecules, at the immediate hydrophilic interface, as determined in a MD simulation by Lee and Rossky.[52]

6.2. THE RELAXING CAGE MODEL

Recently nanosecond time molecular dynamics simulations of supercooled water have been performed. The data have been described with success in terms of the relaxing cage model. The relaxing cage model uses an idea borrowed from mode-coupling theory (MCT) of supercooled liquids. Mode-coupling theory focuses its attention on the "cage effect" in the liquid state that can be pictured as a transient trapping of molecules by their neighbors as a result of lowering of the temperature.[53] In supercooled or interfacial water due to the reduced thermal energy of the molecule and for water due to the formation of a stabler, hydrogen-bonded cage around each molecule, a molecule can translate a substantial distance only by rearranging positions of a large number of molecules around it. Thus the diffusion is strongly coupled to the local structural rearrangements or the structural relaxation. The usual, Markovian, Brownian-like diffusion is no longer valid in the case of supercooled water. The relaxing cage model appears now the more appropriate one to describe the diffusional dynamics in supercooled and interfacial water.[54]

The slow relaxation is characterized by a correlation time τ, related to the lifetime of the cage and a stretched exponent β. $A(Q)$ is the so-called Debye–Waller factor (DWF) which has a Gaussian shape. This implies that the short-time dynamics of interfacial or supercooled water is, to a good approximation, harmonic:

$$A(Q) = \exp(-\frac{1}{3}Q^2a^2) \tag{6}$$

where a is the root mean square vibrational amplitude of water molecules in the cage in which the particle is constrained during its short-time movements. From the results of the MD simulation of the SPC/E water at supercooled temperature, a is found to be about 0.5 Å. Thus for a Q-range between 0 and 1.0 Å^{-1}, applicable to high resolution QENS spectra, the DWF in Eq. 6 is very nearly unity. We can thus write with a good approximation the final intermediate scattering function as

$$F_S(Q,t) = A(Q)\exp\left[-\left(\frac{t}{\tau}\right)^{\beta}\right] \tag{7}$$

The dynamic structure factor of the interfacial water is given by the expression:

$$S_S^{H_2O}(Q,\omega) = \frac{A(Q)}{\pi} \int_0^{\infty} dt \cos \omega t \exp\left[-\left(\frac{t}{\tau}\right)^{\beta} \right] \tag{8}$$

where the Fourier transform is carried out numerically. The theoretical model is convoluted by the resolution of the spectrometer.

In the following the relaxing cage model has been applied in the case of water from fully hydrated Vycor and for hydration water of a C-phycocyanin protein.

6.3. EXAMPLES

6.3.1. Water molecules in hydrated Vycor[55]

The quasi-elastic neutron scattering experiments have been performed at the High Flux Reactor of the Institut Laue Langevin (Grenoble) using the IN5 time of flight spectrometer. The energy resolution (FWHM) at the elastic position was 10 μeV and the Q range covered was from 0.15 to 0. 99Å^{-1} (using 10 Å neutrons). The sample consists of H_2O-hydrated Vycor rectangular plates of thickness 1.9 mm and surface area of 32×36 mm^2. Vycor plates are thin enough to ensure that neutron transmission with water inside is 90%. The procedure of fitting the spectra is described into details in Ref. 55. In particular, the structural relaxation rate and the stretch exponent β are obtained. Figure 10 shows a plot of the structural relaxation rate as a function of Q in log–log scale for 100% hydrated sample at five temperatures, three of the lower ones are supercooled. It looks obvious that the structural relaxation rate has a power law dependence on Q given by

$$1/\tau \propto Q^{\gamma} \tag{9}$$

The values of exponent γ are given in the inset for each temperature. The Q-dependence of the stretch exponent β for the same sample shows that values of β are significantly below unity for large Q, but approaching unity for Q less than 0.1 Å^{-1}. One can infer from these results that even at the highest temperature measured, the hydrodynamic limit is probably reached for Q only below 0.1 Å^{-1}.

In the case of a stretched exponential relaxation, one can calculate an average relaxation time, $\bar{\tau}$ that sums up in one-parameter the two Q-dependent τ and β parameters. $\bar{\tau}$ represents the area below the stretched exponential curve and can be calculated with the following relation:

$$\bar{\tau} = \int_0^{\infty} \mathrm{d}t \exp\left[-\left(\frac{t}{\tau}\right)^{\beta}\right] = \frac{\tau}{\beta}\Gamma\left(\frac{1}{\beta}\right) \tag{10}$$

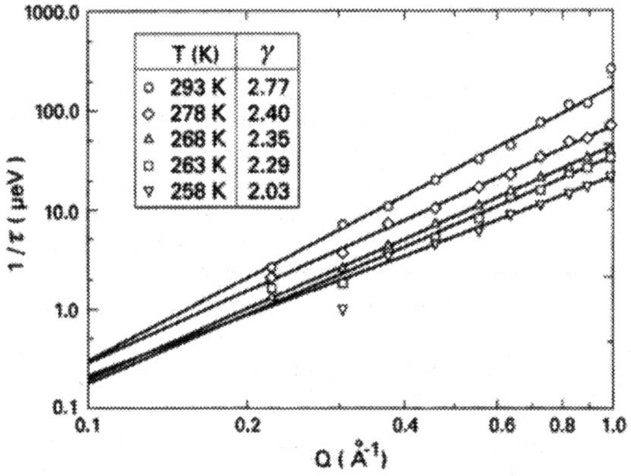

Figure 10. Evidence of a power law dependence of the structural relaxation rate on Q (Eq. 9), measured with 10 μeV resolution, in the 100% hydrated sample. This graph shows a log–log plot of $1/\tau$ vs Q. Within a Q range of 0.1–1 $Å^{-1}$ the exponent γ, which is the slope of the solid lines joining the data points, seems to be a constant. The corresponding data from MD simulation[54] show that the exponent γ is Q-dependent, approaching 2 as Q goes below 0.1 $Å^{-1}$.

Figure 11 is a log–log plot of the average relaxation time $\bar{\tau}$ as a function of Q. One sees that the average relaxation time has, within this Q range, a power law dependence

$$\bar{\tau} \approx Q^{-\gamma'} \tag{11}$$

with an exponent γ' approximately equal to 2 at room temperature, similar to the simple diffusion case. One may then define an average diffusion constant by the relation

$$\bar{\tau} = 1/\bar{D}Q^2 \tag{12}$$

and use it to estimate the average diffusion coefficient $\bar{D}$. From Fig. 10, for 100% hydrated sample at 293 K, we get $\gamma' = -1.95$, $\bar{\tau} = 944$ ps at Q=0.1 $Å^{-1}$. Substituting these two numbers into the above relation, we get $\bar{D} = 1.1\times10^{-5}$ cm^2/s, compared to the measured self-diffusion constant for bulk water at this temperature,[40] that is $D = 2.0\times10^{-5}$ cm^2/s. We thus arrive at a ratio $D/\bar{D} = 1.8$, agreeing with the estimate of Lee and Rossky.[52] We may also say that as far as the single-particle dynamics is concerned, water in Vycor at 293 K behaves as that in a bulk water at 273 K (20° below). As the temperature goes below the freezing point, the exponent γ' becomes less than 2, indicating a deviation from simple diffusion.

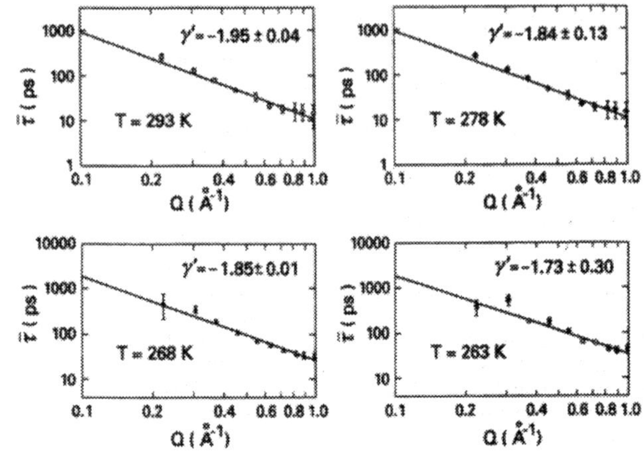

Figure 11. Average relaxation time $\overline{\tau}$ [defined in Eq. 10] for the 100% hydrated sample plotted against Q in log–log scales. It can be seen that the slope is approximately –2 at room temperature.

6.3.2. Water molecules in hydration water of protein[56]

Relaxational dynamics of water molecules at the surface of a C-phycocyanin protein[56] has been studied by high resolution quasi-elastic neutron scattering.

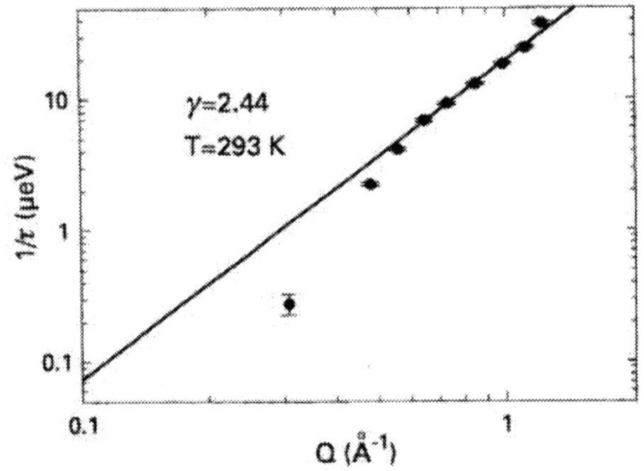

Figure 12. Log–log plot of $1/\tau$ vs Q, at T=293 K. Evidence of a power law dependence on the structural rate $1/\tau$ on Q.

The neutron quasi-elastic spectra are well described by the relaxing cage model. Figure 12 shows a plot of the structural relation rate as a function of Q on a linear scale for the interfacial water at 293 K. It looks obvious that the structural relation rate has a power law dependence on Q given by $1/\tau \propto Q^{\gamma}$. We obtain a value for $\gamma = 2.44 \pm 0.05$. The power law behavior is consistent with experimental results on hydrated Vycor, where a similar Q-dependence is found with $\gamma=2.51$ at 298 K and $\gamma=2.21$ at 278 K. We can also compare the correlation times obtained from these two experiments: $\tau=35.6$ ps at $Q = 0.98$ Å^{-1}; in this case while in case of fully hydrated Vycor, at the same Q-value, $\tau=5$ ps, at 293K. This is in agreement with the fact that when decreasing the level of hydration at a hydrophilic surface, the correlation time of water molecules increases.[54] We can calculate an average diffusion coefficient $\overline{D}$ according to Eq. 12. We find the value of $\overline{D}=0.15\times10^{-5}$ cm^2/s. This value is smaller than the self-diffusion coefficient of bulk water (2.3×10^{-5} cm^2/s) at the same temperature. It is also smaller than the average diffusion coefficient of water from fully hydrated Vycor ($\overline{D}=1.1\times10^{-5}$ cm^2/s, at 293 K).[55]

We have shown that the translational motions of interfacial water near the protein surface can be well described by an α-relaxation model. This model allows us to analyze in detail the slow dynamics of water; the same model has also been used for water in hydrated porous Vycor glass. However, for an equivalent hydration rate of protein, relaxation times have been found to be longer than in Vycor.

7. Conclusion

The recent results from neutron scattering experiments[55–58] and molecular dynamics simulations[54] establish clearly the existence of α-relaxation in supercooled or interfacial water suggesting that the dynamics of supercooled water can be described in terms of the relaxing cage model. The relaxation times of interfacial water exhibit a power law dependence on the wavevector Q. The average diffusion coefficient for water at protein surface is ten times lower than that of bulk water. This confirms that there is a retardation of water molecules at the protein surface, in good agreement with results of water at the surface of hydrophilic model systems.

References

1. Bellissent-Funel, M.-C. and Teixeira, J. (1998) Structural and dynamic properties of bulk and confined water in: L. Rey and J.C. May (Eds.), *Freeze-Drying/ Lyophilization of Pharmaceutical and Biological Products*, (Marcel Dekker, New York) pp. 53–77.
2. Teixeira, J.J. (1993) *Phys. IV*, **3**, C1 163 and references therein.
3. Ball, P. (1999) *H_2O: a Biography of Water*, (Weidenfeld and Nicholson, London).
4. Narten, A.H. and Levy, H.A. (1972) *Water : A comprehensive treatise* (Franks F., ed.), vol. 1 (Plenum Press, New York, London) p. 311.
5. Egelstaff, P.A. (1983) *Adv. Chem. Phys.* **53**, 1.
6. Okhulkov, A.V., Demaniets, Yu.N., and Gorbaty, Yu.E. (1994) *J. Chem. Phys.* **100**, 1578.

7. Page, D.I. (1972) *Water: A Comprehensive Treatise* (Franks F., Ed.), vol. 1 (Plenum Press, New York, London) p. 333.
8. Chen, S.H. and Teixeira, J.J. (1985) *Adv. Chem. Phys.* **64**, 1.
9. Dore, J.C. (1985) *Water Science Reviews* (Franks F., Ed.), vol. 1 (Cambridge University Press) p. 3.
10. Bellissent-Funel, M.-C. (1991) *Hydrogen Bonded Liquids*, J.C. Dore and J. Teixeria, (eds.) vol. 329, NATO ASI Series C, (Kluwer Academic Publishers) p. 117.
11. Wu, A.Y., Whalley, E., and Dolling, G. (1982) *Mol. Phys.* **47**, 603.
12. Neilson, G.W. and Cummings, S. (1984) *Revue Phys. Appl.* **19**, 803.
13. Bellissent-Funel, M.-C. and Bosio, L. (1995) *J. Chem. Phys.* **102**, 3727.
14. Bellissent-Funel, M.-C. (1995) *Physical Chemistry of Aqueous Systems*, H.J. White, Jr, J.V. Sengers, D.B. Neumann, and J.C. Bellows (Eds.) (Begell House, New York), p. 332.
15. Dore, J.C., Garawi, M., and Bellissent-Funel, M.-C. (2004) *Special issue of Molecular Physics* **102**, 2015–2035.
16. Stillinger, F.H. and Rahman, A. (1974) *J. Chem. Phys.* **61**, 4973.
17. Matsuoka, O., Clementi, E., and Yoshimine, M. (1976) *J. Chem. Phys.* **64**, 1351.
18. Corongiu, G. and Clementi, E. (1992) *J. Chem. Phys.* **97**, 2030.
19. Poole, P.H., Sciortino, F., Essmann, U., and Stanley, H.E. (1992) *Nature* **360,** 324.
20. Poole, P.H., Essmann, U., Sciortino, F., and Stanley, H.E. (1993) *Phys. Rev. E* **48**, 4605.
21. Tanaka, H. (1996) *J. Chem. Phys.* **105**, 5099.
22. Harrington, S.T., Zhang, R., Poole, P.H., Sciortino, F., and Stanley, H.E. (1997) *Phys. Rev. Lett.* **78**, 2409.
23. Madura, J.D., Petit, B.M., and Calef, D.F. (1988) *Mol. Phys.* **64**, 325.
24. Hobbs, P.V. (1974) *Ice Physics* (Clarendon, Oxford).
25. Chowdhury, M.R., Dore, J.C., and Wenzel, J.T. (1982) *J. Non Cryst. Solids* **53**, 247.
26. Mishima, O., Calvert, L.D., and Whalley, E. (1985) *Nature* (London) **314**, 76.
27. Bosio, L., Johari, G.P., and Teixeira, J. (1987) *Phys. Rev. Lett.* **87**, 460.
28. Bizid, A., Bosio, L., Defrain, A., and Oumezzine, M. (1987) *J. Chem. Phys.* **87**, 2225.
29. Bellissent-Funel, M.-C., Teixeira, J., and Bosio, L. (1987) *J. Chem. Phys.* **87**, 2231.
30. Bellissent-Funel, M.-C., Bosio, L., Hallbrucker, A., Mayer, E., and Sridi-Dorbez, R. (1992) *J. Chem. Phys.* **97**, 1282.
31. Loerting, T., Salzman, C., Kohl, I., Meyer, E., and Hallbrucker, A. (2001) *Phys. Chem. Chem. Phys.* **3**, 5355–5357.
32. Lovesey, S.W. (1990) Theory of neutron scattering from condensed matter, 3rd edition, Oxford Univ. Press., NY.
33. Bellissent-Funel, M.-C., Teixeira, J., Bosio, L., and Dore, J.C. (1989) *J. Phys.: Cond. Matter* **1**, 7123.
34. Boutron, P. and Alben, R. (1975) *J. Chem. Phys.* **62**, 4898.

35. Bellissent-Funel, M.-C. and Bosio, L.J. (1995) *Chem. Phys.* **102** (9), 3727.
36. Bellissent-Funel, M.-C., Buontempo, U., Petrillo, C., and Ricci, F.P. (1989) *Phys. Rev. A* **40**, 7346.
37. Stanley, H.E. and Teixeira, J. (1980) *J. Chem. Phys.* **73**, 3404.
38. Okhulkov, A.V., Demaniets, Yu.N., and Gorbaty, Yu.E. (1994) *J. Chem. Phys.* **100**, 1578.
39. Teixeira, J., Bellissent-Funel, M.-C., Chen, S.H., and Dianoux, A.J. (1985) *Phys. Rev. A* **31**, 913.
40. Springer, T. (1972) Quasi-elastic neutron scattering for the investigation of diffusive motions in solids and liquids, *Springer Ser. Modern Phys*. **64**, 64.
41. Bellissent-Funel, M.-C., Zanotti, J.M., and Chen, S.H. (1996) *Far. Disc.* **103**, 281 and references therein.
42. Hawkins, R.K. and Egelstaff, P.A. (1980) *Clays and Clay Minerals* **28**, 19.
43. Soper, A.K., (1991) *Hydrogen-Bonded Liquids*, J.C. Dore and J. Teixeira (Eds.), (Kluwer Academic Publishers) Vol. **329**, P. 147.
44. Dore, J.C., Coveney, F., and Bellissent-Funel, M.-C. (1992) *Recent Developments Developments in the Physics of Fluids*, W.S. Howells and A.K. Soper (Adam Hilger Publisher). p. 299.
45. Christenson, H.K. (2001) *J. Phys.: Condens. Matter* **13**, R 95.
46. General information on Vycor Brand Porous thirsty glass, n° 7930, Corning Glass Works, is available from OEM Sales Service, Box 5000, Corning, NY 14830, USA.
47. Bellissent-Funel, M.-C., Bosio, L., Lal, J., et al., (1993) *J. Chem. Phys.* **98**, 4246.
48. Bruni, F., Ricci, M.A., and Soper, A.K. (1998) *J. Chem. Phys.* **109**, 1478.
49. Zanotti, J.-M., Bellissent-Funel, M.-C., and Chen, S.-H. (2005) *Europhys. Lett.* **71**, 1–7.
50. Bellissent-Funel, M.-C., Chen, S.H., and Zanotti, J.-M. (1995) *Phys. Rev. E* **51**, 4558.
51 Van der Maarel, private communication.
52. Lee, S.H. and Rossky, P.J. (1994) *J. Chem. Phys.* **100**, 3334.
53. Gotze,W. and Sjogren, L. (1992) *Rep. Prog. Phys.* **55**, 241.
54. Zanotti, J.-M., Bellissent-Funel, M.-C., and Chen, S.H. (1999) *Phys. Rev. E* **59**, 3084.
55. Chen, S. H., Liao, C., Sciortino, F., Gallo, P., and Tartaglia, P. (1999) *Phys. Rev. E* **59**, 6708.
56 Dellerue, S. and Bellissent-Funel, M.-C. (2000) *Chem. Phys.* **258**, 315–325
57. Chen, S.H. (1999) *Hydration Processes in Biology: Theoretical and Experimental Approaches* (M.-C. Bellissent-Funel Ed.) (IOS Press Publishers.) Vol. 305, and references therein.
58. Bellissent-Funel, M.-C., Longeville, S., Zanotti, J.-M., and Chen, S.H. (2000) PRL **85**, 3644.

RECENT PROGRESSES IN UNDERSTANDING OF WATER INTERACTING WITH BIOMOLECULES, AND INSIDE LIVING CELLS AND TISSUES

ROBERT C. FORD*
Faculty of Life Sciences, University of Manchester, Manchester M60, 1QD, UK
JICHEN LI
School of Physics, University of Manchester, Manchester M60b 1QD, UK

Abstract. Recent inelastic and quasi-elastic neutron scattering measurements of water in cell preparations has provided information on the interfacial (or bound) water molecules. The experiments show that the interfacial water molecules can be readily distinguished from those in the bulk (bulk water), especially using inelastic neutron scattering data over the 20–130 meV range. Studies of intact biological systems – whole cells and tissues – demonstrated the feasibility of using these methods to assess the degree of interfacial water and their potential for monitoring physiological changes. Here we also describe the effect of heat shock and osmotic stress on yeast and E. coli cells, and show that the interfacial water content increases with elevated osmolarity and heat shock, and decreases under hypoosmotic conditions.

Keywords: Neutron scattering, water, interfacial water, cells, biomolecules.

1. Introduction

Cells within living tissues are secluded environments which are separated from the external medium by a cell membrane and often a cell wall. The environment inside the cell is characterized by extremely high macromolecule concentrations[1–4] and water represents about 70% or less of the total cellular mass. Levels of protein molecules alone of around 300 mg/ml have been reported *in-vivo* for eukaryotic cells. Perhaps surprisingly, *in-vitro* studies of biological processes rarely take place at concentrations above 1 mg/ml of

* To whom correspondence should be addressed. Robert Ford, Faculty of Life Sciences, University of Manchester, Manchester M60 1QD, UK, e-mail: robert.ford@manchester.ac.uk

S. J. Rzoska and V. A. Mazur (eds.), Soft Matter under Exogenic Impacts, 433–446.

biomolecules, and more often at the microgram/ml level. The effects of molecular crowding *in-vivo* have therefore often been ignored by biochemists and molecular biologists. Molecular crowding phenomena have been described by theoretical models and by *in-vitro* experiments using molecular crowding agents such as polyethylene glycol to simulate high biomolecule levels,[1–4] but there are few descriptions of experimental studies of crowding effects *in-vivo.* An exception is in the measurement of diffusional motion of biomolecules in cells, which has been carried out mostly using fluorescent-tagged biomolecules, and it has been possible to compare the diffusion of macromolecules within a cell with their motion in a dilute aqueous environment using, e.g., fluorescent recovery after photobleaching (FRAP).[3] Such measurements report that diffusional motion for macromolecules in cells is ~25% of that in dilute solution, although larger discrepancies in the diffusional rates have been reported for some systems such as the green fluorescent protein (GFP), a small globular protein commonly employed as a fluorescent tag and which has an intrinsic fluorescence.[3] Molecular crowding phenomena have so far been largely explained in terms of the excluded volume effect, as postulated by Minton and others,[1–4] and this can successfully predict many of the effects of molecular crowding *in-vivo*. However, it is widely acknowledged that other physical factors might also need to be considered. For example, we have described altered properties of the so-called "interfacial" water molecules which lie within a few hydration shells of biomolecules, and which will comprise a significant fraction of the total solvent molecules under highly crowded conditions in the cell. These interfacial waters show changes in their biophysical properties as detected by inelastic neutron scattering (IINS).[5,6] Energy shifts and broadening of peaks in the librational (60–120 meV) band have been reported as well as shifts in the water O–H stretching band (~420 meV).[7,8] Moreover the complete disappearance (or extensive broadening) of two molecular optic peaks at 28 and 38 meV (associated with water–water hydrogen bonding in bulk water)[7–10] was reported for interfacial water interacting with biomembranes and DNA. Although these were the first studies of interfacial water by inelastic neutron scattering, previous biophysical measurements have also pointed towards altered properties in these proximal shells of water molecules. For example quasi-elastic neutron scattering measurements show a reduced diffusion dynamics versus bulk water,[11–14] and atomic-resolution (~1–2 Å) X-ray crystallography studies of biomolecules show that "bound" water molecules display clathrate-cage structures[15–18] with water–water and water–biomolecule interactions over two water shells distant from the surface of the biomolecule. In general, about one ordered water molecule is observed in protein crystal structures for every amino acid residue present in the protein. Water is able to cross the cell wall and membrane relatively easily, and hence cells must maintain an osmotic equilibrium with the external environment in order to avoid damage due to cell swelling and rupture under hypoosmotic conditions or conversely cell shrinkage and dehydration under hyperosmotic conditions. Osmotic balance is maintained by complex regulatory systems that respond on short (seconds to minutes) and long (hours to days) timescales. Short timescale response systems involve the rapid closing and opening of proteinaceous channels in the cell membrane that sense and modulate cell volume changes. Longer timescale response modes involve upregulation or downregulation of metabolic pathways for, e.g., the synthesis of osmoregulatory small molecules such as sucrose, betaines, etc. Most micro-organisms live under conditions where the external medium is at a lower osmotic potential than the cell interior (hypoosmotic conditions) and hence must cope with a tendency for water influx into the cell. The mechanical properties of the cell wall in these organisms is crucial for resisting the hydrostatic pressure induced by the osmotic imbalance.

2. Methods

E. coli (strain BL21 (D3)) cells were inoculated into LB sterile medium at 310 K and grown overnight in 2 l conical flasks. Cells were harvested by centrifugation in a Sorvall centrifuge at 3000×g for 10 min at 277 K. Cells were osmotically stressed by resuspending the cells in NaCl solutions of various concentrations buffered with 5 mM Tris-hydroxymethyl aminomethane, pH7.5. After incubation for 10 min at 298 K, the cells were harvested by centrifugation, as above. Wet cell paste was transferred into a 50 ml Falcon tube and stored at 277 K overnight prior to transport, in ice, to the neutron source. Bakers' yeast (Saccharomyces cerevisiae w.t.) was obtained commercially as a hydrated paste but treatment of the cells was otherwise exactly as described above for E. coli cells. IINS data recording and treatment were as described earlier.[5,19,20]

3. Results

3.1. THEORETICAL CONSIDERATION

Cells are composed of a range of different biomolecules with a range of sizes.[1] An average g lobular p rotein that was in solution in a cell would have a typical molecular mass of about 50 kDa, and would probably resemble a sphere of diameter ~45 Å with a somewhat rough surface. At a protein concentration of 300 mg/ml, then this average 50 kDa protein would associate with about 30% of the total cellular water, assuming that there are two layers of interfacial water extending out from its surface (see below). This calculation is simply based on geometrical considerations and treats water molecules and protein molecules as hard spheres of a given diameter. The treatment ignores any water bound within the structure of the protein, which will exist, but will be a smaller component than the water at the surface. If the interfacial water effect extends to four layers out from the surface of the hypothetical protein, then over 70% of the total water in the cell is affected by the presence of this hypothetical protein. Interfacial water molecules are readily discernible from the bulk water in our IINS measurements, and this has allowed the experimental estimation of the interfacial water as a fraction of total water. With these measurements an amount corresponding to approxi–mately two shells of interfacial water molecules surrounding DNA and protein biomolecules was determined. With membranes, interfacial water was detected corresponding to a layer up to four water molecules thick on each side of the membrane. A conclusion from these experiments and considerations is that the interfacial water content in living cells or in *in-vitro* studies of pure components is highly dependent on the size and shape of the biomolecules. These effects are illustrated in Fig. 1 and 2 which show the relationship for size and shape, respectively. In the first figure, the protein concentration (in terms of mass/volume) is kept constant at 300 mg/ml, and the overall radius of a hypothetical spherical protein molecule (of partial specific volume 0.73 $cm^3 g^{-1}$) is varied. The interfacial water fraction increases rapidly with decreasing average radius in the crowded environment of a theoretical cell. This simplifies the real situation in the cell, but is nevertheless useful to illustrate the importance of smaller biomolecule components with higher surface area:volume ratios for determining the interfacial water fraction. In the second figure, the protein concentration is kept constant at 300 mg/ml, as is the overall size of the hypothetical molecule.

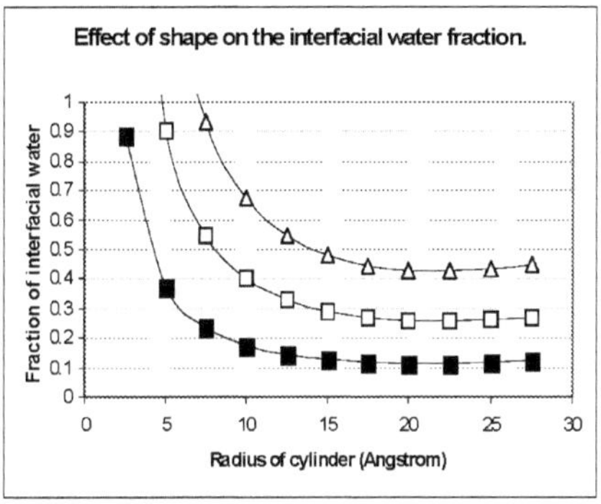

Figure 1. Theoretical calculation of the interfacial water component in a cell as a fraction of the total water at 300 mg/ml protein, and with spherical proteins of variable radius. The three curves show calculations for one, two, and three shells of interfacial water molecules (crosses, diamond and squares respectively).

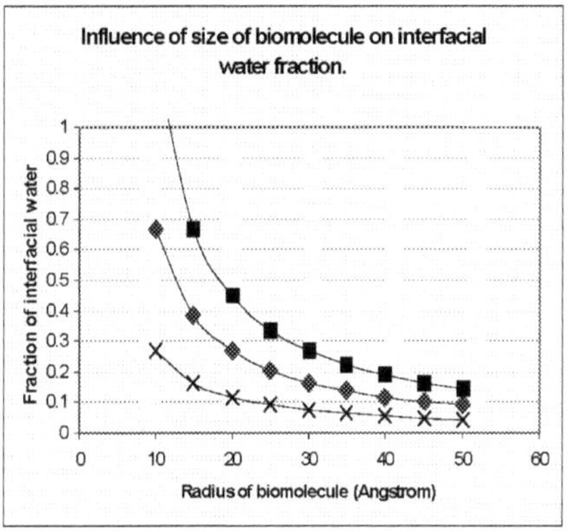

Figure 2. Theoretical calculation of the interfacial water as a fraction of total water in a cell at 300 mg/ml protein and with a cylindrical-shaped protein of mass 50 kDa. The radius of the hypothetical protein is allowed change from 2.5 to 27.5 Å, hence changing its shape to a thin rod at low values for the radius. The three curves show calculations for one, two, and three shells of interfacial water molecules (filled squares, squares and triangles respectively).

In this case the shape of the protein molecule is assumed to be cylindrical and its radius is varied to produce disk-shaped molecules at ~30 Å radius, through roughly globular molecules (radius 20 Å), to highly rod-shaped molecules with a radius of ~2.5 Å. The latter extreme value would represent an unfolded polypeptide chain of ~500 amino acid residues and length ~2500 Å. In this extreme situation, with denatured protein at a concentration of 300 mg/ml, then interfacial water would account for almost the entire water complement in the cell using only one shell of water molecules around the peptide chain (solid squares in Fig. 2). One might predict from such a graph that even small levels of unfolded polypeptides would increase the interfacial water levels significantly in the cell.

3.2. INESLASTIC NEUTRON SCATTERING IN THE ENERGY TRANSFER REGION 3–150 meV

IINS data for water interacting with small globular proteins, DNA biomembranes, and small solutes such as sugars has been obtained.[5,6,19,20] In Fig. 3 a comparison of the bulk and interfacial water spectra is presented as well as a spectrum displaying a mixture of the two. An interfacial water signal dominates at low hydration levels for all these various samples, characterized by a broadening of the librational band, especially the loss of the sharp lower energy edge at 67 meV that is present in the bulk water signal.[5] The degree of interfacial water at specific hydration levels varies between different solutes. For example, a globular protein such as chymotrypsin, at a hydration level of 0.7 g water/g chymotrypsin, then the bulk water:interfacial water ratio is approximately 0.65:1 (see Fig. 4). At similar hydrations levels of DNA a similar ratio of bulk:interfacial water would be observed,[5] but in contrast, proteolipid membranes would show a much larger bulk water signal and a smaller interfacial water signal. The explanation for the latter difference has been simplistically explained in terms of the overall geometry of the biomolecules /biomembranes. DNA can be modeled as a ~2 nm diameter cylinder, (and chymotrypsin as a 3.5 nm diameter sphere), whereas the proteolipid membranes resemble ~8 nm thick planar sheets that are several hundreds of nanometres wide. The latter structure therefore has a much lower surface area to volume ratio, and will bind interfacial water to a lower extent on a unit mass basis, as implied by the theoretical calculations shown in Fig. 2. Clearly other factors due to the surface properties of the macromolecules are also likely to influence the degree of interfacial water, but this geometrical approximation is relatively robust. If we use a model for chymotrypsin as a 35 Å diameter sphere,[21] then the interfacial water signal in Fig. 4 would correspond to about 1–2 layers (shells) of water molecules surrounding the protein. This corresponds closely to the extent of interfacial water determined for both DNA (1–2 layers) and proteolipid membrane samples (2–4 layers).[5] A good example of the robustness of this geometrical treatment of the interfacial water effect is a study by IINS of the librational band before and after a shape change to a biomolecule. This can be achieved by, e.g., denaturation of a protein by heating, but its shape changes from a compact globular shape to an extended coil. Figure 4 shows IINS spectra recorded for two samples of chymotrypsin at a hydration level of ~40 g H_2O per 100 g protein, or 70 g water/100 g protein for both native and denatured (unfolded) chymotrypsin. The native protein shows a significant interfacial water signal at this low hydration level but also displays a small bulk water signal as indicated by the sharp edge at 67 meV, corresponding to about 60% of the total water for the more hydrated sample, and about 75% of total water for the less hydrated one. In contrast, the denatured chymotrypsin spectra are entirely composed of the interfacial water component, and there is no detectable signal due to bulk water, even in the more hydrated sample at 70 g water/100 g protein. This is a clear indication of the importance of shape and conformational integrity in determining the extent of interfacial water in a biological system. As expected, an unfolded protein binds interfacial water to a much greater degree than the native, globular protein.

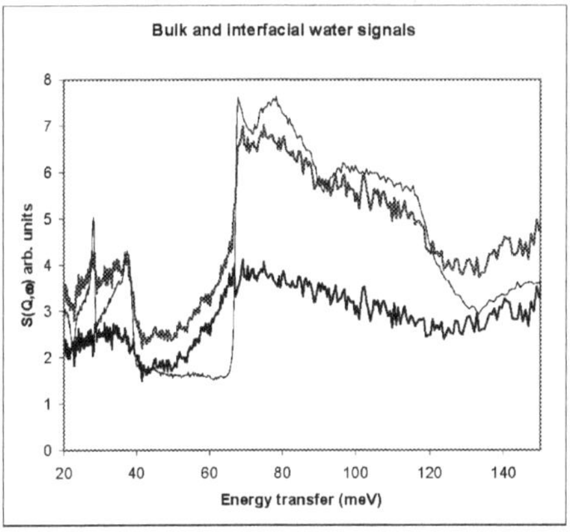

Figure 3. Examples of IINS spectra for bulk water (ice Ih – noise-free upper line), interfacial water (noisy lower line), and a sample displaying a mixture of bulk and interfacial water (noisy upper trace). Note that the very sharp edges in the bulk water spectrum at 67, 28, and 37 meV are absent from the interfacial water spectrum.

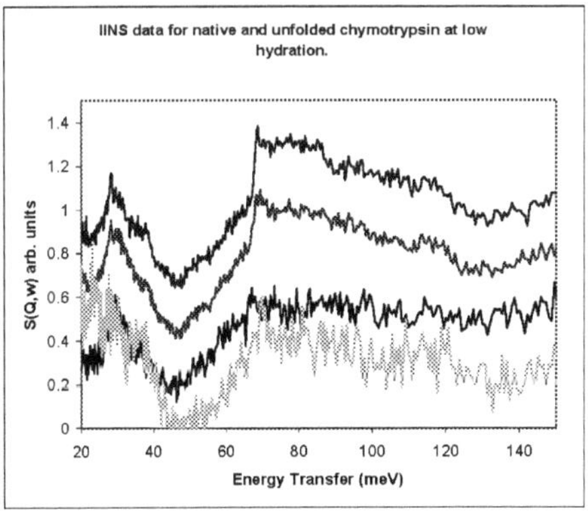

Figure 4. IINS spectra recorded for chymotrypsin, a small protein under different hydration levels and in the native and unfolded (denatured) states. From bottom to top: Unfolded chymotrypsin at 70 g water/100 g protein (gray trace). Unfolded chymotrypsin at 40 g water/100 g protein. Native chymotrypsin at 40 g water/100 g protein. Native chymotrypsin at 70 g water/100 g protein.

The protein remains the same in terms of its atomic composition,

3.3. INELASTIC NEUTRON SCATTERING MEASUREMENTS OF LIVING CELLS AND TISSUES

IINS spectroscopy of living cells and tissues has been carried out for the first time recently, and has been reported in detail elsewhere.[19,20] The complexity of biological material is daunting, but even the most complex tissues contain essentially one major component – water (at a level of about 70%). Although IINS can identify the water component efficiently, the nature of biological material means that there are differing water populations. IINS cannot readily distinguish between intracellular and extracellular water, nor can it presently differentiate organellar compartments within the cell. Even for simple biological samples consisting of single-celled organisms, it is inevitable that part of the bulk water signal will arise from the extracellular growth medium which is part of the cell paste retrieved after filtration or centrifugation. For the more complex tissues of multicellular organs and organisms, the intracellular and extracellular compartments (such as blood vessels) of the tissue cannot be discriminated. Similarly, organelles (compartments) inside cells vary considerably in their water content, for example plants cells have a low total protein content because they contain large water-filled organelles termed vacuoles or tonoplasts. In addition, since the IINS measurements are performed at cryogenic temperatures, then it could be argued that freezing could affect the ratio of bulk and interfacial water if ice Ih crystals form by depleting the interfacial water. It has been proposed that slow external ice crystal growth during freezing may extract water from cells, leading to dehydration.[22] Rapid freezing by plunging into liquid nitrogen should minimize this problem, although experiments where very slow freezing rates were employed did not significantly affect the interfacial water signals in samples (Ford et al., unpublished data). For a unicellular organism (Chlamydomonas) two different media for freezing were assessed, one of which allowed retention of cell viability after the freezing of cells. There was no evidence from these experiments that the cryo-protectant medium could influence the interfacial water signal nor that retention of cell viability during freezing was strongly linked to interfacial water levels. Figure 5 shows the detection of interfacial water in different cells and tissues. The shape of the spectrum in the librational region is remarkably consistent, and presumably reflects the averaging over many different types of macromolecules in the cells.

TABLE 1. Interfacial water content of various cells and tissues. Samples recorded more than once are denoted #A and #B. Y50 and Y42 indicate the heat shock temperature in °C for yeast paste samples.

Sample	Interfacial water (%)	Sample	Interfacial water (%)
Yeast 2.5 OsM	37.9	Chlamydomonas #A	11.7
Yeast 1.25 OsM	33.7	Chlamydomonas #B	7.9
Yeast 0.3 OsM	31.2	Beetroot tuber	20.2
Yeast 25 mOsM	26.9	Spinach leaves	6.4
Yeast paste #A	31.1	Bovine cardiac muscle #A	28.3
Yeast Y50 #B	29.6	Bovine cardiac muscle #B	27.6
Yeast Y42 #B	25.3	E. coli 2.5 OsM	46.9
Yeast paste #B	23.5	E. coli control	38.8

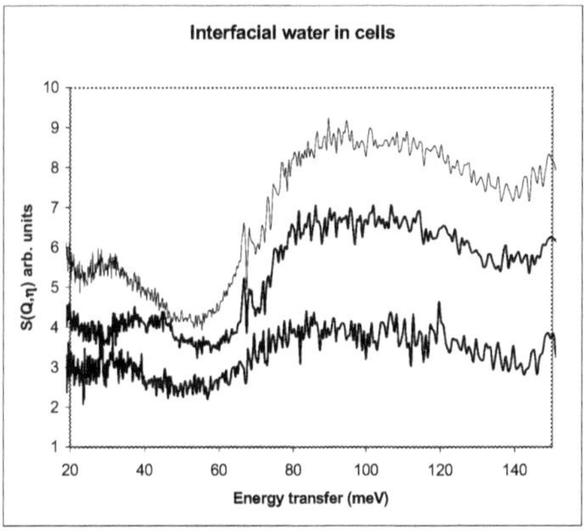

Figure 5. Interfacial water signals detected by IINS in cells and tissues after subtraction of the bulk water signal. Lower trace – Chlamydomonas cells (single-celled green algae). Center trace – beetroot (Beta vulgaris) tuber tissue. Upper trace – bovine cardiac muscle tissue.

These data imply that IINS detection of interfacial water may not at present provide detailed diagnostic information about the cellular composition. However large variation in the magnitude of the interfacial water signal was observed between different cell types, as summarized in Table 1. As expected, organisms with large water-filled organelles such as plants and the single-celled algae have low interfacial water signals and high bulk water signals. Protein-rich tissues such as cardiac muscle and yeast cells showed high interfacial water signals up to about 30% of the total water.

3.4. HEAT SHOCK

Figures 1, 2, and 4 clearly show that simple geometrical considerations (size and shape) can strongly determine the fraction of total cellular water that is interfacial. However, a reasonable question would be whether such factors are relevant to the *in-vivo* situation. Initially, we have focussed on studies of purified biological components, but have extended this to a biological phenomenon which could greatly influence the size and shape of protein molecules in the cell. This is the effect of elevated temperature, or "heat shock". This does not refer to the complete killing of the cell by heat (essentially cooking it), which whilst inevitably affecting size and shape of proteins is hardly of physiological importance. Rather, heat shock refers to a mild elevation of temperature such that some stress to cellular function (such as protein folding or stability) is encountered, but the cell survives this treatment and subsequently recovers full activity. This type of stress is frequently encountered by cells, and may be employed as a strategy to deal with infectious agents.

Figure 6 shows IINS spectra of the interfacial water in control and heat-shocked yeast cells after subtraction of the bulk water component. As before, there is no significant change in the overall shape of the librational peak, but there is a noticeable increase in the interfacial water signal which rises by about 25% after heat shock. The IINS data show that heat shock or denaturation causes an increase in the interfacial water component in both pure biomolecules and also for whole cells. For the pure protein samples (chymotrypsin), the interpretation of a shape change due to denaturation seems to fit the theoretical behavior observed in Fig. 2. For the data for whole cells, two interpretations of the data seem reasonable which relate to the theoretical consideration in both Figs 1 and 2 (size and shape). Firstly heat shock is expected to cause local unfolding of some proteins in the cell[23–26] and this would lead to a much greater surface area available for interaction with the solvent for those proteins that denature. This local unfolding may not necessarily involve the whole polypeptide chain, but could apply to only part of the chain which may have a higher tendency to disorder. Secondly, heat shock could also give rise to the dissociation of some macromolecular complexes (ordered aggregates of several separate proteins) into their constituent proteins. This would lead to a similar increase in solvent-interacting surface area due to the exposure of previously buried residues that were at subunit interfaces in the complexes. A surprisingly small decrease in the average diameter of complexes in the cell could readily explain the changes in interfacial water that we observe. With two water shells, a change of average sphere diameter from ~46 Å to 37 Å would theoretically account for the experimentally determined change of interfacial water from 23.5% to 30% of total water after strong heat shock in yeast cells.

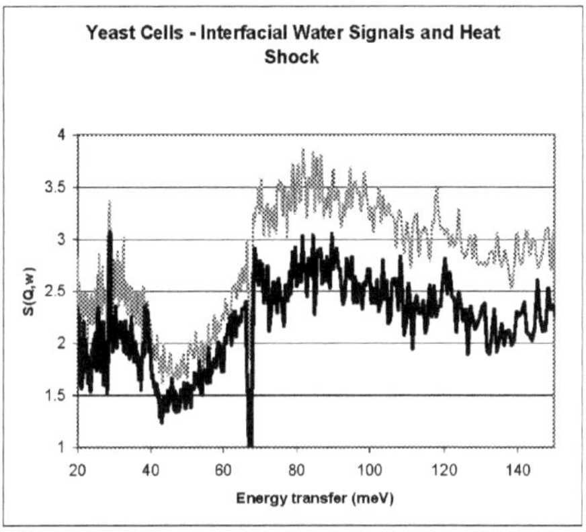

Figure 6. Effect of heat shock on the interfacial water signal and content in yeast cells. Bottom trace – yeast cells (control). Upper trace – yeast cells after heat shock at 323 K. The signals are normalized to the total water signal. The spike at 67 meV is an artefact remaining after subtraction of the bulk water signal due to small errors in the time-of-flight measurement between different runs.

3.5. OSMOTIC SHOCK

Figure 7 shows the raw IINS spectra for control E. coli cell paste; similar live cells which have been osmotically stressed with 2.5 OsM NaCl and dehydrated E. coli cells. Subtraction of a pure water IINS signal is possible, revealing the contribution due to the interfacial water, plus a small contribution due to protons in the non-water molecules present in the sample (Fig. 7, right panel). An indication of the contribution due to non-water molecules can be discerned and is quite minor compared to the water contribution. The broad peaks at ~30 and ~170meV in Fig. 7 appear to be almost entirely due to the non-volatile components in the E. coli sample, and probably arise from $–CH_3$ torsional motions (~30meV band) and from $–CH_2$ and $–NH_2$ wagging, twisting, and bending motions (~170 meV band) in DNA, protein, carbohydrate, and lipid present in the cells.[5] Subtraction of these components is possible with complete removal of these broad peaks, implying that the spectrum for the dehydrated sample gives a reasonably close approximation to the contribution of these components in the hydrated spectrum of the sample.

Bacterial cells treated with a high osmotic strength medium show an increased interfacial water signal (Fig. 7) compared to untreated cells. We estimate that about 47% of the total water in these osmotically stressed E. coli samples is interfacial, somewhat higher than for other cells and tissues measured with this technique.[5,19,20]

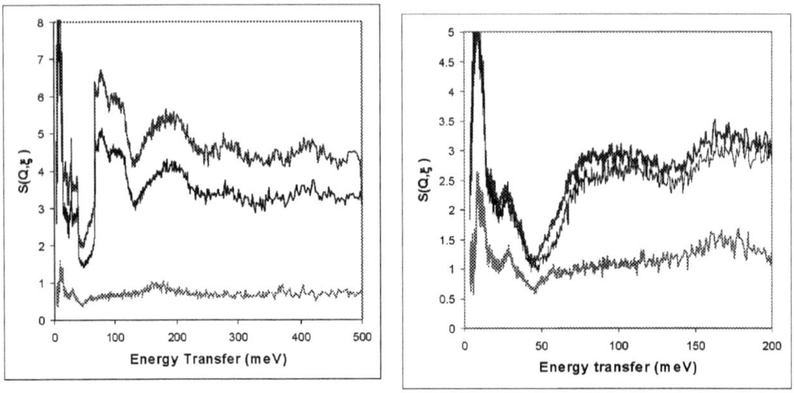

Figure 7. (Above left) IINS data in the 1–500 meV energy transfer range for E. coli cells: dry (bottom/gray), control cells (center), and after 2.5OsM hypertonic shock (upper trace). (Above right) after subtraction of the bulk water signal (same sequence of traces).

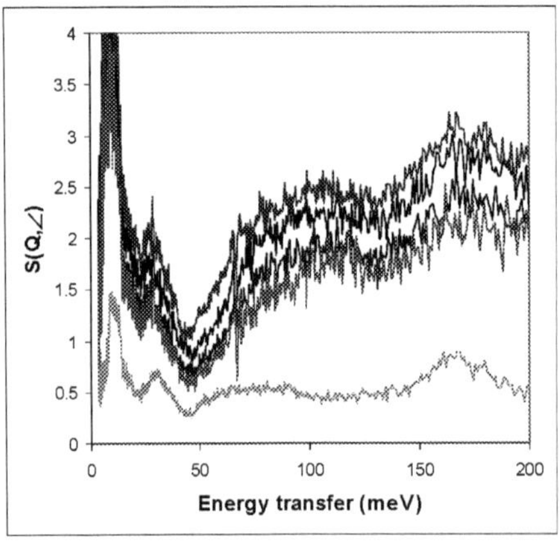

Figure 8. Osmotic effect on the interfacial water IINS signal associated with yeast cells: From bottom to top: dried yeast cells; 25 mOsM; 0.3 mOsM; 1.25 OsM; and 2.5 OsM.

These workers defined "bound" water as the limiting amount of cytoplasmic water that remains at high osmolarity created by high external NaCl concentrations. The bacterium was shown to have a bound water ratio of ~0.5 g/g dry weight and a total cytoplasmic water content of ~2.3 g/g dry weight under normal conditions. This suggests that bound water corresponds to about 22% of the total cell water. This value is close to those determined by IINS measurements of cardiac tissue (28%) and yeast (31–24%). Interfacial water molecules are also measured in X-ray crystallography studies of proteins and other biomolecules. These waters occupy discrete positions in the crystal lattice, i.e., they are not disordered and cannot be considered as "bulk" solvent. These ordered water molecules are discernible in electron density maps obtained from crystals studied at low temperature as well as at room temperature. The mass of bound water can vary, but typically is about 10%–20% of the protein mass in the crystals. Such a level would not account for the "bound" water observed by Record and co-workers of 50% dry mass,[27,28] nor would it match the interfacial water levels observed by IINS. Thus interfacial water will probably include both ordered water as well as disordered water in the hydration shells surrounding proteins.

The conclusion from these data is that the interfacial water is about 30% of total water in the cell and is likely to be highly significant for cellular reactions and processes. Our data do indicate that tissues still contain a significant fraction of bulk water that is indistinguishable in terms of its vibrational properties from ice Ih.

4. References

1. Ellis, J. R (2001) *Curr. Opin. Struct. Biol.* **11**, 114–119.
2. Burg, M. C. (2000) *Cell Physiol. Biochem.* **10**, 251–256.
3. Verkman, A. S. (2002) *TIBS* **27**, 27–33.
4. Hall, D. and Minton, A.P. (2003) *Biochim. Biophys. Acta* **1649**, 127–139.
5. Ruffle, S. V., Michalarias, I., Li, J. C., and Ford, R. C. (2002) *J. Amer. Chem. Soc.* **124**, 565–569.
6. Beta, I. A., Kolesnikov, A. I., Michalarias, I., Wu, G. L., Ford, R. C., and Li, J. C. (2003) *Can. J. Phys.* **81**, 367–371.
7. Michalarias, I., Beta, I. A., Ford, R. C., Ruffle, S. V., and Li, J. C. (2002) *Appl. Phys. A* **74**, 1242–1244.
8. Michalarias, I., Beta, I. A., Li, J. C., Ruffle, S. V., and Ford, R. C. (2002) *J. Mol. Liq.* **101**, 19–26.

9. Li, J. C. (1996) *J. Chem. Phys.* **105**, 6733–6755.
10. Li, J. C. and Tomkinson, J. (1999) In molecular dynamics from classical to quantum methods, *Theoretical and Computational Chemistry.* (Eds. P.B. Balbuena and J. M. Seminario), Vol. **7**. pp. 471–512.
11. Beta, I. A., Michalarias, I., Li, J. C., Ford, R. C., and Bellissent-Funel, M.-C. (2003) (2003) *Chem. Phys.* **292**, 451–454.
12. Dellerue, S. and Bellissent-Funel, M. C. (2000) *Chem. Phys.* **258**, 315–325.
13. Russo, D., Baglioni, P., Peroni, E., and Texeira, J. (2003) *Chem. Phys.* **292**, 235–245.
14. Pieper, J., Charalambopoulou, G., Steriotis, T., Vasenkov, S., Desmedt, A. and Lechner, R.E. (2003) *Chem. Phys.* **292**, 465–476.
15. Esposito, L., Vitagliano, L., Sica, F., Sorrentino, G., Zagari, A., and Mazzarella, L. (2000) *J. Mol. Biol.* **297**, 713–792.
16. Clark, G. R., Squire, C. J., Baker, L. J., Martin, R. F., and White, J. (2000) *Nucl. Acids Res.* **28**, 1259-1265.
17. Soler-Lopez, M., Malinia, L., and Subirana, J. A. (2000) *J. Biol. Chem.* **275**, 23034–23044.
18. Goupil-Lamy, A. V., Smith, J. C., Yunoki, J., Parker, S. F., and Kataoka, M. (1997) (1997) *J. Am. Chem. Soc.* **119**, 9268–9273.
19. Ford, R. C., Ruffle, S. V., Ramirez-Cuesta, A., Michalarias, I., Beta, I. A., Miller, A., and Li, J. C. (2004) *J. Amer. Chem. Soc.* **126**, 4682–4688.
20. Ford, R. C., Ruffle, S. V., Michalarias, I., Beta, I. A., Ramirez-Cuesta, A., and Li, J. C. (2004) *J. Mol. Recogn.* **17**, 505–511.
21. Mathews, B. W., Sigler, P. B., Henderson, R., Blow, D. M. (1967) *Nature* **214**, 652-656.
22. Lovelock, J. E. (1953) *Biochim. Biophys. Acta* **11**, 28–36.
23. Minton, A. P. (2000) *Biophys. J.* **78**, 101–109.
24. Eggers, D. K. and Valentine, J. S. (2001) *Protein Sci.* **10**, 250–261.
25. Fitter, J. (2003) *Chem. Phys.* 292, 405–411.
26. Tarek, M., Neumann, D. A. and Tobias, D. J. (2003) *Chem. Phys.* **292**, 435–443.
27. Record, M. T., Courtenay, E. S., Cayley, D. S., and Guttman, H. J. (1998) *TIBS* **23**, 143–147.
28. Cayley, D. S., Guttman, H. J., and Record, M. T. (2000) *Biophys. J.* **78**, 1748–1764.

SELF-ASSEMBLY OF POLYPEPTIDES. THE EFFECT OF THERMODYNAMIC CONFINEMENT

G. FLOUDAS* and P. PAPADOPOULOS
University of Ioannina, Dept. of Physics and Biomedical Research Institute BRI-FORTH

Abstract. Despite common belief the most well-known polypeptide secondary structure (α-helical) is of rather low persistence. We have investigated the effect of the thermodynamic confinement produced by the unlike segments of nano-phase separated block copolymers on the persistence length of α-helices. The thermodynamic field give rise to a partial annihilation of helical defects. This suggests that thermodynamic confinement can be employed as a means of secondary structure perfection in phase separated polypeptide block copolymers.

Keywords: Polypeptides, block copolymers, confinement, α-helices, persistence length.

1. Introduction

The self-assembly of polypeptides (i.e., macromolecular sequences of certain amino acids) have been studied in detail with a variety of "structural" techniques such as nuclear magnetic resonance (NMR), infrared (FTIR), circular dichroism (CD), wide-angle X-ray scattering (WAXS).[1,2] These techniques have shown that polypeptides form tertiary structures composed from α-helices and β-sheets as the primary secondary motifs stabilized, respectively, by intra- and inter-molecular correlations. In recent years there has been a strive toward the design of macromolecules that combine these peptidic secondary motifs with the nanometer scale self-assembly of block copolymers.[3–6] Of special interest are systems with biocompatibility and biofunctionality. In this respect, the synthesis of block copolymers based on polypeptides can produce new biomaterials with potential applications in drug delivery and tissue engineering. One of the main concerns in these systems is how the thermodynamic field (normally described by the product of the interaction parameter χ and the total

* To whom correspondence should be addressed.

S. J. Rzoska and V. A. Mazur (eds.), Soft Matter under Exogenic Impacts, 447–455.

degree of polymerization N) imposed by the incompatibility of the blocks forming the copolymer[7] affect the peptide secondary motifs. It was shown recently that interfacial mixing is a key factor in controlling the appearance of β-sheets in low molecular weight polypeptides.[8] In the present work we report on the effect of the thermodynamic confinement on the persistence length of α-helical secondary structures. The investigated polypeptide is poly(γ-benzyl-L-glutamate) (PBLG) – a well-known α-helical polypeptide[2] – that forms one of the blocks of the copolymer. For the other block we have employed both synthetic macromolecules (based on polyethylene glycol (PEG) and polydimethyl siloxane (PDMS)) as well as another polypeptide (polyglycine). The results presented herein show a substantial effect of the thermodynamic confinement on the persistence length of α-helical secondary structures.

2. Systems

Testing the effect of thermodynamic confinement on the persistence length of peptide secondary structures requires different block copolymers with a variety of interactions ranging from weakly segregated (WSL: weak-segregation limit) to strongly segregated blocks (SSL: strong-segregation limit). First, a series of triblock copolymers of PBLG-*b*-PEG-*b*-PBLG with PBLG volume fractions in the range $0.2 \leq f_{PBLG} \leq 0.96$ have been investigated.[8] This system combines a peptidic block with the ability to form secondary structures with a synthetic block (PEG) having the tendency to crystallize by chain-folding. Second, a series of triblock copolymers of PBLG-*b*-PDMS-*b*-PBLG with PBLG volume fractions in the range $0.49 \leq f_{PBLG} \leq 0.96$.[9] This system combines the peptidic block with a flexible synthetic macromolecule (PDMS). Third, a series of model (i.e., nearly monodisperse) copolypeptides of PBLG-*b*-PGly with PBLG volume fractions in the range $0.67 \leq f_{PBLG} \leq 0.97$ and molecular weights in the range from 2×10^4 to 3×10^4 g/mol were employed.[10] The latter combines two peptidic blocks with different secondary motifs (α-helices and β-sheets) within the nanodomains of the copolymers. The complete molecular characterization of all samples are given in the suggested references.

3. Methods

For the investigation of the phase state a combination of "structural" techniques was used: small-angle X-ray scattering (SAXS) for the domain spacing, WAXS, NMR, FTIR and differential scanning calorimetry (DSC) for the peptide secondary structure. For the dynamics associated with the persistence length of secondary structures, dielectric spectroscopy (DS) was employed. The dielectric measurements were made at different temperatures in the range 113–413 K, at atmospheric pressure, and for frequencies in the range from 10^{-2} to

10^6 Hz using a Novocontrol BDS system composed of a frequency response analyzer (Solartron Schlumberger FRA 1260) and a broadband dielectric converter. The complex dielectric permittivity $\varepsilon^*=\varepsilon'-1\varepsilon''$, where ε' is the real and ε'' is the imaginary part, is generally a function of frequency ω, temperature T, and pressure P. The analysis of the complex dielectric function has been made using the empirical equation of Havriliak and Negami:

$$\frac{\varepsilon^*(\omega,T)-\varepsilon_\infty(T)}{\Delta\varepsilon(T)}=\frac{1}{[1+(i\omega\tau_{HN}(T))^\alpha]^\gamma}, \tag{1}$$

where $\Delta\varepsilon(T)$ is the relaxation strength of the process under investigation, τ_{HN} is the relaxation time of the equation and α, γ $(0<\alpha,\ \alpha\gamma\leq 1)$ describe the symmetrical and asymmetrical broadening of the distribution of relaxation times. At lower frequencies ε'' rises due to the conductivity ($\varepsilon''=\sigma/(\omega\varepsilon_f)$, where σ is the dc conductivity and ε_f the permittivity of free space). The conductivity contribution has also been taken into account during the fitting process. The measured ε'' spectra have been used for the analysis except at high temperatures where the derivative of ε' has been employed ($d\varepsilon'/d\ln\omega \sim -(2/\pi)\varepsilon''$). Since ε' is not affected by the conductivity, this method is very useful in fitting relaxation processes which are hidden under the conductivity, provided that the system is free from polarization. Figure 1 gives some representative spectra for bulk PBLG for different temperatures. Three dielectrically active mechanisms can be seen at temperatures above the glass temperature T_g; the primary (α-) process associated with the liquid-to-glass dynamics, a weak intermediate process associated with the relaxation of amorphous chains and a third very intense process associated with the relaxation of the α-helical secondary structure. In this work we are mainly concerned with the dynamics of this slower mode.

In addition to the dielectric permittivity the electric modulus (M*) representation was used:

$$M^*=\frac{1}{\varepsilon^*}=M'+iM'' \tag{2}$$

The latter representation is very sensitive to the presence of slow processes as well as to the process due to ionic mobility.

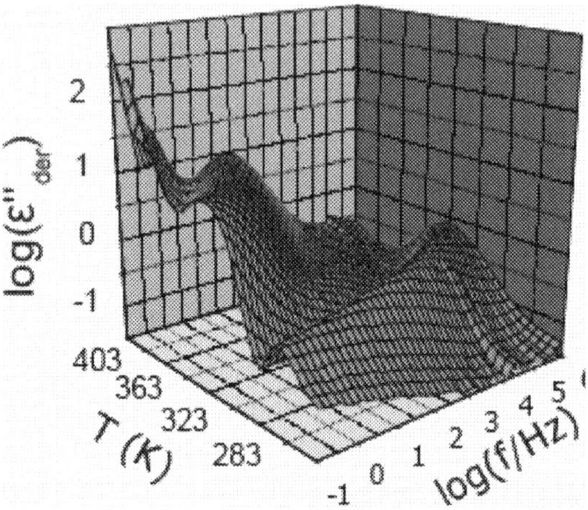

Figure 1. Representative dielectric loss data from PBLG in a 3D representation. Three processes can be identified corresponding to the α-process, the intermediate, and the slower process associated, respectively, with the liquid-to-glass dynamics, the dynamics of amorphous chains and the dynamics of the secondary (α-helical) structures.

4. Results and discussion

Block copolymer segregation is influenced[7] by (i) the segment–segment interaction parameter χ, describing the free energy cost per monomer of contacts between unlike monomers and (ii) the total degree of polymerization N, affecting the translational and configurational entropy of chains. Since the entropic and enthalpic contributions to the free energy scale as N^{-1} and χ, respectively, the phase state of block copolymers is discussed in terms of the product χN. Other effects, such as the conformational asymmetry, copolymer architecture, and fluctuation effects can strongly influence the phase state and have been discussed in detail in the literature.[7] The results on the phase state for the PBLG-*b*-PEG-*b*-PBLG triblock copolymer system are shown in Fig. 2. In the construction of the phase diagram the mean-field predictions[11] have been used for the disordered phase structure factor giving rise to $\chi \sim 0.32 \pm 0.05$. The figure displays regions that are thermodynamically mixed and other regions with more symmetric compositions where the block copolymers are nano-phase separated.

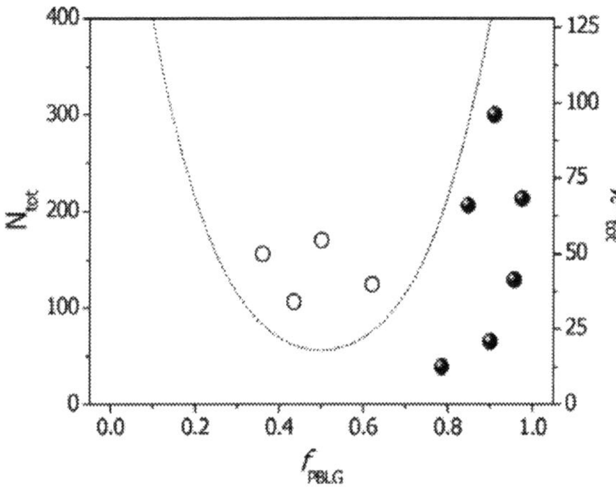

Figure 2. Schematic phase diagram of the PBLG-*b*-PEG-*b*-PBLG triblock copolymer systems. The disordered (filled symbols) and ordered (open symbols) phases are shown separated by a phase boundary calculated from MFT.[11]

Subsequently, the dynamics of the secondary structure (α-helix) have been obtained from DS from the slower and more intense process associated, in principle, with the end-to-end relaxation of dipoles in a perfect α-helical structure.[12] DS is very sensitive not only to the presence of α-helices, but also to their persistence length. In this respect, we have recently shown that, contrary to the expectation born out by the static experiments and common belief, α-helical polypeptides cannot be considered as ideal rigid-rods and that a model of "broken" rods is closer to reality.[13] This is based on the fact that there exist two processes above T_g that are slower than the segmental (α-) process;[14] an "intermediate" process with a strong molecular weight dependence reflecting the relaxation of amorphous parts of PBLG chains, and a "slow" process which reflects the relaxation of α-helical defects. The persistence length ξ of the α-helical structure can be extracted from the intensity of the slower DS process. In a study of PBLG as a function of molecular weight, ξ was found in the range of 1–2 nm, i.e., comprising about 10–20 monomers.[5]

The effect of the thermodynamic field on the persistence length of PBLG is discussed with respect to Figs. 3, 4, and 5 for the PBLG-*b*-PDMS-*b*-s. PBLG, PBLG-*b*-PEG-*b*-PBLG, and PBLG-*b*-PGly copolymers, respectively.

In the PBLG-*b*-PDMS-*b*-PBLG system, the PBLG "slow" process in the block copolymers is shifted to longer times by about one decade. In addition, the intensity of the "slow" process is enhanced. Both trends are a strong indication that the α-helical persistence length of PBLG increases in the presence of PDMS despite the lower lateral coherence of helices as revealed by WAXS. To make the comparison more quantitative, we plot in Fig. 3 the effective dipole moment corresponding to the "slow" process for the three copolymers.

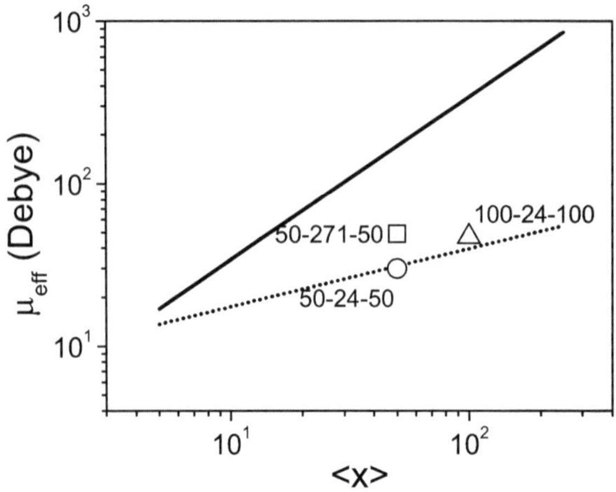

Figure 3. Effective dipole moment of PBLG plotted as a function of the degree of polymerization in the triblock copolymers PBLG-*b*-PDMS-*b*-PBLG. The solid line gives the expected effective dipole moment for a perfect α-helical structure. The symbols correspond to the different copolymers; up triangle: $PBLG_{100}$-*b*-$PDMS_{24}$-*b*-$PBLG_{100}$, square: $PBLG_{50}$-*b*-$PDMS_{271}$-*b*-$PBLG_{50}$, circle: $PBLG_{50}$-*b*-$PDMS_{24}$-*b*-$PBLG_{50}$.

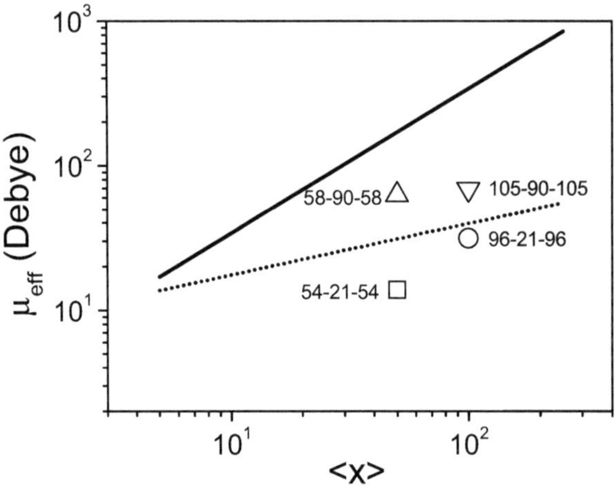

Figure 4. Calculated effective dipole moment of PBLG plotted as a function of the degree of polymerization in the triblock copolymers PBLG-*b*-PEO-*b*-PBLG. The solid line gives the expected effective dipole moment for a perfect α-helical structure. The symbols correspond to the different copolymers; up triangle: $PBLG_{58}$-*b*-PEG_{90}-*b*-$PBLG_{58}$, down triangle: $PBLG_{105}$-*b*-PEG_{90}-*b*-$PBLG_{105}$, square: $PBLG_{54}$-*b*-PEG_{21}-*b*-$PBLG_{54}$, circle: $PBLG_{96}$-*b*-PEG_{21}-*b*-$PBLG_{96}$.

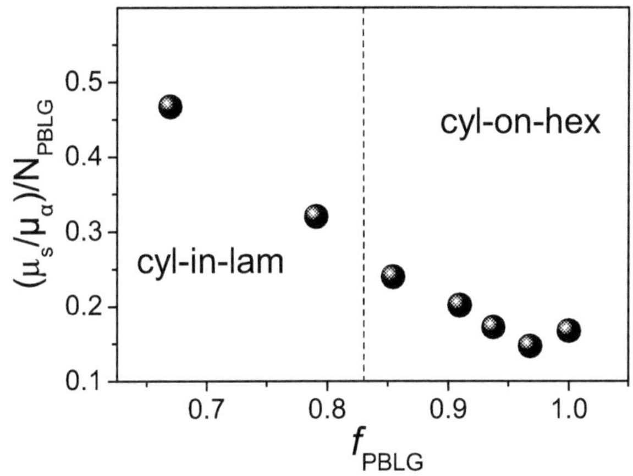

Figure 5. Ratio of the effective dipole moments corresponding to the "slow" and segmental (α-) processes of the PBLG block for the PBLG-*b*-PGly copolymers as well as the pure PBLG. Due to the small differences in the molecular weight of the PBLG block the ratios are normalized by the average degrees of polymerization. Notice the increase of the effective dipole moment with increasing PGly volume fraction suggesting the annihilation of a part of helical defects. The vertical line separates the two nanodomain morphologies.

The dipole moment was calculated using the Buckingham equation that describes rod-like molecules in solution

$$\frac{Nfg\mu^2}{3\varepsilon_0 k_B T} = \frac{(2\varepsilon_s + 1)(\varepsilon_s - n^2)}{2\varepsilon_s + n^2} - \frac{(2\varepsilon_\infty + 1)(\varepsilon_\infty - n^2)}{2\varepsilon_\infty + n^2}, \tag{3}$$

where f is a value related to the geometry of the molecule ($f \rightarrow 2/3$ for an infinitely long rod), N is the number density of dipoles, g is the Kirkwood–Fröhlich correlation factor between neighboring dipoles and ε_S and ε_∞ are the permittivity values at low and high frequencies, respectively, and n is the refractive index. The dashed line shows the respective values for a series of PBLG homopolymers and the solid line the theoretical value for an ideal α-helix. The effective dipole moments of the two block copolymers with high PBLG content ($PBLG_{50}$-*b*-$PDMS_{24}$-*b* $PBLG_{50}$ and $PBLG_{100}$-*b*-$PDMS_{24}$-*b*-$PBLG_{100}$) show almost no difference from the PBLG homopolymer with the same molecular weight. The $PBLG_{50}$-*b*-$PDMS_{271}$-*b*-$PBLG_{50}$ case with the higher PDMS content, however, is different; the dipole moment is much higher than the homopolymer (by about a factor of 2) although still lower than the theoretical value. Understanding this effect requires some knowledge of the block copolymer phase state. For strongly segregated copolymers (i.e., $\chi N \rightarrow \infty$), chains are strongly stretched and can be envisioned as grafted polymer brushes with a domain period scaling as $d \sim N^{2/3}\chi^{1/6}$, i.e., with a strong N-dependence.[7] Following these ideas, the thermodynamic confinement and chain stretching should be more pronounced in the $PBLG_{50}$-*b*-$PDMS_{271}$-*b*-$PBLG_{50}$ copolymer as compared to the $PBLG_{50}$-*b*-$PDMS_{24}$-*b*-$PBLG_{50}$ because of the higher N and possibly by the more symmetric composition that drives the copolymer deeper into the phase-segregated region.

In the PBLG-*b*-PEG-*b*-PBLG triblock copolymer system (Fig. 4), there is an increase in the effective dipole moment for the $PBLG_{58}$-*b*-PEG_{90}-*b*-$PBLG_{58}$ (f_{PBLG}=0.85) and $PBLG_{105}$-*b*-PEG_{90}-*b*-$PBLG_{105}$ (f_{PBLG}=0.91) and a decrease to a value below the bulk PBLG for the $PBLG_{54}$-*b*-PEG_{21}-*b*-$PBLG_{54}$ (f_{PBLG}=0.96) and $PBLG_{96}$-*b*-PEG_{21}-*b*-$PBLG_{96}$ (f_{PBLG}=0.98) copolymers. Both can de discussed with respect to the location of the particular copolymers in the phase diagram (Fig. 2); the latter are well within the disordered state whereas the former within the WSL. Interestingly, even weak segregation can influence the persistence length of α-helices.

In the PBLG-*b*-PGly copolypeptides (Fig. 5), the effective dipole moment is a strong function of the PBLG content; increasing for the more symmetric compositions. This chain stretching gives rise to the partial annihilation of helical defects and to the more persistent helical sequences as observed experimentally. This suggests that thermodynamic confinement can be employed as a means of secondary structure perfection in phase separated polypeptide block copolymers.

5. Conclusion

It is well known that external stimuli (i.e., a generalized thermodynamic force such as electric field, magnetic field, shear, etc.) can induce order in systems with a tendency for self-assembly. We have shown that the thermodynamic field created by the unlike segments forming the nanodomains of block copolymers produce a similar effect on the secondary structures of polypeptides. This suggests that thermodynamic confinement can be employed as a means of secondary structure perfection in polypeptides.

Acknowledgments

We are indebted to professors H.-A. Klok (EPFL), N. Hadjichristidis (UoA), and H. Iatrou (UoA) for the synthesis of the model copolymers.

References

1. Walton, A.G., Blackwell, J. (1973) In *Biopolymers*, Academic Press, New York.
2. Block, H. (1983) In *Poly(benzyl-L-glutamate) and Other Glutamic Acid Containing Polymers*, Gordon and Breach Science Publisher, New York.
3. Klok, H.-A., Langenwalter, J.F., Lecommandoux, S. (2000) *Macromolecules* **30**, 7819.
4. Schlaad, H., Smarsly B., Losik, M. (2004) *Macromolecules* **37**, 22210.
5. Wong, M.S., Cha, J.N., Choi, K.-S., Deming, T.J., Stucky, C.D. (2002) *Nano Lett.* **2**, 583.
6. Aliferis, Th., Iatrou, H., Hadjichristidis, N. (2004) *Biomacromolecules* **5**, 1653.
7. Hadjichristidis, N., Pispas, S., Floudas, G. (2002) In *Block Copolymers: Synthetic Strategies, Physical Properties and Applications*, J. Wiley & Sons Inc., New York.
8. Floudas, G., Papadopoulos, P., Klok, H.-A., Vandermeulen, G.W.M., Rodriguez-Hernandez, J. (2003) *Macromolecules* **36**, 3673.
9. Papadopoulos, P., Floudas, G., Schnell, I., Lieberwirth, I., Nguyen T.Q., Klok, H.-A. (2005) *Biomacromolecules*, **7**, 618.
10. Papadopoulos, P., Floudas, G., Schnell, I., Aliferis, T., Iatrou, H., Hadjichristidis, N. (2005) *Biomacromolecules* **6**, 2352.
11. Mayes, A.M., de la Cruz, M.O. (1989) *J. Chem. Phys.* **91**, 7228.
12. Hartmann, L., Kratzmüller, T., Braun, H.-G., Kremer, F. (2000) *Macromol. Rapid Commun.* **21**, 814.
13. Papadopoulos, P., Floudas, G., Klok, H.-A., Schnell, I., Pakula, T. (2004) *Biomacromolecules* **5**, 8.
14. Papadopoulos, P., Floudas, G., Schnell, I., Klok, H.-A., Aliferis, T., Iatrou, H., Hadjichristidis, N. (2005) *J. Chem. Phys.* **122**, 224906.

COULOMB LIQUIDS UNDER ELECTRIC FIELD – APPLICATION OF A NEW COMPUTER SIMULATION METHOD

E. S. YAKUB
Computer Science Dept., Odessa State Economic University
8 Preobrazhenskaya St., Odessa, 65029, Ukraine
E-mail: yakub@oseu.odessa.ua

Abstract. Basics and some applications of a new method for computation of Coulomb forces in Monte Carlo or molecular dynamics simulation of highly non-ideal disordered systems like strongly coupled plasma and high-temperature ionic liquids is discussed. This method, based on angular averaging of Ewald sums over all orientations of the reciprocal lattice vector, eliminates periodicity artifacts imposed by conventional Ewald scheme under conditions of computer simulation and provides much faster computation of electrostatic interactions in simulation of disordered condensed systems. Application of this approach to study of solvation effect on transport properties of ionic liquids under strong electric field is discussed. Diffusion coefficients and external mobilities of ions are determined and analyzed in relationship to structure characteristics of the melt. External ionic mobilities of heavy actinide ions in their different oxidation states in lithium chloride–potassium chloride eutectic melt at high temperature are determined using ionic model and Fumi–Tosi potentials. Negative values of external mobilities of multiple charged actinide ions are found. In order to explain this effect, radial distribution functions, coordination numbers, and dynamic structure characteristics of solvated ions are determined and analyzed. Concentration dependency of the apparent charges of $Li^{(+)}$, $K^{(+)}$, and $Cl^{(-)}$ was studied at constant density in the whole range of Li concentrations from pure LiCl to dilute solution of LiCl in KCl. It was found that the apparent charge of $Li^{(+)}$ ion at low concentrations also becomes negative. The explanation for this effect based on structure and stability of solvation shells discussed.

Keywords: Coulomb liquids, molecular dynamics, mobility, apparent charge.

1. Introduction

Molecular dynamics (MD) and Monte Carlo (MC) computer simulations are playing now an increasing important role in predicting properties of matter under extreme conditions that are hard to investigate experimentally or calculate by theory. Despite of many simulations performed an essential

S. J. Rzoska and V. A. Mazur (eds.), Soft Matter under Exogenic Impacts, 457–465.

problem within such approach remains how to combine an accurate account of long-ranged Coulomb forces with periodic boundary conditions. Conventional Ewald summation procedure, used now throughout, imposes a heavy processor load, which actually determines the limits in feasibility of such simulations and restricts the number of particles in the main MD or MC cell by few hundred. Obviously, the larger the number of charged particles in the main cell, the more acute the problem of the effective evaluation of the electrostatic contribution is. In studies of dilute ionic solutions where the total number of particles in the cell is large, this problem becomes crucial.

Moreover, when applied to simulation of ionic liquids, plasmas and other disordered Coulomb systems, Ewald procedure invokes a non-isotropic electric field having the cubic symmetry of the crystalline lattice composed of main cells as elementary units. It results in artificial 'crystalline field' effects in simulation of molten salts and other spatially uniform Coulomb systems which may be important, especially in *ab-initio* studies with small simulation cells.[1]

Recent modification of the Ewald scheme[1] based on angular averaging of Ewald sums over all orientations of the reciprocal lattice eliminates the periodicity artifacts imposed by conventional Ewald scheme and provides much faster computation of electrostatic energy in computer simulations of disordered condensed systems.

This approach results in simple expression for the total Coulomb energy for N point charges Z_i ($i=1,\ldots,N$) placed in the main cubic cell of the size L:[1]

$$U_N^{(C)} = -\sum_{i=1}^{N} \frac{3Z_i^2}{16\pi\varepsilon_0 r_m} + \sum_{1\le i<j\le N} \phi_{ij}^{(C)}, \tag{1}$$

where

$$\phi_{ij}^{(C)} = \begin{cases} \dfrac{Z_i Z_j}{4\pi\varepsilon_0 r_{ij}}\left\{1+\dfrac{1}{2}\left(\dfrac{r_{ij}}{r_m}\right)\left[\left(\dfrac{r_{ij}}{r_m}\right)^2-3\right]\right\} & r_{ij} < r_m \\ 0 & r_{ij} \ge r_m \end{cases} \tag{2}$$

are effective potentials of electrostatic interaction and r_m is radius of the volume-equivalent sphere of the main cell. This method is more accurate and much faster than the conventional Ewald scheme. It allows for extensive simulations of ionic systems, containing up to few thousand of particles in the main cell, even on PC. New MD simulation software, implementing the above-mentioned approach, was recently developed[2] and tested on simple model systems like one-component plasma and primitive ionic model of electrolyte.

The purpose of the present work is an application of this new method in study of non-ideality effects on ionic mobility and diffusion in extremely diluted ionic solutions under external electric field. Simulations of such systems require big simulation cells, containing up to several thousand of charged particles. Here we report the results of MD simulation of kinetic behavior of heavy actinide ions in their different oxidation states dissolved in eutectic LiCl–KCl melt. Molten salts, especially alkali chloride mixtures, play an important role in industrial chemical processes such as electrolytic reprocessing of spent metallic fuels in a high-temperature molten salt bath, enrichment of the fission products by electromigration method, etc.

At the same time study of thermodynamic, structural, and transport properties of these systems are not only of technological importance, but have also essential theoretical significance, because molten salts are textbook examples of highly non-ideal Coulomb fluids. Many conventional relations such as Nernst–Einstein equation cannot be directly applied to highly non-ideal solutions like molten salts. Theoretical study of dilute solutions in molten salts is much more difficult than that of non-electrolyte solutions, mainly due to the strong long-ranged Coulomb interaction. There is still no rigorous theoretical approach able to predict thermodynamic and transport properties of such strongly coupled Coulomb systems as molten salts are.

We applied the new MD simulation software, implementing the above-mentioned approach[2] to predict high-temperature behavior of dilute solutions of some actinide ions ($Am^{(2+)}$, $Am^{(3+)}$) in lithium chloride–potassium chloride eutectic melt. Pressure, energy, structure characteristics (radial particle–particle and charge distributions, coordination numbers, etc.), velocity autocorrelation functions, as well as mean square displacements of ions are computed. The last two properties are used in calculations of diffusion coefficients. Along with equilibrium simulations we determined also electrical conductivity and ionic mobilities by applying an external electric field and monitoring the migration velocities v_α of all ions along its direction. Some of these results are presented and discussed below.

TABLE 1. Comparison of MD diffusion coefficients (10^{-5}cm^{2}s^{-1}) with predictions of Stokes–Einstein (SE) equation using experimental viscosity[3] and ionic radii $R(Li^{(+)}) = 0.816$, $R(K^{(+)}) = 1.463$, $R(Cl^{(-)}) = 1.587$ (10^{-10} m).[4]

LiCl, %	T,K	$Li^{(+)}$		$K^{(+)}$		$Cl^{(-)}$	
		MD	SE	MD	SE	MD	SE
0	1400			13.6	12.8	12.4	11.8
100	1073	9.5	10.2			5.0	5.2

2. Results

Structure and transport properties of ions in lithium chloride–potassium chloride melt are studied using Nosè–Hoover NVT/NPT ensemble dynamics,[5] and Verlet–Stroemer algorithm.[6] We used a simulation cell containing from 1000 ions in the case of pure LiCl–KCl melt and up to 8788 ions for solution of actinides in it. Dilute solutions of Americium ions having valence of two and three (0.15 mol % of each $Am^{(2+)}$ and $Am^{(3+)}$) in LiCl–KCl eutectic melt (58 mol % of LiCl) are considered.

Pure ionic model and Fumi–Tosi potentials[4] are utilized. This model is relatively simple and widely used for simulation of such systems as molten salts.[7] Potential parameters for actinide ions are determined from their ionic radii (1.3 Å for $Am^{(2+)}$ and 0.9 Å for $Am^{(3+)}$) estimated on the basis known properties of actinide compounds.[8] Coulomb contributions to inter-ionic forces are described by Eqs. (1) and (2). Self-diffusion coefficients are calculated from mean square displacements at zero electric field according to Einstein expression and from velocity autocorrelation functions using Kubo formula. Both values agree within estimated error (~10%) and are in reasonable agreement with Stokes–Einstein equation using (extrapolated) experimental viscosity data[3] and effective ionic radii as only parameters. Some results are presented in Table 1.

External ionic mobilities are computed in simulations of electromigration process in LiCl–KCl melt under external electric field by monitoring the migration velocity of different ions along this field. External mobility u_α, was determined as a proportionality coefficient between the migration velocity v_α of an ion and field strength E:

$$v_\alpha = u_\alpha E.$$

To facilitate the simulation we applied a rather high (~10^8 Vm^{-1}) external field (which at the same time amounts only few percents of the mean squared internal field at the position of an ion). After a quite long relaxation process (more than 100 ps) following switching-on the electric field, the migration velocities stabilize and mean values of the average mobilities are determined (see Fig. 1). The most striking results found were negative values of the actinide mobilities. That means the positively charged cations migrate against the external electrical field, in the anode direction. Analyzing ion–ion radial distribution functions and stability of the solvation shells one can reveal the reason for this effect. The charge distribution around heavy ions is shown in Fig. 2. It consists of a number of alternating solvation shells.

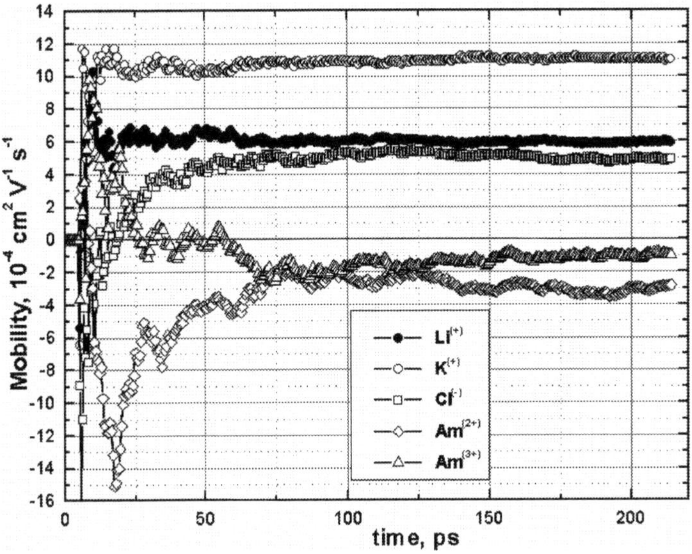

Figure 1. Predicted ionic mobilities as functions of simulation time (1 ps = 10^{-12} s) at high temperature (T=1400 K) and field strength (E=10^8 Vm^{-1}).

If these shells are stable enough they produce electrophoretic drag of the central ion and it behaves as an complex ion with an apparent charge Z_α^*, which may be positive or negative depending on relative contribution of different shells. Value of the apparent charge Z_α^* can be found from MD simulation data on ionic mobility and diffusion coefficient using the modified Nernst-Einstein relation:

$$u_\alpha = \left(\frac{Z_\alpha^*}{Z_\alpha}\right)\frac{D_\alpha}{kT}. \tag{3}$$

The apparent ionic charge Z_α^* is more convenient for interpretation of the ionic mobility, because it is less sensitive to variations of density and temperature. At simulation conditions the apparent charge was found to be about –0.3 for $Am^{(3+)}$ and near –1.2 for $Am^{(2+)}$.

Along with sizes and total charges of the solvation shells (Fig. 2), the most important is stability of shells, characterized by average times spent by an ion within some shell or other. Following the movement of anions forming first solvation shells and different ions in the second shell one may study their stability. We found that numbers of ions left both in first and second shells as

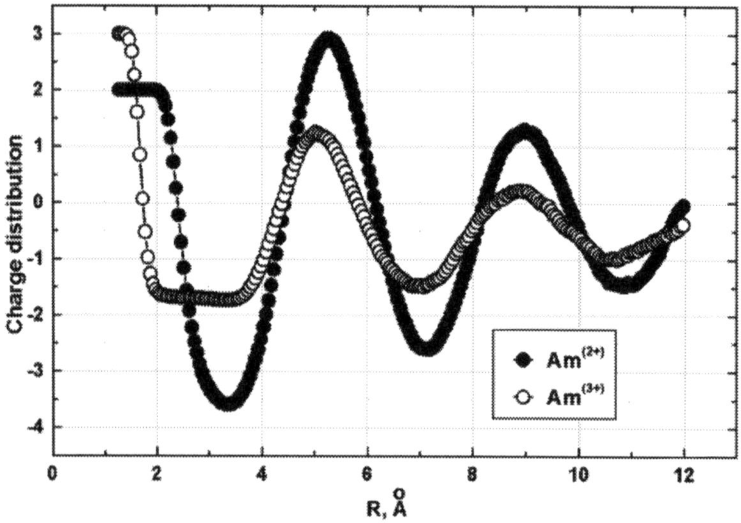

Figure 2. Charge distribution (in units of elementary charge) around $Am^{(2+)}$ and $Am^{(3+)}$ ions in eutectic LiCl–KCl melt at 1400 K.

functions of time obey, like radioisotopes, the exponential decay law with different values of half-life $\tau_{1/2}$.

Within ionic model at simulation conditions both $Am^{(3+)}$ and $Am^{(2+)}$ have stable first solvation shells composed from $Cl^{(-)}$ ions: $\tau_{1/2}$ ($Am^{(3+)}$)~70 ps with coordination number ~4.8; and $Am^{(2+)}$ ($\tau_{1/2}$ ~7 ps) with larger coordination number ~5.7. The total charge in the second coordination shell of $Am^{(3+)}$ is about +3 when that of $Am^{(2+)}$ is about +6. All ions in second shells of both $Am^{(2+)}$ and $Am^{(3+)}$ behave alike ($\tau_{1/2} \sim 3$ ps). These times are close to characteristic diffusion times $\tau_D = R_i^2 \ln 2/(D_\alpha \pi^2)$ one can expect from simple diffusion decay[9] of the second coordination sphere (τ_D~1.5 ps for a sphere of a radius $R_i \sim 5$ Å composed from ions having diffusion coefficient $D_\alpha = 1.2 \cdot 10^{-4} cm^2 s^{-1}$).

The simplest way to explain the solvation effect on mobility is to assume that all ions in solvation shells can be subdivided into 'free' and 'bound', and only last ones contribute to electrophoretic drag of solvated ion. Estimating fractions of 'bound' ions as $(1-\tau_D/\tau_{1/2})$, one can find estimations of apparent charge as a sum of charges of solvated ion and total charges of solvation shells multiplied by fractions of 'bound' ions in this shells in reasonable agreement with observed values. This simple considerations give an qualitative understanding of negative apparent charges of $Am^{(2+)}$ and $Am^{(3+)}$ ions in chloride melts and may be used in development of a model for quantitative estimation of apparent charge of solvated ions.

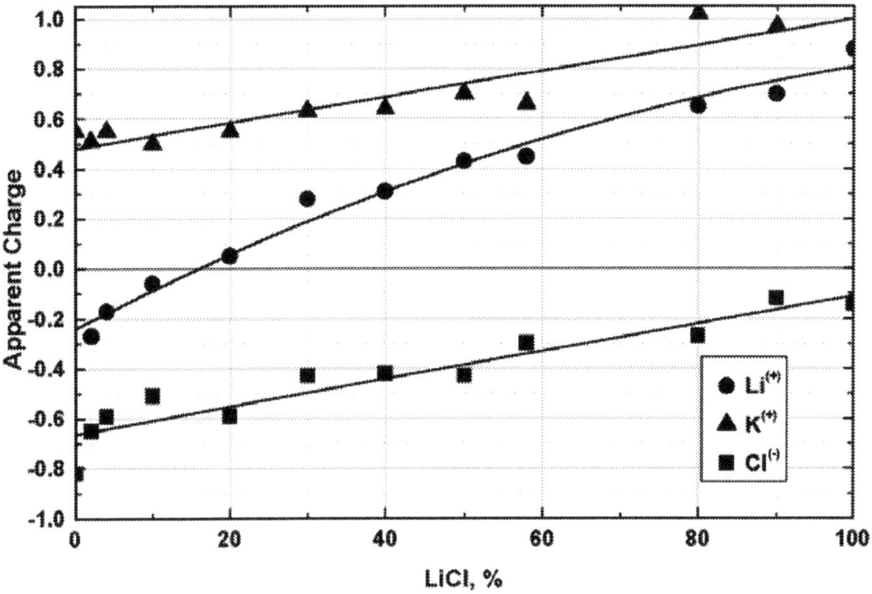

Figure 3. Concentration dependencies of the ionic apparent charge in LiCl–KCl melt. Symbols represent MD simulation results, solid lines are guides for eye.

To clarify the question if such a peculiarity as negative mobility is characteristic only for multiply charged ions, or this is quite a common feature in diluted solutions of all ions having small size, we carried out an additional set of simulations of three-component LiCl–KCl ionic liquid.

Diffusion coefficients, external ionic mobilities, and structure characteristics are determined at constant temperature (T=1400 K), well above the melting point of KCl, and fixed density (1.22 g cm^{-3}) in the whole range of concentration from pure LiCl to dilute solution of Li in KCl. Apparent charges of $Li^{(+)}$, $K^{(+)}$, and $Cl^{(-)}$ as functions of LiCl concentration are presented in Fig. 3.

3. Discussion

Results of simulation obtained in this work need to be explained and interpreted in more detail from theoretical point of view. It has been known for long that behavior of ionic mobility in LiCl–KCl salt cannot be described within conventional Debye–Hückel theory or Nernst–Einstein approximation due to strong Coulomb non–ideality.

A rather coherent theory of Fuoss–Onsager[10] takes into account two important effects: relaxation of ionic atmosphere and electrophoretic drift. Both effects

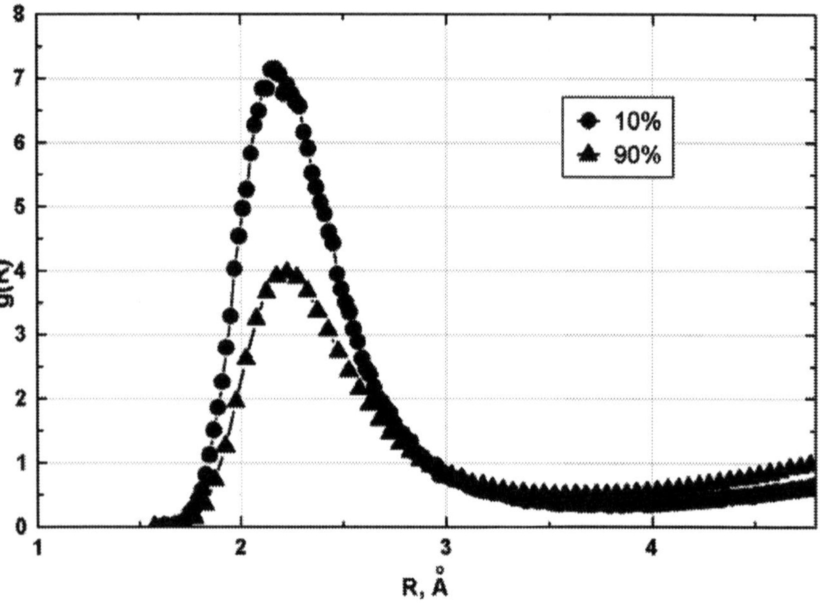

Figure 4. LiCl radial distribution function at two concentrations of LiCl: 10% (circles) and 90% (triangles).

reduce the absolute values of ionic mobility. Unfortunately, Fuoss–Onsager theory is valid only for dilute electrolyte solutions and provides non-ideality corrections to ionic mobility at very low Coulomb coupling parameter $\Gamma << 1$.

According to our results, as well as to and many other simulations, the charge distribution around an ion dissolved in molten salt consists of a number of alternating solvation shells (Fig. 2) – a striking contrast to the Debye–Hűckel limit. Mobilities of all ions are largely affected by electrophoretic drift of solvation shells and their values, predicted by Nernst–Einstein equation, are normally essentially higher than measured. This effect significantly reduces the magnitude of apparent ionic charge Z^*_α, defined by Eq. (3), and may even change its sign, when small ions are present at low concentrations[1] as in the cases of $Am^{(2+)}$, $Am^{(3+)}$, or $Li^{(+)}$. Electrophoretic drift depends both on co-ordination number and stability of the first and second solvation shells. Trajectories of individual anions forming solvation shells have been studied in this work for a number of ions dissolved in LiCl–KCl melt and the mean 'solvation time' in the first and second shells was determined. This time in first and second solvation shells of different ions was found to obey the exponential decay law with half-lives being in general case sufficiently different.

For heavy ions in their higher oxidation states these half-lives differ essentially in the first solvation shell, which was found to be much more stable

[1] Negative apparent charge of solvated Li ions at low concentrations was experimentally discovered many years ago.[11]

than the second one. The concentration dependence of D($Cl^{(-)}$) at conditions of simulation was found to be rather weak, D($K^{(+)}$) and D($Li^{(+)}$) increase (about 20% and 40%, respectively) with increase of LiCl concentration from 0% to 100%.

In contrast to diffusion, the concentration dependence of mobility is more pronounced. Mobility of cations increase, mobility of $Cl^{(-)}$ decreases with increase of LiCl concentration. The most interesting feature of the concentration behavior of $Li^{(+)}$ mobility is its negativity at small concentrations of $Li^{(+)}$ (Fig. 3). Radial distribution functions in LiCl–KCl melt at different compositions reveal essential changes in nearest-neighbor structure at low concentrations of $Li^{(+)}$. In Fig. 4 two radial distribution functions are compared: one at low (10%) and other at high (90%) concentration of LiCl. One can see essential change in coordination number in first shell. At low concentrations this number is about 3 or 3.5. More detailed examination reveals formation of an relatively stable planar complex $Li^{(+)}(Cl^{(-)})_3$ which are probably responsible for negative mobility of lithium ions at low concentrations.

There is also another point, which may be important here, namely, the different apparent charge Z^*_α of $Cl^{(-)}$ at low and high concentrations of Li in LiCl–KCl mixture (see Fig. 3). $Cl^{(-)}$ anions, being more 'negative' in pure KCl provide higher electrophoretic drift of $Li^{(+)}$ at low concentrations. This drift, in our opinion, should be taken into account in quantitative explanation of negative values of $Z^*_\alpha(Li^{(+)})$ at low concentrations of LiCl.

Acknowledgment

The part of this investigation was performed during author's stay as a visiting scientist at Joint Research Center of European Commission, Institute for Transuranium Elements, Karlsruhe, Germany.

References

1. Yakub, E. and Ronchi, C. (2003) *J. Chem. Phys.* **119**, 11556.
2. Yakub, E. and Ronchi, C. (2005) *J. Low Temp. Phys.* **139**, 633.
3. Borcan, R. and Zuca, S. (1970) *Rev. Roum. Chim.* **15**, 1681.
4. Fumi, F. G. and Tosi, M. P. (1964) *J. Phys. Chem. Solids* **25**, 31.
5. Hoover, W. G. (1985) *Phys. Rev. A* **31**, 1695.
6. Allen, M. P. and Tildesley, D. J. (1989) *Computer Simulation of Liquids* (Oxford: Clarendon Press).
7. Sangster, M. J. L. and Dixon, M. (1976) *Adv. Phys.* **25**, 247.
8. Cotton, S. (1991) *Lanthanides and Actinides* (Macmillan Education, London).
9. Crank, J. (1964) *The Mathematics of Diffusion* (Oxford, Clarendon Press).
10. Fuoss, R. and Onsager, L. (1957) *J. Phys. Chem.* **61**, 668.
11. Blander M. (ed.) (1964) *Molten Salt Chemistry* (New York u.a.: Interscience), p. 775.

SOLVATION EFFECTS IN NEAR-CRITICAL POLAR FLUIDS

AKIRA ONUKI
Department of Physics, Kyoto University, Kyoto 606-8502, Japan

Abstract. A Ginzburg–Landau theory is presented for polar binary mixtures containing ions, which takes account of the solvation and image interactions between ions and solvent molecules. The ion distributions and the electric potential are calculated around an interface with finite thickness. The surface tension is shown to increase in the presence of ions. Critical behavior and phase separation are also investigated, which are much affected even by a small amount of ions. We find that there arise long-range attractive interactions among ions mediated by the critical fluctuations. Under strong solvation conditions they can dominate over the Coulomb interaction in the range shorter than the correlation length.

Keywords: Critical mixtures, solvation effects, strong electric field.

1. Introduction

In usual electrolyte theories, ions interact via the Coulombic potential in a fluid with a homogeneous dielectric constant ε. However, ε is strongly inhomogeneous particularly in the presence of mesoscopic structures in aqueous solutions. Furthermore, in most of the physics literature, the microscopic molecular interactions between ions and solvent molecules are not explicitly considered. Around a microscopic ion such as Na^+ or Cl^- in a polar fluid, the ion–dipole interaction gives rise to a solvation (hydration) shell composed of a number of solvent molecules (those of the more polar component for a mixture).[1,2] In this work, the resultant solvation free energy per ion will be called the solvation chemical potential and will be written as $\mu^i_{\rm sol}$ where i represents the ion species. It is important that $\mu^i_{\rm sol}$ strongly depends on the solvent density n or the composition ϕ for binary mixtures, with its typical values much larger than the thermal energy k_BT. As a classic result, Born,[3] took into account polarization around an ion to obtain the formula $(\mu^i_{\rm sol})_{\rm Born} = Z_i^2e^2/2\varepsilon R^i_{\rm ion}$, where ε is the linear dielectric constant, Z_ie is the ion charge, and $R^i_{\rm ion}$ is the microscopic ion radius (Born radius).[1] Notice that ε can strongly depend on the density or the composition. However, this formula is not a good approximation when the dielectric constant is largely increased in the vicinity of ions, for example, in the presence of a hydration shell in aqueous solutions[4] or in gaseous water.[5]

S. J. Rzoska and V. A. Mazur (eds.), Soft Matter under Exogenic Impacts, 467–479.

We consider an interface between a polar phase α and a less polar phase β, across which there arises a difference of $\mu^i_{\rm sol}$ because of its density or composition dependence:

$$\Delta\mu^i_{\alpha\beta} = \mu^{i\beta}_{\rm sol} - \mu^{i\alpha}_{\rm sol}, \tag{1.1}$$

where $\mu^{iK}_{\rm sol}$ ($K = \alpha, \beta$) are the bulk values of the solvent chemical potential of species i in the two phases. The difference of the solvation free energies $\Delta G^i_{\alpha\beta} (= N_i \Delta\mu^i_{\alpha\beta}$ with N_i being the ion number of species i) between two phases have been called the standard Gibbs transfer energy in electrochemistry (usually measured in units of kJ per mole).[4,6] Using data of $\Delta G^i_{\alpha\beta}$ on water + nitrobenzene at room temperatures,[6] we estimate $\Delta\mu^i_{\alpha\beta}/k_{\rm B}T$ as 13.6 for Na^+, 10.6 for K^+, 11.3 for Br^-, and 7.46 for I^- with α being a water-rich phase. As another aspect, $\Delta\mu^i_{\alpha\beta}$ is an important parameter dramatically influencing the nucleation process. That is, when a polar fluid in a less polar phase β is driven into in a metastable state, the solvation shell around an ion can serve as a seed of a critical droplet of a more polar phase α.[7] For water at $T = 0.6T_c \cong 390\,{\rm K}$, an approximate value of $\Delta\mu^i_{\alpha\beta}/k_{\rm B}T$ was recently estimated to be about 60 for $Z^2 e^2/k_{\rm B} T \sigma = 106$ with $\sigma = 3.1$ Å between gaseous water (phase β) and liquid water (phase α),[5] which much reduces the nucleation barrier in the presence of ions.

Here we will present a simple Ginzburg–Landau theory accounting for the solvation effects and will present a number of predictions.[5, 8–10]

2. Ginzburg-Landau free energy

Let us consider a polar binary mixture containing a small amount of salt composed of two species of charges. The volume fraction of the more polar component A will be written as $\phi(\boldsymbol{r})$. The ions are composed of two species with charges $Q_1 = Z_1 e$ and $Q_2 = -Z_2 e$. Their densities are $n_1(\boldsymbol{r})$ and $n_2(\boldsymbol{r})$. Here ϕ, n_1, and n_2 are smooth variables coarse–grained on the microscopic level. We will set up a Ginzburg-Landau free energy $F = F_{\rm ch} + F_{\rm im} + F_{\rm e}$ composed of three parts.

For simplicity, the chemical part $F_{\rm ch}$ is written as

$$F_{\rm ch} = \int d\boldsymbol{r} \left\{ f(\phi) + \frac{1}{2} C |\nabla\phi|^2 + \sum_{i=1,2} \left[k_{\rm B} T n_i \ln n_i + n_i \mu^i_{\rm sol}(\phi) \right] \right\}, \tag{2.1}$$

where the free energy density f is assumed to be of the Bragg–Williams form,

$$f = \frac{k_B T}{a^3}\left[\phi \ln \phi + (1-\phi)\ln(1-\phi) + \chi\phi(1-\phi)\right], \tag{2.2}$$

a is the molecular diameter assumed to be the same for the two components, and χ is the interaction parameter dependent on the temperature T. In the absence of ions the mean field critical values of χ and ϕ are $\chi_c = 2$ and $\phi_c = 1/2$, respectively, while the critical value of χ is reduced with increasing the ion densities as in Eq. (3.6) below. The coefficient C of the gradient term is of order $k_B T / a$ and we will use a simple form $C = k_B T_\chi / a$ in our numerical results to follow. As regards the composition dependence of $\mu^i(\phi)$ we assume

$$\mu^i_{\rm sol}(\phi) = \mu^i_{\rm c} - k_B T g_i(\phi - \phi_c) \qquad (i = 1, 2). \tag{2.3}$$

The first term on the right hand side is an irrelevant constant when the ion numbers are conserved quantities. It follows the relation $\Delta\mu^i_{\alpha\beta} = k_B T g_i \Delta\phi$ with $\Delta\phi = \phi_\alpha - \phi_\beta$. In aqueous solutions, the coupling constants g_i are positive for hydrophilic ions and can much exceed unity for small ions such as Li^+ or Na^+ (typically of order 10 from the experimental values of the Gibbs transfer energy). They should be even larger for small multivalent ions such as Mg^{2+} or Al^{3+}. It is worth noting that g_i should be negative for hydrophobic organic ions.[4] In the Born approximation we find $(g_i)_{\rm Born} = Z_i^2 e^2 \varepsilon_1 / 2k_B T \varepsilon_c^2 R^i_{\rm ion}$, where ε_c is the dielectric constant at the critical point and $\varepsilon_1 = \partial\varepsilon / \partial\phi$.

The $F_{\rm im}$ arises from a deformation of the self-interaction of ions due to inhomogeneous ε and is of the form,

$$F_{\rm im} = \int d\boldsymbol{r}(Z_1^2 n_1 + Z_2^2 n_2)\mu_{\rm im}. \tag{2.4}$$

The chemical potential contribution $\mu^i_{\rm im} = \delta F_{\rm im} / \delta n_i = Z_i^2 \mu_{\rm im}$ are the so-called image potentials. If an ion of species i located in phase α is close to an interface, we obtain[11,12]

$$\mu^i_{\rm im} = \frac{\varepsilon_\alpha - \varepsilon_\beta}{2(\varepsilon_\alpha + \varepsilon_\beta)\varepsilon_\alpha} \cdot \frac{e^2 Z_i^2}{|z|}, \tag{2.5}$$

where ε_α and ε_β are the dielectric constants of the two bulk phases and $|z|$ is the distance of the ion to the interface. For electrolyte Onsager and Samaras argued that the other ions give rise to a decaying factor $\exp(-2\kappa|z|)$ in the right hand side of Eq. (2.5), where κ is the Debye–Hückel wave number in phase α. Around a planar interface, where all the quantities vary along the z direction, the following integral form is convenient:

$$\mu_{\rm im}(z) = 2\pi B_0 \int \frac{dz'}{z-z'} e^{-2k|z-z'|} \frac{d}{dz'}\phi(z'), \tag{2.6}$$

where B_0 is an appropriate constant. It is equal to $e^2\varepsilon_1 / 8\pi\varepsilon_{\rm c}^2$ near the critical point. In the numerical calculations to follow, we assume that κ in Eq. (2.6) is the local value determined by $n_1(z)$ and $n_2(z)$.

The $F_{\rm e}$ is the electrostatic contribution. If the charge densities at the boundaries are fixed, it is written as

$$F_{\rm e} = \int dr \frac{\varepsilon(\phi)}{8\pi} \boldsymbol{E}^2, \tag{2.7}$$

where $\boldsymbol{E} = -\Delta\Phi$ is the electric field and the electrostatic potential Φ satisfies

$$-\nabla\cdot\varepsilon(\phi)\nabla\Phi = 4\pi\rho. \tag{2.8}$$

Here $\rho = e(Z_1 n_1 - Z_2 n_2)$ is the charge density. The dielectric constant of the mixture is assumed to depend on the concentration as

$$\varepsilon(\phi) = \varepsilon_0 + \varepsilon_1\phi, \tag{2.9}$$

where ε_0 and ε_1 are positive constants. Then notice the general relations,

$$\frac{\delta F_{\rm e}}{\delta\phi} = -\frac{\varepsilon_1}{8\pi}\boldsymbol{E}^2, \quad \frac{\delta F_{\rm e}}{\delta n_1} = Z_1 e\Phi, \quad \frac{\delta F_{\rm e}}{\delta n_2} = -Z_2 e\Phi. \tag{2.10}$$

The linear composition dependence of $\mu^{i}_{\rm im}(\phi)$ and $\varepsilon(\phi)$ are adopted to gain the physical consequences in the simplest manner and should not be taken too seriously.

For our model the grand potential density ω is defined by

$$\omega = f + \frac{C}{2}|\nabla\psi|^2 - h\psi + \sum_{i=1,2} n_i\left[k_B T \ln n_i + \mu^i_{\rm sol}(\phi) + Z^2_i\mu_{\rm im} - \mu_i\right] + \frac{\varepsilon \boldsymbol{E}^2}{8\pi}, \quad (2.11)$$

where h, μ_1, and μ_2 are constants. At equilibrium, the grand potential $\Omega = \int d\boldsymbol{r}\omega = F - \int d\boldsymbol{r}(h\psi + \mu_1 n_1 + \mu_2 n_2)$ should be minimized and then the chemical potentials,

$$\delta F/\delta n_1 = \mu_1, \quad \delta F/\delta n_2 = \mu_2, \quad \delta F/\delta\phi = h, \quad (2.12)$$

become homogeneous. With the aid of Eq. (2.10) the ion densities are expressed as

$$n_1 = n_1^0 \exp\left[-(\mu^1_{\rm sol} + Z_1 e\Phi + Z^2{}_1\mu_{\rm im})/k_B T\right],$$

$$n_2 = n_2^0 \exp\left[-(\mu^2_{\rm sol} + Z_2 e\Phi + Z^2{}_2\mu_{\rm im})/k_B T\right], \quad (2.13)$$

where n_1^0 and n_2^0 are constants. Near a planar interface all the quantities depend only on z and the surface tension γ is expressed as

$$\gamma = \int dz[\omega(z) - \omega_\infty]. \quad (2.14)$$

We may generally prove that $\omega(z)$ tends to a common constant ω_∞ as $z \to \pm\infty$.[10]

3. Planar interface

We make the equilibrium equations dimensionless measuring the space in units of a. There appears a parameter A representing the strength of the charges,

$$A = \pi e^2 / 4a\varepsilon_c k_B T. \quad (3.1)$$

The normalized charge densities are $c_1 = a^3 n_1$ and $c_2 = a^3 n_2$ in the following. The interface structure is then determined by the six dimensionless parameters, χ, g_1, g_2, $\varepsilon_1 / \varepsilon_c$, A, and the bulk ion concentration $c_{1\alpha} = a^3 n_{1\alpha}$ in phase α.

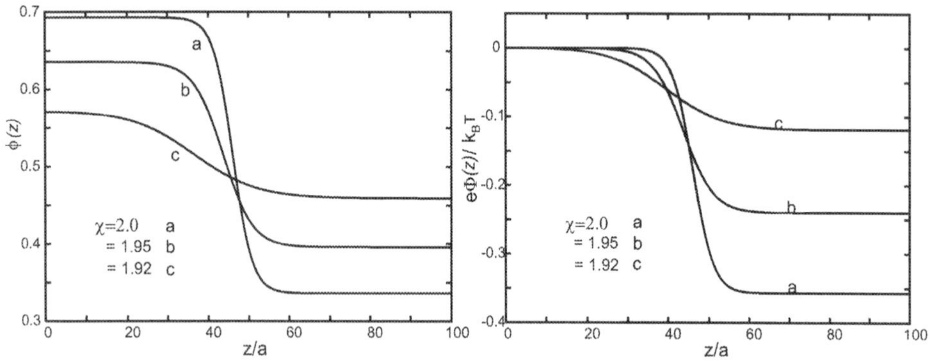

Figure 1. Composition $\phi(z)$ (left) and normalized electric potential $e\Phi(z)/k_BT$ (right) near an interface with varying $\chi = 2$, 1.95, and 1.92 for $g_1 = 4$, $g_2 = 2$, and $\sqrt{n_{1\alpha}n_{1\beta}} = 0.01a^{-3}$.

We assume that there are significant ion densities in the two phases such that the bulk screening lengths $1/\kappa_\alpha$ and $1/\kappa_\beta$ in the two phases are shorter than the cell length L. Far from the interface, the charge density $\rho(z)$ tends to zero and the electric potential $\Phi(z)$ tends to constants, Φ_α and Φ_β. The difference $\Phi_\alpha - \Phi_\beta$ is called the Galvani potential difference in electrochemistry.[4,6] Since the ion densities in the bulk phases n_{iK} satisfy $Z_1 n_{1K} = Z_2 n_{2K}$ ($K = \alpha, \beta$) and $\mu^i_{\rm im}(z)$ vanish far from the interface, we obtain

$$\begin{aligned}\Phi_\alpha - \Phi_\beta &= \frac{1}{(Z_1+Z_2)e}[\Delta\mu^2_{\alpha\beta} - \mu^1_{\alpha\beta}] \\ &= \frac{k_BT}{(Z_1+Z_2)e}(g_1 - g_2)\Delta\phi. \end{aligned} \tag{3.2}$$

The first line is written in terms of $\Delta\mu^i_{\alpha\beta}$ in Eq. (1.1), where we define the solvation chemical potentials in the bulk, $\mu^{1K}_{\rm sol} = \mu^1_{\rm sol}(\phi_K)$ and $\mu^{2K}_{\rm sol} = \mu^2_{\rm sol}(\phi_K)$. The second line holds under Eq. (2.3) with $\Delta\phi = \phi_\alpha - \phi_\beta$. The potential difference arises in the asymmetric case ($g_1 \neq g_2$) only and vanishes in the symmetric case ($g_1 = g_2$). We also find that the ion densities in the bulk two phases are related by

$$\frac{n_{1\beta}}{n_{1\alpha}} = \frac{n_{2\beta}}{n_{2\alpha}} = \exp\left(-\frac{\varepsilon_r}{k_BT}\right). \tag{3.3}$$

The degree of the ion-density reduction is represented by the energy ε_r written as

$$\varepsilon_r = \frac{Z_2}{Z_1 + Z_2}\Delta\mu^1_{\alpha\beta} + \frac{Z_1}{Z_1 + Z_2}\Delta\mu^2_{\alpha\beta}$$

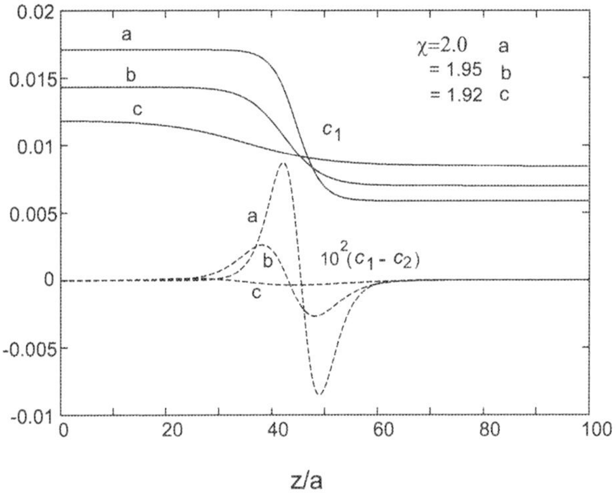

Figure 2. Normalized ion density $c_1(z) = a^3 n_1(z)$ and normalized charge density $c_1(z) - c_2(z) = a^3(n_1(z) - n_2(z))$ (multiplied by 100) with varying χ with the same parameter values as in the previous figure.

$$= \frac{k_B T}{Z_1 + Z_2}(Z_2 g_1 + Z_1 g_2)\Delta\phi. \tag{3.4}$$

For example, if $\Delta\mu^1_{\alpha\beta}/k_B T = \Delta\mu^2_{\alpha\beta}/k_B T = 10$ and 60, the ion reduction factor becomes $e^{-10} = 2.4 \times 10^{-4}$ and $e^{-60} = 0.67 \times 10^{-26}$ respectively, in the monovalent case. In the latter case the ion density is virtually zero in phase β. The Debye–Hückel wave numbers in Eq. (4.2) satisfy

$$k_\beta/k_\alpha = (\varepsilon_\alpha/\varepsilon_\beta)^{1/2}\exp(-\varepsilon_r/2k_B T). \tag{3.5}$$

If $\varepsilon_r / k_B T$ is considerably larger than unity, we have $n_{1\beta} << n_{1\beta}$, $n_{2\beta} << n_{2\alpha}$, and $1/\kappa_\beta << 1/\kappa_\alpha$. On the other hand, as we approach the critical point, $\Delta\phi$ and $\varepsilon_r/k_B T$ eventually become small, where the heterogeneity in the ion distributions is weak. If g_1 and g_2 are large, this crossover occurs very close to the critical point.

We now give numerical results for the monovalent case $Z_1 = Z_2 = 1$. In this article we set $A = 4$ and $\varepsilon_1/\varepsilon_c = 0.8$. In Fig. 1 we show the composition $\phi(z)$ and the normalized electric potential $e\Phi(z)/k_BT$ (taken to be zero in phase α) near an interface, where $g_1 = 4$ and $g_2 = 2$. Here the critical point is approached as χ is decreased, where the ion-density-dependent critical value of χ is.[10]

$$\chi_c^{\text{ion}} = 2 - \frac{1}{2}[(g_1Z_2 + g_2Z_1)/(Z_1 + Z_2)]^2 a^3\langle n\rangle, \tag{3.6}$$

where $n = n_1 + n_2$. In Fig. 1 we fix the product $n_{1\alpha}\, n_{1\beta} = 10^{-4}a^{-6}$, where $\chi_c^{\text{ion}} = 1.91$ for $Z_1 = Z_2 = 1$. The electric potential is confirmed to satisfy Eq. (3.2). Figure 2 displays the ion density $c_1(z)$ and the normalized charge density $c_1(z) - c_2(z)$. The ion densities are reduced in phase β as in Eq. (3.3). An electric double layer at the interface diminishes as the critical point is approached. It is convenient to choose $\bar{n} \equiv \sqrt{n_\alpha n_\beta}$ $(= 2\sqrt{n_{1\alpha}n_{1\beta}}$ in the monovalent case) as the parameter representing the degree of ion doping.

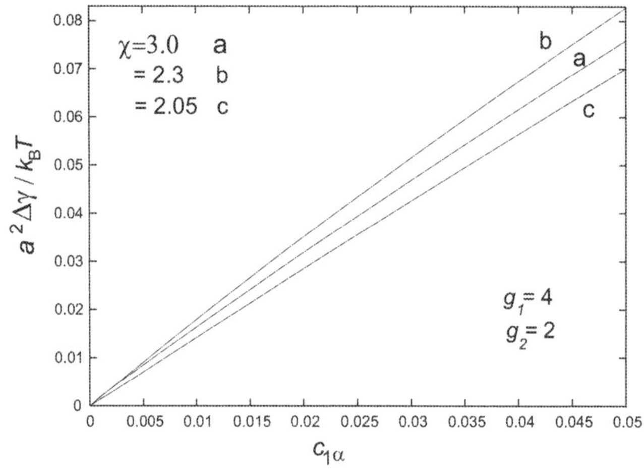

Figure 3. Normalized excess surface tension $a^2\Delta\gamma/k_BT$ vs the normalized ion density $c_{1\alpha}$ in phase α.

Over a rather wide parameter region in our model including the cases in Fig. 1, the bulk concentrations in the two phases are expressed as.[10]

$$\phi_\alpha + \phi_\beta - 1 \cong [(g_1 Z_2 + g_2 Z_1)/(Z_1 + Z_2)]^3 a^3 \bar{n}/16, \qquad (3.7)$$

$$\Delta\phi = \phi_\alpha - \phi_\beta \cong [3(\chi - \chi_c^{\text{ion}})/2]^{1/2}. \qquad (3.8)$$

The first line is the shift of the critical composition multiplied by 2. The second line is the usual mean field expression for the average order parameter difference.

The surface tension γ can be calculated from Eq. (2.14). In Fig. 3 we show the normalized excess surface tension $a^2\Delta\gamma/k_B T$ vs. $c_{1\alpha}$ at $A = 4$, where $\Delta\gamma = \gamma - \gamma_0$ with γ_0 being the surface tension without ions. For $\chi = 3$, 2.3, and 2.05, we find $a^2\gamma_0/k_B T = 0.498$, 0.103, and 0.0773, respectively. While these γ_0 values are very different, we recognize that $\Delta\gamma$ increases roughly linearly with increasing the ion density, in agreement with experiments.[11,12] Figure 4 shows that the surface tension γ itself tends to zero as χ approaches χ_c^{ion} in Eq. (3.6). Here the

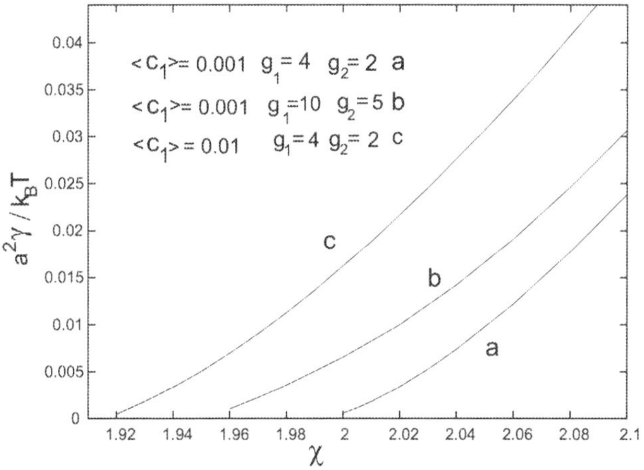

Figure 4. Normalized surface tension $a^2\gamma / k_B T$ as a function of χ. It tends to zero as χ tends to its critical value χ_c^{ion} in Eq. (3.6).

interface is located at the middle of the cell and the average normalized ion density $\langle c_1 \rangle$ is fixed.

4. Fluctuations in one-phase states and interactions among ions

In our previous papers we examined the structure factor $S(q) = \left\langle |\phi_q|^2 \right\rangle$ of the composition fluctuations with wave number q in one-phase states. It can be strongly affected even by a small amount of ions for large solvation coefficients

g_i .[8–10] One salient effect is the shift of the critical temperature shown in Eq. (3.6). Here we are interested in the effective interactions among ions mediated by the critical fluctuations.

We consider small plane-wave fluctuations in a one-phase state, where the inhomogeneity in the dielectric constant may be neglected ($\varepsilon = \varepsilon_c$) to leading order. The fluctuation contributions to F in the bilinear order are written as

$$\delta F = \int_q \left[\frac{1}{2}(f'' + Cq^2)|\phi_q|^2 + \frac{2\pi}{\varepsilon_c q^2} + |\rho_q|^2 + k_B T \sum_{i=1,2} \left(\frac{|n_{iq}|^2}{2n_i} - g_i n_{iq} \phi_q^* \right) \right], \tag{4.1}$$

where $\int_q \cdots = (2\pi)^{-3} \int d_q \cdots$ denotes the integration over the wave vector $\boldsymbol{q}$ and $f'' = \partial^2 f / \partial \phi^2$ is the small coefficient near the criticality. The correlation length is written as $\xi = (C/f'')^{1/2}$. The $\phi_{\boldsymbol{q}}$, $n_{i\boldsymbol{q}}$, and $\rho_{\boldsymbol{q}}$ are the Fourier transformations of $\phi(\boldsymbol{r})$, $n_i(\boldsymbol{r})$, and $\rho(\boldsymbol{r}) = e(Z_1 n_1\ \boldsymbol{r} - Z_2 n_2(\boldsymbol{r}))$, respectively. The n_i in the last term of Eq. (4.1) are the average ion densities satisfying $Z_1 n_1 = Z_2 n_2$ from the overall charge neutrality. The image interaction is relevant only in the presence of strong inhomogeneity in ε and is negligible for small fluctuations.

Elimination of the ion density fluctuations yields the composition structure factor in the mean field theory,

$$\frac{k_B T}{S(q)} = f'' - \Delta r_{\text{ion}} + Cq^2 \left[1 - \frac{\gamma_p^2 k^2}{k^2 + q^2} \right], \tag{4.2}$$

where κ is the Debye–Hückel wave number defined by

$$k^2 = 4\pi (Z_1^2 n_1 + Z_2^2 n_2) e^2 / \varepsilon_c k_B T. \tag{4.3}$$

In accord with Eq. (3.6) the ion-induced shift Δr_{ion} is given by

$$\Delta r_{\text{ion}} = k_B T (Z_2 g_1 + Z_1 g_2)^2 (n_1 + n_2) / (Z_1 + Z_2)^2 \tag{4.4}$$

For the monovalent case $Z_1 = Z_2 = 1$ we find $\Delta r_{\text{ion}} = (k_B T / 16\pi \ell_B)$ $(g_1 + g_2)^2 k^2{}_D$, where $\ell_B = e^2 / \varepsilon_c k_B T$ is the Bjerrum length. There appears a dimensionless parameter,

$$\gamma_{\mathrm{P}} = (k_{\mathrm{B}}T / 4\pi C \ell_{\mathrm{B}})^{1/2} \,|\, g_1 - g_2 \,|\, /(Z_1 + Z_2), \tag{4.5}$$

which is independent of the ion density and represents the strength of asymmetry in the ion-induced polarization between the two components. As discussed previously, a Lifshitz point can be reached for $\gamma_{\mathrm{p}} = 1$, formation of a mesoscopic phase can be predicted for $\gamma_{\mathrm{p}} > 1$, and the usual critical point exists for $\gamma_{\mathrm{p}} < 1$.[8–10,13]

We may eliminate the critical fluctuations in F in Eq. (4.1) by setting

$$C(\xi^{-2} + q^2)\phi_q = k_{\mathrm{B}}T \sum_{i=1,2} g_i n_{iq}. \tag{4.6}$$

For example, the composition deviation $\delta\phi_i(r) = (k_{\mathrm{B}}T\, g_i/4\pi C)e^{-r/\xi}/r$ is induced around an ion of species i,[9] where r is the distance from the ion center. In one-phase states we thus obtain attractive interactions among the ions:

$$\begin{aligned} \delta F &= \int_q \left[\frac{2\pi}{\varepsilon_c q^2} |\rho_q|^2 + \sum_{i,j} \left(\frac{k_{\mathrm{B}}T}{2n_i} \delta_{ij} |n_{jq}|^2 - \frac{(k_{\mathrm{B}}T)^2 g_i g_j}{2C(\xi^{-2} + q^2)} n_{iq} n^*_{jq} \right) \right] \\ &= \int d\boldsymbol{r} \sum_i \frac{k_{\mathrm{B}}T}{2n_i} \delta n_i^2 + \frac{1}{2} \int d\boldsymbol{r} \int d\boldsymbol{r}' \sum_{i,j} V_{ij}(|\boldsymbol{r} - \boldsymbol{r}'|) \delta n_i(\boldsymbol{r}) \delta n_j(\boldsymbol{r}') \end{aligned} \tag{4.7}$$

The second line is the expression in the real space with δn_1 and δn_2 being the ion density deviations. The effective interaction potentials $V_{ij}(r)$ are expressed as

$$V_{ij}(r) = Q_i Q_j \frac{1}{\varepsilon_c r} - \frac{(k_{\mathrm{B}}T^2)}{4\pi C} - g_i g_j \frac{e^{-r/\xi}}{r}, \tag{4.8}$$

where $Q_1 = Z_1 e$ and $Q_2 = -Z_2 e$ are the ion charges. Among the ions of the same species ($i = j$), the second attractive term dominates over the first Coulomb repulsive term in the range $\lesssim r \lesssim \xi$ when

$$g^2_{\ i} > 4\pi C Z^2_{\ i} e^2 / \varepsilon_c (k_{\mathrm{B}}T)^2 \sim 4\pi Z_i^2 \ell_{\mathrm{B}} / a \tag{4.9}$$

As long as the system remains in a one-phase state with $\bar{r} > \Delta r_{\mathrm{ion}}$, however, ξ is shorter than the length,

$$\xi_{\rm ion} = (C/\Delta r_{\rm ion})^{1/2} \sim [(Z_1+Z_2)/Z_1 g_2 + Z_2 g_1)](an)^{-1/2}, \qquad (4.10)$$

where use has been made of Eq. (4.4). For $g_1 \sim g_2 \sim 10$ we have $\xi_{\rm ion} \sim \kappa^{-1}$.

It is of great interest whether or not Eq. (4.10) really holds in experiments. For example, if we set $g_i = 13$, $Z_i = 1$, and $a = 3$ Å, Eq. (4.10) holds for $\ell_B < 40$ Å. Physically, there should be a tendency of aggregation of ions under the condition (4.10). This effect might explain a number of observations of microheterogeneities in near-critical binary mixtures containing salt.[14,15]

5. Summary

We summarize our results:

(i) The solvation effects are taken into account with the composition-dependent solvation chemical potentials in Eq. (2.3). The resultant interaction in the free energy F is bilinear and characterized by dimensionless parameters g_i dependent on the ion species i. A shift of the coexistence curve of the composition follows from this coupling as in Eq. (3.6) or Eq. (4.4). Intriguing is the asymmetric case $g_1 \neq g_2$, where a potential difference arises at an interface. In one-phase states this asymmetry can lead to a peak in the structure factor of the composition fluctuations at an intermediate wave number. There can even be a mesoscopic phase,[13] which can possibly occur in the case of strongly asymmetric salt with large $g_1 - g_2$ (say, salt composed of small cations and relatively large anions with distinctly different solvation powers).[9] Notice that g_2 can be negative for hydrophobic ions.[4]

(ii) The image interaction is shown to arise from general inhomogeneous dielectric constant and is expressed in the integral form in Eq. (2.6). It can dominate over the solvation interaction only for $A \sim \ell_B/a >> 1$, at very small ion densities, and far below the critical point.[10]

(iii) The dielectric constant in Eq. (2.4) is dependent on the composition and is inhomogeneous. Though we have treated a moderate case of $\varepsilon_1 / \varepsilon_0 = 4/3$, the ratio $\varepsilon_1/\varepsilon_0$ can be of order 80 in aqueous solutions. Such very strong composition-dependence should lead to various effects not well recognized so far.

(iv) We have derived the attractive interactions among ions mediated by the critical fluctuations as in the second term of Eq. (4.9). We should then investigate how they can produce large-scale structures under Eq. (4.10) near the criticality.[14,15] It is also of interest how charged colloidal particles interact in polar fluids in the presence of strong solvation effects.

We mention related problems:

(i) There can be an electric double layer and a potential difference at an interface in general charged systems, including low-molecular-weight fluids (as in this paper), complex fluids, gels, and even liquid metals.

(ii) Wetting transitions should be greatly influenced by ions.[16] If the wetting layer is more polar than the outer fluid, ions can even be confined within the layer.

(iii) Phase separation processes in ionic systems should also be investigated. Ions are more strongly segregated than the solvent composition for large g_i. We already examined the effect of a very small amount of ions on nucleation in a metastable fluid in a less polar phase.[5]

(iv) Recently we examined solvation effects of charged particles in liquid crystals.[17] In nematic states the dielectric tensor is anisotropically dependent on the director orientation and, as a result, the electric field of ions distorts the orientation over long distances and sometimes create nanometer scale defects.

References

1. J.N. Israelachvili, *Intermolecular and Surface Forces* (Academic Press, London, 1991).
2. Y. Marcus, *Ion Solvation* (Wiley, New York, 1985).
3. M. Born, *Z. Phys.* **1**, 45 (1920).
4. T. Osakai and K. Ebina, *J. Phys. Chem. B* **102**, 5691 (1998).
5. H. Kitamura and A. Onuki, *J. Chem. Phys.* **123**, 124513 (2005).
6. Le Quoc Hung, *J. Electroanal. Chem.* **115**, 159 (1980).
7. J.J. Thomson, *Conduction of Electricity through Gases* (Cambridge University Press, Cambridge, 1906), Section 92.
8. A. Onuki, in *Nonlinear Dielectric Phenomena in Complex Liquids*, NATO Science Series II: **157**, edited by S.J. Rzoska (Kluwer Academic, Dordrecht, 2004).
9. A. Onuki and H. Kitamura, *J. Chem. Phys.* **121**, 3143 (2004).
10. A. Onuki, to be published in *Phys. Rev. E.*
11. L. Onsager and N.N.T. Samaras, *J. Chem. Phys.* 2, 528 (1934).
12. Y. Levin and J.E. Flores-Mena, *Europhys. Lett.* 56, 187 (2001).
13. V.M. Nabutovskii, N.A. Nemov, and Yu.G. Peisakhovich, *Phys. Lett. A*, **79**, 98 (1980); *Sov. Phys. JETP* **52**, 111 (1980) [*Zh. Eksp. Teor. Fiz.* **79**, 2196 (1980)].
14. J. Jacob, M.A. Anisimov, J.V. Sengers, A. Oleinikova, H. Weingärtner, and A. Kumar, *Phys. Chem. Chem. Phys.* **3**, 829 (2001); A.F. Kostko, M.A. Anisimov, and J.V. Sengers, *Phys. Rev. E* 70, 026118 (2004).
15. M. Wagner, O. Stanga, and W. Schröer, *Phys. Chem. Chem. Phys.* **6**, 580 (2004).
16. N.A. Denesyuk and J.-P. Hansen, *J. Chem. Phys.* **121**, 3613 (2004).
17. A. Onuki, *J. Phys. Soc. Jpn.* **73**, 511 (2004).